CRC SERIES IN NUTRITION AND FOOD

Editor-in-Chief

Miloslav Rechcigl, Jr.

Handbook of Nutritive Value of Processed Food
Volume I: Food for Human Use
Volume II: Animal Feedstuffs

**Handbook of Nutritional Requirements
in a Functional Context**
Volume I: Development and Conditions of
Physiologic Stress
Volume II: Hematopoiesis, Metabolic Function, and
Resistance to Physical Stress

Handbook of Agricultural Productivity
Volume I: Plant Productivity
Volume II: Animal Productivity

Handbook
of
Nutritive Value
of
Processed Food

Volume II
Animal Feedstuffs

Miloslav Rechcigl, Jr., Editor
Nutrition Advisor and Director
Interregional Research Staff
Agency for International Development
U.S. Department of State

CRC Series in Nutrition and Food
Miloslav Rechcigl, Jr., Editor-in-Chief

CRC Press, Inc.
Boca Raton, Florida

Library of Congress Cataloging in Publication Data

Main entry under title:

Handbook of nutritive value of processed food.

(CRC series in nutrition and food)
Bibliography: p.
Includes index.
CONTENTS: v. 1. Food for human use.—v. 2. Animal feedstuffs.
1. Food—Composition—Collect works. 2. Feeds—Composition—Collected works. 3. Food industry and trade—Collected works. 4. Feed processing—Collected works. 5. Nutrition—Collected works. 6. Animal nutrition—Collected works. I. Rechcigl, Miloslav. II. Series.

TX551.H264   641.1   80-21652
ISBN 0-8493-3951-0 (v. 1)
ISBN 0-8493-3953-7 (v. 2)

Direct all inquiries to CRC Press, Inc., 2000 N.W. 24th Street, Boca Raton, Florida, 33431.

© 1982 by CRC Press, Inc.

International Standard Book Number 0-8493-3951-0 (Volume I)
International Standard Book Number 0-8493-3953-7 (Volume II)

Library of Congress Number 80-21652
Printed in the United States

PREFACE
CRC SERIES IN NUTRITION AND FOOD

Nutrition means different things to different people, and no other field of endeavor crosses the boundaries of so many different disciplines and abounds with such diverse dimensions. The growth of the field of nutrition, particularly in the last 2 decades, has been phenomenal, the nutritional data being scattered literally in thousands and thousands of not always accessible periodicals and monographs, many of which, furthermore, are not normally identified with nutrition.

To remedy this situation, we have undertaken an ambitious and monumental task of assembling in one publication all the critical data relevant in the field of nutrition.

The *CRC Series in Nutrition and Food* is intended to serve as a ready reference source of current information on experimental and applied human, animal, microbial, and plant nutrition presented in concise tabular, graphical, or narrative form and indexed for ease of use. It is hoped that this projected open-ended multivolume compendium will become for the nutritionist what the *CRC Handbook of Chemistry and Physics* has become for the chemist and physicist.

Apart from supplying specific data, the comprehensive, interdisciplinary, and comparative nature of the *CRC Series in Nutrition and Food* will provide the user with an easy overview of the state of the art, pinpointing the gaps in nutritional knowledge and providing a basis for further research. In addition, the series will enable the researcher to analyze the data in various living systems for commonality or basic differences. On the other hand, an applied scientist or technician will be afforded the opportunity of evaluating a given problem and its solutions from the broadest possible point of view, including the aspects of agronomy, crop science, animal husbandry, aquaculture and fisheries, veterinary medicine, clinical medicine, pathology, parasitology, toxicology, pharmacology, therapeutics, dietetics, food science and technology, physiology, zoology, botany, biochemistry, developmental and cell biology, microbiology, sanitation, pest control, economics, marketing, sociology, anthropology, natural resources, ecology, environmental science, population, law politics, nutritional and food methodology, and others.

To make more facile use of the series, the publication has been organized into separate handbooks of one or more volumes each. In this manner the particular sections of the series can be continuously updated by publishing additional volumes of new data as they become available.

The Editor wishes to thank the numerous contributors many of whom have undertaken their assignment in pioneering spirit, and the Advisory Board members for their continuous counsel and cooperation. Last but not least, he wishes to express his sincere appreciation to the members of the CRC editorial and production staffs, particularly President Bernard J. Starkoff, Earl Starkoff, Sandy Pearlman, Pamela Woodcock, Lisa Levine Eggenberger, John Hunter, and Amy G. Skallerup for their encouragement and support.

We invite comments and criticism regarding format and selection of subject matter, as well as specific suggestions for new data which might be included in subsequent editions. We should also appreciate it if the readers would bring to the attention of the Editor any errors or omissions that might appear in the publication.

Miloslav Rechcigl, Jr.
Editor-in-Chief

PREFACE
HANDBOOK OF NUTRITIVE VALUE OF PROCESSED FOOD

Industrial as well as home processing can bring about profound changes in the chemical composition of food and its nutritive value. While a variety of treatments may be detrimental, certain modifications of foodstuffs can actually improve their digestibility and biological value. The effect on specific nutrients also varies and may differ depending on the type of treatment and food used. The purpose of this handbook is to provide a systematic and critical treatment of these questions. This publication should be an invaluable tool to food technologists, dieticians, and nutritionists, as well as to livestock producers and persons engaged in production, processing, and formulation of animal feeds. It should also be of great interest to conscientious consumers concerned about the quality and wholesomeness of food products.

Miloslav Rechcigl, Jr.

THE EDITOR

Miloslav Rechcigl, Jr. is a Nutrition Advisor and Chief of Research and Methodology Division in the Agency for International Development.

He has a B.S. in Biochemistry (1954), a Master of Nutritional Science degree (1955), and a Ph.D. in nutrition, biochemistry, and physiology (1958), all from Cornell University. He was formerly a Research Biochemist in the National Cancer Institute, National Institutes of Health and subsequently served as Special Assistant for Nutrition and Health in the Health Services and Mental Health Administration, U.S. Department of Health, Education and Welfare.

Dr. Rechcigl is a member of some 30 scientific and professional societies, including being a Fellow of the American Association for the Advancement of Science, Fellow of the Washington Academy of Sciences, Fellow of the American Institute of Chemists, and Fellow of the International College of Applied Nutrition. He holds membership in the Cosmos Club, the Honorary Society of Phi Kappa Pi, and the Society of Sigma Xi, and is recipient of numerous honors, including an honorary membership certificate from the International Social Science Honor Society Delta Tau Kappa. In 1969, he was a delegate to the White House Conference on Food, Nutrition, and Health and in 1975 a delegate to the ARPAC Conference on Research to Meet U.S. and World Food Needs. He served as President of the District of Columbia Institute of Chemists and Councillor of the American Institute of Chemists, and currently is a delegate to the Washington Academy of Sciences and a member of the Program Committee of the American Institute of Nutrition.

His bibliography extends over 100 publications including contributions to books, articles in periodicals, and monographs in the fields of nutrition, biochemistry, physiology, pathology, enzymology, molecular biology, agriculture, and international development. Most recently he authored and edited *Nutrition and the World Food Problem* (S. Karger, Basel, 1979), *World Food Problem: a Selective Bibliography of Reviews* (CRC Press, 1975), and *Man, Food and Nutrition: Strategies and Technological Measures for Alleviating the World Food Problem* (CRC Press, 1973) following his earlier pioneering treatise on *Enzyme Synthesis and Degradation in Mammalian Systems* (S. Karger, Basel, 1971), and that on *Microbodies and Related Particles, Morphology, Biochemistry and Physiology* (Academic Press, New York, 1969). Dr. Rechcigl also has initiated a new series on *Comparative Animal Nutrition* and was Associated Editor of *Nutrition Reports International.*

ADVISORY BOARD MEMBERS

CONTRIBUTORS

W. M. Beeson, Ph.D.
Emeritus Professor of Agriculture
Purdue University
West Lafayette, Indiana

A. W. M. Brooymans, Ph.D.
Quaker Europe
Brussels, Belgium

Julián A. Buitrago
Livestock Research Coordinator
ELANCO
Cali, Colombia

Leo V. Curtin, Ph.D.
Senior Vice President
NAMOLCO Inc.
Willow Grove, Pennsylvania

P. J. de Wet
Professor, Department of Sheep and
  Wool Science
University of Stellenbosch
Stellenbosch, South Africa

Bernard J. Francis
Animal Feeds Section
Overseas Development Administration
Tropical Products Institute
London, England

James C. Fritz
Division of Nutrition
U.S. Food and Drug Administration
Washington, D.C.

Henry L. Fuller, Ph.D.
Professor of Poultry Nutrition
The University of Georgia
Athens, Georgia

Guillermo G. Gómez
Head, Cassava Utilization
Centro Internacional de Agricultura
  Tropical
Cali, Colombia

G. F. W. Haenlein, Ph.D.
Professor of Animal Science and
  Agricultural Biochemistry
University of Delaware
Newark, Delaware

Olav Herstad
Lic. Agric.
Department of Poultry and Fur Animal
  Science
The Agricultural University of Norway
Ås, Norway

M. L. Kakade, Ph.D.
Manager, Biochemistry
Cargill, Inc.
Minneapolis, Minnesota

J. O. L. King, Ph.D., F.R.C.V.S.
Professor of Animal Husbandry
The University of Liverpool
Veterinary Field Station
Liverpool, England

V. Friis Kristensen
National Institute of Animal Science
Copenhagen, Denmark

T. L. J. Lawrence, Ph.D.
Reader in Animal Husbandry
The University of Liverpool
Veterinary Field Station
Liverpool, England

J. L. L'Estrange, Ph.D.
Department of Agricultural Chemistry
  and Soil Science
Belfield, Dublin, Ireland

James G. Linn, Ph.D.
Extension Dairy Specialist
University of Minnesota
St. Paul, Minnesota

A. L. Livingston
Project Leader
Feedstuffs Research Unit
United States Department of Agriculture
Western Regional Research Center
Berkeley, California

J. C. MacRae, Ph.D.
Energy Metabolism Department
The Rowett Research Institute
Bucksburn, Aberdeen, Scotland

Peter McDonald, Ph.D., D.Sc.
Reader, The University of Edinburgh
Edinburgh, Scotland

W. R. McManus, Ph.D., MAIAS
Associate Professor, Animal Nutrition
  and Physiology
School of Wool and Pastoral Sciences
The University of New South Wales
Kensington, Australia

James M. McNab, Ph.D.
Agricultural Research Council
Poultry Research Centre
Roslin, Scotland

T. W. Perry, Ph.D.
Professor of Animal Science
Purdue University
West Lafayette, Indiana

J. H. B. Roy, Ph.D., D.Sc.
Feeding and Metabolism Department
National Institute for Research in Dairying
University of Reading
Shinfield, Reading, England

Milton L. Scott, Ph.D.
Chairman, Department of Poultry Science
Cornell University
Ithaca, New York

Geoffrey R. Skurray, Ph.D.
School of Food Sciences
Hawkesbury Agricultural College
Richmond, New South Wales, Australia

K. Ross Stevenson
Parliamentary Assistant
Minister of the Environment
Government of Ontario
Toronto, Ontario, Canada

M. L. Sunde, Ph.D.
Chairman, Poultry Science Department
University of Wisconsin-Madison
Madison, Wisconsin

H. L. A. Tarr
Environment Canada
Fisheries and Oceans
West Vancouver Laboratory
West Vancouver, British Columbia,
  Canada

M. van Schothorst, Ph.D.
Nestlé Products Technical Assistance
  Co. Ltd.
La Tour de Peilz, Switzerland

J. F. Wood
Animal Feeds Section
Overseas Development Administration
Tropical Products Institute
London, England

L. G. Young, Ph.D.
Professor, Animal and Poultry Science
University of Guelph
Guelph, Ontario, Canada

DEDICATION

To my inspiring teachers at Cornell University—Harold
H. Williams, John K. Loosli, the late Richard H.
Barnes, the late Clive M. McCay, and the late Leonard
A. Maynard. And to my supportive and beloved
family—Eva, Jack, and Karen.

TABLE OF CONTENTS

Volume I

TABLE OF CONTENTS

Volume II

Specific Processes

EFFECT OF HEAT TREATMENT ON THE NUTRITIONAL VALUE OF FEEDS

M. L. Sunde

INTRODUCTION

Obtaining the most feeding value from our grains and protein sources is probably the most important feed formulation problem today. This problem also involves the study of methods of improving the assimilation of the nutrients from materials not normally considered suitable for animals we value as food. How can the nutrients from the feedstuffs be most efficiently utilized? In feeding a laying hen, Bird,[1] by using simultaneous equations, determined that about 45% of the cost of the diet was for energy and 45% for protein. The other 10% was attributed to vitamins, minerals, and other additives.

This report will consider the effect of heat on materials used as animal feeds and will stress the effect of various heating processes on the efficiency of utilization of our protein and carbohydrate sources. Because of the increased cost of energy, only the minimum amount of heat should be used.

In the first quarter of the 20th century, many farms had steam cookers to prepare feed for their pigs. Experiments at many locations[2] showed that this practice did not improve performance and was no longer recommended. In the mid-1940s, it was possible to produce a "properly cooked soybean meal".[3] Early experiments by Vestal and Shrewsbury[4] and by Hayward et al.[5] clearly showed that soybeans should be heated for young animals. By the early 1960s some researchers and some farmers were again applying heat on the farm but in a more sophisticated way, and they were again producing a high-quality soybean product. Some researchers were also recommending heating, pelleting, wafering, and cooking of grains, completing the cycle begun 50 years before.

Although the intent here is to discuss primarily the effect of heat on soybean protein, where the greatest impact appears to be, some data will deal with the effects of heat on grains and other products.

Heating affects both proteins and carbohydrates, and the presence of moisture generally tends to increase the effectiveness of the heating process. Proteins are partially denatured by this process. Heating interrupts the orderly protein structure and increases digestion efficiency in the young monogastric animal.

The starches in grains are gelatinized by heat processing. If almost complete rupture of all starch granules occurs, weight gains and efficiency of gains are reduced.[6] Hale et al.[7] reported that milo or barley, properly heated and moisturized prior to flaking, will improve gains in cattle by about 10%. Starch digestion was improved considerably in in vitro tests.[8] Corn responds in a similar way for large livestock. Several persons have indicated that similar results can be obtained with poultry, but the author is not aware of any substantiating data.

EFFECT OF HEATING ON SOYBEAN MEAL

Heating affects several aspects of soybean utilization by chickens. Laying hens are not affected as quickly or as much as younger birds. Several workers, including Carver et al.[9] and Fisher et al.,[10] have reported that raw soybeans or soybean meal would do as well for laying hens as properly heated soybean meal. Adding methionine improved performance of laying hens, but did not, in the report by Salman and McGinnis[11] or

Table 1

EFFECT OF SUPPLEMENTING RAW SOYBEAN MEAL WITH
METHIONINE ON LAYER PERFORMANCE

Addition to raw soy	Production (%)	Egg size (g)	Feed consumption (g)
None	51	49	87
+ 0.1% methionine	70	53	113
+ 0.2% methionine	72	54	112
+ 0.3% methionine	78	55	111
+ 0.4% methionine	71	56	107
+ 0.5% methionine	80	55	114
Heated soy and methionine	86	57	125

From Salman, A. J. and McGinnis, J., *Poult. Sci.*, 47, 247, 1968. With permission.

Table 2

EFFECT OF HEAT
TREATMENT OF SOYBEANS
ON METABOLIZABLE ENERGY

	Metabolizable energy	
Heat treatment (107°C)	Soybean flakes (kcal/g)	Soybean (kcal/g)
None	1.94	2.84
5 min	2.70	3.48
10 min	2.76	3.36
40 min	2.86	3.40
60 min	2.77	3.48
120 min	2.42	3.70

From Hill, F. W. and Renner, J., *J. Nutr.*,
80, 375, 1963. With permission.

Latshaw and Clayton[12] permit performance equal to that of the heated soybean meal (Table 1).

This means that one does not need to be as concerned about feeding meals with somewhat less heat processing than normal to laying hens. However, some heating of soybeans is of value for adult chickens as well as for the young.

The use of 30% soybean meal is quite common for chicks because of the increased need for protein. With turkey poults, it is not unusual to use 50% soybean meal. Therefore, more effects on growth and efficiency could logically be expected, even if all ages could handle the raw soybean meal to a fairly satisfactory degree. Saxena et al.[13] found that the young bird cannot digest an "active protein fraction in the raw soybean" because of the lack of a specific enzyme and that a part of this protein is absorbed, and that this part produces the pancreatic hypertrophy observed in young chicks. The destruction of the trypsin inhibitor has also been proposed as a factor.[14] Faba beans when heated to 121°C for 20 min result in improved feed efficiency.[15]

The heating of soybeans also affects other nutrients. Metabolizable energy (ME) is increased with heat for both the full-fat bean and the fat-extracted bean (Table 2). Only the 120-min heat treatment with the flakes reduced the ME. Fat absorption from the full-fat beans increased from 81 to 94%. As indicated earlier, carbohydrate utili-

Table 3
EFFECT OF PELLETING ON
UTILIZATION OF SOYBEANS

	Average gain, 0—51 days (g)		Feed/gain	
	Mash	Pellet	Mash	Pellet
Control	1231	1242	1.90	1.87
Autoclaved	1174	1217	1.98	1.85
Extruded	1160	1297	2.00	1.83
Infrared	1209	1197	1.91	1.82

From White, C. L., Greene, D. E., Waldroup, P. W., and Stephenson, E. L., *Poult. Sci.,* 44, 1180, 1967. With permission.

tion was also increased by the heating process. However, these improvements probably are not as great as those obtained from the effects on the proteins and amino acid availability.

Evaluation techniques using enzymatic vs. acid hydrolysis of soybean meals heated in differing degrees by Reisen et al.[17] and Ingram et al.[18] clearly showed that heating increased amino acid availability.

Overall growth responses have been recorded by many research workers. Wood et al.[19] showed that nitrogen efficiency increased from 21% to 26 to 28% when heat was applied in a sufficient degree to soybean protein. Weight gains were also increased considerably. Saxena et al.[20] and Nesheim and Garlich[21] also reported decreased nitrogen digestibility with raw soybeans.

OTHER METHODS OF HEATING FEEDSTUFFS

Pelleting all or a portion of the diet also appears to affect the utilization of diet. Table 3 shows some of the Arkansas data.[22] In almost every instance, pelleting improved gain and feed conversion. This effect was present regardless of heat treatment. Because of the effect on density of the extrusion process, the pelleting effect was primarily control of density, part cell disruption, and part heat effect. Pelleting and subsequent crumbling had their greatest effect on baby pheasants during the first week.[23]

Pelleting of diets for growing swine will produce similar results. The results of Baird[24] are shown below.

	Meal	Pelleted
Average daily gain (kg)	0.69	0.72
Feed/gain	3.71	3.42

Extruded beans can be prepared on the farm from whole soybeans with somewhat elaborate equipment. A preconditioning chamber that heats the beans is sometimes involved. The beans are then forced through an extruder die under high pressure. When the soybean product leaves the die, the great reduction in pressure ruptures cells. The product is quite acceptable and permits good growth of chicks (Table 4). Some portable units that can be attached to the power takeoff on a farm tractor are also used. The weight gains of birds fed meals prepared by either process are generally lower than those of birds fed the control soybean meal. Feed efficiency is equally good with the control beans or the extruded product. The fairly complete rupturing of the cells makes the oil readily available in the extruded beans. In some studies the extruded

Table 4
EFFECT OF EXTRUDING AND INFRARED
COOKERY ON SOYBEANS FOR POULTRY

	Average gain, 7—28 days (g)	Feed/gain
Control	467	1.61
Raw flakes	324	2.35
Extruded beans	444	1.59
Infrared-cooked beans	429	1.70

From White, C. L., Greene, D. E., Waldroup, P. W., and Stephenson, E. L., *Poult. Sci.,* 44, 1180, 1967. With permission.

Table 5
EFFECT OF INFRARED-COOKED FULL-FAT
SOYBEANS ON TURKEY POULTS

	Weight at 3 weeks[a] (g)	Feed/gain
Soybean meal (46%)	432, a[b]	1.51
Full-fat beans (55%)	406, b	1.53

[a]   Five groups of 20 on each diet.
[b]   Letters that are different are significantly different at 5% levels.

From Shen, T. F., Bird, H. R., and Sunde, M. L., *Poult. Sci.,* 49, 1738, 1970. With permission.

beans have not permitted good gains until the diet is pelleted to increase the density and to possibly decrease the time spent eating.

Another method of treatment involves the infrared cooking of soybeans. This is done by moving the beans past an infrared cell fueled by propane gas. A number of these units are in operation in Wisconsin. The beans come out of the machine still whole, but the skin is cracked if the moisture in the bean is near 15%. Experimental lots were obtained from two of these for evaluation at the Wisconsin station, and the diets were described by Shen et al.[25] Basically, they were the same except that choice white grease was added to the soybean meal diet to keep the energy levels comparable. The diet was comprised of 55% infrared-cooked soybeans for the day-old turkey poults. The poults were reared in electrically heated batteries.

Weight gains were slightly better for the regular soybean meal than for the full-fat beans (Table 5).

A second experiment, in which methionine was also added as a variable (Table 6), was conducted with the other sample of the beans. Again, with or without added methionine, the regular soybean meal supported slightly better growth. The differences were significant; feed efficiency followed the growth pattern rather closely.

More recent data by Roberts[26] shown in Table 7 also illustrate this point. As the percentage of unheated soybean flakes was increased and properly processed, soybean meal decreased, body weight, and feed efficiency decreased.

Work with swine (Table 8) has indicated that the cooked soybeans need to be heated to at least 132°C to permit maximum daily gains.[27]

The proper heating of the infrared cooked full-fat soybeans was considered next from a chemical determination standpoint. Farmers and small feed dealers use a quick

Table 6
EFFECT OF METHIONINE ADDITIONS TO SOY DIETS

	Weight at 3 weeks[a] (g)	Feed/gain
Soybean meal (45%)	411, a[b]	1.49
Soybean meal (45% + 0.2% methionine)	490, b	1.36
Full-fat beans (55%)	378, c	1.62
Full-fat beans (55% + 0.2% methionine)	448, d	1.39

[a] Two or three groups of 20 on each diet.
[b] Letters that are different are significantly different at 5% level.

From Shen, T. F., Bird, H. F., and Sunde, M. L., *Poult. Sci.,* 49, 1738, 1970. With permission.

Table 7
INFLUENCE OF PROPORTIONS OF PROCESSED AND UNHEATED SOYBEAN MEAL ON CHICK PERFORMANCE

Soybean meal		Chick weight		Urease activity	Dye binding
Heated	Unheated	(g)	Feed/gain	(mg/g)	(mg/g)
100	0	223	2.10	37	3.64
80	20	209	2.18	84	3.18
60	40	184	2.46	177	2.75
40	60	176	2.54	228	2.18
20	80	147	2.60	403	1.86
0	100	145	2.88	526	1.33

From Roberts, R., *J. Am. Oil Chem. Soc.,* 53, 302, 1976. With permission.

Table 8
EFFECT OF HEATING ON SOYBEANS FOR PERFORMANCE OF GROWING SWINE

Temperature	Average daily gain (kg)	Feed/gain
115°C	0.50[b]	4.80[b]
132°C	0.62[c]	4.16[c]
143°C	0.66[c]	3.82[d]
Commercial soybean meal	0.66[c]	3.75[d]

Note: Letters that are different are significant at the 5% level of probability. Heating to 132°C seemed to be sufficient for weight gains but more heat was needed for best feed efficiency.

From Seerley, R. W., Emberson, J. W., McCampbell, H. C., Burdick, D., and Grimes, L. W., *J. Anim. Sci.,* 39, 1082, 1974. With permission.

evaluation method that involves thoroughly mixing ten level teaspoons of ground soybean meal and one teaspoon of commercial urea. This is placed in a pint jar (commonly used for canning purposes) and stirred. Five teaspoons of water are then added, the mixture is again stirred, and the lid is screwed or clamped on. After 15 to 30 min the sample is sniffed for presence of ammonia. No ammonia indicates that a sufficient

Table 9
DYE ABSORPTION OF SOYBEANS AS
AFFECTED BY HEAT TREATMENT[23,28]

Defatted raw soybeans	
No treatment	2.59[a]
30 min at 15 lb/in.² steam pressure	2.94
50 min at 140°C	3.44
Commercial soybean meal	
No treatment	3.77
90 min at 15 lb/in.² steam pressure	4.36
Infrared-cooked full-fat beans	
Batch 1	3.77
Batch 2	4.09

[a]   Milligram Cresol Red per gram of meal.

amount of heat was applied. This method will not indicate overheating. The technique of Olomucki and Bornstein,[28] however, shows both underheating and overheating. The results are shown in Table 9. Olomucki and Bornstein reported that values greater than 3.8 and less than 4.3 are considered to be properly heated. When values are less than 3.8 the meal is considered underheated, and when over 4.3 the meal is overheated. This is also shown in Table 5.

EFFECT OF HEAT ON OTHER FEEDSTUFFS

Other grains are also heated by various mechanical devices. McConnell et al.[29] have reported on their experiments and the reports of other workers. With swine, the effects were greatest with soybean meal and barley.

Sunflower meal is another plant protein used in livestock feeding. It is a better source of methionine than soybean meal but is lower in lysine. Heat also improves this product (Table 10).

Recent data by Lawrence[31] indicates improvements for the pig by roasting either whole soybeans or rapeseeds.

METHIONINE AND PROCESSING OF ANIMAL PRODUCTS

The response to methionine (Table 6) was unexpected because the basal diet contained 3% meat scrap and 3% fish meal. Calculations on the methionine and cystine indicate a sufficient amount to meet the National Research Council (NRC) figures without additions.[32] Actual analysis also showed the calculated analysis of methionine and cystine to be accurate. The problem apparently was in availability of the methionine and cystine from meat scrap and fish meal. Further evaluation seemed desirable. This was done by enlisting the cooperation of several experiment stations in our area. The diets, with and without 0.1% methionine, were mixed for all experiment stations at one location. In addition, the same formula plus the same sample of methionine were mixed at each location. All poults were obtained from one source the same day. The data on growth are shown in Table 11. When the control feed (mixed at one location) was fed at any of the four stations, growth varied from 433 to 478 without methionine addition. A response was obtained with methionine in seven out of eight comparisons.

A possible explanation is that the meat scrap had been overprocessed to reduce the chances of salmonella contamination. It has also been suggested that formaldehyde

Table 10
EFFECT OF PROCESSING TEMPERATURE AND L-LYSINE SUPPLEMENTATION ON UTILIZATION OF SUNFLOWER MEAL (SFM) BY THE GROWING RAT

Temperature	Rat weight (g)	Added lysine to meal (%)	SFM No heat	SFM 100°C
			Rat weight	
			(g)	(g)
0°C	42.8	—	—	—
75°C	41.2	0	51	54
100°C	54.5	0.17	80	76
115°C	41.2	0.34	90	95
127°C	38.8	0.51	79	98
Commercial soybean meal	157.6			

Note: Near optimum weight gain was obtained with unheated SFM with 0.34% L-lysine, but better gains were obtained with the heated product. Overheating can also be a problem with this meal.

From Amos, H. E., Burdick, D., and Seerley, R. W., *J. Anim. Sci.,* 40, 90, 1975. With permission.

Table 11
EFFECT OF LOCALLY PURCHASED INGREDIENTS ON TURKEY GROWTH

	Local[a] (g)	Local + 0.1% methionine (g)	Control[a] (g)	Control + 0.1% methionine (g)
Station A	523	518	461	473
Station B	508	518	478	497
Station C	426	437	433	466
Station D	368	374	451	470

[a] See text for description.

may be used during the processing of the fish meal, which could reduce the availability of the sulfur amino acids. Even the most sterile product is not much good nutritionally if the nutrients are unavailable (Table 12).

Weight gains with local mix were not good at one station (Wisconsin). The wire screens quickly showed the effects: droppings were loose and sticky, and gains were low even at 1 week. Samples were available for analysis. The sample absorbed only 3.5 mg of dye per gram of the commercial soybean meal. It was underheated. The data suggest that in spite of knowing how to prepare soybean meal, problems still exist. As processes are altered, amino acid availability is affected. It is necessary for nutritionalists to continually reevaluate ingredients.

SUMMARY

Proper heating of protein sources and cereal grains will result in better availability of nutrients. The availability of the fat in soybeans is increased by proper heating.

Table 12
EFFECT OF LOCALLY PURCHASED INGREDIENTS ON POULTRY FEED EFFICIENCY

	Local[a] (feed/gain)	Local + 0.1% methionine (feed/gain)	Control[a] (feed/gain)	Control + 0.1% methionine (feed/gain)
Station A	1.63	1.49	1.64	1.51
Station B	1.61	1.54	1.58	1.57
Station C	1.54	1.55	1.56	1.55
Station D	1.92	1.80	1.58	1.52

[a]   See text for description.

This results in increased ME. Cereal grains, such as corn, barley, and milo, are partially gelatinized by the heating process.

Flaking, grinding, and pelleting all cause physical disruption of cells and some heating that affects availability of nutrients.

Heating soybeans by the expeller process or the heating following solvent extraction both destroy the trypsin inhibitor or a possible active protein fraction in raw soybeans. The heating increases the ME and amino acid availability.

The use of an extruder process on the farm also produces a very good product when properly done. The product loses density and is improved by pelleting. The use of the infrared cooker also produces a good product under controlled conditions. When the processes are compared directly, the commercially prepared meals usually produce slightly higher gains when fed to poults.

Commercially prepared soybean meals do vary in the amount of heating, and the growth response can vary considerably.

REFERENCES

1. Bird, H. R., Planning manufactured food for efficient egg production, *Feed Age*, 5, 36—39, 1955.
2. Morrison, F. B., *Feeds and Feeding*, 20th ed., Morrison, Ithaca, N.Y., 1945, 1—62.
3. Hayward, J. W., Fifty years of soybean meals, *Feedstuffs*, 42(46), 22—24, 1970.
4. Vestal, C. M. and Shrewsbury, C. L., The nutritive value of soybeans with preliminary observations on the quality of pork produced, in *Proc. 25th Annual Meeting Am. Soc. Anim. Prod.*, Cornell University, Ithaca, N.Y., 1933, 127—130.
5. Hayward, J. W., Steenbock, H., and Bohstedt, G., The effect of heat as used in the extraction of soybean oil upon the nutritive value of the protein, *J. Nutr.*, 11, 219—234, 1936.
6. Bohstedt, G., Criteria for food preparation — Grinding, flaking, wafering, cooking, in *Beef Cattle Science Handbook*, Vol. 4, Agriservices Foundation, Clovis, Calif., 1967.
7. Hale, W. R., Cuitun, L., Saba, W. J., Taylor, B., and Theurer, B., Effect of steam processing and flaking milo and barley on performance and digestion by steers, *J. Anim. Sci.*, 25, 392—396, 1966.
8. Osman, H. F. Theurer, B., Hale, W. H. and Mehen, S. M., Influence of grain processing on *in vitro* enzymatic starch digestion of barley and milo, *Proc. West Sect. Am. Soc. Anim. Sci.*, 25, 593, 1966.
9. Carver, J. S., McGinnis, J., McClary, C. F., and Evans, R. J., The utilization of raw and heat-treated soybean meal for egg production and hatchability, *Poult. Sci.*, 25, 399, 1946.
10. Fisher, H., Johnson, D., Jr., and Ferdo, S., The utilization of raw soybean meal protein for egg production in the chicken, *J. Nutr.*, 61, 611—621, 1957.
11. Salman, A. J. and McGinnis, J., Effect of supplementing raw soybean meal with methionine on performance of layers, *Poult. Sci.*, 47, 247—251, 1968.
12. Latshaw, J. D. and Clayton, P. C., Raw and heated full fat soybeans in laying diets, *Poult. Sci.*, 55, 1268—1272, 1976.

13. Saxena, H. C., Jensen, L. S., and McGinnis, J., Influence of age on utilization of raw soybean meal by chickens, *J. Nutr.*, 80, 391—396, 1963.
14. Kunitz, M., Crystalline soybean trypsin inhibitor. I. General properties, *J. Gen. Physiol.*, 30, 291, 1947.
15. Campbell, L. D. and Marquardt, R. R., *Poult. Sci.*, 56, 442—448, 1977.
16. Hill, F. W. and Renner, R., Effects of heat treatment on the metabolizable energy value of soybeans and extracted soybean flakes for the hen, *J. Nutr.*, 80, 375—380, 1963.
17. Reisen, W. H., Clandinin, D. R., Elvehjem, C. A., and Cravens, W. W., Liberation of essential amino acids from raw properly heated and overheated soybean oil meal, *J. Biol. Chem.*, 167, 143—150, 1947.
18. Ingram, G. R., Ricson, W. W., Cravens, W. W., and Elvehjem, C. A., Evaluating soybean oil meal protein for chick growth by enzymatic release of amino acids, *Poult. Sci.*, 28, 898—902, 1949.
19. Wood, A. S., Summers, J. D., Moran, E. T., Jr., and Pepper, W. F., The utilization of unextracted, raw and extruded full-fat soybeans by the chick, *Poult. Sci.*, 50, 1392—1399, 1971.
20. Saxena, H. C., Jensen, L. S., and McGinnis, J., Protein metabolism in chicks fed raw soybean meal, *Poult. Sci.*, 42, 788—790, 1963.
21. Nesheim, M. C. and Garlich, J. D., Digestibility of unheated soybean meal for laying hens, *J. Nutr.*, 88, 187—192, 1966.
22. White, C. L., Greene, D. E., Waldroup, P. W., and Stephenson, E. L., The use of unextracted soybeans for chicks. I. Comparison of infrared cooked, autoclaved, and extruded soybeans, *Poult. Sci.*, 44, 1180—1185, 1967.
23. Sunde, M. L. and Bird, H. R., The niacin requirement of the young Ringneck pheasant, *Poult. Sci.*, 36, 34—42, 1967.
24. Baird, D. M., Influence of pelleting swine diets on metabolizable energy, growth and carcass characteristics, *J. Anim. Sci.*, 36, 516—521, 1973.
25. Shen, T. F., Bird, H. R., and Sunde, M. L., Cooked full-fat soybeans for turkey poults, *Poult. Sci.*, 49, 1738—1740, 1970.
26. Roberts, R., Protein products and hulls for animal feeds, *J. Am. Oil Chem. Soc.*, 53, 302—304, 1976.
27. Seerley, R. W., Emberson, J. W., McCampbell, H. C., Burdick, D., and Grimes, L. W., Cooked soybeans in swine and rat diets, *J. Anim. Sci.*, 39, 1082—1091, 1974.
28. Olomucki, E. and Bornstein, S., The dye absorption test for the evaluation of soybean meal quality, *J. Assoc. Off. Agric. Chem.*, 43, 440, 1960.
29. McConnell, J. C., Skelley, G. C., Handlin, D. L., and Johnstone, W. E., Corn, wheat, milo and barley with soybean meal or roasted soybeans and their effect on feedlot performance, carcass traits and pork acceptability, *J. Anim. Sci.*, 41, 1021—1030, 1975.
30. Amos, H. E., Burdick, D., and Seerley, R. W., Effect of processing temperature and L-lysine supplementation on utilization of sunflower meal by the growing rat, *J. Anim. Sci.*, 40, 90—95, 1975.
31. Lawrence, T. L. J. Effects of micronization on the digestibility of whole soya beans and rapeseeds for the growing pig, *Anim. Feed Sci. Technol.*, 3, 179—182, 1978.
32. National Research Council-National Academy of Sciences, Nutrient Requirements of Poultry, National Academy of Sciences, Washington, D.C., 1971.

EFFECT OF PROCESSING ON NUTRIENT CONTENT OF FEEDS: FREEZING*

J. C. MacRae

INTRODUCTION

Although the freeze storage of perishable human foodstuffs has grown over the last 30 to 40 years into a worldwide and multibillion dollar industry, the practice has found little application in the preservation of animal feedstuffs. In recent years it has been used quite extensively in research to preserve fresh temperate herbages of varying quality for feeding at a later date to ruminant animals, but its only significant commercial application has been in mink farming.

Following earlier studies in the United Kingdom and Canada which suggested that freezing freshly cut, high-quality temperate pasture had little effect on its intake[1] or apparent digestion[2] by sheep, or on the nitrogen or energy metabolism of cattle,[3] there was a widespread acceptance of the idea that data from experiments made with freeze-stored herbage could be readily extrapolated to fresh herbage.[4,5] This led to the development of large-scale cold storage facilities[6] at several research centers, e.g., the Grasslands Research Institute, Hurley, U.K.; the Ruakura Research Centre, New Zealand; and the Prospect Laboratories, Sydney, Australia. More recent studies have demonstrated, however, that freezing high-quality herbage does alter the physicochemical composition of the material to an extent likely to affect its nutritive value to the ruminant. Although they have not been studied in detail, there is no reason to assume that other changes known to occur during the freezing of vegetable foodstuffs for human consumption do not also occur in herbage.

Traditionally, feeds for ranch mink have been mainly fresh or frozen, high-quality protein materials derived from other mammals. In the past, horse meat was favored feed, but byproducts of the meat, fish, and poultry industries are used currently in combination with cereal, cereal grain, and grain byproducts.[7] Feed manufacturers are now starting to produce completely dehydrated feeds which do not require refrigeration.

This review will be limited to a discussion of mechanisms associated with the freezing of plant and animal tissues based on (1) observations made in experimental nutrition studies and (2) published literature dealing with deteriorative changes in the storage of feed for ranch mink.

PHYSICOCHEMICAL REACTIONS ASSOCIATED WITH FREEZING

Despite innumerable publications on freezing of animal and vegetable tissues, relatively few studies have been made of the causal mechanisms associated with the changes which take place on freezing and thawing. Readers are referred to articles by Love[8] and Jansen[9] for a detailed discussion of the known physicochemistry of freezing as it affects human foodstuffs; the present discussion will attempt to deal only with the mechanisms governing reactions which are relevant to the freeze storage of herbage and of mink diets.

As biological materials are frozen, ice crystals form in the extracellular space. At normal freezer temperatures (−10 to −30°C) there is little intracellular freezing; instead water is drawn out of the cell. The size of these crystals is inversely proportional to

---

* Article submitted 1976.

the freezing rate,[10] so slower rates of freezing give more opportunity for physical damage to the cell brought about by the formation of large ice crystals. Ice crystal formation outside the cell is accompanied by a concentration of the solutes within the cell, not unlike the process which takes place during heat dehydration. It is mainly this physical process which sets up conditions where physicochemical changes can occur, either by altering the environmental conditions within the cell (e.g., solute concentrations or pH) or simply by bringing enzymes and substrates into closer proximity.

Changes in Texture — Protein Denaturation

Protein denaturation is the term given to the process which involves the irreversible precipitation of protein out of solution. It leads to a reduction in water-binding capacity, which results in drip loss upon thawing and a toughening of animal tissues during subsequent thawing and cooking. Protein denaturation takes place in the freezing of both plant and animal tissues but appears (at least on the evidence of the amount of study it has been given) to be more important in fish muscle tissues.

It is still not clear if the phenomenon is largely a physical effect brought about by chemical reactions by an overconcentration of protein as water is withdrawn from the cell, possibly causing the formation of increased numbers or increased strength of bonds between the constituent myofibrillar proteins.[11] Evidence from studies on fish protein denaturation supports the significance of chemical reactions, in that free fatty acids appear to cause denaturation of the actomysin components of fish muscle.[8] Indeed, in fatty fish such as herring and halibut which contain considerable amounts of neutral lipids, these lipids appear to protect the protein by absorbing some of the released free fatty acids during cold storage.[8] Alternatively, cells of marine fish of the gadoid species (cod, haddock, whiting) contain a chemical, trimethylamine oxide, which tends to be enzymically broken down into formaldehyde and dimethylamine upon cold storage.[12] Formaldehyde is known to form acid-reversible cross-linkages with amino and amide groups of protein, rendering them insoluble at nonacidic pH values,[13] and this could cause considerable denaturation in these species.

Probably because freezing in itself has little effect on the quality of meat as assessed by taste and other organoleptic tests,[14] protein denaturation in meat and poultry products has been studied less. However, the slow freezing out of water from intercellular spaces, a decrease in the degree of binding of water by protein molecules, and changes in their tertiary and quaternary structures have been observed;[15] drip losses in thawing of meat products can contain some 11 to 15% protein.

Production of Off-Flavor — Oxidative Rancidity

The causes of off-flavoring are numerous and complex. Off-flavors can develop in any frozen tissue containing appreciable amounts of lipid material. In human foodstuffs, oxidative rancidity most commonly occurs in animal tissues, because the blanching pretreatments used in the preparation of frozen vegetable products apparently inactivate the enzyme responsible for the most common type of off flavor production.[16]

Fats undergo two types of change in the process of off-flavor development. The first is a simple lipase hydrolysis of the fat into its component parts, glycerol, and free fatty acids. This reaction does not give rise to serious problems in frozen meat products, but it is important in milk products. The predominant source of off-flavor from fat breakdown products in frozen foods is an oxidative reaction. Figure 1 illustrates the chain reactions involved when unsaturated fats are frozen in the presence of oxygen. The first product of oxidation, the hydroperoxides, do not themselves impart rancidity. They can, however, give rise to a whole series of reactions, the most important of which leads to fission products (see Figure 1). The rancid odor and flavor in

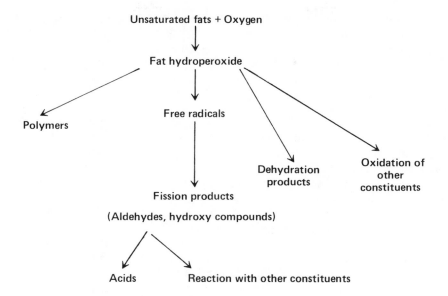

Unsaturated fats + Oxygen

Fat hydroperoxide

Polymers

Free radicals

Dehydration products

Oxidation of other constituents

Fission products

(Aldehydes, hydroxy compounds)

Acids   Reaction with other constituents

FIGURE 1.   Reactions involved in rancidity formation. (From Jansen, E. F., in *Quality and Stability in Frozen Foods,* van Arsdel, W. B., Copley, M. J., and Olsen, R. L., Eds., John Wiley & Sons, New York, 1969, 19. With permission.)

food is mainly due to these volatile aldehydes and aldehyde derivatives. Aldehydes will also react with amino acids (Maillard reaction) to form colored compounds such as those observed in the "rusting" of stored frozen fish.

The oxidative processes can be speeded up in frozen fish and meat by the presence of accelerators such as organic catalysts like hemoglobin and trace metal ions, e.g., copper and iron. Fortunately, certain naturally occurring substances such as tocopherol and vitamin E are antioxidants. These interrupt the chain reaction and so delay the onset of rancidity. Synthetic antioxidants have also been developed. The use of tocopherol supplements for poultry diets[17] and synthetic antioxidants in the ice used for glazing fish[18] are examples of practical attempts to prevent oxidative rancidity in freeze-stored products.

Production of off flavoring compounds in freeze-stored plant tissue follows the reactions shown in Figure 1. However, these are at least three related control mechanisms which can initiate these reactions.[19] Autoxidation, the slowest type, does not require any enzyme action, as it is catalyzed by light and free radicals. When chlorophyll is also present a sensitized photooxidation occurs at a much faster rate. In addition, lipoxygenase (also called lipoxidase) will catalyze the primary reaction of oxidation of linoleic acid c18:2 (which makes up some 10 to 15% of leaf lipids[20]) and, to a lesser extent, linolenic acid c18:3 (60% of leaf lipids[20]).[21]

It is the destruction of lipoxygenase in the blanching process which prevents serious off-flavor production in frozen vegetables.[16]

Changes in Color

When green plant tissues are stored frozen at temperatures above −20°C, there is a pronounced color change from the rich green of the fresh material to a much duller green. The green color is due to the photosynthetic pigments, chlorophyll a and b. During cold storage a hydrolysis reaction can occur, whereby the magnesium ions contained in chlorophyll a and b are replaced by hydrogen ions to form an olive-colored pigment, pheophytin a and b.[9] A similar kind of reaction occurs on the surface of freeze-stored beef, but in this case the color change is due to oxidation. Myoglobin, a purplish-red muscle pigment (very similar to the hemoglobin of blood) becomes oxygenated to the red oxymyoglobin in cut, fresh meat. However, when meat is freeze stored, the ferrous ion of the oxymyoglobin is oxidized and metmyoglobin, a red brown pigment, is formed.[22]

Losses of Vitamins

Significant losses of vitamins can occur when vegetable and animal tissues are freeze stored for human consumption.[23] These losses are caused mainly by enzymic or chemical oxidation, and they can be alleviated to a considerable extent in most vegetable products by blanching before freezing; losses of vitamin C and thiamin in blanched peas, green beans, and sprouts are usually less than 20 to 25 and 10%, respectively.[24-26] Chemical oxidation is caused mainly by products of fat oxidation, and these products have been reported to destroy the fat-soluble vitamins A, D, E, and K as well as certain B vitamins in freeze-stored meat products.[27]

CHANGES IN THE PHYSICOCHEMICAL COMPOSITION OF FREEZE-STORED HERBAGE

Although erratic changes in the carbohydrate and nitrogen components of fresh-frozen leaf tissue stored at −20°C had been reported earlier,[28] it was only when considerable invertase activity (breakdown of sucrose to glucose and fructose) and reduction in nitrogen solubility were found to accompany the freeze storage of grass and clover[29] that animal nutritionists began to examine the possible implications of such changes on the nutritive value of the material.

The magnitude of the physicochemical changes reported in a subsequent investigation,[30] where herbages of differing quality were analyzed freshly cut, frozen, and thawed for periods of 3, 6, 12, and 24 hr is illustrated in Table 1 and Figure 2. There was little change in the chemical composition of the poor quality grass, possibly because this material was collected from a hill site in late winter after several periods of sub-zero temperatures, but the composition of the spring-grown, higher quality grass and clover was altered considerably. As invertase hydrolyzed sucrose there was an increase in glucose and a corresponding doubling of total reducing sugar (glucose plus fructose) (Table 1); some 30 to 50% of the total hexose sugar present in the freshly cut herbage was accounted for in this way as hydrolyzable sucrose. Total hexose and total nitrogen contents remained unchanged throughout the freezing and thawing, but the solubility of the protein-nitrogen fraction, extracted from the tissue by grinding in pH 7 phosphate buffer at 0°C, was reduced by up to 50% on freezing (Figure 2). This latter observation is very similar to those made when fish products are frozen.[31] The concentrations of nonprotein nitrogen, extracted by the phosphate buffer, and of cell wall carbohydrates and the in vitro digestibility of the herbages showed little change.

Other workers have reported a similar marked reduction in soluble nitrogen but little change in hot water-extracted carbohydrate in freeze-stored mixed pasture.[32]

Although there is no information which bears directly on the point, it is unlikely that the hydrolytic changes which occur in the carbohydrate components of frozen herbage (Table 1) have any significant effect on the nutritive value of the herbage when fed to ruminant animals, since this reaction is one of the first steps in the digestive process when the animals consume the material.

On the other hand, the considerable reduction in nitrogen solubility brought about by freezing (Figure 2) may have a considerable influence on the nutritive value of frozen herbage. Ruminant digestion is characterized by a pregastric microbial fermentation which takes place predominantly in the reticulorumen. The rumen microorga-

Table 1

CHANGES IN THE CONCENTRATIONS (mg/g DM) OF TOTAL HEXOSE (T), REDUCING SUGAR (R), AND GLUCOSE (G) IN THE ALCOHOL-EXTRACTED "SOLUBLE SUGARS" UPON FREEZING AND THAWING HERBAGES[30]

| | Species | | | | | | | | | | | |
| Sampling times | Agrostis-Festuca poor | | | Agrostis-Festuca good | | | Ryegrass | | | Clover | | |
	T	R	G	T	R	G	T	R	G	T	R	G
Fresh	37	3.4	1.5	84	12.4	3.7	243	40.7	15.6	109	39.4	21.7
Frozen	23	6.2	2.8	99	13.8	10.9	282	113.3	55.1	108	70.9	28.6
Thawed												
3 hr	23	12.5	6.3	91	26.4	19.0	252	106.8	59.9	105	78.0	34.7
6 hr	52	25.1	11.1	81	35.1	19.9	253	155.6	71.6	124	73.9	42.6
12 hr	36	18.7	8.9	92	47.9	22.2	232	150.3	66.5	117	82.3	39.7
24 hr	36	14.9	5.0	77	37.0	17.2	292	166.3	74.8	127	79.6	44.8
Weighted S.E.	±1.0	±0.60	±0.23	±4.7	±1.89	±0.63	±6.9	±5.07	±1.69	±4.0	±4.44	±1.45

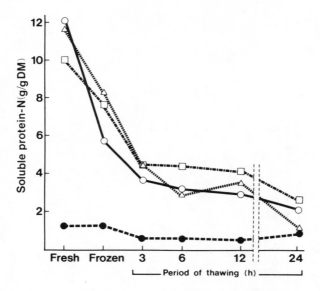

FIGURE 2.   Changes in soluble protein-nitrogen content upon freezing and thawing of poor quality *Agrostis-Festuca* (● — — — — ●) and of high-quality *Agrostis-Festuca* (○ —————— ○), perennial ryegrass (□ — · — · — □), and clover (Δ — — — — - Δ) herbages. (From MacRae, J. C., Campbell, D. R., and Eadie, J., *J. Agric. Sci.,* 84, 125, 1975. With permission.)

nisms degrade dietary protein to peptides and amino acids, which are largely deaminated to ammonia.[33] This ammonia is either utilized along with amino acids and shorter chain peptides for microbial protein synthesis (protein which will ultimately become available to the host animal when the microbes leave the rumen and are enzymically digested in the abomasum — the true stomach — and small intestine) or it is absorbed directly from the rumen as ammonia, passing via the portal blood to the liver. Here it is converted to urea and ultimately excreted from the body via the kidneys into the urine. The proportion of ammonia lost to the animal in this way will depend on the diet, but the loss is considerable in high-nitrogen diets such as fresh herbage (3.5 to 4.5% nitrogen). On such diets there is a rapid production of ammonia immediately after feeding. This rate of ammonia production exceeds the capacity of the microbes to utilize it for protein production. As a result, considerable amounts (up to 25% of the dietary nitrogen intake) of the ammonia can be absorbed directly from the rumen.[34] Freezing reduces the solubility of the nitrogen in such herbage, and it is postulated that this reduces the rate of rumen ammonia production immediately after feeding and gives rise to a more stable pattern of ammonia production over time. This, in turn, leads to more efficient utilization of the ammonia by the microorganisms and less direct absorption.[35]

   While much of the foregoing argument is at the present time circumstantial and speculative, Beever and co-workers[36] recently observed that, when frozen perennial ryegrass was fed to sheep, they obtained a 26% increase in microbial protein production and a 15% increase in the amount of amino acids entering the small intestine with the frozen herbage compared to freshly cut material.[36] Microbial protein production per 100 g of organic matter digested within the rumen (a good indication of the efficiency of nitrogen digestive processes within the rumen) was 10.2 g for the frozen herbage but only 7.7 g for the same herbage fed freshly cut, while values for Y(ATP or grams microbial mass per mole ATP used) were 8.1 and 6.3, respectively.[48] Although the microbial production values were some 30% higher on the frozen herbage diet than those obtained with the fresh herbage, they were still lower than other values obtained with artificially dried herbages[37] where nitrogen solubility was reduced even further.

If, as has been postulated, the productive performance of animals given fresh pasture is limited by the amount of amino N absorbed from the small intestine,[35] then freeze storing of herbage, which will enhance the utilization of its nitrogen component, will lead to an improvement in its nutritive value.

Unfortunately, few investigations have been carried out dealing with other changes occurring during the freeze storage of herbage. A darkening in color when pasture is frozen has been reported,[6] and personal observation would suggest that this is a common phenomenon under the conditions associated with large-scale freezing and storage of herbage. It is very probably caused by a breakdown of cholorophyll pigments to pheophytin compounds (noted earlier).

Research into human foodstuffs would suggest that blanching is a necessary prerequisite to cold storage of plant materials if enzyme oxidation of fatty acids and trace nutrients is to be kept to a minimum. There is no attempt made to blanch forages prior to freezing, so it might be expected that these reactions can proceed to some extent. Indeed it might be expected that off-flavoring could occur in freeze-stored herbage because there is considerable lipoxygenase activity in leaf tissue.[38] The fact that ruminant animals will eat similar quantities of fresh and frozen herbages,[16] suggests that oxidative degradation of unsaturated fatty acids is in some way inhibited in frozen herbage, possibly by agents such as α-tocopherol normally located in cell mitochondria, or that off-flavors do not lead to any significant reduction in voluntary intake. These questions require further study.

Losses of water-soluble vitamins (vitamin C and the B vitamins) and of the fat-soluble vitamin K will be of little nutritional consequence to herbivores as these can be synthesized in sufficient quantities to meet the animal's requirements by the microorganisms of the digestive tract. However, there is a need for studies on losses of vitamins A, D, and E following the cold storage of forages, deficiencies of which can cause problems in ruminant animals.

PHYSICOCHEMICAL CHANGES IN FREEZE-STORED MINK DIETS

Freezing and thawing can damage the ingredients of mink diets. Mink feeds (e.g., unwashed chicken offal) characteristically have inherently high bacterial counts and require immediate cold storage. Short-term storage at temperatures above freezing probably has no important effect on nutrient quality, but if such storage is prolonged, psychrotrophic organisms multiply and eventually cause spoilage.[39] There has been little research centered specifically on mink diets but, since the feed ingredients are basically similar to those of human diets, the changes that occur during the cold storage of mink diets have been fairly well established.

The main deteriorative changes that can occur during the freezing of meat, poultry, and fish byproducts are protein denaturation and oxidative rancidity of fats. Protein denaturation results in a toughening of the meat and a reduction in the water-binding capacity. There is little evidence that denaturation of protein in freeze-stored material has any serious effect on its nutritive value to the mink. Kolakowski and co-workers[41] reported a lower availability of lysine (when measured chemically[40]) in frozen as compared to fresh minced flesh of Baltic herring, but there is little evidence to suggest that such differences are common. Indeed, the digestibility values of fresh and frozen fish products are very similar. Drippings of thawed ranch mink diet ingredients contain small but significant amounts of protein and various vitamins,[15,42] but these are generally fed along with the thawed product.

Probably the most serious effect on the nutritive value of mink rations occurs when poor freezing and storage procedures allow oxidative rancidity to proceed, i.e., the oxidative breakdown of unsaturated linkages of fatty acids to form peroxides, aldehydes, and other derivatives (see earlier discussion). The reactions destroy essential fatty acids and other nutrients, especially fat-soluble vitamins A, D, E, and K, as well as some of the B vitamins including thiamine, folic acid, pantothenic acid, pyridoxine, biotin, and vitamin B_{12}.[27]

If feedstuffs containing oxidized fat are fed to mink, a disease condition known as "yellow fat", "steatitis", or "watery-hide disease" can result.[43] Affected animals exhibit anorexia, progressive weakness, paralysis, and a high mortality rate. If the disease occurs during the furring period, the developing under fur is often produced without pigment, i.e., "cotton fur", a specific symptom of iron deficiency.[44] Fortunately, the symptoms are easily prevented or alleviated by the addition of vitamin E or other antioxidant materials to the ration.[7] The feeding of freeze-stored fish of the gadiod species (Pacific hake, *Merlucius productus*, and whiting, *Merlucius bilinearis*) can also produce "cotton fur".[45] Affected animals are small, thin, anemic, and lacking in fur pigment. It is thought that formaldehyde (450 to 600 ppm in raw, frozen Pacific hake) produced on freeze storage as a breakdown product of thimethylamine oxide[12] (see earlier discussion) reduces the absorption of dietary iron by the mink. Parenteral administration of iron restores normal body iron levels, but the addition of iron to the diet is not effective.[46] Heat processing (88°C for 5 min) inactivates the depressing effects of formaldehyde on iron absorption, possibly by coupling the formaldehyde with free amino groups of protein in a Maillard-type reaction.[47]

CONCLUSIONS

It would appear from the limited data available that freeze storage of plant and animal tissues to be used as livestock feed can have significant effects on the physico-chemical composition and on the subsequent nutritional value of the material. It is doubtful whether economic considerations will ever allow freeze storage to have any widespread practical application in the feedstuffs industry. However, it is obvious that, where feeds so treated are to be used in research or in livestock production, further work on the characterization of changes occurring during cold storage and the nutritional implications of these changes to the animals concerned will need to be undertaken.

REFERENCES

1. Pigden, W. J., Pritchard, G. I., Winter, K. A., and Logan, V. S., Freezing — a technique for forage investigations, *J. Anim. Sci.*, 20, 796—801, 1961.
2. Raymond, W. F., Harris, C. E., and Harker, V. G., Studies on the digestibility of hergate. II. Effect of freezing and cold storage of herbage on its digestibility by sheep, *J. Br. Grassl. Soc.*, 8, 315—320, 1953.

3. Ekern, A., Blaxter, K. L., and Sawers, D., The effect of artificial drying and of freezing on the energy value of pasture herbage, in *Energy Metabolism,* Blaxter, K. L., Ed., Academic Press, New York, 1965, 217—224.

4. Rattray, P. V. and Joyce, J. P., The utilization of perennial ryegrass and white clover by young sheep, *Proc. N.Z. Soc. Anim. Prod.,* 29, 102—113, 1969.

5. Beever, D. E., Thomson, D. J., Pfeffer, E., and Armstrong, D. G., The effect of drying and ensiling grass on its digestion in sheep, *Br. J. Nutr.,* 26, 123—134, 1971.

6. Hutton, J. B., Hughes, J. W., Bryant, A. M., Pluck, L. J., and Taylor, R. E. C., Evaluation of high moisture forages with ruminants. I. Equipment and techniques for harvesting, processing and storing high-moisture forages, *N.Z. J. Agric. Res.,* 18, 37—43, 1975.

7. Stout, F. M., Freezing and storage of fish, poultry, and meat, *Symp. on Effect of Processing on the Nutritional Value of Feeds,* National Research Council-National Academy of Sciences, Washington, D.C., 1972, 383—392.

8. Love, R. M., The freezing of animal tissue, in *Cryobiology,* Meryman, H. T., Ed., Academic Press, London, 1966, 317—401.

9. Jansen, E. F., Quality-related chemical and physical changes in frozen foods, in *Quality and Stability in Frozen Foods,* van Arsdel, W. B., Copley, M. J., and Olsen, R. L., Eds., John Wiley & Sons, New York, 1969, 19—42.

10. Kaminarskaya, A. K., Character of crystallisation in animal tissue during freezing under different conditions, *Bull. Inst. Int. Froid,* 2 (Suppl.), 229—236, 1972.

11. Connell, J. J., Changes in amount of myosin extractable from cod flesh during storage at $-14°$, *J. Sci. Food Agric.,* 13, 607—617, 1962.

12. Amano, K. and Yamada, K., Formaldehyde formation from trimethylamine oxide by the action of pyloric caeca of cod, *Bull. Jpn. Soc. Sci. Fish,* 30, 639—645, 1964.

13. Ferguson, K. A., Hemsley, J. A., and Reis, P. J., The effect of protecting dietary protein from microbial degradation in the rumen, *Aust. J. Sci.,* 30, 215—217, 1967.

14. Jul, M., Quality and stability of frozen meats, in *Quality and Stability in Frozen Foods,* van Arsdel, W. B., Copley, M. J., and Olsen, R. L., Eds., John Wiley & Sons, New York, 1969, 191—216.

15. Gorna, M., An attempt to determine changes occurring in sarcoplasm proteins due to the long term frozen storage of beef, *Bull. Inst. Int. Froid,* 2(Suppl.), 241—249, 1972.

16. Rhee, K. S. and Watts, B. M., Lipid oxidation in frozen vegetables in relation to flavour change, *J. Food Sci.,* 31, 675—679, 1966.

17. Mecchi, E. P., Pool, M. F., Nonaka, M., Klose, A. A., Marsden, S. J., and Lillie, R. J., Further studies on tocopherol content and stability of carcass fat of chickens and turkeys, *Poult. Sci.,* 35, 1246—1251, 1956.

18. Tarr, H. L. A., Control of rancidity in fish flesh. II. Physical and chemical methods, *J. Fish. Res. Board Can.,* 7, 237—247, 1948.

19. Hitchcock, C. and Nichol, B. W., *Plant Lipid Biochemistry,* Academic Press, London, 1971, 223—235.

20. Garton, G. A., Lipid metabolism in herbivorous animals, *Nutr. Abst. Rev.,* 30, 1—16, 1960.

21. Holman, R. T., Egwim, P. O., and Christie, W. W., Substrate specificity of soya bean lipoxidase, *J. Biol. Chem.,* 244, 1149—1151, 1969.

22. Ramsbottom, J. M. and Koonz, C. G., Freezer storage temperature as related to drip and to color in frozen-defrosted beef, *Food Res.,* 6, 571—580, 1941.

23. Cook, D. J., The nutritional value of frozen foods. II. The composition of frozen foods, *Bull. Br. Nutr. Found.,* No. 12, 42—56, 1974.

24. Morrison, M. H., The vitamin C and thiamin contents of quick frozen peas, *J. Food Technol.,* 9, 491—500, 1974.

25. Morrison, M. H., The vitamin C content of quick frozen green beans, *J. Food Technol.,* 10, 19—28, 1975.

26. Abrams, C. I., The ascorbic acid content of quick frozen brussel sprouts, *J. Food Technol.,* 10, 203—213, 1975.

27. Barnes, R. H., Clausen, M., Rusoff, J. J., Hanson, H. T., Swenseid, M. E., and Burr, G. O., The nutritive characteristics of rancid fats, *Arch. Sci. Physiol.,* 2, 313—328, 1948.

28. Perkins, H. J., Note on chemical changes occurring in freeze-dried and fresh-frozen wheat leaves during storage, *Can. J. Plant Sci.,* 41, 689—691, 1961.

29. MacRae, J. C., Changes in chemical composition of freeze-stored herbage, *N. Z. J. Agric. Res.,* 13, 45—50, 1970.

30. MacRae, J. C., Campbell, D. R., and Eadie, J., Changes in the biochemical composition of herbage upon freezing and thawing, *J. Agric. Sci.,* 84, 125—131, 1975.

31. Cowie, W. P. and Mackie, I. M., Examination of the protein extractability method for determining cold-storage protein denaturation in cod, *J. Sci. Food Agric.,* 19, 696—700, 1968.

32. **Bryant, A. M. and Newth, R. P.,** Evaluation of high-moisture forages with ruminants. II. Changes in the chemical composition of herbage after freezing and thawing, *N. Z. J. Agric. Res.,* 18, 375—378, 1975.

33. **Annison, E. F. and Lewis, D.,** *Metabolism in the Rumen,* Methven, London, 1962, 92—118.

34. **MacRae, J. C. and Ulyatt, M. J.,** Quantitative digestion of fresh herbage by sheep. II. The sites of digestion of some nitrogenous constituents, *J. Agric. Sci.,* 82, 309—319, 1974.

35. **MacRae, J. C.,** Utilisation of the protein of green forage by ruminants at pasture, in *From Plant to Animal Protein — Reviews in Rural Science II,* Sutherland, T. M., McWilliam, J. R., and Leng, R. A., Eds., University of New England Publishing Unit, Armidale, Australia, 1976.

36. **Beever, D. E., Cammell, S. B., and Wallace, A.,** The digestion of fresh, frozen and dried perennial ryegrass, *Proc. Nutr. Soc.,* 33, 73A-74A, 1974.

37. **Hogan, J. P. and Weston, R. H.,** Quantitative aspects of microbial protein synthesis in the rumen, in *Physiology of Digestion and Metabolism in the Ruminant,* Phillipson, A. T., Ed., Oriel Press, Newcastle-upon-Tyne, 1970, 474—485.

38. **Holden, M.,** Lipoxidase activity of leaves, *Phytochemistry,* 9, 507—512, 1970.

39. **Rolfe, E.,** Characteristics of preservation processes as applied to proteinaceous foods, in *Proteins as Human Foods,* Lawri, R. A., Ed., AVI Publishing, Westport, Conn., 1970, 107—125.

40. **Carpenter, K. J.,** The estimation of the available lysine in animal protein foods, *Biochem. J.,* 77, 604—610, 1960.

41. **Kolakowski, E., Fik, M., and Karminska, S.** Investigations into changes in the protein nutritive value of frozen fish sausages produced from fresh and frozen minced flesh, *Bull. Inst. Int. Froid,* 2(Suppl.), 59—63, 1972.

42. **Schweigert, B. S. and Lushbough, C. H.,** The effects of commercial processing on the nutrient composition of meat, poultry and fish products, in *Nutritional Evaluation of Food Processing,* Harris, R. S. and Von Loesecke, H., Eds., AVI Publishing, Westport, Conn., 1971, 261—304.

43. **Lalor, R. J., Loeschke, W. L., and Elvehjem, C. A.,** Yellow fat in the mink, *J. Nutr.,* 45, 183—188, 1951.

44. **Stout, F. M., Oldfield, J. E., and Adair, J.,** Fur abnormalities in mink, in *Progress Reports of the Mink Farmers Research Foundation,* Mink Farmers Research Foundation, Milwaukee, 1963.

45. **Stout, F. M., Oldfield, J. E., and Adair, J.,** Nature and cause of the "cotton-fur" abnormality in mink, *J. Nutr.,* 70, 421—426, 1960.

46. **Stout, F. M., Oldfield, J. E., and Adair, J.,** Aberrant iron metabolism and the "cotton-fur" abnormality in mink, *J. Nutr.,* 72, 46—52, 1960.

47. **Stout, F. M., Adair, J., and Oldfield, J. E.,** Feeding Pacific hake to mink, *U.S. Fish Wildl. Serv. Circ.* No. 332, 149—152, 1970.

48. **Beever, D. E., Cammel, S. B., and Wallace, A.,** unpublished data.

EFFECT OF PROCESSING ON NUTRITIVE VALUE OF FEEDS: DEHYDRATION

A. L. Livingston

INTRODUCTION

The preservation of farm crops for use in animal feeds by drying to a low moisture level so as to inactivate enzymes and microorganisms and to prevent spoilage is an ancient practice. For centuries, solar heat was the chief source of energy for the drying of plant materials on the farm. Excessive nutrient and dry matter losses during the harvesting and drying of forages,[1,2] particularly those losses due to rain damage,[3] prompted the development of artificial drying or dehydration of forage crops.[4,5] Comparison of the nutrient quality of sun-dried and artificially dehydrated forages found the latter to be superior for many animal feeds, particularly dairy[6,7] and poultry[8-10] feeds where β-carotene and vitamin E are essential nutrients.

Initially, dehydrators were concerned with the preservation of provitamin A (largely β-carotene) and vitamin E (α-tocopherol) during the dehydration of freshly cut forages.[11,12] However, other primary feed uses, such as pigmentation of poultry products,[13,14] and improved methods of analyses to determine digestibility of protein[15,16] and fiber[17] coupled with animal feeding trials,[18] have led to new emphasis on those enzymatic and chemical analyses that can be related to animal digestibility.

Not all forage and feed ingredients used in the preparation of feed will be described. However, the effect of dehydration on the principal nutrients as well as some minor ones will be described in this review.

Although several makes of dehydrators are successfully used to dry forages and other crops, this study shall not endeavor to describe advantages of the different makes. Figure 1 presents a typical forage crop dehydrator.

CAROTENOIDS

The loss of carotenoids during the drying of forage plants has been found to be from three sources: enzymes,[19,20] oxidation,[22-24] and isomerization.[25-27] Most of the enzymatic losses begin following cutting and continue until the plant is sufficiently dry that the enzymes are inactivated; this might be expected at about 35% moisture.[28,29]

The most serious losses during drying would be through oxidation. Table 1 shows the losses found during the dehydration of alfalfa. Carotene was generally found to be more stable during drying than xanthophyll. The mono- and diepoxide xanthophylls, neoxanthin, and violaxanthin, were less stable than lutein. Although both contribute to the total xanthophyll, neither were found by Kuzmicky et al.[31] to be poultry pigmenters. Table 2 shows the effect of dehydration on the isomerization of the principal poultry pigmenting xanthophyll, lutein. The preparation of a leaf protein concentrate from fresh forage afforded an enriched carotenoid source in which the stability of the carotenoids in the concentrate was found by Witt et al.[32] to be improved over that of the dehydrated meal (Table 3).

Dehydration of other plant materials[33-35] including wastes from fruit (Table 4), vegetables (Table 5), and grass (Table 6) has been undertaken to prevent environmental pollution and to ascertain their economic potential as feeds. The stability of the nutrients, particularly xanthophyll, during dehydration would seem to be most affected by product meal moisture (Figure 2). Careful meal moisture control including the use of infrared moisture sensing devices[36] would seem to be very important for the production of dehydrated meals rich in carotenoids.

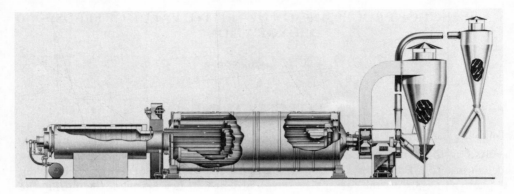

FIGURE 1.   A typical triple pass forage dehydrator. (Photo courtesy of The Heil Co.)

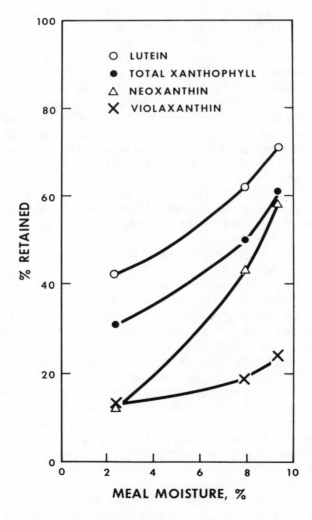

FIGURE 2.   Effect of dehydrated meal moisture on the re-
tention of xanthophyll. (Adapted from Livingston, A. L.,
Knowles, R. E., and Kohler, G. O., *U.S. Dept. of Agric. Tech.
Bull.*, No. 1414, 1, 1970.)

Table 1

STABILITY OF XANTHOPHYLL AND CAROTENE DURING INDUSTRIAL-SCALE ALFALFA DEHYDRATION

Dryer temperature (°C)		Time in dryer (min)	Moisture of meal (%)	Content in fresh alfalfa (mg/kg)					Retained in dehydrated meal (%)				
At inlet	At outlet			Carotene	Total xanthophyll	Lutein	Neoxanthin	Violaxanthin	Carotene	Total xanthophyll	Lutein	Neoxanthin	Violaxanthin
Industrial Arnold® Trial 1													
871	149	3—5	9.2	378	831	610	92	116	91	61	71	58	24
871	154	4—6	7.8	353	736	488	92	150	76	50	62	43	19
871	166	5—7	2.3	353	798	591	82	121	82	36	42	12	13
Stearns-Roger® Trial 1													
454	116	13—17	8.3	411	876	614	97	169	94	63	62	50	17
482	132	8—10	9.5	407	905	581	116	213	87	60	79	42	16
482	149	4—6	9.9	411	861	590	102	169	88	64	80	47	20
482	166	5—7	5.9	363	822	576	106	140	97	55	67	32	21
Stearns-Roger® Trial 2													
216	121	3—5	12.2	349	784	529	97	160	100	72	84	85	35
438	121	4—6	7.1	411	861	580	102	179	83	56	67	62	16
538	121	4—6	2.5	339	808	537	92	179	86	41	53	37	13
482	135	3—5	7.1	445	972	638	97	242	85	57	70	63	16
493	135	2—4	3.1	445	1005	643	111	252	67	40	50	40	13
528	135	2—4	1.5	363	847	555	92	198	79	37	50	26	13

Adapted from Livingston, A. L., Knowles, R. E., and Kohler, G. O., *U.S. Dep. Agric. Tech. Bull.*, No. 1414, 1, 1970.

Table 2

ISOMERIZATION OF LUTEIN DURING ALFALFA DEHYDRATION

Dryer temperature at outlet (°C)	Moisture of meal (%)	Fresh alfalfa		Dehydrated alfalfa	
		Total lutein content (mg/kg)	Content of isomers[a] (%)	Total lutein content (mg/kg)	Content of isomers[a] (%)
Industrial Arnold®					
149	9.2	610	13	431	20
154	7.8	488	14	301	25
166	2.3	591	14	250	36
Pilot Arnold®					
132	2.8	538	10	305	19
149	1.6	534	13	252	23
166	1.5	552	18	214	45
Stearns-Roger®					
135	7.1	638	12	449	20
135	3.1	643	10	324	25
135	1.5	555	18	276	32

[a]   Percent isomers calculated as percent of total lutein absorbance at 475 mμ.

Adapted from Livingston, A. L., Knowles, R. E., and Kohler, G. O., *U.S. Dep. Agric. Tech. Bull.*, No. 1414, 1, 1970.

Table 3
CAROTENOID STABILITY IN PRO-XAN DRUM DRIED TO DIFFERENT MOISTURE CONTENTS AND STORED IN OPEN CONTAINERS AT 38°C

Sample	% Moisture	Initial (Mg/kg) Carotene	Initial (Mg/kg) Xanthophyll	% Carotene retained 4 Weeks Untreated	4 Weeks Treated[a]	12 Weeks Untreated	12 Weeks Treated[a]	% Xanthophyll retained 4 Weeks Untreated	4 Weeks Treated[a]	12 Weeks Untreated	12 Weeks Treated[a]
PRO-XAN[b]											
Batch 1	7	746	1076	56	75	25	56	69	79	42	68
	9	719	1041	59	73	27	52	70	79	46	69
	11	726	1074	57	67	28	47	69	78	47	68
	13	786	1080	57	67	29	47	69	77	48	69
Batch 2	8	460	629	42	74	12	46	60	83	27	56
	9	458	605	40	73	13	44	56	82	28	59
	11	473	640	42	72	13	45	56	79	28	58

[a] Samples treated with 0.125% ethoxyquin.
[b] Samples prepared from unwashed concentrate containing about 25% water-solubles.

Reprinted with permission from Witt, S. C., Spencer, R. R., Bickoff, E. M., and Kohler, G. O., *J. Agric. Food Chem.*, 19, 162, 1971. Copyright by the American Chemical Society.

Table 4
DEHYDRATION OF PIMENTO FRUIT WASTE

Meal sample and dryer temperature at outlet	Moisture of meal (%)	Carotene (mg/kg)	Xanthophyll (mg/kg)	Protein (%)	Fat (%)	Fiber (%)
Trial 1, 121°C						
Freeze-dried whole pimento	9.7	358	1280	13.8	7.28	36.5
Dehydrated whole pimento	13.7	323	1064	10.4	4.62	59.2
Trial 2, 149°C						
Freeze-dried whole pimento	9.36	259	1138	14.7	7.30	48.4
Dehydrated whole pimento — 1	3.01	255	942	10.0	5.64	53.6
Dehydrated whole pimento — 2	2.90	244	896	10.8	5.77	49.0
Trial 3, 127°C						
Freeze-dried pressed pimento	4.55	279	1191	10.9	7.96	50.4
Dehydrated pressed pimento — 1	4.53	258	956	10.8	7.84	25.4
Dehydrated pressed pimento — 2	3.56	282	925	11.9	9.48	44.5
Dehydrated pressed skins	3.00	294	1114	8.8	4.99	61.9
Dehydrated pimento seed	8.41	—	<100	15.9	16.72	38.6

From Livingston, A. L., Knowles, R. E., and Kohler, G. O., *J. Sci. Food Agric.*, 25, 483, 1974. With permission.

VITAMIN E

In forage plant materials the chief source of vitamin E activity is α-tocopherol.[12] The α-tocopherol content of dehydrated or sun-cured alfalfa was found by Charkey et al.[38] to be variable depending upon the source. Although α-tocopherol has been found to be an excellent antioxidant in milk, it is subject to oxidative losses during forage dehydration. Table 7 shows the stability of α-tocopherol during alfalfa dehydration, while Table 8 shows the subsequent storage stability of the same dehydrated meals. Included among the stored meals are the comparable meals which were freeze-dried. The α-tocopherol in these meals was much more stable during storage than that of any of the dehydrated meals, demonstrating heat damage during dehydration to the natural antioxidant system in alfalfa.

AMINO ACIDS

Dehydrated forage plant meals, particularly alfalfa, are important sources of amino acids for animal feeds. Much of the heat damage to amino acids and proteins during dehydration has been attributed to sugar-amino acid interactions of the Maillard type.[39,40] Dehydrator operating conditions, including inlet and outlet temperatures as well as meal moisture, would seem to be critical factors in determining amino acid stability during dehydration.[41] Table 9 presents the stability of the amino acids during pilot and industrial-scale alfalfa dehydration. The essential amino acids, lysine, methionine, and cystine, underwent the most severe losses, particularly when the alfalfa was dried to levels of less than 3% meal moisture. Figure 3 presents the correlation between meal moisture and four amino acids.

Table 5
DEHYDRATION OF WASTE CAULIFLOWER LEAF

Moisture-Free Basis

Meal sample and dryer temperature at outlet	Weight (%)	Moisture of meal (%)	Carotene (mg/kg)	Xanthophyll (mg/kg)	Protein (%)	Fat (%)	Fiber (%)
Trial 1, 135°C							
Freeze-dried whole plant	—	2.9	190.0	398.0	24.6	4.06	13.5
Dehydrated whole plant	100.0	6.1	176.8	280.1	24.1	3.06	13.7
Dehydrated leaf fraction	53.8	8.1	256.2	420.3	28.1	3.64	12.5
Dehydrated stem fraction	46.2	17.0	65.6	119.4	21.3	2.43	14.5
Pressed, freeze-dried whole plant	—	2.9	184.7	373.5	23.0	3.48	14.6
Pressed, dehydrated whole plant	100.0	8.8	185.4	298.5	24.1	3.11	14.8
Pressed, dehydrated leaf fraction	58.2	4.9	238.7	375.0	26.0	2.87	12.5
Pressed stem fraction	41.8	16.2	67.9	119.3	20.7	2.25	15.9
Trial 2, 121°C							
Freeze-dried whole plant	—	3.1	265.3	648.2	26.2	4.52	11.8
Dehydrated whole plant	100.0	15.3	244.1	413.4	26.7	4.10	13.2
Dehydrated leaf fraction	55.5	7.2	383.7	599.0	31.0	4.34	11.1
Dehydrated stem fraction	44.5	24.3	19.8	42.2	19.2	2.58	14.8
Pressed, freeze-dried whole plant	—	3.7	235.3	506.0	23.5	4.18	14.5
Pressed, dehydrated whole plant	100.0	12.1	230.8	420.0	24.2	3.47	15.1
Pressed, dehydrated leaf	64.4	8.0	334.0	620.0	26.9	4.16	14.1
Pressed, dehydrated stem	35.6	21.9	84.0	176.0	19.3	2.42	16.0

Table 6
XANTHOPHYLL AND CAROTENE STABILITY DURING TURF GRASS DEHYDRATION

Moisture-Free Basis

Trial	Outlet temperature of dryer (°C)	Dryer time (min)	Fan speed (r/min)	Moisture of meal (%)	Carotene (mg/kg)		Total xanthophyll (mg/kg)		Nonepoxide xanthophyll (mg/kg)	
					Freeze dried	Dehydrated	Freeze dried	Dehydrated	Freeze dried	Dehydrated
1	113	3	1800	2.6	420	420	1094	857	854	599
2	132	3	1800	1.2	548	544	1248	830	840	747
		2.5	2100	1.7	528	511	1225	976	828	796
	121	3	1800	1.5	532	528	1235	1023	832	772
		2.5	2100	2.2	544	524	1324	1186	882	798
3	121	2.5	2100	2.4	438	419	1120	876	787	708
	113	2.5	2100	3.4	447	424	1100	979	776	732
	103	2.5	2100	4.5	465	427	1245	1107	830	747

Reprinted with permission from Livingston, A. L., Knowles, R. E., Page, J., and Kohler, G. O., *J. Agric. Food Chem.*, 19, 951, 1971. Copyright by the American Chemical Society.

Table 7

STABILITY OF α-TOCOPHEROL AND RELATED COMPOUNDS DURING ALFALFA DEHYDRATION

Dryer	Outlet temp of dryer (°C)	Dehydrated meal moisture (%)	α-Tocopherol (mg/100 g)[a]			Related reducing compounds[a] (mg/100 g)[a]		
			Fresh freeze-dried meal	Dehydrated meal	Loss (%)	Fresh freeze-dried meal	Dehydrated meal	Loss (%)
Arnold®	149	9.2	18.1	17.1	5	9.3	10.9	Increase
Stearns-Roger® (Trial 1)	166	2.3	22.8	18.1	21	13.9	13.6	2
	121	12.2	18.8	13.7	27	10.1	5.0	50
		7.1	17.5	12.9	26	11.5	3.8	67
		2.5	20.6	13.8	33	9.4	7.9	16
Stearns-Roger® (Trial 2)	135	7.1	19.6	18.7	5	14.7	9.2	35
		3.1	19.0	18.1	5	13.9	11.0	21
		1.5	21.9	17.8	19	9.1	9.1	None

[a]   Dry basis, average of duplicate analysis.

Reprinted with permission from Livingston, A. L., Nelson, J. W., and Kohler, G. O., *J. Agric. Food Chem.*, 16, 492, 1968. Copyright by the American Chemical Society.

Table 8
STABILITY OF α-TOCOPHEROL AND RELATED REDUCING
COMPOUNDS DURING STORAGE[a] OF ALFALFA

Sample	Meal moisture (%)	α-Tocopherol (mg/100)[b]			Related reducing compounds (mg/100 g)[b]	
		Initial	12 weeks	Loss (%)	Initial	12 weeks
Dehydrated meals						
1	12.2	13.7	3.7	73	5.0	8.7
2	9.2	17.2	4.9	72	10.9	12.9
3	7.1	12.9	3.7	71	3.8	9.1
4	2.5	13.8	6.3	54	7.9	12.1
5	2.5	13.8	9.0	35	7.9	21.2
6	8.3	18.1	7.6	55	13.6	18.4
Freeze-dried meals						
1	8.4	18.8	14.5	23	10.1	21.4
2	5.4	17.5	16.9	3	11.5	20.7
3	5.4	20.6	19.3	6	9.4	25.6
4[c]	5.4	20.6	19.7	4	9.4	25.0
5	3.3	18.1	13.1	28	9.3	25.3

[a] Stored 12 weeks at 32°C.
[b] Dry basis, average of duplicate analyses.
[c] Added 0.150% ethoxyquin.

PROTEINS

Excessive heating of plant proteins in the presence of carbohydrates or lipids has been shown to result in loss of digestibility of the proteins.[42-46]

Booth et al.[47] showed a correlation between in vivo and in vitro digestibility of alfalfa leaf protein concentrate. Saunders et al.[48] demonstrated a decrease in the in vitro crude protein digestibility with increasing dehydrator outlet temperatures (Table 10). This decrease in the in vitro digestibility could be correlated with the xanthophyll content of leaf protein concentrates as well as dehydration temperatures (Figure 4). The protein of alfalfa meal dehydrated to a high moisture content has been found to be more efficiently utilized by baby pigs than the comparable alfalfa dehydrated to a low moisture meal (Table 11). In this study[49] the high-moisture alfalfa meal was comparable to an ideally freeze-dried alfalfa meal or a corn-soy basal.

SOLUBLE CARBOHYDRATES AND STARCH

Following the cutting of forage and other plant materials, plant enzymes might be expected to act upon starch and the soluble carbohydrates, converting these to simple sugars and ultimately to CO_2, thus leading to respiratory losses.[50] These losses may continue until the plant material is sufficiently dry to inactivate the enzymes.[51,52] Rapid dehydration of the plant material following cutting could therefore prevent dry matter losses of as much as 5 to 8%.[53,54] During dehydration the reaction between carbohydrates and protein or amino acids may result in losses of soluble carbohydrates as well as amino acids, while simultaneously decreasing the digestibility of protein.[41,55-57] Link[28] found that drying beet and corn leaves at 65 to 80°C gave higher carbohydrates analyses than drying at lower than 65 or at 98°C. The higher temperature resulted in

Table 9

RETENTION OF AMINO ACIDS DURING ALFALFA DEHYDRATION

Dehydrated meal (Grams of amino acid per 16 g N)

	Dehydrator								Lyophilized controls (average of three samples)
	Pilot Arnold®			Stearns-Roger®			Industrial Arnold®		
Outlet temp (°C)	130	148	166	130	148	166	148	166	—
Meal moisture (%)	2.8	1.8	1.5	9.5	9.9	5.9	9.2	2.3	—
Amino acid									
Lysine	4.78	4.36	3.41	5.25	5.38	4.60	4.88	3.25	6.11
Histidine	2.05	2.07	1.95	2.19	2.15	2.14	2.20	1.98	2.31
Ammonia	2.17	1.76	1.96	2.06	2.37	2.23	2.59	2.08	2.27
Arginine	4.68	4.50	4.52	4.82	4.74	4.47	4.72	4.02	5.04
Aspartic acid	11.42	10.48	9.86	11.19	11.07	10.51	11.31	10.92	11.45
Threonine	4.46	4.33	4.34	4.37	4.43	4.47	4.37	4.51	4.66
Serine	4.66	4.46	4.32	4.52	4.89	4.50	4.53	4.53	4.79
Glutamic acid	10.27	9.67	9.78	9.93	10.03	9.98	10.21	10.79	10.31
Proline	4.23	4.20	4.17	4.22	4.94	4.26	4.35	4.73	4.65
Glycine	5.29	5.04	5.21	5.07	5.07	5.23	5.15	5.73	5.22
Alanine	5.69	5.31	5.45	5.40	5.47	5.64	5.52	5.87	5.76
Valine	6.05	5.82	5.88	5.83	6.40	5.77	6.05	6.62	6.11
Isoleucine	5.09	4.84	4.92	4.81	4.88	5.00	4.95	5.34	5.05
Leucine	7.91	7.59	7.63	7.27	7.76	7.81	7.81	8.33	7.66
Tyrosine	3.33	3.37	3.19	3.42	3.50	3.40	3.38	3.44	3.39
Phenylalanine	5.30	5.96	5.12	5.21	5.22	5.35	5.25	5.46	5.24
Methionine	1.66	1.56	1.43	1.74	1.74	1.66	1.27	1.25	1.79
Cystine	1.08	1.08	1.02	1.18	1.19	1.05	1.13	0.92	1.27
% Nitrogen recovery	82.9	77.9	77.2	82.0	83.4	81.7	85.1	81.4	87.1

Adapted with permission from Livingston, A. L., Allis, M. H., and Kohler, G. O., *J. Agric. Food Chem.*, 19, 947, 1971. Copyright by the American Chemical Society.

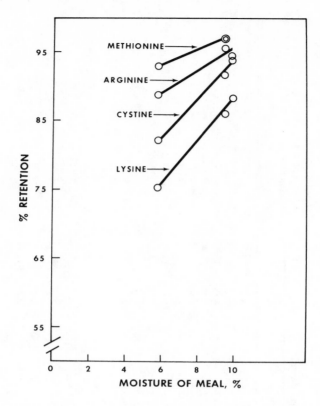

FIGURE 3.   Correlation of amino acid content and meal moisture during forage dehydration. (Reprinted with permission from Livingston, A. L., Allis, M. H., and Kohler, G. O., *J. Agric. Food Chem.,* 19, 947, 1971. Copyright by the American Chemical Society.

Table 10
IN VITRO CRUDE PROTEIN DIGESTIBILITY OF DEHYDRATED ALFALFA AFTER DRYING AT DIFFERENT TEMPERATURES[48] OR FREEZE-DRIED

Alfalfa and drying conditions	Crude protein digestibility (N × 6.25) (%)
Leaf, dried at 121°C outlet temperature	79.0
Leaf, dried at 130°C outlet temperature	77.4
Leaf, dried at 143°C outlet temperature	73.9
Leaf, freeze-dried	85.7
Whole alfalfa, dried at 121°C outlet temperature	83.2
Whole alfalfa, dried at 130°C outlet temperature	81.8
Whole alfalfa, dried at 143°C outlet temperature	78.1
Whole alfalfa, freeze-dried	85.7
Whole alfalfa, dried at 135°C	80.4
Whole alfalfa, recycled gas dried at 135°C	84.8
Whole alfalfa, freeze-dried	87.2
Whole alfalfa, dried at 135°C	84.0
Whole alfalfa, dried at 80°C	90.2
Whole alfalfa, freeze-dried	90.0

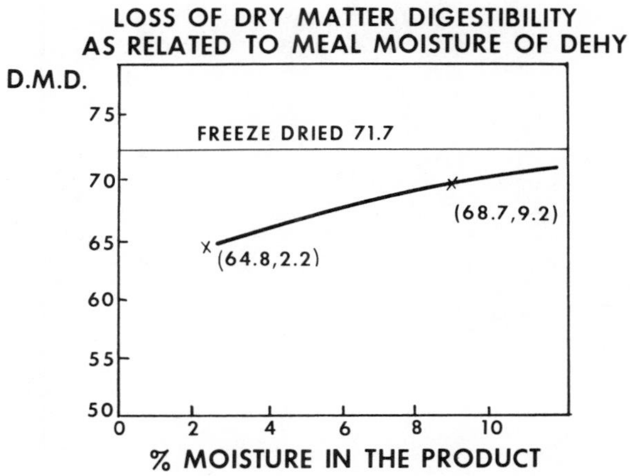

LOSS OF DRY MATTER DIGESTIBILITY
AS RELATED TO MEAL MOISTURE OF DEHY

FIGURE 4.   Loss of dry matter digestibility during forage dehydration. (Reproduced from *Effect of Processing on the Nutritional Value of Feeds,* page 320, with the permission of the National Academy of Sciences, Washington, D.C.)

carmelization of the sugars, while the lower temperature resulted in enzymatic losses. Hathout[58] found that 33 to 70% of the reducing sugars and 14 to 66% of the amino acid nitrogen was lost during alfalfa dehydration (Table 12).

CELL WALL CARBOHYDRATES

The digestibility of total fiber or cell wall constituents such as lignin, cellulose, and hemicellulose may vary widely between ruminants and nonruminants. Due to enzymes and microflora in the rumen, ruminants generally are more efficient in the digestion of cellulose and hemicellulose.[59-61] Therefore, increases in poorly digestible cell wall carbohydrates during dehydration might be expected to more adversely affect nonruminants than ruminants. Van Soest has demonstrated that the increased yield of poorly digestible lignin and fiber during the drying of forages at temperatures above 50°C was due to the production of lignin via the nonenzymic browning reaction.[62] He suggested that the lignin and nitrogen content of acid-detergent fiber be used as a sensitive assay for heat damage in feeds.

Goering and Van Soest[63] found a high correlation between acid-detergent fiber insoluble nitrogen (ADIN) and pepsin digestion when investigating nonenzymic browning in heated forage plant materials. The ADIN content of dehydrated alfalfa was recently found by Goering et al.[64] and Goering and Lindahl[65] to be related to the dehydrator outlet temperature (Table 13). In these studies sheep fed alfalfa dehydrated at 180°C consumed less feed and gained more slowly than those fed alfalfa dehydrated at 120 to 160°C. The ADIN/N values for the meals dehydrated at 160°C or less were about equal to mildly heated, barn-dried hay, whereas the alfalfa meals dried at the high temperature (180°C) had ADIN/N values more than double those of the meals dried at lower temperatures. Additional alfalfa dehydration animal feeding trials would be valuable in more precisely determining the dehydrator outlet temperature or meal moisture limits which adversely affect nutrient quality for sheep. These results could then be carefully correlated with chemical analyses such as ADIN/N or with in vitro, enzymatic digestibility results.

Similarly, Kohler et al.[66] found that the dry matter digestibility of dehydrated alfalfa

Table 11

EFFECT OF DEHYDRATION MOISTURE LEVELS IN ALFALFA MEAL ON FEED QUALITY FOR
BABY PIGS

			Treatments[a]			
Criterion	Opaque-2 corn-soy basal[b]	Low-moisture alfalfa[c] (meal form)	Low-moisture alfalfa[c] (pellet form)	High-moisture alfalfa[c] (meal form)	High-moisture alfalfa[c] (pellet form)	Freeze-dried alfalfa[c] (meal form)
Av daily gain (lb)	0.53	0.46	0.48	0.51	0.53	0.48
Av daily feed intake (lb)	1.12	1.10	1.17	1.10	1.17	1.03
Gain:feed ratio	0.46	0.42	0.42	0.45	0.47	0.47

[a]　Data based on the average of four pigs per pen; four pens per treatment.
[b]　Sixteen percent pure protein opaque-2 corn-soybean meal basal diet.
[c]　Alfalfa meals furnished 50% of the supplemental protein or approximately one third of the total dietary protein.

From Livingston, A. L., Knowles, R. E., Kohler, G. O., and Peo, E. R., Jr., *Proc. 12th Tech. Alfalfa Conf.*, American Dehydrators Association, Overland Park, Kan., 1974, 19. With permission.

Table 12
LOSS OF AMINO ACID NITROGEN AND REDUCING SUGAR DURING ALFALFA DEHYDRATION

Dehydrator outlet temperature (°C)	Lost amino acid nitrogen (%)	Lost reducing sugar (%)
149	66	77
127	37	62
105	9	34
82	14	33

Adapted from Hathout, M. K., Ph.D. thesis, North Carolina State College, Raleigh, 1961, 79. With permission.

Table 13
GROWTH AND METABOLISM OF SHEEP FED RATIONS CONTAINING ALFALFA HAY OR DEHYDRATED ALFALFA

Component	Barn-dried hay	Dehydrator outlet temperature (°C)		
		120	145	180
DMD (%)	54	62	60	56
ND (%)	46	53	50	42
NR (%)	4.9	6.3	4.8	5.1
ADIN/total N (%)	7.7	7.1	7.6	18.9
Av daily feed intake (g/day)	736	821	803	707
Av daily gain (g/day)	145	160	162	136
Feed:gain ratio	5.1	5.3	5.0	5.4

Adapted from Goering, H. K., Menear, J., and Lindahl, I. L., *J. Dairy Sci.,* 57(Abstr.), 621, 1974. With permission.

could be correlated with the dehydrator outlet temperature and product meal moisture. High drying temperatures were found by Delic et al.[67] to cause a decrease in hemicellulose content and an increase of cellulose and lignin contents during industrial alfalfa dehydration (Table 14).

OTHER VITAMINS

Certain of the water-soluble vitamins, particularly the B group, have been found to be relatively stable during dehydration.[67] However, vitamin C has been found to be almost completely destroyed, even during rather mild dehydration conditions.

Table 14
INFLUENCE OF THE DRYING TEMPERATURES ON CHANGES IN THE CONCENTRATION OF EXAMINED CONSTITUENTS IN ALFALFA HERBAGE

(Percent Dry Weight)

Treatment	Moisture	Hemicellulose	Cellulose	Lignin
Fresh alfalfa	81.20	12.96	21.10	10.60
Heat-dried alfalfa				
600°C	7.48	11.80	22.37	12.50
700°C	7.09	11.62	23.40	12.16
800°C	7.88	10.27	25.90	12.93

Reprinted with permission from Delic, I., Milic, B., and Vlahovic, M., *J. Agric. Food Chem.*, 19, 254, 1971. Copyright by the American Chemical Society.

REFERENCES

1. Hoglund, C. R., Harvest and storage losses for several forage handling systems, *Mich. State Univ. Agric. Exp. Econ. Rep.*, No. 947, 1—16, 1964.
2. Weaver, J. M., Jr. and Wylie, E., Drying hay in the barn and testing its feeding value, *Tenn. Agric. Exp. Sta. Bull.*, No. 170, 1939.
3. Shepherd, J. B., Wiseman, H. G., Ely, R. E., Melin, C. G., Sweetman, W. J., Gordon, C. H., Schoeleber, L. G., Wagner, R. E., Campbell, L. E., Roane, G. D., and Hosterman, W. H., Harvest and nutrients losses during hay-making, *U.S. Dep. Agric. Tech. Bull.*, No. 1079, 1—18, 1954.
4. Bechdel, S. I., Clyde, A. W., Cromer, C. O., and Williams, P. S., Dehydrated and sun-cured hay, *U.S. Dep. Agric. Tech. Bull.*, No. 396, 1—23, 1940.
5. Silker, R. E., Development and use of artificial dehydration of forage crops in the United States, in *6th Int. Grass Congr.*, Vol. 2, Pennsylvania State University, University Park, 1952, 1753.
6. Eaton, H. D., Doldge, K. L., Mochrie, R. D., and Avampato, J. E., Field-cured and field-baled alfalfa hay versus artificially dried and chopped and pelleted alfalfa hays as a source of carotene and roughage for Guernsey and Holstein calves, *J. Dairy Sci.*, 55(1), 98—105, 1952.
7. Eaton, H. D., Carpenter, C. A., Caverno, R. J., Johnson, R. E., Elliott, F. I., and Moore, L. A., A comparision of U.S. No. 2 field-cured and field-baled alfalfa hay, *J. Dairy Sci.*, 34, 124—135, 1951.
8. Russell, W. C., The effect of the curing process upon the vitamin A and D content of alfalfa, *J. Biol. Chem,*. 85, 289—297, 1930.
9. Adamstone, F. B., Histologic comparison of the brains of vitamin A deficient and vitamin E deficient chicks, *Arch. Pathol.*, 43, 301—312, 1947.
10. Harris, P. L., Swanson, W. J., and Hickman, K. C. Effect of tocopherol supplementation on the output of vitamin A, carotene and fat in dairy cows, *J. Nutr.*, 33, 411—427, 1947.
11. Thompson, C. R., Bickoff, E. M., Van Atta, G. R., Kohler, G. O., Guggolz, J., and Livingston, A. L., Carotene stability in alfalfa as affected by laboratory and industrial scale processing, *U.S. Dep. Agric. Techn. Bull.*, No. 1232, 1—27, 1960.
12. Livingston, A. L., Nelson, J. W., and Kohler, G. O., Stability of alphatocopherol during alfalfa dehydration and storage, *J. Agric. Food Chem.*, 16, 494—495, 1968.
13. Sullivan, T. W., Holleman, K. A., Kingan, J. R., and Skinner, J. L., Color makes $ difference, *Nebr. Exp. Stn. Q.*, 19—21, Fall 1962.
14. Guenthner, E., Carlson, C. W., Olson, O. E., Kohler, G. O., and Livingston, A. L., Pigmentation of egg yolks of xanthophylls from corn, marigold, alfalfa and synthetic sources, *Poult. Sci.*, 52, 1787—1798, 1973.

15. Tilley, J. M. A. and Terry, R. A., Two-stage technique for the digestion of forage crops, *J. Br. Grassl. Soc.,* 18, 104—111, 1963.
16. Saunders, R. M. and Kohler, G. O., In vitro determination of protein digestibility in wheat mill feeds for monogastric animals, *Cereal Chem.,* 49(1), 98—103, 1972.
17. Van Soest, P. J., The use of detergents in the analysis of fibrous feeds. II. A rapid method for the determination of fiber and lignin, *J. Assoc. Off. Agric. Chem.,* 46, 825—829, 1963.
18. Guggolz, J., Saunders, R. M., Kohler, G. O., and Klopfenstein, T. J., Enzymatic evaluation of processes for improving agricultural wastes for ruminant feeds, *J. Anim. Sci.,* 33, 167—170, 1971.
19. Griffith, R. B. and Thompson, C. R., Production and utilization of alfalfa, *Bot. Gaz.,* 111, 165—166, 1949.
20. Booth, V. H., *J. Sci. Food Agric.,* 11, 8—11, 1960.
21. Walsh, K. A., and Hauge, S. M., Carotene: factors affecting destruction in alfalfa, *J. Agric. Food Chem.,* 1, 1001—1004, 1953.
22. Wall, M. E. and Kelley, E. G., Stability of carotene and vitamin A in dry mixtures, *Ind. Eng. Chem.,* 43, 1146—1150, 1951.
23. Thompson, C. R., Stability of carotene in alfalfa meal: effect of antioxidants, *Ind. Eng. Chem.,* 42, 922—925, 1950.
24. Livingston, A. L., Bickoff, E. M., and Thompson, C. R., Effect of added animal fats and antioxidant on stability of xanthophyll concentrates in mixed feeds, *J. Agric. Food Chem.,* 3, 439—443, 1955.
25. Thompson, C. R., Bickoff, E. M., and Maclay, W. D., Formation of stereoisomers of beta-carotene in alfalfa, *Ind. and Eng. N.Y.,* 43, 126—129, 1951.
26. Knowles, R. E., Livingston, A. L., Nelson, J. W., and Kohler, G. O., Xanthophyll and carotene storage stability in commercially dehydrated and freeze-dried alfalfa, *J. Agric. Food Chem.,* 16, 654—658, 1968.
27. Zechmeister, L. and Tuzon, P., Isomerization of carotenoids, *J. Biochem.,* 32, 1305—1311, 1938.
28. Link, K. P., Effects of the methods of dessication on the carbohydrates of plant tissues, *J. Am. Chem. Soc.,* 47, 470—476, 1925.
29. Link, K. P. and Tottingham, W. E., Effects of the method of dessication on the carbohydrates of plant tissue, *J. Am. Chem. Soc.,* 45, 439—443, 1923.
30. Livingston, A. L., Knowles, R. E., and Kohler, G. O., Xanthophyll, carotene and alpha-tocopherol stability in alfalfa as affected by pilot and industrial-scale dehydration, *U.S. Dep. Agric. Tech. Bull.,* No. 1414, 1—14, 1970.
31. Kuzmicky, D. D., Kohler, G. O., Livingston, A. L., Knowles, R. E., and Nelson, J. W., Broiler pigmentation of neoxanthin and violaxanthin relative to lutein, *Poult. Sci.,* 48, 326—330, 1969.
32. Witt, S. C., Spencer, R. R., Bickoff, E. M., and Kohler, G. O., Carotenoid storage stability in drum dried Pro-Xan, *J. Agric. Food. Chem.,* 19, 162—165, 1971.
33. Livingston, A. L., Knowles, R. E., and Kohler, G. O., Processing of pimento waste to provide a pigment source for poultry feed, *J. Sci. Food Agric.,* 25, 483—490, 1974.
34. Livingston, A. L., Knowles, R. E., Page, J., Kuzmicky, D. D., and Kohler, G. O., Processing of cauliflower leaf waste for poultry and animal feed, *J. Agric. Food Chem.,* 20, 277—281, 1972.
35. Livingston, A. L., Knowles, R. E., Page, J., and Kohler, G. O., Turf grass dehydration, *J. Agric. Food Chem.,* 19, 951—953, 1971.
36. Livingston, A. L., Knowles, R. E., Amella, A., Kohler, G. O., Arnold, W. L., and Smith, K. D., Alfalfa wilting and dehydration: changes in nutrient composition during industrial drying, *Feedstuffs,* 48(6), A8—A12, 1976.
37. Blaylock, L. G., Richardson, L. R., and Pearson, P. B., The riboflavin, pathothenic acid, niacin and folic acid content of fresh, dehydrated and field cured alfalfa, *Poult. Sci.,* 29, 692—695, 1950.
38. Charkey, L. W., Pyke, W. E., Kano, A., and Carlson, R. E., Carotene and tocopherol content of dehydrated and sun-cured alfalfa meals, *J. Agric. Food Chem.,* 9, 70—74, 1961.
39. Folk, J. E., Reactions of glucose with lysine, *Arch. Biochem. Biophys.,* 61, 150—151, 1956.
40. Halery, S. and Guggenheim, K., The biological availability of heated wheat gluten-glucose mixtures, *Arch. Biochem. Biophys.,* 44, 211—217, 1953.
41. Livingston, A. L., Allis, M. H., and Kohler, G. O., Amino acid stability during alfalfa dehydration, *J. Agric. Food Chem.,* 19, 947—950, 1971.
42. Beauchene, R. E. and Mitchell, H. L., Effect of temperature of dehydration on proteins of alfalfa, *J. Agric. Food Chem.,* 5, 762—765, 1957.
43. Buchanan, R. A., *Br. J. Nutr.,* 23, 533—537, 1969.
44. Evans, R. J. and McGinnis, J., The influence of autoclaving soybean oil meal on the availability of cystine and methionine for the chick, *J. Nutr.,* 31, 449—461, 1946.
45. Hurrell, R. F., Carpenter, K. J., Sinclair, W. J., Otterburn, M. S., and Asquith, R. S., Mechanism of heat damage in proteins. VII. The significance of lysine-containing isopeptides and of lanthionine in heat proteins, *Br. J. Nutr.,* 35, 383—395, 1976.

46. Israelsen, M., Influence of drying conditions on quality and quantity artificially dried crops, *Medd. Grommelsofdelingen,* 14, 297—308, 1973.

47. Booth, A. N., Saunders, R. M., Connor, M. A., and Kohler, G. O., In vivo and in vitro protein digestibility of alfalfa and concentrates, in *Proc. 11th Tech. Alfalfa Conf.,* ARS-74-60, Agricultural Research Service, U.S. Department of Agriculture, Washington, D.C., 1971, 51—57.

48. Saunders, R. M., Livingston, A. L., and Kohler, G. O., Effect of drying temperatures on nutritional value of dehydrated lucerne meal and protein concentrate, in *Proc. 1st Int. Green Crop Drying Congr.,* E. E. Plumridge, Linton, Cambridge, 1973, 309—317.

49. Livingston, A. L., Knowles, R. E., Kohler, G. O., and Peo, E. R., Jr., Nutritional changes during alfalfa wilting and dehydration, in *Proc. 12th Tech. Alfalfa Conf.,* American Dehydrators Association, Overland Park, Kan., 1974, 19—23.

50. Melvin, J. R. and Simpson, B., Chemical changes and respiratory drift during the air drying of ryegrass, *J. Sci. Food Agric.,* 14, 228—234, 1963.

51. Wylam, C. B., Analytical studies on the carbohydrates of grasses and clovers. III. Carbohydrate breakdown during wilting and ensilage, *J. Sci. Food Agric.,* 4, 527—531, 1953.

52. Laidlaw, R. A. and Wylam, C. B., Carbohydrates of grasses and clovers. II. Preparation of grass samples for analysis, *J. Sci. Food Agric.,* 3, 494—496, 1953.

53. Greenhill, W. L., The respiration drift of harvested pasture plants during drying, *J. Sci. Food Agric.,* 10, 495—499, 1959.

54. Ekelund, S., Carbohydrates in hay, *Lantbrukshoegsk. Ann.,* 16, 179—181, 1949.

55. Livingston, A. L., Kuzmicky, D. D., and Kohler, G. O., Amino acid contents of dehydrated alfalfa meals, *Feedstuffs,* 44(5), A4—A5, 1972.

56. Lea, C. H. and Hannan, R. S., Biochemical and nutritional significance of the reaction between proteins and reducing sugars, *Nature (London),* 65, 438—439, 1950.

57. Kretovich, V. L. and Tokareva, R. K., Interaction of amino acids and sugars at high temperatures, *Biokhimiya,* 13, 508—515, 1948.

58. Hathout, M. K., Effect of Drying Temperatures on Chemical Composition of Alfalfa, Ph.D. thesis, North Carolina State College, Raleigh, 1961, 1—80.

59. van Soest, P. J., Symposium on nutrition and forage and pastures: new chemical procedures, *J. Anim. Sci.,* 23, 838—845, 1964.

60. Gaillard, B. D. E., The relationship between cell-wall constituents of roughages and the digestibility of dry matter., *J. Agric. Sci.,* 59, 369—373, 1962.

61. van Soest, P. J., Development of a comprehensive system of feed analysis and its application to forages, *J. Anim. Sci.,* 26, 119—128, 1967.

62. van Soest, P. J., Use of detergents in analysis of fibrous feeds. III. Study of effects of heating and drying on yield of fiber and lignin in forages, *J. Assoc. Off. Agric. Chem.,* 48, 785—790, 1965.

63. Goering, H. K., and Van Soest, P. J., Effect of moisture, temperature and pH on the relative susceptibility of forages, *J. Dairy Sci.,* 50(Abstr.), 989—990, 1967.

64. Goering, H. K., Menear, J., and Lindahl, L. L., Growth and nitrogen metabolism of sheep food alfalfa dehydrated at different temperatures, *J. Dairy Sci.,* 57(Abstr.), 621, 1974.

65. Goering, H. K. and Lindahl, I. L., Growth and metabolism of sheep fed rations containing alfalfa, *J. Dairy Sci.,* 58(Abstr.), 759, 1975.

66. Kohler, G. O., Livingston, A. L., and Saunders, R. M., Effect of processing on the nutritional value of dehydrated meal and other forages, in *Effect of Processing on the Nutritional Value of Feeds,* National Academy of Sciences, Washington, D.C., 1973, 311—325.

67. Delic, I., Milic, B., and Vlahovic, M., Influence of temperature on changes of some constituents in alfalfa meal and formation of glycosamines, glycoproteins, and melanoidins, *J. Agric. Food Chem.,* 19, 254—256, 1971.

EFFECT OF PROCESSING ON NUTRIENT CONTENT OF FEEDS: ENSILING

Peter McDonald

INTRODUCTION

Silage is the material produced by the fermentation of a crop of high moisture content. Ensilage is the name given to the process, and the container, if used, is called the silo. The fermentation is controlled by encouraging the rapid development of anaerobic conditions and maintaining anaerobiosis by effective sealing of the silo. The naturally occurring lactic acid bacteria ferment the water-soluble carbohydrates (WSC) in the crop mainly to lactic acid which reduces the pH to about 4.0. Such a lactate type of silage will preserve satisfactorily as long as oxygen is excluded from the silo. Almost any crop can be preserved as silage although the most common crops used are the grasses, whole cereal crops such as corn (*Zea mays* L.), and legumes such as alfalfa (*Medicago sativa* L.) and red clover (*Trifolium pratense* L.)[1]. It is common practice to chop the crop with a forage harvester prior to ensiling, a process which releases the plant sap allowing more rapid microbial growth[2] and improves silage dry matter (DM) intake.[3-5]

Under certain conditions silages other than the lactate type may be produced. Silages can be classified into six types (including lactate).

LACTATE SILAGE

In this, the most common type of silage produced from grasses and corn, lactic acid bacteria have dominated the fermentation. Table 1 lists those lactic acid bacteria most commonly found on fresh herbage and in silage. These microorganisms, which can be classified into the two types, homofermentative and heterofermentative, ferment the WSC (mainly glucose, fructose, sucrose, and fructans) present in the crop to lactic acid and other products[6] (Table 2). Some pentoses may be liberated from partial hydrolysis of hemicelluloses and these can also be fermented to lactic acid.

Although lactic acid bacteria may frequently be scarce on fresh crops,[7] counts of these organisms after forage harvesting are frequently high. This is illustrated in the results of an experiment shown in Table 3.

The composition of a typical lactate ryegrass silage is shown in Table 4. Lactate silages are characterized by a low pH (approximately 3.7 to 4.2), a high concentration of lactic acid in the DM (approximately 80 to 120 g/kg), with smaller amounts of formic, acetic, propionic, and butyric acids. Variable quantities of mannitol and ethanol, derived from the activities of heterofermentative lactic acid bacteria and yeasts, are present as well.

In addition to the WSC, other compounds, e.g., the plant organic acids, citric and malic, act as substrates for fermentation to lactic acid and other products such as acetoin and 2,3-butanediol (Table 5).

The buffering capacity (Bc) values of lactate silage, expressed as milliequivalents of alkali required to change the pH of macerates from 4 to 6 per kilogram DM, are high.[13] The Bc values for fresh grasses normally range from 250 to 400, while legumes such as clovers and alfalfa are higher from 500 to 600.[13,14] In lactate silages Bc values are usually three to four times those in original forage, due mainly to the high concentration of fermentation acids and their salts[13] (Table 6).

The cell wall constituents of the crop (fiber) may increase during ensilage depending upon the losses of nutrients in gaseous form or via the effluent. The flow of effluent is related to the DM content of the ensiled crop (Figure 1).[15]

The nitrogenous components of lactate silages are mainly in a nonprotein, soluble form in contrast to those present in fresh forage crops, where some 75 to 85% of the nitrogen is present as protein.[16] Rapid and extensive proteolysis occurs immediately after a crop is harvested and this continues in the silo until a low pH is attained. The main nitrogenous components of red clover and a lactate silage made from it are shown in Table 7.

In lactate silages about 50% net breakdown of the protein can take place[8] and most of this proteolysis occurs during the first week of ensiling.[18] Studies with ensiled corn have shown that proteolytic enzyme activity declines rapidly from ensiling to a non-measurable level after 5 days, although considerable changes in relative properties of amino acids occur[19] (Table 8).

In a study with a red clover lactate silage,[17] the overall changes in amino acids present in proteins, peptides, and free amino acids indicated complete breakdown of arginine as well as appreciable losses of glutamic acid, aspartic acid, serine, and threonine (Table 9).

The hydrolysis of proteins during ensiling appears to be nonselective in that residual protein is of similar amino acid composition to the original and there is no preferential liberation of any of the acids.[20] The catabolism of amino acids, and in particular their deamination and decarboxylation is influenced by the rate of pH fall reflecting the rapidity with which the lactic acid bacteria dominate the fermentation and inhibit the growth of proteolytic bacteria, such as clostridia.[21]

The lactic acid bacteria are considered to have a limited ability to deaminate amino acids although L-serine and L-arginine are broken down by these organisms to ornithine and acetoin, respectively.[8] The NH_3-N content, an indication of the extent of amino acid deamination[22] is usually <110 g/kg total N in lactate silages.[8]

There is strong evidence to suggest that the high soluble N content of silages is a disadvantage in terms of its subsequent utilization by the ruminant and several workers have shown low retentions of N in ruminants on diets of silage compared with fresh herbage or hay.[23-25] The highly degraded nature of the N components of ensiled materials, combined with very low levels of WSC, can result in high NH_3-N concentrations in rumen liquor. In one study with six grass lactate silages,[12] concentrations of ruminal NH_3-N varied from 195 to 450 mg NH_3-N/ℓ and a highly significant correlation between peak concentration and the nonprotein-N and NH_3-N contents of the silages was found. Poor utilization of N can occur when ruminal NH_3-N concentration exceeds 150 mg/ℓ.[26]

As a result of the extensive fermentation changes, and, in particular, those affecting the WSC which lead to the formation of high energy compounds such as ethanol, the gross energy (GE) concentration of lactate silages increases significantly[27,28] as is demonstrated in Table 10.

The changes are also reflected in the metabolizable energy (ME) values of the silages after deductions of fecal, urinary, and methane energy losses.[29] There is no evidence to indicate that in the production of lactate silages digestibility is altered, and fecal energy losses are similar to those obtained with parent materials.[12] Similarly, methane losses differ little from those obtained with fresh herbage, and range from 0.06 to 0.09 (CH_4E/GE).[30,31]

There is some evidence that urine energy losses are slightly higher in animals given lactate silages compared with parent materials,[12] but these differences are small compared with the increases in GE during ensiling and consequently ME values of lactate silages are likely to be higher than those of original herbage as is seen in Table 11.

The efficiency of utilization of the ME of fresh forage ranges from 0.60 to 0.75 for maintenance (km) and 0.28 to 0.52 for growth (kg).[30,34,35] The km values for lactate silages do not appear to differ markedly from values obtained with fresh herbage or similar material conserved as hay.[12] In one study consisting of 280 balance trials, it was concluded that km and kℓ (efficiency of utilization of ME for lactation) did not differ for hays and silages of similar ME value.[36]

The voluntary DM intake (DMI) of high-moisture lactate silages is lower than that of fresh or dried herbage.[37-40]. Negative correlations between DMI and titratable acidity, total volatile fatty acids and acetic acid contents in silages have been established.[41,42] The action of lactic acid per se on DMI is less clear although some workers have demonstrated a depressing effect.[43,44] A clearer relationship between the DM content of the ensiled crop and DMI has been established. Such relationships have been shown for grass,[40] alfalfa,[45] corn,[46] and sorghum.[47]

ACETATE SILAGE

Under certain ill-defined conditions, acetic acid-producing bacteria may dominate the fermentation. Such acetate-type silages have been reported in studies with tropical grass species.[48-50] These silages rarely appear in temperate crops although they have been reported in hand harvested grass samples ensiled in laboratory silos[8] and in an autumn cut of ryegrass ensiled in 3-t capacity plastic silos.[10] In spite of their low lactic acid content, these silages appear to be stable. The composition of a typical acetate silage is shown in Table 12.

There is very little information concerning the nutritional value of high acetate silages although the negative correlation of acetic acid content of silage and DMI,[41,51] and the results of intraruminal infusion experiments[52-54] would suggest that the DMI of acetate silages would be low.

BUTYRATE SILAGE

If a stable pH value has not been achieved in silage, the saccharolytic clostridia, present on the original crop as spores, will proliferate, fermenting lactic acid and any residual WSC to butyric acid and causing a rise in pH.[55] The less acid tolerant proteolytic clostridia usually then become active leading to a further increase in pH caused by the production of ammonia.[55] Table 13 lists the main clostridial bacteria associated with silage.

Clostridia require moisture for active growth and with crops of DM contents (g/kg) approximately 150, even the achievement of a pH value of 4.0 may not inhibit their activity. With crops of DM 250 g/kg, clostridial activity is likely to be minimal. Crops low in WSC, such as orchard grass (*Dactylis glomerata* L.) or heavily buffered crops such as the legumes, alfalfa, and the clovers, are liable to produce butyrate silages unless suitably pretreated with an additive. Clostridia produce a range of fermentation products which are shown in Table 14.

The composition of a typical butyrate silage is given in Table 15. Butyrate silages are characterized by having pH values normally within the range 5 to 6 with low levels of both lactic acid and WSC. Butyric acid is usually the dominant fermentation acid present although acetic acid levels may also be relatively high. Normally, although not always, the amino acid-fermenting bacteria have also been active in this type of silage; consequently the amount of ammonia-N is high. This type of fermentation is undesirable for several reasons. Nutrient losses will be high because of the large evolution of gaseous products (carbon dioxide, hydrogen, and ammonia). The DMI of these silages is low, and there is a close negative correlation between DMI and the concentration of ammonia-N,[41] although the exact components of butyrate silages reponsible for reduced DMI have not been identified. There is no specific information available relating to the utilization of the energy of butyrate silages.

WILTED SILAGE

Prewilting a crop prior to ensiling restricts fermentation increasingly as DM content increases (Table 16). In such silages, clostridial activity is slight although some lactic acid bacterial growth occurs, even in herbage wilted to 500 g DM/kg. Total fermentation acids and Bc are reduced although wilting, unless rapid, does not prevent proteolysis occurring. Deamination of amino acids is minimal. Table 17 gives the composition of a silage made from ryegrass wilted to a DM of 308 g/kg. Restricting fermentation results in little change to GE (Table 10). Small reductions in digestibility may occur.[57,58] Methane and urine energy losses are likely to be similar to those obtained by feeding original herbage, consequently ME values of wilted silages resemble more closely those of parent materials (Table 11).

Although considerable proteolysis occurs in the production of these silages ,there is evidence for reduced ruminal ammonia concentrations[59] and better nitrogen retention,[60] compared with unwilted silages. The beneficial effects on DMI of increasing the DM content of forages prior to ensiling have been well documented,[12,43,56,61] and are illustrated in Figure 2.

CHEMICALLY RESTRICTED SILAGE

Sugar-rich compounds, such as molasses, have been used as silage additives for many years. Such additives are classed as stimulants and encourage the production of lactate silages. A detailed classification of these silage additives is shown in Table 18. Recently attention has been focused on the use of chemicals which restrict or inhibit fermentation and in particular on formic acid and formalin.

Formic acid restricts fermentation increasingly with level of application and with increasing DM content[62] (Table 19). With low DM formic acid-treated crops, effluent flow may be increased.[63] The beneficial effects of formic acid on the fermentation characteristics of "difficult" crops such as alfalfa and grasses low in WSC have been well established.[64,65] Improvements in animal performance and DMI have also been demonstrated.[66-68] There is evidence that the direct application of formic acid to freshly harvested crops reduces the extent of proteolysis.[12] Formalin is usually applied on its own or mixed with an acid. At low concentrations of pure formalin application (2.5 g/kg fresh herbage) butyrate silages may be produced[69] (Table 20).

At intermediate levels (approximately 8 g/kg fresh herbage) fermentation is restricted and improvements in DMI and animal performance can be demonstrated[70] (Table 21).

At high levels of application, >20 g/kg fresh herbage, severe restrictions in DMI and DM digestibilities occur.[71] Formaldehyde is known to form strong chemical bonds with protein and application of this chemical to the crop at harvesting prevents proteolysis.

The degree of protection depends upon the species and protein content of the crop as well as on the amount applied. At moderate levels of application, formalin treated silages have resulted in a greater entry of amino acids into the small intestine.[72]

Formalin is more effective as a fermentation inhibitor when applied with an acid.[73] The effect of a formalin/formic acid mixture (3:1 w/w) applied to ryegrass at 10 g/kg fresh herbage on the composition of the silage is shown in Table 22. Typical effects of such additives are low levels of fermentation acids, high protein N levels, and low concentrations of ammonia-N.

DETERIORATED SILAGE

The continuous infiltration of air during the storage period in the silo leads to the growth of a range of aerobic microorganisms which break down the organic matter to form composted material unfit for feeding to animals. Such waste material is commonly found in the surface and sides of silage made in bunker and clamp silos. This deterioration process will also commence during the feeding period when silage is exposed for varying periods of time to atmospheric oxygen. The breakdown of the organic matter, and especially the soluble fermentation products and residual WSC, is brought about initially by yeasts and bacteria followed by the growth of molds. The factors governing rate of deterioration are unknown but the extent of breakdown may, over a 10-day period, range from virtually nil to about one third of the DM.[74] Silages, in which fermentation has been restricted either by wilting or by use of normal silage additives, are inclined to be less stable than those in which fermentation has been unrestricted.[70,75] Corn silage appears to be particularly susceptible to aerobic deterioration.[76] Higher volatile fatty acids such as propionic, butyric, and caproic[75,83] appear to improve the aerobic stability of silages (Tables 23 and 24). The use of such additives, however, does not prevent oxidation but merely retards it.[84]

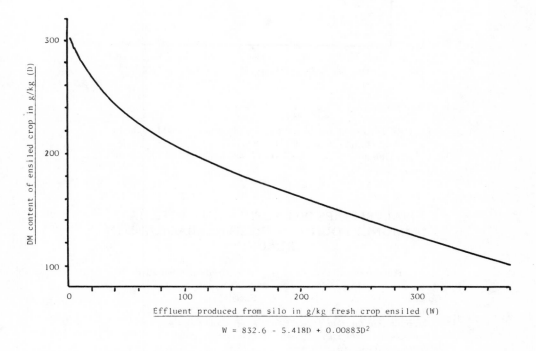

$$W = 832.6 - 5.418D + 0.00883D^2$$

FIGURE 1. Relationship between the DM content of an ensiled crop and the effluent production.[15]

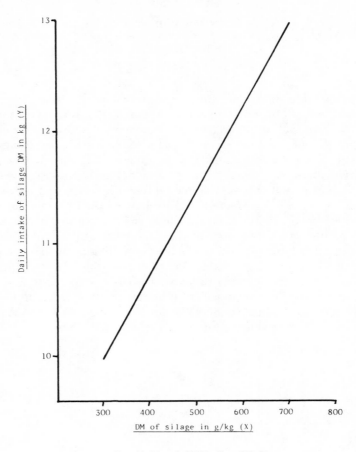

$$Y = 11.54 + 0.00736 \ (X - 511.3)$$

FIGURE 2.   Effect of DM content of alfalfa silage on DM intake by dairy cows. (From Gordon, C. H., Derbyshire, J. C., Jacobson, W. C., and Humphrey, J. L., *J. Dairy Sci.*, 48, 1062, 1965. With permission.)

Table 1
MAIN SPECIES OF LACTIC ACID BACTERIA COMMONLY FOUND ON FRESH HERBAGE AND IN SILAGE[7]

Homofermentative	Heterofermentative
Coccus	*Coccus*
Streptococcus faecalis	*Leuconostoc mesenteroides*
Streptococcus faecium	*Leuconostoc dextranicum*
Pediococcus acidilactici	*Leuconostoc cremoris*
Pediococcus cerevisiae	
Pediococcus pentosaceus	
Rod	*Rod*
Lactobacillus plantarum	*Lactobacillus brevis*
Lactobacillus curvatus	*Lactobacillus fermentum*
Lactobacillus casei	*Lactobacillus buchneri*
Lactobacillus coryniformis subsp. *coryniformis*	*Lactobacillus viridescens*

Table 2
MAIN PRODUCTS OF FERMENTATION OF SUGARS BY LACTIC ACID BACTERIA[6]

Homofermentative Lactic Acid Bacteria

Glucose $\longrightarrow$ 2 Lactic acid

Fructose $\longrightarrow$ 2 Lactic acid

Xylose $\longrightarrow$ Lactic acid + acetic acid

Arabinose $\longrightarrow$ Lactic acid + acetic acid

Heterofermentative Lactic Acid Bacteria

Glucose $\longrightarrow$ Lactic acid + ethanol + CO_2

3 Fructose $\longrightarrow$ Lactic acid + acetic acid + CO_2 + 2 mannitol

2 Fructose + glucose $\longrightarrow$ Lactic acid + acetic acid + CO_2 + 2 mannitol

Table 3
MICROBIAL COUNTS (NUMBER OF ORGANISMS PER GRAM FRESH MATERIAL) ON FRESH GRASS AND SILAGE SAMPLES[7]

	Total count	Lactic acid bacteria
Uncut grass	5.9×10^6	<100
Forage harvested grass	2.5×10^8	4.9×10^5
Grass at silo	3.0×10^8	8.3×10^4
Silage (after 189 days)	3.9×10^4	7.0×10^3

Table 4
TYPICAL DRY MATTER COMPOSITION OF A LACTATE RYEGRASS SILAGE[9]

(g/kg)

Total N	23
PN[a]	235
NH_3-N[b]	78
WSC[c]	10
Glucose	2
Fructose	3
Fructans	1
Formic acid	Trace
Acetic acid	36
Propionic acid	2
Butyric acid	1
Lactic acid	102
Ethanol	12
Mannitol	41

Note: Silage at pH 3.9; 190 g DM/kg buffering capacity 1120 meq alkali per kg DM (Reference 8).

[a] PN = Protein N as g/kg total N.

[b] NH_3-N = Ammonia N as g/kg total N.

[c] WSC = Water-soluble carbohydrates.

Table 5
FERMENTATION OF CITRIC AND MALIC ACIDS BY HOMO- AND HETEROFERMENTATIVE LACTIC ACID BACTERIA[8]

Citric acid $\longrightarrow$ 2 Acetic acid + formic acid + CO_2

2 Citric acid $\longrightarrow$ 2 Acetic acid + acetoin + CO_2

2 Citric acid $\longrightarrow$ 3 Acetic acid + lactic acid + 3 CO_2

Malic acid $\longrightarrow$ Lactic acid + CO_2

Malic acid $\longrightarrow$ Acetic acid/ethanol + formic acid + CO_2

2 Malic acid $\longrightarrow$ Acetoin + 4 CO_2

Table 6
BUFFERING CAPACITY (Bc) OF FRESH HERBAGES AND SILAGES[9,13]

	Original herbage		Silage	
	DM (g/kg)	Bc (meq/kg DM)	DM (g/kg)	Bc (meq/kg DM)
Ryegrass				
Fresh	178	350	184	1245
Wilted	323	320	309	935
Fresh + formic acid[a]	145	340	152	1110
Wilted + formic acid[b]	360	350	337	580
Red clover				
Fresh	150	578	—	1471
Wilted	318	491	—	763

[a] 2.2 g formic acid per kg fresh grass.

[b] 3.3 g formic acid per kg wilted grass.

Table 7
NITROGENOUS COMPOSITION OF RED CLOVER HERBAGE AND SILAGE[17]

	Per total N (g/kg)		Per total NPN[a] (g/kg)	
	Herbage	Silage	Herbage	Silage
Ammonia-N	10	144	42	256
Amide-N	75	Trace	313	Trace
Nitrate-N	25	10	104	17
Amino-N	43	250	181	445
Peptide-N	44	Nil	183	Nil
Protein-N	760	439	—	—
Total	957	844	823	719

[a] NPN = Non protein N.

From Ohshima, M., *Mem. Fac. Agric., Kagawa Univ.*, 26, 1, 1971. With permission.

Table 8
FREE AMINO ACIDS IN CORN PLANT WATER-SOLUBLE NITROGEN BEFORE AND AFTER ENSILING FOR 20 DAYS

Amino acids	g/kg Total amino acids	
	Fresh	Silage
Lys	39	75
His	20	49
Thr	26	45
Val	14	66
Met	3	17
Ile	9	33
Leu	9	75
Phe	8	35
Asp + AspNH$_2$	138	89
Glu	355	60
Ala	147	106

From Bergen, W. G., Cash, E. H., and Henderson, H. E., *J. Anim. Sci.*, 39, 629, 1974. With permission.

Table 9
RECOVERIES OF AMINO ACIDS N (PROTEIN, PEPTIDE, AND FREE AMINO ACID N) DURING THE ENSILING OF RED CLOVER FOR 80 DAYS

Amino acid	Recovery (g/kg)
Arg	0
Glu + Asp	341
Ser	527
Thr	669
Pro	732
Phe	761
Gly	819
Leu	941
Val	1000
Ile	1000
Ala	1507

From Ohshima, M., *Memoirs Fac. Agric., Kagawa Univ.*, 26, 1, 1971. With permission.

Table 10a
GROSS ENERGY (GE) VALUES OF
GRASSES AND SILAGES[12]

(MJ/kg DM)

	No	GE	SE
Grasses	18	18.3	±0.16
Lactate silages	18	20.0	±0.25
Grasses	7	18.4	±0.17
Wilted silages	7	19.1	±0.15
Grasses	7	18.7	±0.17
Chemically restricted silages	7	20.0	±0.36

Table 10b
GROSS ENERGY (GE) VALUES OF SOME
HERBAGE AND SILAGE CONSTITUENTS

(MJ/kg)

Glucose	15.64	Citric acid	10.33
Fructose	15.70	Malic acid	10.00
Sucrose	16.51	Lactic acid	15.16
Cellulose	17.49	Acetic acid	14.60
Crude fiber	17.60	Propionic acid	20.76
Ether extract	38.53	Butyric acid	25.51
		Mannitol	16.73
		Ethanol	29.80

Table 11
DIGESTIBILITY AND ME VALUES OF RYEGRASS AND RYEGRASS
SILAGES USING SHEEP[32]

		Silage			
Value	Grass	Lactate[a]	Wilted[b]	Chemically restricted and wilted[c,d]	SE
DM digestibility	0.784	0.794	0.752	0.776	±0.0133
OM digestibility	0.797	0.809	0.768	0.788	±0.0128
N digestibility	0.752	0.782	0.723	0.784	±0.0150
ME (MJ/kg DM)	11.6	13.6	11.4	12.0	±0.33

[a]   pH = 3.94; DM = 186 g/kg; lactic acid = 102 g/kg DM.
[b]   pH = 4.18; DM = 316 g/kg; lactic acid = 59 g/kg DM.
[c]   pH = 4.39; DM = 336 g/kg; lactic acid = 43 g/kg DM.
[d]   Formic acid applied 3.3 g/kg fresh herbage.

Table 12
TYPICAL DRY MATTER COMPOSITION OF AN ACETATE RYEGRASS SILAGE[10]

(g/kg)

Total N	47
PN[a]	440
NH$_3$-N[b]	128
WSC[c]	3
Mannitol	2
Formic acid	Nil
Acetic acid	97
Propionic acid	7
Butyric acid	2
Lactic acid	34
Ethanol	8

Note: Silage at pH 4.8; buffering capacity of 1090 meq alkali per kg DM (Reference 8); 176 g DM/kg.

[a] PN = Protein N as g/kg total N.
[b] NH$_3$-N = Ammonia N as g/kg total N.
[c] WSC = Water-soluble carbohydrates.

Table 13
SOME EXAMPLES OF CLOSTRIDIAL BACTERIA ASSOCIATED WITH BADLY PRESERVED SILAGE[55]

Lactate fermenters	Amino acid fermenters	Others
Clostridium butyricum	*Clostridium sporogenes*	*Clostridium sphenoides*
C. tyrobutyricum	*C. bifermentans*	*C. skatol*
C. paraputrificum		*C. perfringens*

Table 14
SOME EXAMPLES OF CLOSTRIDIAL FERMENTATION[6]

Organic acids

2 Lactic acid → Butyric acid + 2 CO$_2$ + 2 H$_2$

Amino acids

Coupled oxidation — reduction reactions (Stickland)

Alanine + 2 glycine → 3 acetic acid + 3 NH$_3$ + CO$_2$

Deamination

3 Alanine	⟶	2 propionic acid + acetic acid + 3 NH$_3$ + CO$_2$
Valine	⟶	isobutyric acid + NH$_3$ + CO$_2$
Leucine	⟶	isovaleric acid + NH$_3$ + CO$_2$

Decarboxylation

Histidine	⟶	histamine
Lysine	⟶	cadaverine
Arginine	⟶	ornithine → putrescine
Tryptophan	⟶	tryptamine
Tyrosine	⟶	tyramine
Glutamic acid	⟶	γ-amino butyric acid
Aspartic acid	⟶	β-alanine
Phenylalanine	⟶	β-phenylethylamine

Table 15
TYPICAL DRY MATTER COMPOSITION OF A BUTYRATE ORCHARD GRASS (COCKSFOOT) SILAGE[11]

(g/kg)

Total N	36
PN[a]	353
NH N[b]	246
WSC[c]	5
Formic acid	Nil
Acetic acid	24
Propionic acid	Nil
Butyric acid	35
Lactic acid	1

Note: Silage pH 5.2; 170 g DM/kg.

[a] PN = Protein N as g/kg total N.
[b] NH_3-N = Ammonia N as g/kg total N.
[c] WSC = Water-soluble carbohydrates.

Table 16
EFFECT OF PREWILTING ON GRASS SILAGE COMPOSITION, DIGESTIBILITY, AND INTAKE BY CATTLE[56]

	Silage			
	1	2	3	4
DM (g/kg)	190	273	323	432
pH	3.95	4.14	4.28	4.48
DM digestibility	0.740	0.710	0.690	0.700
DMI (g/kg $W^{0.75}$)	70	80	85	83
ME (MJ/kg DM)	10.6	10.2	9.8	9.9

Composition of Silage DM (g/kg)

Total N	22.6	22.0	22.0	21.2
Protein N (total N)	470	457	445	482
WSC	11	34	33	55
Lactic acid	71	55	45	32
Acetic acid	32	14	7	9
Butyric acid	28	30	27	11

From Jackson, N. and Forbes, T. J., *Anim. Prod.*, 12, 591, 1970. With permission.

Table 17
TYPICAL DRY MATTER
COMPOSITION OF A
WILTED RYEGRASS
SILAGE[9]

(g/kg)

Total N	23
PN[a]	289
NH$_3$-N[b]	83
WSC[c]	48
Glucose	16
Fructose	14
Fructans	1
Formic acid	Nil
Acetic acid	24
Propionic acid	Trace
Butyric acid	Trace
Lactic acid	59
Ethanol	6
Mannitol	36

Note: Silage pH, 4.2; 308 g DM/kg; buffering capacity of 890 meq alkali per kg DM.

[a] PN = Protein N as g/kg total N.
[b] NH$_3$-N = Ammonia N as g/kg total N.
[c] WSC = Water-soluble carbohydrates.

Table 18
CLASSIFICATION OF MAIN COMPOUNDS USED AS
SILAGE ADDITIVES

Silage Additives

Stimulants		Inhibitors	
Direct	**Indirect**	**Sterilizers**	**Acids**
Lactic acid bacteria	Glucose	Bacitracin	Mineral acids (AIV)
	Sucrose	Formalin	Formic acid
	Molasses	Sodium	Propionic acid
	Beet pulp	metabisulfite	Calcium formate/
	Corn		sodium nitrite
	Barley meal		Benzoic acid
			Phosphoric acid

Table 19
THE EFFECT OF FORMIC ACID ON RYEGRASS
SILAGE COMPOSITION[a]

The Effect on pH and WSC at Different Ryegrass DM Levels[62]

Grass DM (g/kg)	Silage			
	pH		WSC[b] (g/kg DM)	
	Untreated	Acid-treated[c]	Untreated	Acid-treated[c]
150	3.7	3.8	6	53
200	4.0	3.9	10	69
250	4.2	4.1	16	90
300	4.3	4.3	22	115
350	4.4	4.5	31	148

The Effect of Different Levels of Formic Acid on Ryegrass Ensiled at Same DM Level (200 g/kg)

Formic acid level (g/kg)		pH of grass[d]	WSC in silage[b] (g/kg DM)
Fresh	DM		
1	5	5.3	29
2	10	5.0	58
3	15	4.7	100
4	20	4.5	151

[a]   Harvested at 50% ear emergence.
[b]   Calculated from regression: $\log (WSC + 1) = -0.0893 + 0.0603F - 0.00077F^2 - 0.00199 \, DM$ when F = formic acid applied to grass (g/kg).
[c]   Treated with formic acid (11.5 g/kg DM).
[d]   Calculated from regression $100/y = 17.4 + 0.3F - 0.003F^2$ when y = pH.

Table 20
THE COMPOSITION OF SILAGES MADE FROM RYEGRASS (*L. perenne*) WITH THE ADDITION OF FORMALIN AT ENSILING

Quantity of Formalin (g/kg Fresh Herbage) Added

	0	2.5	4.9	9.8	19.8
pH	3.9	5.3	4.3	5.5	5.3
NH$_3$-N[a]	79	363	77	25	20

DM Composition (g/kg)

WSC	27	7	53	202	200
Acetic acid	29	58	11	6	5
Propionic acid	Nil	19	Nil	Nil	Nil
Butyric acid	Nil	78	Nil	Nil	Nil
Lactic acid	137	Nil	50	10	5

[a]   NH$_3$-N = ammonia N as g/kg total N.

From Wilkins, R. J., Wilson, R. F., and Woolford, M. K., *Vaextodling*, 29, 197, 1974. With permission.

Table 21
THE EFFECT OF FORMALIN[a] PRETREATMENT ON THE COMPOSITION AND NUTRITIVE VALUE TO SHEEP OF GRASS/CLOVER SILAGE

	Silage	
	Untreated	Formalin treated[a]
DM (g/kg)	179	196
pH	5.1	5.6
DM digestibility	0.602	0.585
DMI (g/kg W$^{0.75}$)	48.3	74.4
N digestibility	0.695	0.623
Liveweight gain (g/day)	38.3	104.0

Composition of silage DM (g/kg)

Total N	41	39
NH$_3$-N (TN)	145	65
Lactic acid	23	8
Acetic acid	46	12
Propionic acid	2	Trace
Butyric acid	11	Trace

[a]   Formalin applied at 8.3 g/kg fresh forage.
[b]   NH$_3$-N = Ammonia N as g/kg total N.

From Barry, T. N. and Fennessy, P. F., *N.Z. J. Agric. Res.*, 15, 712, 1972. With permission.

Table 22
TYPICAL DRY MATTER COMPOSITION OF A CHEMICALLY RESTRICTED[a] RYEGRASS SILAGE[6,12]

(g/kg)

Total N	34
PN[b]	740
NH$_3$-N[c]	30
WSC[d]	133
Ethanol	4
Formic acid	4.1
Acetic acid	10
Propionic acid	0.3
Butyric acid	1
Lactic acid	26

Note: Silage at pH 5.1; buffering capacity of 560 meq alkali per kg DM; 212 g DM/kg.

[a] Grass pretreated with formalin: formic acid (3:1 w/w) mixture, 10 g/kg.
[b] PN = Protein N as g/kg total N.
[c] NH$_3$-N = Ammonia N as g/kg total N.
[d] WSC = Water-soluble carbohydrates.

Table 23
AEROBIC DETERIORATION OF CORN SILAGES

The pH Values During Aerobic Deterioration of Corn Silages Treated with Organic Acids[a]

	Days of aerobic exposure					
	0	2	14	22	29	36
Control	3.93	4.08	5.80	7.95	—[b]	—[b]
5 P	4.05	3.95	5.20	5.95	4.90	5.05
10 P	4.53	4.38	3.93	4.25	4.64	4.64
20 P	4.23	4.28	4.18	4.18	4.20	4.20
5 F	4.38	4.35	4.18	4.30	4.48	5.30
10 F	4.45	4.85	4.30	5.90	4.60	4.65
20 F	4.35	4.70	4.15	5.05	7.05	—[b]
5 P + F	4.43	4.20	4.05	4.73	5.05	4.70
10 P + F	4.38	4.28	4.35	5.05	5.29	4.06
20 P + F	3.60	3.58	4.13	4.15	4.14	4.40
5 P + A	3.95	3.90	4.15	4.50	5.45	5.10
10 P + A	4.30	4.15	5.15	5.00	5.05	5.25
20 P + A	4.33	4.20	4.23	4.34	4.30	4.23

Table 23 (continued)
AEROBIC DETERIORATION OF CORN SILAGES

The pH Values During Aerobic Deterioration of Corn
Silages Treated with Organic Acids[a]

		Days of aerobic exposure			
0	2	14	22	29	36

Average Types of Platable Fungal Colonies at Ensiling and During Aerobic Deterioration (Relative Proportion)

	Yeast	*Geotrichum*	*Aspergillus* sp.	*Penicillium* sp.
At ensiling	0.717[e]	0.173[c]	0.001[c]	0.0005
On opening	0.084[c]	0.352[d]	0.505[d]	0.0007
14 Days exposure	0.653[e]	0.267[d]	0.072[c]	Nil
36 Days exposure	0.385[d]	0.164[c]	0.443[d]	Nil

[a]   Acids applied (5, 10, or 20 g/kg fresh corn) P = propionic, F = formic, P + F (0.6 propionic + 0.4 formic), P + A (0.8 propionic + 0.2 acetic).
[b]   Complete spoilage.
[c,d,e] Means not sharing the same superscript are different (P <0.01).

From Britt, D. G., Huber, J. T., and Rogers, A. L., *J. Dairy Sci.,* 58, 532, 1975. With permission.

Table 24
THE BIOCHEMICAL AND MICROBIOLOGICAL CHANGES DURING 7 DAY AEROBIC EXPOSURE OF TWO SILAGES[81]

	Lactate silage[a]		Chemically restricted silage[a,b]	
	On opening	After 7 days	On opening	After 7 days
Chemical composition				
pH	4.3	7.4	5.3	5.5
WSC in DM (g/kg)	86	13	191	191
Lactic acid in DM (g/kg)	69	7	7	3
NH_3-N in DM (g/kg)	1.6	4.1	0.3	0.3
Microbiological counts				
Yeasts (per gram silage)	4.4×10^4	2.8×10^7	<10	$<10^2$
Molds (per gram silage)	4.3×10^2	9.1×10^2	3.8×10^4	3.0×10^5

[a]   Made from ryegrass/red clover mixture.
[b]   Formalin + caproic acid (30/1 mixture) @ 10g/kg.

REFERENCES

1. Watson, S. J. and Nash, M. J., *Conservation of Grass and Forage Crops,* Oliver and Boyd, Edinburgh, 1960, 3—8.
2. Gibson, T., Stirling, A. C., Keddie, R. M., and Rosenberger, R. F., Bacteriological changes in silage as affected by laceration of the fresh grass, *J. Appl. Bacteriol.,* 24, 60—70, 1961.
3. Murdoch, J. C., The effect of length of silage on its voluntary intake by cattle, *J. Br. Grassl. Soc.,* 20, 54—58, 1965.
4. Dulphy, J. P. and Demarquilly, C., Effect of harvest machine and fineness of chop on the feeding value of silages, *Ann. Zootech.,* 22, 199—217, 1973.
5. Dulphy, J. P., Bechet, G., and Thomson, E., Influence of the physical structure and quality of conservation of grass silages on their voluntary intake by sheep produced from grass and corn, *Ann. Zootech.,* 24, 81—94, 1975.
6. Whittenbury, R., McDonald, P., and Bryan-Jones, D. G., A short review of some biochemical and microbiological aspects of ensilage, *J. Sci. Food Agric.,* 18, 441—444, 1967.
7. McDonald, P., Trends in silage making in *Microbiology in Agriculture, Fisheries and Food,* Skinner, F. A. and Carr, J. G., Eds., Academic Press, New York, 1976, 109—123.
8. McDonald, P. and Whittenbury, R., The ensilage process in *Chemistry and Biochemistry of Herbage,* Vol. 3, Buteer, G. W. and Bailey, R. W., Eds., Academic Press, New York, 1973, 33—60.
9. Henderson, A. R., McDonald, P., and Woolford, M. K., Chemical changes and losses during the ensilage of wilted grasses treated with formic acid, *J. Sci. Food Agric.,* 23, 1079—1087, 1972.
10. Henderson, A. R. and McDonald, P., The effect of delayed sealing on fermentation losses during ensilage, *J. Sci. Food Agric.,* 26, 653—667, 1975.
11. McDonald, P., Stirling, A. C., Henderson, A. R., and Whittenbury, R., Fermentation studies on wet herbage, *J. Sci. Food Agric.,* 13, 581—590, 1962.
12. McDonald, P. and Edwards, R. A., The influence of conservation on digestion and utilisation of forages by ruminants, *Proc. Nutr. Soc.,* 35, 201—211, 1976.
13. Playne, M. J. and McDonald, P., The buffering constituents of herbage and of silage, *J. Sci. Food Agric.,* 17, 264—268, 1966.
14. Carpintero, M. C., Holding, A. J., and McDonald, P., Fermentation studies on lucerne, *J. Sci. Food Agric.,* 20, 677—681, 1969.
15. Zimmer, E., The influence of prewilting on nutrient losses, particularly on the formation of fermentation gas, *Tagungsber. Dtsch. Landwirtschaftswiss. Berl.,* 92, 37—47, 1967.
16. Macpherson, H. T. and Slater, J. S., γ amino-n-butyric, aspartic, glutamic and pyrrolidonecarboxylic acids. Their determination and occurrence in grass during conservation, *Biochem. J.,* 71, 654—659, 1959.
17. Ohshima, M., Studies on nutritional nitrogen from red clover silage, *Mem. Fac. Agric., Kagawa Univ.,* 26, 1—68, 1971.
18. De Vuyst, A., Verack, W., Vanbelle, M., and Jadim, V., The amino acid and protein composition of green forages and the degradation of this protein during ensilage. Action of additives on this degradation, *Z. Tierphysiol. Tierernaehr. Futtermittelkd.,* 27, 82—99, 1971.
19. Bergen, W. G., Cash, E. H., and Henderson, H. E., Changes in nitrogenous compounds of the whole corn plant during ensiling and subsequent effects on dry matter intake by sheep, *J. Anim. Sci.,* 39, 629—637, 1974.
20. Kemble, A. R. and Macpherson, H. T., Mono-amino monocarboxylic acid content of preparations of herbage protein, *Biochem. J.,* 58, 44—46, 1954.
21. Macpherson, H. T. and Violante, P., The influence of pH on the metabolism of arginine and lysine in silage, *J. Sci. Food Agric.,* 17, 128—130, 1966.
22. Hughes, A. D., The nonprotein nitrogen composition of grass silages. II. The changes occurring during the storage of silage, *J. Agric. Sci., Camb.,* 75, 421—431, 1970.
23. Forbes, T. J. and Irwin, J. H. D., The use of barn-dried hay and silage in fattening young beef cattle, *J. Br. Grassl. Soc.,* 23, 299—305, 1968.
24. Waldo, D. R. and Moore, L. A., Growth, intake, and digestibility from formic acid silage versus hay, *J. Dairy Sci.,* 52, 1609—1616, 1969.
25. Durand, M., Zelter, S. Z., and Tisserand, J. L., Effect of some methods of preservation on the N efficiency of lucerne in sheep, *Ann. Biol. Anim. Biochim. Biophys.,* 8, 45—67, 1968.
26. Ciszuk, P. and Eriksson, S., Ammonia formation in the rumen of sheep fed on grass, clover, or lucerne preserved in various ways, *Swed. J. Agric., Res.,* 3, 13—20, 1973.
27. McDonald, P., Henderson, A. R., and Ralton, I., Energy changes during ensilage, *J. Sci. Food Agric.,* 24, 827—834, 1973.
28. Alderman, G., Collins, F. C., and Dougall, H. W., Laboratory methods of predicting feeding value of silage, *J. Br. Grassl. Soc.,* 26, 109—111, 1971.

29. McDonald, P., Edwards, R. A., and Greenhalgh, J. F. D., *Animal Nutrition,* Longman, London, 1975, 198—222.
30. Ekern, A., Blaxter, K. L., and Sawers, D., The effect of artificial drying on the energy value of grass, *Br. J. Nutr.,* 19, 417—434, 1965.
31. Wainman, F. W., First Report of the Feedstuffs Evaluation Unit, Rowett Research Institute, Scotland, 1975, 1—55.
32. Donaldson, E. and Edwards, R. A., Feeding value of silage: silages made from freshly cut grass, wilted grass, and formic acid treated wilted grass, *J. Sci. Food Agric.,* 27, 536—544, 1976.
33. Ekern, A. and Sundstol, F., Energy utilisation of hay and silages by sheep, *Eur. Assoc. Anim. Prod. Publ.,* 11, 231, 1974.
34. Corbett, J. L., Langlands, J. P., Mcdonald, I., and Puller, J. D., Comparison by direct animal calorimetry of the net energy values of an early and a late season growth of herbage, *Anim. Prod.,* 8, 13—27, 1966.
35. Graham, N. McC., Effect of artificial drying and of freezing on energy value of herbage, *Eur. Assoc. Anim. Prod. Publ.,* 11, 231—242, 1965.
36. Van Es., A. J. H., Utilisation of metabolisable energy by cows given silage, pelleted hay, or straw or long hay with their concentrate, in *Proc. 3rd Gen. Meeting of the European Grassland Federation,* Braunschweig, W. Germany, 1969, 275—281.
37. Moore, L. A., Thomas, J. W., and Sykes, J. F., The acceptability of grass/legume silage by dairy cattle, *Proc. 8th Int. Grassland Cong.,* Reading, England, 1960, 701—704.
38. Campling, R. C., Factors affecting the voluntary intake of grass, *Proc. Nutr. Soc.,* 23, 80—88, 1964.
39. Jarrige, R., Demarquilly, C., and Dulphy, J. P., The voluntary intake of forages, Väextodling, 28, 98—106, 1974.
40. Wilkins, R. J., The nutritive value of silages, in *University of Nottingham Nutrition Conf. for Feed Manufacturers,* Swan, H. and Lewis, D., Eds., Butterworths, London, 1974, 167—189.
41. Wilkins, R. J., Hutchinson, K. J., Wilson, R. F., and Harris, C. E., The voluntary intake of silage by sheep. I. Inter-relationship between silage composition and intake, *J. Agric. Sci. Camb.,* 77, 531—537, 1971.
42. Demarquilly, C., Chemical composition, fermentation characteristics, digestibility, and intake of forage silages: changes in relation to the initial green forage, *Ann. Zootech.,* 22, 1—35, 1973.
43. Thomas, J. W., Moore, L. A., Okamoto, M., and Sykes, J. F., A study of factors affecting rate of intake of heifers fed silage, *J. Dairy Sci.,* 44, 1471—1483, 1961.
44. McLeod, D. S., Wilkins, R. J., and Raymond, W. F., The voluntary intake by sheep and cattle of silages differing in free acid content, *J. Agric. Sci. Camb.,* 75, 311—319, 1970.
45. Gordon, C. H., Derbyshire, J. C., Jacobson, W. C., and Humphrey, J. L., Effects of dry matter in low moisture silage on preservation, acceptability, and feeding value for dairy cows, *J. Dairy Sci.,* 48, 1062—1068, 1965.
46. Foss, D. C., Niedermeier, R. P., Baumgardt, B. R., and Lance, R. D., Digestibility and intake corn, oats, and sorghum silages, *J. Dairy Sci.,* 44, 1175, 1961.
47. Ward, G. M., Boren, F. W., Smith, E. F., and Brethour, J. R., Relation between dry matter content and dry matter consumption of sorghum silage, *J. Dairy Sci.,* 49, 399—402, 1966.
48. Catchpoole, V. R. and Williams, W. T., The general pattern in silage fermentation in two subtropical grasses, *J. Br. Grassl. Soc.,* 24, 317—324, 1969.
49. Catchpoole, V. R., Laboratory ensilage of *Setaria sphaceleta* cv. Nandi and *Chloris gayena* cv. Pioneer at a range of dry matter contents, *Aust. J. Exp. Agric. Anim. Husb.,* 12, 269—273, 1972.
50. Catchpoole, V. R. and Henzell, E. F., Silage and silage-making from tropical herbage species, *Herb. Abstr.,* 41, 213—221, 1971.
51. Brown, D. C. and Radcliffe, J. C., Relationship between intake of silage and its chemical composition and *in vitro* digestibility, *Aust. J. Agric. Res.,* 23, 25—33, 1972.
52. Rook, J. A. F., Balch, C. C., Campling, R. C., and Fisher, L. J., The utilisation of acetic, propionic, and butyric acids by growing heifers, *Br. J. Nutr.,* 17, 399—406, 1963.
53. Ulyatt, J. M., The effect of intraruminal infusions of volatile fatty acids on food intake of sheep, *N.Z. J. Agric. Res.,* 8, 397—408, 1965.
54. De Vuyst, A., Vanbelle, A. M., and Deswysen, A., Effect of sodium acetate on the fermentation in the rumen and on the appetite of sheep, *Z. Tierphysiol. Tierernaehr. Futtermittelkd.,* 32, 279—288, 1974.
55. Gibson, T., Clostridia in silage, *J. Appl. Bacteriol.,* 28, 56—62, 1965.
56. Jackson, N. and Forbes, T. J., The voluntary intake by cattle of four silages differing in dry matter content, *Anim. Prod.,* 12, 591—599, 1970.
57. Harris, C. E., Raymond, W. F., and Wilson, R. F., The voluntary intake of silage, in *Proc. 10th International Grassland Congr. Helsinki,* 1966, 564—567.

58. Alder, F. E., McLeod, D. St.L., and Gibbs, B. G., Comparative feeding value of silages made from wilted and unwilted grass and grass/clover herbage, *J. Br. Grassl. Soc.*, 199—206, 24, 1969.

59. Durand, M., Zelter, S. Z., and Tisserand, J. L., Influence of some conservation techniques on the N efficiency of lucerne in the sheep, *Ann. Biol. Anim. Biochim. Biophys.*, 8, 45—67, 1968.

60. Fatianoff, W., Durand, M., Tisserand, J. L., and Zelter, S. Z., Comparative effects of wilting and of sodium metabisulphite on quality and nutritive value of alfalfa silage, in *Proc. 10th International Grassland Congr., Helsinki*, 1966, 551—557.

61. Hinks, C. E., Edwards, I. E., and Henderson, A. R., Beef production from formic acid-treated and wilted silages, *Anim. Prod.*, 22, 217—224, 1976.

62. Henderson, A. R. and McDonald, P., The effect of formic acid on the fermentation of ryegrass ensiled at different stages of growth and dry matter levels, *J. Brit. Grassl. Soc.*, 31, 47—51, 1976.

63. Candlish, E. and McKirdy, J., Organic acid determination on treated and untreated corn silage, *Can. J. Plant Sci.*, 53, 105—111, 1973.

64. Carpintero, C. and Suarez, A., Formic acid as a preservative of lucerne silage, *Ann. Fac. Vet. Leon, Univer. Oviedo*, 19, 209—218, 1974.

65. Ulvesli, O. and Saue, O., Comparison of different additives used in ensiling forage crops, *Meld. Norg. Landbrukshoegsk.*, 44, 1—31, 1965.

66. Castle, M. E. and Watson, J. N., Silage and milk production, a comparison between grass silages made with and without formic acid, *J. Br. Grassl. Soc.*, 25, 65—71, 1970.

67. Waldo, D. R., Smith, L. W., and Gordon, C. H., Formic acid silage versus untreated silage for growth, *J. Dairy Sci.*, 51, 982, 1968.

68. Brierem, K. and Ulvesli, O., Ensiling methods, *Herb. Abstr.*, 30, 1—8, 1960.

69. Wilkins, R. J., Wilson, R. F., and Woolford, M. K., The effects of formaldehyde on the silage fermentation, *Vaextodling*, 29, 197—201, 1974.

70. Barry, T. N. and Fennessy, P. F., The effect of formaldehyde treatment on the chemical composition and nutritive value of silage. I. Chemical composition, *N.Z. J. Agric. Res.*, 15, 712—22, 1972.

71. Brown, D. C. and Valentine, S. C., Formaldehyde as a silage additive. I. The chemical composition and nutritive value of frozen lucerne, lucerne silage, and formaldehyde treated lucerne silage, *Aust. J. Agric. Res.*, 23, 1093—1100, 1972.

72. Beever, D. E., Thompson, D. J., and Harrison, D. G., Energy and protein transformations in the rumen and the absorption of nutrients by sheep fed forage diets, in *Proc. 12th International Grassland Congr., Moscow*, 1974.

73. Wilson, R. F. and Wilkins, R. J., The use of mixtures of formalin and certain acids to restrict fermentation, *Annual Report of the Grassland Research Institute*, Hurley, England, 1974, 56.

74. Honig, H., Changes and losses by secondary fermentation, *Wirtschaftseigene Futter*, 21, 25—32, 1975.

75. Gross, F. and Beck, T., Investigations into the prevention of aerobic degradation processes after unloading of silage with propionic acid, *Wirtschaftseigne Futter*, 16, 1—14, 1970.

76. Cook, J. E., The use of additives to improve the stability of maize silage, *Bulletin of the Maize Development Association*, 54, 13—16, 1973.

77. Britt, D. G., Huber, J. T., and Rogers, A. L., Fungal growth and acid production during fermentation and refermentation of organic acid treated corn silages, *J. Dairy Sci.*, 58, 532—539, 1975.

78. Leaver, J. D., The use of propionic acid as an additive for maize silage, *J. Br. Grassl. Soc.*, 30, 17—21, 1975.

79. Ohyama, Y. and Masaki, S., Deterioration of silage after opening silo. I. Changes in temperature and chemical composition in some wilted silages, *J. Jpn. Soc. Grassl. Sci.*, 17, 176—183, 1971.

80. Mann, E. M. and McDonald, P., The effect of formalin and lower volatile fatty acids on silage fermentation, *J. Sci. Food Agric.*, 27, 612—616, 1976.

81. Ohyama, Y. and McDonald, P., The effect of some additives on aerobic deterioration of silages, *J. Sci. Food Agric.*, 26, 941—948, 1975.

82. Ohyama, Y., Masaki, S., and Hara, S., Factors influencing aerobic deterioration of silages and changes in chemical composition after opening silos, *J. Sci. Food Agric.*, 26, 1137—1147, 1975.

83. Yamashita, Y. and Yamazaki, A., The conditions for the occurrence of secondary fermentation on wilted silage, *Hokkaido Natl. Exp. Stn. Res. Bull.*, 110, 81—95, 1975.

84. Norgäard-Pedersen, E. J. and Witt, N., The effect of silage additives. I. Investigations on the effect of some silage additives on the oxidation processes, *Tidsskr. Planteavl.*, 77, 415—425, 1973.

EFFECT OF PROCESSING ON NUTRIENT CONTENT OF FEEDS: CHEMICAL PRESERVATION

K. Ross Stevenson

INTRODUCTION

Chemical additives can be used to maintain the quality of feeds or frequently to improve the quality of treated feeds relative to that of similar untreated material. The usefulness of preservatives is usually restricted to high moisture feeds which will deteriorate by natural processes such as micorbial degradation if not properly handled and stored. The two main areas of feed preservation in which the use of chemicals can result in improved feed quality are

1. The reduction or inhibition of plant respiration and/or microbial activity which result in deterioration of high moisture feeds stored under aerobic conditions.
2. To assist with satisfactory fermentation in feeds which are difficult to ensile under natural conditions.

A number of chemical preservatives will be discussed in the following sections. Where possible, the mode of action and the effects on feed quality will be described.

PROPIONIC ACID

Propionic acid or mixtures of propionic acid with some acetic acid have been shown to inhibit seed respiration and microbial activity on many types of high-moisture grain.[1-6] This acid is known to possess stronger fungicidal and fungistatic than bactericidal and bacteriostatic properties.[4] When propionic acid is sprayed on high moisture grain, the acid kills the seed embryo and essentially sterilizes the grain. Thus there is no heating or deterioration of properly treated grain and the high moisture grain can be stored in a similar manner to dry grains. Reinfection of the grain occurs over time and the length of the storage period in which satisfactory feed grain quality can be maintained varies with the treatment rate, moisture content of the grain, and storage conditions, such as ambient temperature and cleanliness. Because propionic acid inhibits mold activity in treated grain, it also has been shown to reduce mycotoxin production in such material.[7]

Many research trials have been conducted to determine the feeding value of acid-treated grains. Experiments have ranged from feeding trials to detailed studies of rates of digestion and energy and nitrogen retentions. In general, the consumption and utilization of treated high moisture grains have been satisfactory. Studies with dairy cattle,[2,8-10] beef cattle,[8,11-14] pigs,[2,3,15] poultry,[16-18] and sheep[15,19] have shown that animals fed treated high moisture grain performed equal to or slightly better than animals fed dried or ensiled grain.

Because the main action of propionic acid is the inhibition of aerobic reactions resulting in feed deterioration and spoilage, there has been considerable interest in the use of this acid in applications to other high moisture feed stuffs. Propionic acid has been shown to reduce heating,[20,21] dry matter loss,[20] and nitrogen content in the acid detergent fiber[21] of silages stored under variable oxygen exclusion. Propionic acid is also effective in reducing after-fermentation, the resulting heating,[22-26] mold,[22,26] and yeast[23,26] growth in high moisture feed materials. Propionic acid has been shown to reduce heating and spoilage in other feed stuffs such as damp hay[27] and wet brewers grain[28] when compared to untreated controls.

When any artificial acidification is added to fresh crop materials prior to ensiling, the acid tends to reduce the extent of the primary lactic acid fermentation, and propionic acid addition has this effect.[22,23,29] Because propionic acid is a weaker acid than formic, propionic has less effect than formic at equal application rates on altering both primary and secondary fermentation processes.[22,29]

FORMIC ACID

Formic acid is used widely in some countries in Europe to improve the fermentation processes in unwilted grass silage. Formic acid, with the lowest equivalent weight and lowest pKa of the volatile fatty acids, is the most effective organic acid in causing a pH reduction in treated crop material. Direct-cut grass and legume forages usually contain insufficient fermentable carbohydrate to enable stable silage to be produced. The addition of formic acid to such forages, causes an immediate reduction in pH of the forage and reduces the levels of lactic acid required to obtain a stable pH value.[30]

The low pH values of the treated forage reduce the extent of secondary fermentation, and thus, inhibit the activity of bacteria which are involved in the production of butyric acid and the breakdown of protein and amino acids with the resulting production of ammoniacal-nitrogen and several longer chain volatile fatty acids.

Considerable research has been conducted in many countries to study the effects of formic acid on feed quality and livestock performance. Much of this research has been reviewed previously.[31,32] Research at the Ruminant Nutrition Laboratory, Beltsville, Md., generally has shown increases in daily gain, efficiency, intake, energy digestibility feed recovery from silos, nitrogen retention, and reductions in ammoniacal-nitrogen.[31-34] The improved intake of treated silage may be related to the reduced ammoniacal-nitrogen content. Formic acid has not been used to any extent in North America because of the high cost of the acid.

Also, formic acid is an effective additive for the ensiling of other materials which are difficult to ensile.[29] Because formic acid acts primarily indirectly by reducing pH rather than by direct biostatic activity, it is much less effective than propionic acid in controlling the breakdown of high moisture feeds under aerobic conditions.[28]

FORMALDEHYDE

Formaldehyde has two important properties which have led to considerable research into its use as a silage and general feed additive. Formaldehyde has bactericidal effects and it also combines with protein to protect the protein from microbial degradation in the silo and in the rumen.

Formaldehyde is available in several forms and concentrations. The forms most often used in feeds are formalin; a solution, containing approximately 37% formaldehyde, and paraformaldehyde; a solid polymer in prill or flake form, containing approximately 92% formaldehyde.

When added to unwilted hay crop forage, formaldehyde significantly reduces the degradation of protein[32,35] and the production of ammoniacal-nitrogen[30] relative to that of untreated silage. Similar results have been obtained in other countries.[32] Feeding trial results indicate that animal performance for formaldehyde treated silage is similar to that of formic treated material. Limiting protein degradation may prove to be important in future research in animal nutrition.

OTHER PRESERVATIVES

Ammonium isobutyrate has shown promise as a preservative for high moisture feeds.[25,26] The action of ammonium isobutyrate appears to be somewhat similar to that of propionic acid.

Ammonia has shown potential as a preservative for high moisture grain.[36] Also, anhydrous ammonia is effective in controlling heating, mold growth, and dry matter loss in damp hay.[37,38] As a result of the biostatic properties of ammonia, nonprotein nitrogen additives in high moisture feedstuffs extend the trough life of such feeds.[24]

REFERENCES

1. **Huitson, J. J.**, Cereals preservation with propionic acid, *Proc. Biochem.*, 3, 31—32, 1968.
2. **Jones, G. M., Donefer, E., and Eliot, J. J.**, Feeding value for dairy cattle and pigs of high moisture corn preserved with propionic acid, *Can. J. Anim. Sci.*, 50, 483—489, 1970.
3. **Young, L. G., Brown, R. G., and Sharp, B. A.**, Proprionic for acid preservation of corn for pigs, *Can. J. Anim. Sci.*, 50, 711—715, 1970.
4. **Singh —Verma, S. B.**, Über den Einsatz der Propionsäure zur Konservierung von industriell hergestellten Mischfuttermitteln sowie von feuchtem Getriede und Mais, *Zentralbl. Baktariol.*, 125, 100—111, 1970.
5. **Stevenson, K. R. and Alexander, J. C.**, Propionic acid for storage of high moisture soybeans, *Can. J. Plant Sci.*, 52, 291—294, 1972.
6. **Saur, D. B. and Burrough, R.**, Efficacy of various chemicals as grain mold inhibitors, *Trans. ASAE*, 17, 557—559, 1974.
7. **Vandegraft, E. E., Hesseltine, C. W., and Shotwell, O. L.**, Grain preservatives: effect on aflatoxin and achratoxin production, *Cereal Chem.*, 52, 79—84, 1975.
8. **Forsyth, J. G., Mowat, D. N., and Stone, J. B.**, Feeding value for beef and dairy cattle of high moisture corn preserved with propionic acid, *Can. J. Anim. Sci.*, 52, 73—79, 1972.
9. **Clark, J. H., Frobish, R. A., Harshbarger, K. E., and Derrig, R. G.**, Feeding value of dry corn, ensiled high moisture corn, and propionic acid treated high moisture corn fed with hay or haylage for lactating dairy cows, *J. Dairy Sci.*, 56, 1531—1539, 1973.
10. **Britt, D. G. and Huber, J. T.**, Preservation of and animal performance on high moisture corn treated with ammonia or propionic acid, *J. Dairy Sci.*, 59, 278—287, 1976.
11. **Tonroy, B. R., Perry, T. W., and Beeson, W. M.**, Dry ensiled high-moisture, ensiled reconstituted high-moisture and volatile fatty acid treated high moisture corn for growing-finishing beef cattle, *J. Anim. Sci.*, 39, 931—936, 1974.
12. **Horton, G. M. J. and Holmes, W.**, Feeding value of whole and rolled acid-treated corn for beef cattle, *J. Anim. Sci.*, 40, 706—713, 1975.
13. **Macleod, G. K., Mowat, D. N., and Curtis, R. A.**, Feeding value for finishing steers and holstein male calves of whole dried corn and of whole and rolled high moisture acid-treated corn, *Can. J. Anim. Sci.*, 56, 43—49, 1976.
14. **Bolsen, K. K., Cox, O. J., and Riley, J. G.**, Influence of organic acids upon feeding of dry, reconstituted and early harvested milo for ruminants, *J. Anim. Sci.*, 39, 286—291, 1974.
15. **Bayley, H. S., Holmes, J. H. G., and Stevenson, K. R.**, Digestion by the pig of the energy and nitrogen in dried, ensiled and organic-acid-preserved corn: with observations on the starch content of digesta samples, *Can. J. Anim. Sci.*, 54, 377—383, 1974.
16. **Longworth, D. M.**, Influence of Moisture and Method of Preservation on the Nutritive Value of Corn to the Chick, unpublished M. Sci. thesis, University of Guelph, Ontario, 1972.
17. **Arends, L. G. and Gehle, M. H.**, Preservation of high moisture corn with volatile fatty acids, *Poult. Sci.*, 51, (Abstr.), 1868—1869, 1972.
18. **Akinola, A. A.**, Corn: Influence of Moisture Content and Method of Preservation on Nutritive Value for the Chick after Long Term Storage, unpublished M. Sci. thesis, University of Guelph, Ontario, 1973.
19. **Harpster, H. W., Long, T. A., and Wilson, L. L.**, A nutritive evaluation of dried, high moisture and acid-treated corn and sorghum grains for sheep, *J. Anim. Sci.*, 41, 1124—1133, 1975.

20. Thomas, J. W., Organic acids for haylage in snow-fence silos, *J. Dairy Sci.,* 59, 1104—1109, 1976.

21. Yu, Y. and Thomas, J. W., Effect of propionic acid and ammonium isobutyrate on preservation and nutritive values of alfalfa haylage, *J. Anim. Sci.,* 41, 1458—1467, 1975.

22. Britt, D. G., Huber, J. T., and Rogers, A. L., Fungal growth and acid production during fermentation and refermentation of organic acid treated corn silages, *J. Dairy Sci.,* 58, 532—539, 1976.

23. Daniel, P., Honig, H., Weise, F., and Zimmer, E., Wirkung von Propionsäure bei der Grünfuttersilierung, *Wirtschaftseigene Futter,* 16, 239—252, 1970.

24. Stevenson, K. R., Stability and bunk life of high moisture corn, in *Proc. High Moisture Grains Symp.,* Oklahoma State University, Stillwater, 1976, 105—111.

25. Goering, H. K. and Gordon, C. H., Chemical aids to preservation of high moisture feeds, *J. Dairy Sci.,* 56, 1347—1351, 1973.

26. Bothast, R. J., Adams, G. H., Hatfield, E. E., and Lancaster, E. B., Preservation of high-moisture corn: a microbiological evaluation, *J. Dairy Sci.,* 58, 386—391, 1975.

27. Knapp, W. R., Holt, D. A., and Lechtenberg, V. L., Propionic acid as a hay preservative, *Agron. J.,* 68, 120—123, 1976.

28. Allen, W. R., Stevenson, K. R., and Buchanan-Smith, J., Influence of additives on short-term preservation of wet brewers grain stored in uncovered piles, *Can. J. Anim. Sci.,* 55, 609—618, 1975.

29. Allen, W. R. and Stevenson, K. R., Influence of additives on the ensiling process of wet brewers grains, *Can. J. Anim. Sci.,* 55, 391—402, 1975.

30. Davidson, T. R., Stevenson, K. R., and Buchanan-Smith, J., Influences of formic acid and formalin on the production of organic acid in direct-cut alfalfa silage, *Can. J. Plant Sci.,* 53, 81—85, 1973.

31. Waldo, D. R. and Derbyshire, J. C., The Feeding Value of Hay Crop Silages, Proc. Int. Silage Res. Conf., Chicago, 1971.

32. Waldo, D. R., Chemical Preservation of Forages, Proc. Cornell Nutrition Conf., Buffalo, N.Y., 1973, 50—58.

33. Waldo, D. R., Keys, J. E., Jr., Smith, L. W., and Gordon, C. H., Effect of formic acid on recovery, intake, digestibility and growth from unwilted silage, *J. Dairy Sci.,* 54, 77—84, 1971.

34. Waldo, D. R., Keys, J. E., Jr., and Gordon, C. H., Preservation efficiency and dairy heifer response from unwilted formic and wilted untreated silages, *J. Dairy Sci.,* 56, 129—134, 1973.

35. Waldo, D. R., Keys, J. E., Jr., and Gordon, C. H., Paraformaldehyde compared with formic acid as a direct-cut silage preservative, *J. Dairy Sci.,* 59, 922—930, 1976.

36. Bothast, R. J., Lancaster, E. B., and Hassetine, C. W., Ammonia kills spoilage molds in corn, *J. Dairy Sci.,* 56, 241—245, 1973.

37. Knapp, W. R., Holt, D. A., and Lechtenberg, V. L., Anhydrous ammonia and propionic acid as hay preservatives, *Agron. J.,* 66, 823—824, 1974.

38. Knapp, W. R., Holt, D. A., and Lechtenberg, V. L., Hay preservation and quality improvement by anhydrous ammonia treatment, *Agron. J.,* 67, 766—769, 1975.

EFFECT OF PROCESSING ON NUTRIENT CONTENT OF FEEDS: ALKALI TREATMENT

V. Friis Kristensen

INTRODUCTION

Cellulose is the most abundant source of organic energy on earth and in man's search for future sources of food, cellulose-rich plant materials may come to play an expanding role. These materials cannot serve directly as food for humans and is utilized only to a very limited degree by simple-stomached animals. Ruminants are through their symbiosis with microorganisms in the rumen capable of utilizing cellulose and hemicellulose as sources of energy and convert them to valuable foods as meat and milk.

Plant cell walls are composed mainly of cellulose, hemicellulose, and lignin. Certain plant materials, for instance wood and the stalks and leaves of mature annual plants, consist almost entirely of cell walls. Ruminants utilize pure cellulose and hemicellulose as sources of energy with the same efficiency as starch.[50,94] However, in the mature plant cell wall these materials are physically and chemically so tightly bound in the cell wall complex that they are more or less unaccessible for the microbes or the microbial enzymes in the rumen. Many woods are practically undigestible for ruminants, and only 35 to 55% of the dry matter in straw is digested. Therefore, there is severe limitation on the use of such materials as feed for ruminants.

In the last part of the 19th century it was discovered that the accessibility of fibrous materials could be increased considerably by treatment with dilute alkali.[60] During the following decades great attempts were made to increase the digestibility of fibrous plant materials by chemical treatment, and a great number of treatment procedures were developed. At the beginning the procedures comprised cooking in a solution of NaOH, eventually under pressure. This method was used to some extent in Germany before and during World War I. Treatment of straw in a weak NaOH solution (1.5-2.0% w/v) at ambient temperature and pressure, a method developed by Beckmann,[8] gained much more importance and was widely used, especially in Norway, during and after World War II.[14,39] Homb[39] made a comprehensive review of the early work on treatment procedures, and it is also presented by Homb et al.[40]

In the Beckmann procedure straw is soaked for 12 to 24 hr in a 1.5% NaOH solution in an amount of about eight times the weight of the straw. Subsequently, the straw is carefully washed in order to remove the excess alkali. The wet material is fed directly. The method involves the use of very great amounts of water, causes a loss of dry matter during washing and gives rise to serious pollution problems. Because of these disadvantages, this method is now used only to a very limited extent.

Since the 1960s, the search for and investigations of new treatment methods have been very extensive. In 1964 Wilson and Pigden[102] described a successful new method, which was based on the use of a much smaller amount (30% of the weight of straw) of more concentrated alkali solutions. The treated material was fed directly either unneutralized or with the excess alkali neutralized by organic acids. Since then a great number of studies on procedures of alkali treatment have been made. NaOH is the chemical most widely used, but the effect of other alkaline chemicals as, e.g., KOH and Ca(OH)$_2$ have also been studied. Besides NaOH, ammonia is the most studied alkaline chemical for treatment of low-quality roughages. Ammonia may be used either anhydrous (liquid or gaseous), aqueous (NH$_4$OH), or supplied in the form of NH$_3$-releasing substances (i.e., NH$_4$HCO$_3$, urea). This chapter reviews results of experiments mainly carried through during the last 20 years. Many technical procedures

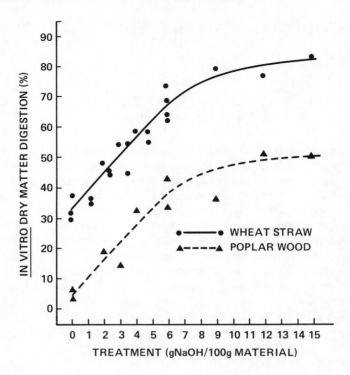

FIGURE 1.   The effect of sodium hydroxide upon the percentage dry matter digestion with rumen microorganisms. (From Wilson, R. K. and Pigden, W. J., *Can. J. Anim. Sci.,* 44, 122, 1964. With permission.)

for treatment of roughages with alkali have been developed. The most important procedures are thoroughly discussed by Jackson,[44] and treatment methods are not covered in this chapter.

EFFECT OF ALKALI TREATMENT ON THE DIGESTIBILITY OF FIBROUS MATERIALS IN VITRO

Treatment with Sodium Hydroxide

Wilson and Pigden[102] found that the in vitro dry matter digestibility (IVDMD) of wheat straw and poplar wood was increased with application of increasing levels of NaOH up to 9% of the air dry weight of straw and wood, but above this level no further increases were obtained (Figure 1). Homb[39] in his review gives results from early German experiments with the Beckmann method conducted by Fingerling in the 1920s. They showed that the greatest increase in the in vivo digestibility was obtained by increasing the amount of NaOH up to 8% of the straw, and a further increase to 12% NaOH increased the digestibility to a smaller extent.

Results from newer studies are given in Figure 2. They cover a variety of fibrous materials and a number of quite different treatment procedures. Of the results in Figure 2, those obtained by Chandra and Jackson[18] are determined by use of the nylon bag technique (suspension in the rumen), while all other determinations were done in vitro by incubation with rumen microorganisms. Jayasuriya and Owen[46] gave the results as digestibility of the organic matter, Gharib et al.[31] as digestibility of NDF, while all other results in Figure 2 are given as digestibility of the dry matter. The NaOH-dosage is given as percent of air dry plant material.

Digestibility in vitro, %

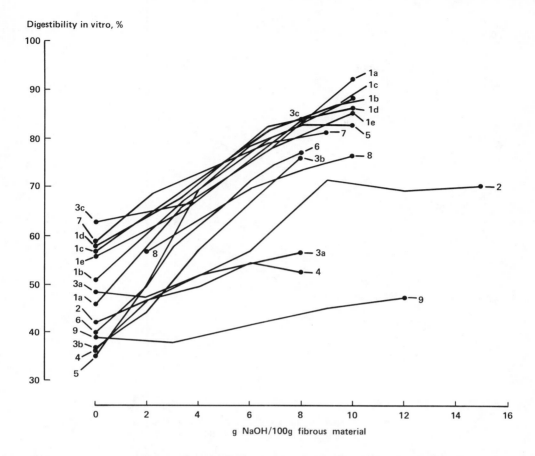

FIGURE 2.   The effect of increasing amounts of NaOH on the digestibility of various fibrous materials in vitro. 1a. Maize cobs, b. wheat straw, c. paddy straw, d. sorghum stover, e. sugar cane tops. Roughages ground, 10 mℓ solution per 10 g roughage. Treatment time 1 hr, dried at 100°C.[18] 2. Wheat straw, ground, 30 mℓ solution per 100 g straw, dried at room temperature.[101] 3a. Alfalfa stem, b. barley straw, c. corn stover. Roughages ground, 2 mℓ solution per 250 mg roughage. Treatment time 24 hr at 23°C.[76] 4. Aspen wood, ground, 0.5% NaOH solution in various solution-to-wood ratios from 4:1 to 16:1. Treatment time 2 hr.[67] 5. Barley straw, chopped, treated in a rotating batch processor with 68% moisture and at 100°C. Maximum digestibility determined after treatment in various times from 4 to 120 min.[79] 6. Means for wheat, rye, oat, and barley straw. Roughages chopped, 100 mℓ solution per 100 g straw. Treatment time 3 days at room temperature.[84] 7. Barley straw, ground, 100 mℓ solution per 100 g straw. Treatment time 24 hr at room temperature.[46] 8. Wheat straw, ground, mixed with dry powdered chemical under addition of steam and treated at different pressures (electro-hydraulic press), temperatures, and times.[48] 9. Poplar bark, ground, 31.3 mℓ of solution per 50 g bark. Averages for varying times and temperatures of treatment.[31]

Figure 2 gives an impression of the effect of the level of NaOH, but the results of different studies may not be directly compared. The levels of digestibility are dependent on the procedure of determination of digestibility, and the level and the rate of increase of digestibility are dependent on the treatment procedure used.

Wilson and O'Shea[101] found that treatment of wheat straw with up to 9% NaOH caused marked increases in digestibility. Higher alkali levels gave no further increases. Donefer et al.[25] found increasing in vitro digestibilities of oat straw by increasing the NaOH level up to 16%, but at a diminishing rate. However, the last step increased the NaOH level from 8 to 16%, and the point, where maximum effect was reached, was possibly not shown. Feist et al.[27] noted maximum digestibility of quaking aspen and red oak at 5 to 6%, and no further increases were obtained at higher levels. At

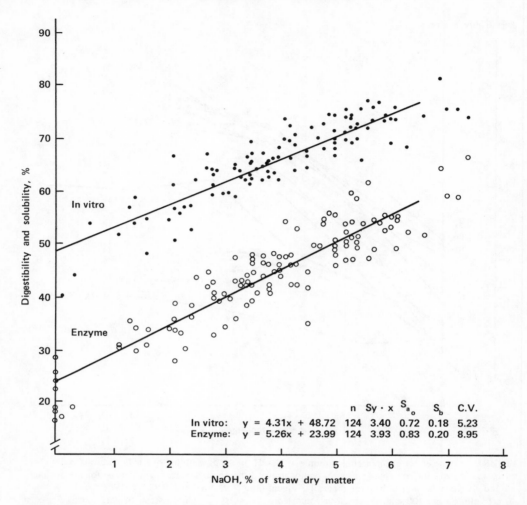

FIGURE 3.   The effect of increasing amounts of NaOH on the in vitro (rumen microorganisms) digestibility of organic matter and the enzyme (cellulase) solubility of dry matter in barley straw.[58] Straw chopped, mixed with a 33 to 35% NaOH solution and pressed through a ring die press.

different temperatures and durations of treatment, IVDMD values of barley straw increased with increased concentration of NaOH up to 8%.[76] No significant differences occured between 8 and 12% NaOH. Chandra and Jackson[18] increased the digestibility of ground maize cobs almost linearly up to the 10% NaOH level and no further increase was obtained at higher levels. Carmona and Greenhalgh[17] obtained increased digestibilities of barley straw by soaking it in solutions with up to 14 g NaOH/100 g straw, though at a greatly diminishing rate from about 7 g NaOH. When spraying NaOH solutions on the straw, the digestibility was increased with up to 8 g NaOH/ 100 g straw, which was the highest amount tested. In studies of Phoenix et al.,[79] the digestibility of barley straw was increased up to 8% NaOH, but at a diminishing rate. Raising the NaOH percentage from 8 to 10 gave no further increase. The IVDMD values of both rye, wheat, oat, and barley straw were increased almost linearly up to 8% NaOH.[84] Increasing the amount of NaOH stepwise from 2.25 to 9.0 g/100 g barley straw increased the in vitro digestibility, but to a progressively smaller extent.[46] Gharib et al.[31] found maximum digestibility of the neutral detergent fiber (NDF) of poplar bark at 12% NaOH, which was the highest level tested. Junker[48] noted a high rate of

Table 1
EFFECT OF TEMPERATURE, DURATION, AND CONCENTRATION OF SODIUM HYDROXIDE ON IN VITRO DRY MATTER DIGESTIBILITY (%) OF BARLEY STRAW

Temp. (°C)	Processing duration (min)	NaOH (% of dry matter)					
		0	2	4	6	8	12
23	0 (hr)	37.9	39.3	41.5	41.2	41.8	38.8
	24 (hr)	37.6	43.8	53.8	63.4	68.0	68.4
60	5	38.4	42.7	51.7	60.4	63.9	66.6
	15	36.4	45.1	54.2	61.9	67.4	68.1
	45	37.9	45.0	55.4	65.2	68.4	67.8
	90	37.7	44.9	56.0	65.1	67.9	68.2
80	5	36.2	45.3	55.8	66.4	72.5	73.1
	15	37.0	45.9	60.1	68.6	72.3	70.0
	45	36.8	49.3	63.2	70.1	73.4	76.2
	90	37.6	51.0	63.5	72.3	76.0	75.6
100	5	37.7	46.6	59.0	69.9	73.9	76.8
	15	37.4	46.4	59.7	69.5	73.7	75.2
	45	37.7	49.1	62.8	72.4	76.8	79.1
	90	37.7	50.6	65.0	73.0	78.0	79.9
130	5	38.6	48.3	63.4	73.5	77.9	80.3
	15	38.2	49.1	65.2	74.5	77.9	80.8
	45	39.6	51.2	66.1	74.3	79.1	81.6
	90	40.8	52.2	66.8	75.3	78.3	81.0

Note: Barley straw ground (1 mm screen), 2 mℓ NaOH solution added to 250 mg straw. Digestibility determined by the two stage rumen microorganisms — acid pepsin procedure.

From Ololade, B. G., Mowat, D. N., and Winch, J. E., *Can. J. Anim. Sci.,* 50, 660, 1970. With permission.

increase of the digestibility of wheat straw up to 6% NaOH. The digestibility was still increased up to 10% NaOH, though at a slower rate. Kristensen et al.[58] determined the digestibility of 124 lots (each constituting 500 to 1000 kg) of straw treated on a pilot plant (Figure 3). Determinations were made both with rumen liquid and a commercial cellulase enzyme preparation ("Onozuka" SS). The results indicated a nearly linear increase in digestibility up to approximately 6% NaOH.

In general, it may be concluded that with increasing rate of application of NaOH, the effect on the digestibility in vitro of a variety of fibrous materials is curvelinear in pattern. The digestibility is increased nearly linearly up to about 6% NaOH, and a substantial further increase is obtained up to 8 to 10% NaOH, though often at a slower rate. Differences in this pattern may possibly be caused by differences in the efficiency of the various treatment procedures used. Also, various plant materials may respond differently. Thus, the digestibility determined by the nylon bag technique of the organic matter in rice hulls was increased with up to 100 g NaOH per 100 g rice hull dry matter.[63]

The rate and extent of reaction of NaOH with the lignocellulose complex of fibrous plant materials are also dependent on other physical factors such as temperature and pressure and adequacy of mixing the NaOH solution with the plant material.

Basically, this process can be expected to behave like other chemical reactions on increased temperature and pressure. Ololade et al.,[76] demonstrated the effect of tem-

Table 2
CALCULATED VALUES OF MAXIMUM IN VITRO
DIGESTIBILITY (IVD) INCREASE (B_o^{-1}) AND INITIAL RATE
OF DIGESTIBILITY INCREASE (B_1^{-1}) WITH NaOH-
TREATMENT[a] OF BARLEY STRAW UNDER VARIOUS
PHYSICAL CONDITIONS

NaOH %	Moisture content %	Temp. °C	Max. IVD %	B_o^{-1} (Maximum digestibility increase) %	B_1^{-1} (Initial rate of digestibility increase) % per min
4	68	44	65.1	30.1	22.0
4	68	56	65.8	30.8	30.6
4	68	75	66.9	31.9	44.3
4	68	100	69.3	34.3	82.8
4	32	100	63.6	28.6	27.9
4	43	100	68.4	33.4	19.7
4	56	100	68.2	33.2	15.9
4	68	100	69.3	34.3	82.8
4	80	100	72.0	37.0	111.3

[a] Barley straw chopped, treated in a rotating batch processor, air cooled and dried at 45°C. Digestibility determined by the rumen microorganisms — acid pepsin procedure.

From Phoenix, S. L., Bilanski, W. K., and Mowat, D. N., *Trans. ASAE*, 15, 1091-1093, 1972. With permission.

perature on the action of NaOH on barley straw (Table 1). The effect of increasing the temperature from 23 to 130°C as measured on the digestibility in vitro was substantial and highest at the highest NaOH concentrations. The effect of treatment time was greatest at lower temperatures and lower NaOH concentrations, which indicates the importance of these factors for the rate of reaction. Treatment for 15 min at 60°C produced similar results as treatment for 24 hr at 23°C.

Phoenix et al.,[79] similarly, reported an increased extent of response to NaOH-treatment at elevated temperatures, but especially, the initial rate of reaction was very much increased (Table 2). At 100°C the initial rate of digestibility increase was nearly four times as high as at 44°C. It is noted that at the lower temperatures, the values may be slightly biased, because it was not possible to stop chemical reactions during the drying process at 45°C. Increasing the temperature beyond 100°C did not improve the in vitro dry matter digestibility, when more than 3% NaOH was used at 68% moisture content.[80] At lower NaOH concentrations, the digestibility was increased somewhat with increasing temperature up to 175°C. Rexen et al.[90] reported increased in vitro digestibility of NaOH-treated barley straw as a result of elevated temperature and pressure, when chopped straw was treated with concentrated NaOH solutions for 1 min at various pressures and temperatures in a hydraulic particleboard press. The effect of increased pressure was highest at higher temperatures and at higher NaOH-concentrations. Bolduan and Piatkowski[11] found that increased temperature increased the percentage of added NaOH bound in straw after treatment.

On the contrary, Gharib et al.[31] did not find any difference in the effect of NaOH treatment of poplar bark at 25, 50, or 75°C for 1 or 20 days, when 0, 3, 6, 9, or 12 g NaOH in 31.3 m*l* H_2O was mixed with 50 g of bark. Junker[48] treated wheat straw mixed with dry alkali in a hydraulic press for various times (1 to 7 min). The digestibility in vitro tended to be reduced after treatment at 150°C as compared to 60°C and at a pressure of 900 bar compared to 400.

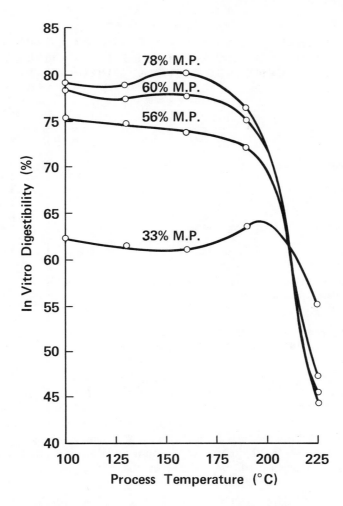

FIGURE 4.   The effect of treatment temperature and moisture percentage (M.P.) on the in vitro digestibility of NaOH-treated barley straw (5% NaOH).[80] Treatment time 30 min in pressure vessels. (From Phoenix, S. L., Bilanski, W. K., and Mowat, D. N., *Trans. ASAE,* 17, 780, 1974. With permission.)

Detrimental effects on the in vitro digestibility may occur, when roughages are exposed to strong reaction conditions. Charring occurred, when barley straw was treated with NaOH at temperatures above 190°C, the IVDMD was severely depressed and the treated material rendered unusable[80] (Figure 4). In high-pressure-temperature treatments of grass straw Guggolz et al.[33] found that with treatment at 28 kg/cm², the digestibility increased with increasing treatment time up to 4 min. When the treatment time was increased to 6 min, the digestibility was decreased compared to the results at 4 min. The digestibility was determined by incubation with a crude cellulase enzyme followed by protease digestion as described by Guggolz et al.[35] It was suggested that the decrease in digestibility possibly is owing to a decelerating process caused by high temperature polymerizing or complexing. Dekker and Richards[24] found corresponding results, when the IVDMD of bagacillo was decreased by pulping (kraft process) for 70 hr compared to 3 and 23 hr.

The effect of alkali treatment may be dependent on the amount of moisture added to straw in the treatment. Donefer et al.[25] added 4, 8, or 16 g NaOH each in 30, 60,

Table 3
INFLUENCE OF AMBIENT TEMPERATURE AMMONIA
TREATMENT[a] ON ENZYME SOLUBILITY AND NITROGEN
CONTENT OF SEVERAL STRAWS

	Enzyme solubility of D.M. %		N-content, %	
Straw	Original	Treated	Original	Treated
Alfalfa	53	62	2.36	3.24
Barley	37	73	0.53	2.01
Bean	52	65	0.91	2.47
Fescue	37	62	0.79	1.50
Oat	33	63	0.43	1.77
Perennial ryegrass	40	65	1.00	1.92
Rice	29	62	0.56	1.32
Wheat	37	62	0.86	2.14

[a]   Treatment with 5.2% aquous NH_3 for 30 days at room temperature (22°C). Enzyme solubility determined by a two stage (crude cellulase + proteinase) technique developed by Guggolz, et al.[35]

From Waiss, A. C., Jr., Guggolz, J., Kohler, G. O., Walker, H. V., Jr., and Garrett, W. N., *J. Anim. Sci.,* 35, 111, 1972. With permission.

or 120 mℓ solution to 100 g oat straw. After 24 hr treatment, the digestibility was substantially increased with increasing dilution rate at all NaOH levels. Phoenix et al.[79] found a slight increase in the IVDMD with increasing moisture content in the straw (Table 2), and in later high-pressure-temperature trials more pronounced effects of increased moisture content were noted[80] (Figure 4). Jayasuriya and Owen[46] similarly reported increased IVDMD by increasing amount of solution from 30, over 60, to 120 mℓ/100 g straw, while no further response was obtained at higher dilution rates. In in vivo digestibility trials with sheep, the digestibility of treated straw was significantly increased by increasing the amount of added solution from 30 to 60 mℓ/100 g straw.[46] The treatments consisted of mixing the solution and straw and allowing the mixture to react for 24 hr at room temperature followed by neutralization. Carmona and Greenhalgh[17] concluded that the concentration, at which the sodium hydroxide was sprayed on the straw, appeared not to influence its effect on digestibility, provided there was sufficient liquid to allow adequate mixing. In this experiment, the treatment consisted of mixing and storing for 24 hr at room temperature and drying.

The amount of water added to the material in NaOH treatments could act in two ways to increase the effect. First, it can provide a more efficient mixing of alkali and straw, and secondly, a higher moisture content could increase the reactivity of alkali with the cell wall constituents. The results do not demonstrate the importance of each of these factors.

Treatment with Ammonia

Ammonia is a slow reacting chemical, but as Waiss et al.[99] stated, a substantial rise in the digestibility of straw can also be obtained at room temperature and atmospheric pressure, if the reaction of ammonia on the straw is allowed to proceed for a longer period of time (Table 3).

Waiss et al.[99] treated rice straw for 1 hr at 160°C with water or NH_4OH solutions. Under these conditions it was found that the enzyme digestibility (crude cellulase + proteinase) was increased substantially at the level of 2.6% NH_3 on the straw. Only a

Table 4
ORGANIC MATTER DIGESTIBILITY IN VITRO (OMD)
AND CRUDE PROTEIN CONTENT (CP) (% of DM) OF
BARLEY STRAW TREATED[a]) WITH VARIOUS
AMOUNTS OF AMMONIA, TREATMENT TIMES AND
TREATMENT TEMPERATURES

Treatment		NH$_3$- and water-dosage per 100 g straw (87% D.M.)					
		3.4 g NH$_3$ 3 g water		4.4 g NH$_3$ 9 g water		5.9 g NH$_3$ 16 g water	
Temp. °C	Time days	OMD	CP	OMD	CP	OMD	CP
15	3	51	8.5	57	9.4	63	9.9
	7	51	8.6	59	10.3	64	10.6
	14	55	9.4	60	10.1	65	10.7
30	3	61	9.7	63	10.4	66	10.8
	7	63	10.3	67	10.8	67	11.1
	14	68	11.1	69	10.9	66	11.2
45	3	67	11.3	69	11.4	66	11.8
	7	67	11.2	66	11.6	65	11.9
	14	67	11.1	67	11.9	66	12.1
55	3	66	11.1	65	11.4	65	11.9
	7	66	12.0	63	11.7	64	12.2
	14	65	11.9	62	11.9	65	12.1
Untreated straw		38	3.1	—	—	—	—

[a] 100 g of straw moistened with water and placed in 1-liter glass jars; anhydrous, gaseous NH$_3$ added. Digestibility determined by two stage in vitro technique.

From Waagepetersen, J. and Thomsen, K. V., *Anim. Feed. Sci. Technol.*, 2, 134, 1977. With permission.

very slight improvement was produced by 5.2% NH$_3$ over 2.6%. At room temperature, 5.2% NH$_3$ gave an appreciable effect over the 2.6% level. Waagepetersen and Thomsen[104] also found interactions between NH$_3$ levels, treatment temperatures, and treatment times, when barley straw was treated with anhydrous, gaseous ammonia (Table 4). At the lowest temperature (15°C) the levels of NH$_3$ (3.4, 4.4, and 5.9%) had a pronounced effect on the in vitro digestibility. At 30°C the amount of NH$_3$ had an effect only at the shortest treatment times. At higher temperatures, the digestibility was not increased by increasing the amount of NH$_3$ above 3.4%.

The treatment time (3, 7, or 14 days) had an effect only at the lowest temperatures and NH$_3$ levels. At 15°C, a longer period of time than used in this experiment, would be necessary to attained maximum digestibility. At 30°C maximum digestibility was reached after 14 days with 3.4% NH$_3$ and after 3 days with 5.9% NH$_3$.

Similar results were obtained by Becker and Pfeffer[7] (Table 5). Maximum effect was reached at a shorter time (within 2 weeks) at a higher temperature (20°C). With long time treatment (8 weeks) the temperature had only minor effects. At 20°C the amount of NH$_3$ increased the effect on digestibility only to the 3 to 4% NH$_3$ level. At 2°C the digestibility was rising in the whole range of NH$_3$ concentrations up to 6%.

Sundstøl et al.[94f] reported a pronounced effect on in vitro digestibility of oat straw of increasing the ammonia level from 1 to 2.5% and some effect of increasing further to 4% NH$_3$ when treating with gaseous ammonia for 4 weeks. No significant effect was found by increasing to 5.5% NH$_3$. A pronounced effect of increasing the temperature from −20°C to +25°C when treating with 3.4% NH$_3$ was noted.

Table 5
DRY MATTER DIGESTIBILITY IN VITRO (DMD) AND CRUDE PROTEIN CONTENT (CP) OF WHEAT STRAW TREATED[a] WITH VARIOUS AMOUNTS OF AMMONIA, TREATMENT TIMES AND TREATMENT TEMPERATURES[7]

Treatment time	2 weeks		4 weeks		6 weeks		8 weeks	
Temperature, °C	20	2	20	2	20	2	20	2
DMD, %								
g NH₃/100 g straw								
1	51	52	51	51	54	50	52	52
2	57	55	55	53	61	54	61	55
3	62	55	60	56	61	55	64	62
4	63	58	62	57	61	58	71	64
5	64	50	63	58	58	58	62	65
6	68	62	61	59	61	60	66	69
Untreated	46	—	—	—	—	—	—	—
CP, % of DM								
1	7.0	7.0	7.6	5.8	7.8	5.2	7.2	6.6
2	9.3	8.6	9.3	7.9	10.1	8.1	9.2	7.9
3	11.2	8.5	10.0	8.9	10.8	7.9	10.7	8.3
4	10.2	9.2	10.4	9.2	10.8	9.4	12.8	9.8
5	10.8	8.8	12.3	9.8	10.8	9.2	10.8	9.6
6	11.2	10.0	11.8	8.9	11.9	10.0	12.3	10.3
Untreated	2.9	—	—	—	—	—	—	—

[a]   10 kg wheat straw in plastic bags treated with gaseous ammonia. Digestibility determined by two stage in vitro technique.

From Becker, K. and Pfeffer, E., *Das Wirtschaftseigene Futter,* 23, 84, 1977. With permission of DLG-Verlag, Frankfurt.

Waiss et al.[99] noted an improved enzyme digestibility of ammonia treated straw with increasing moisture content of the straw. The rate of increase for added water was highest at lower moisture levels. Sundstøl et al.[94f] reported increasing in vitro digestibility of ammonia treated straw with increasing moisture content up to 50%.

Treatment with ammonia increases the N-content of feeds. From Tables 3, 4, and 5 it can be seen that the crude protein content of straw may be increased to 10 to 12%. This means that a maximum of 1 to 1.5% of N from ammonia is bound in the treated straw. The crude protein content of aspen wood was increased from 0.5 to 9% by NH_3 treatment.[6] Oji et al.[74] found that the crude protein content of corn stover increased from 8.8 to 17.1 and 20.9% after treatment with 3 and 5% NH_3, respectively. Some of this nitrogen is present in a loose volatile form and is evaporated during drying at elevated temperatures.

Not only is the rate of reaction slower for NH_3 than for NaOH, but also the maximum effect seems to be less. The presented results show maximum in vitro digestibility coefficients between 60 and 70 for NH_3 treated straw and between 70 and 80 or more for NaOH treated straw. Maximum effect seems to be obtained at about the same level of added chemical, if measured on a molar basis, as 3 to 4% NH_3 is equivalent to about 7 to 9% NaOH. Percentages are in most of the cited experiments measured on an air dry basis.

Treatment with Other Alkaline Chemicals

Rounds et al.[92] found that KOH was equally effective as NaOH in increasing the in vitro digestibility of corn cobs, when tested on an equimolar basis, and similar results were obtained by Anderson and Ralston.[3] Ca(OH)₂ on the other hand, appeared to be ineffective when allowed to react for only 24 hr after addition of the chemical together with 30 g water per 100 g cobs.[92] However, when corn cobs were treated with mixtures of NaOH and Ca(OH)₂, the chemicals added with water to increase the moisture content to 60% and allowed to react for 24 days, the feeding value, as measured in experiments with lambs, was as high or higher than the value of cobs treated with NaOH alone. Treatment with 3% NaOH + 1% Ca(OH)₂ (on dry matter basis) and 2% NaOH + 2% Ca(OH)₂ tended to give higher weight gain and feed efficiency than treatment with 4% NaOH. Equal intakes and digestibilities were obtained by treating corn stover with 2% NaOH + 2% Ca(OH)₂ (50% H₂O), 3% NH₃ (30% H₂O), or 5% NH₃ (30% H₂O) for 30 days at ambient temperature[74] (% alkali on dry matter basis).

Based on these results it may be concluded that KOH and NaOH can be used with equal efficiency for increasing the digestibility of low-quality roughages. Animal performance data have shown very promising results of treating corn stover and corn cobs with mixtures of NaOH and Ca(OH)₂. It is indicated that Ca(OH)₂ reacts slowly, and a longer period of time is needed for reaction. There is no information about the effect of the amount of water added together with Ca(OH)₂. Ca(OH)₂ is an interesting chemical in this context, because it is rather cheap; it lowers the level of sodium in the treated feed and supplies calcium to the ration. More experiments with Ca(OH)₂ are needed.

EFFECT OF ALKALI TREATMENTS ON THE CHEMICAL COMPOSITION OF FIBROUS MATERIALS

Results of the effects of alkali treatment on the content of cell wall constituents in different fibrous materials are shown in Table 6. The values are given as percentages of dry matter. Thus, when the alkali is sprayed on, and the material not washed, the ash content is increased about 1 percentage unit for each percent NaOH, and the content of organic constituents proportionately decreased. It is apparent that the content of cell wall constituents (CWC) in straw and corn stover has been reduced as a result of spray treatment, and this reduction is greater than that caused by the increased ash content. The effects on the contents of ADF, lignin, and cellulose are small and varying. There is a difference between plant materials, as the cell wall content of alfalfa stems and poplar bark respond only slightly to alkali spray treatment.

The reason for the decrease in the content of cell wall constituents is that an increased proportion of the dry matter is rendered soluble by mild alkali treatment and is measured as neutral detergent soluble cell content in the analysis. Ololade et al.[76] found that the content of water-solubles was increased from 14 to 20% in alfalfa stems, from 9 to 26% in barley straw, and from 19 to 30% in corn stover as a result of spray treatment with 8% NaOH. Phoenix et al.[80] increased the water-solubility of dry matter in barley straw from 5 to 10%, in untreated straw to 30 to 35%, and in straw treated with 8% NaOH. These solubles together with some fine particulate matter are lost, when the soaking and washing procedure is used. Homb[39] concluded that 20% of the dry matter of straw was lost, when the Beckmann method of alkali treatment was used. Saxena et al.[93] reported that approximately 25% of the dry matter in oat straw was lost in the soaking and washing process of alkali treatment. In experiments done by Carmona and Greenhalgh[17] it was noted that 20 to 30% of the organic matter in barley straw was lost, when treatment (soaking and washing) was done with normal levels of NaOH.

Table 6
CHANGES IN CONTENT OF CELL WALL CONSTITUENTS OF VARIOUS ROUGHAGES FOLLOWING ALKALI TREATMENT

Roughage	Treatment	Percent of dry matter				Ref.
		CWC[a]	ADF[b]	Lignin	Cellulose	
Alfalfa stem	Untreated	71.7	53.8	14.8[c]	39.2	76
	8% NaOH, sprayed	70.1	56.1	15.0[c]	41.2	
Barley straw	Untreated	81.9	54.0	11.1[c]	39.3	
	8% NaOH, sprayed	70.4	54.0	10.0[c]	41.1	
Corn stover	Untreated	73.1	38.9	4.8[c]	33.1	
	8% NaOH, sprayed	65.7	40.8	5.1[c]	35.5	
Barley straw	Untreated	—	—	14.3[d]	—	83
	6% NaOH, sprayed	—	—	13.5[d]	—	
Straw	Untreated	—	—	13.2[d]	41.5	11
	6% NaOH, sprayed	—	—	12.9[d]	41.1	1
Oat straw	Untreated	—	—	—	40.1	
	8% NaOH, sprayed	—	—	—	34.8	
	3% NH₄OH, sprayed	—	—	—	39.0	
Corn stover	Untreated	81.7	49.9	7.9[e]	—	53
	5% NaOH, sprayed	61.5	45.3	7.4[e]	—	
Alfalfa stem	Untreated	68.4	52.0	11.6[e]	—	
	4% NaOH, sprayed	69.2	54.5	12.4[e]	—	
Barley straw	Untreated	84.7	51.7	8.0[e]	43.6	58
	5.9% NaOH, sprayed	66.8	49.4	8.1[e]	41.3	
Poplar bark	Untreated	62.3	54.3	22.0[e]	32.3	31
	12% NaOH, sprayed	54.1	49.6	16.5[e]	33.1	
Oat straw	Untreated	73.4	56.5	10.5[e]	—	93
	12% NaOH, soaked, washed	80.2	67.7	9.5[e]	—	
Barley straw	Untreated	80.4	54.0	7.1[e]	43.2	17
	9% NaOH, soaked, washed	85.8	64.0	7.9[e]	55.9	
Wood[f]	Untreated	—	80.3	18.1[e]	—	41
	6% NaOH, soaked, washed	—	89.6	18.8[e]	—	

[a]   CWS = cell wall constituents.
[b]   ADF = acid detergent fiber.
[c]   Permanganate lignin.
[d]   Modified detergent method.
[e]   Acid detergent lignin.
[f]   Average values for wood from poplar, alder and fir, and sludge (a byproduct from pulping).

Therefore, the effects of the soaking and washing procedure on the chemical composition of roughages are quite different from the effects of the spraying process (Table 6). The contents of CWC, ADF, and cellulose are increased as a result of the loss of soluble constituents. The content of silica is reduced by the alkali treatment and washing. Saxena et al.[93] noted that the content of silica in oat straw decreased from 2.7 to 0.2% of the dry matter. The silica content of the dry matter in rice hulls was reduced from 23 to 10% after treatment with 20 g NaOH per 100 g DM and washing.[63]

The percentage of lignin in the soaked and washed material remains almost unchanged. The reason for this is that part of the lignin is solubilized by the alkali treatment and lost with the washing water. The proportion lost may be of the same order of magnitude as that of total dry matter loss. Homb[39] found that 33% of the lignin was lost in the Beckmann procedure of treatment. The content of soluble, acid precipitable lignin in straw is increased by NaOH spray treatment[83,84] (Table 7). Some of the

Table 7
CONTENT OF TOTAL LIGNIN AND OF
SOLUBLE, ACID PRECIPITABLE
LIGNIN[a] IN DIFFERENT KINDS OF
UNTREATED AND NaOH TREATED[b]
STRAW (% OF DRY MATTER)[84]

Treatment % NaOH	Rye	Wheat	Oat	Barley
% Total Lignin				
Untreated	19.78	18.79	15.39	15.83
% Soluble, Acid Precipitable Lignin				
0	1.25	1.40	2.43	0.79
2	5.06	5.30	5.44	4.06
4	7.70	6.38	9.75	6.49
5	7.33	8.66	12.05	7.92
6	9.13	9.72	13.72	9.68
8	12.38	12.76	13.95	11.30

[a] Straw was extracted by cooking in distilled water and filtered. HCl was added to the filtrate followed by centrifugation, washing to neutrality, drying and weighing.
[b] NaOH spray treatment, 100 mℓ solution/100 g straw reacted for 3 days.

From Piatkowski, B., Bolduan, G., Zwierz, P., and v. Lengerken, J., *Arch. Tierernähr.*, 24, 515, 1974. With permission.

soluble, acid precipitable matter may be constituents other than lignin. Müller and Bergner[71] found 7.5% soluble, acid precipitable dry matter in oat straw after strong ammonia treatment. From determinations of the methoxyl content of total lignin and of the acid precipitable fraction, it was calculated that lignin constituted approximately 60% of the acid precipitable dry matter.

The increased solubility of lignin would account for part of the reduction in the cell wall content of spray treated straw, as lignin is dissolved in the neutral detergent used to determine the cell wall content. However, the greater part of this reduction is due to increased solubility of hemicellulose. Thus, mild alkali treatment increases the solubility of hemicellulose, lignin, and silica, while the solubility of cellulose is not altered.

MECHANISM OF ACTION OF ALKALI ON FIBROUS MATERIALS

The main reasons for the low availability of carbohydrates in fibrous materials are suggested to be the following:

1. Chemical bonding by cross-links between polymers of the ligno-cellulosic complex and by intermolecular hydrogen bonds
2. Crystallinity of cellulose
3. Physical barriers of lignin and silica

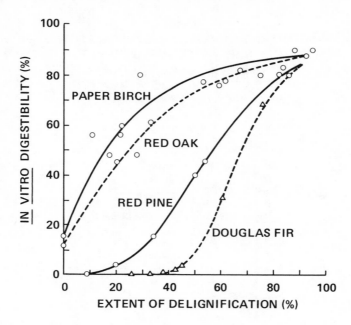

FIGURE 5.   Relationship between in vitro digestibility and extent of delignification for kraft pulps made from four species of wood. Pulps obtained by cooking in pulping liquor (liquor to wood ratio 5:1) with varying concentrations of sodium hydroxide and sodium sulfide (kraft process) and varying time and temperature. (From Baker, A. J., *J. Anim. Sci.*, 36, 770, 1973. With permission.)

Extensive studies on the chemical and physical factors limiting the availability of these carbohydrates and of the mechanism of digestibility increase by alkali treatment have been done on wood substances.[5,6,27,67,94d,95]

Lignin is an inert substance, which is practically indigestible for rumen microorganism, and it is an important factor limiting the digestion of holocellulose. Strongly delignified wood pulp has a high digestibility (80 to 90% for organic matter in vivo) and feeding value for ruminants, and the digestibility is dependent on the degree of delignification.[42,92a] Baker[5] studied the digestibility in vitro of four species of wood with varying contents of lignin obtained by different degrees of pulping. The results (Figure 5) confirmed that the digestibility depends upon the extent of delignification measured as the percent of lignin removed from the original wood. However, they also indicated that softwoods increase more slowly in digestibility, as lignin is removed, than do hardwoods. At high lignin contents, the pulps from hardwoods were more digestible than pulps from softwoods. Baker[5] found these results to be in excellent accordance with in vivo digestibilities of softwood and hardwood pulps obtained by Saarinen et al.[92a] using rams. The results of Saarinen et al.[92a] were obtained by pulps made by ten different pulping methods. Thus, it is indicated that the digestibility is dependent on the extent of delignification regardless of the pulping method used.

Later, Baker et al.[6] calculated from the same data the degree of delignification required to attain 60% in vitro digestibility of the four species of wood (Table 8). Again, these data indicate a strong correlation between response in digestibility and lignin content; response here being measured as the degree of delignification needed to achieve a certain level of digestibility. The same seems to be true for mild alkali treatment. Thus, the lignin content of legumes (e.g., alfalfa stem) is higher than that of grass straws, and the response to mild alkali treatment is lower for legumes than for

Table 8
DEGREE OF DELIGNIFICATION REQUIRED TO ATTAIN
60% IN VITRO DIGESTIBILITY OF WOOD PULPS

Wood	Required delignification (%)	Lignin in original wood (%)	Lignin in pulp (%)
Paper birch	25	20	21
Red oak	35	23	20
Red pine	65	27	14
Douglas fir	73	32	13

Cellulose Technology Research

From Baker, A. J., Millett, M. A., and Satter, L. D., *Cellulose Technology Research,* ACS Symp. Ser. 10, Turbak, A. F., Ed., American Chemical Society, Washington, D.C., 1975, 94. With permission.

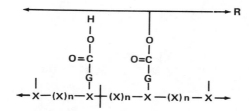

FIGURE 6.   Schematic structure of a hardwood xylan chain with a free carboxyl group and an esterified carboxyl group functioning as a cross-link between the xylan chain and some other polymeric unit. (From Tarkow, H. and Feist, W. C., *Cellulases and Their Applications,* Advances in Chemistry Ser. No. 95, Hajny, G. T. and Reese, E. T., Eds., American Chemical Society, Washington, D.C., 1969, 208. With permission.)

grasses (see later section). A similar pattern was found from mild alkali treatments of different species of wood.[27]

But, the differences between various species show that there is no simple relationship between the lignin content and the availability of carbohydrates. This may be due to differences in the lignin-carbohydrate associations, the constitution of lignin, or the lignin distribution in the cell walls.

The mechanism of digestibility increase by treatment with dilute alkali has been discussed by Tarkow and Feist.[95] Cellulosic materials swell in aqueous NaOH. The true additional swelling beyond the normal swelling of waterlogged material can be measured by the increase in the fiber saturation point measuring the total water of swelling.

The increased swelling capacity is caused by chemical changes brought about by the alkali. 4-O-Methyl-D-glucuronic acid is present on every 6 to 9 anhydro xylan residues along the xylan chains in hardwood. These uronic acid residues act as intermolecular cross-links by forming ester groups with alcoholic components on another xylan chain or in the lignin (Figure 6). The xylan chain also contains acetyl groups. Both the glucuronic acid and the acetyl ester groups are saponified by NaOH, thus breaking these intermolecular cross-links. Values for free carboxyl content can be measured by calcium ion exchange. By measuring the amount of NaOH held by NaOH treated wood

Table 9
DISTRIBUTION OF NaOH IN NaOH-TREATED SUGAR MAPLE

		Computed content of NaOH			
NaOH concentration %	In lumen mols/100 g	Held by carboxyl on xylan chain mols/100 g	Held as acetate mols/100 g	Total including only 15% held as an acetate[a] mols/100 g	Total observed NaOH content mols/100 g
1	0.033	0.022	0.110	0.071	0.077
4	0.133	0.022	0.110	0.171	0.191

[a]   In a supplementary experiment it was found that under the treatment conditions used (5-min period in the treating solution) saponification of the acetate occurred and 85% of the resulting acetate diffused out of the wood.

Cellulases and Their Applications

From Tarkow, H. and Feist, W. C., *Cellulases and Their Applications,* Advances in Chemistry Ser. No. 95, Hajny, G. T. and Reese, E. T., Eds., American Chemical Society, Washington, D.C., 1969, 203. With permission.

Table 10
CHANGE IN FREE CARBOXYL CONTENT AND FIBER SATURATION POINT (FSP) FOLLOWING NaOH TREATMENT[a]

Species	Uronic acid, %	Total carboxyl content[b] meq/100 g	Free carboxyl content[c] meq/100 g	Relative free carboxyl content, %	FSP %
Sugar maple					
Original	4.40	25.0	7.4	30	42
NaOH-treated	4.10	23.2	22.4	97	87
Aspen					
Original	4.20	23.9	7.2	30	52
NaOH-treated	3.64	20.7	18.3	89	100
Beech					
Original	4.84	27.5	7.8	28	41
NaOH-treated	4.01	23.0	22.9	99	84

[a]   1.5% NaOH-solution, 3 hr at room temperature.
[b]   Computed from uronic acid using uronic acid molecular weight of 176.
[c]   Measured by calcium ion exchange.

Cellulases and Their Applications

From Tarkow, H. and Feist, W. C., *Cellulases and Their Applications,* Advances in Chemistry Ser. No. 95, Hajny, G. T. and Reese, E. T., Eds., American Chemical Society, Washington, D.C., 1969, 203, 207. With permission.

it was found that this amount was in agreement with the computed content of NaOH, the computation being based on the content of carboxyl groups as measured by calcium ion exchange (Table 9). In a later work[27] computations based on the same principles showed that the calculated amount of NaOH necessary to saponify acetate plus uronic esters was also in agreement with the observed NaOH requirement for achieving maximum digestibility of hardwoods.

Tarkow and Feist[95] compared the free carboxyl content with the fiber saturation point of hardwoods (Table 10). The results indicated a close correlation between these factors. Together, these observations support the hypothesis that cleavage of uronic-

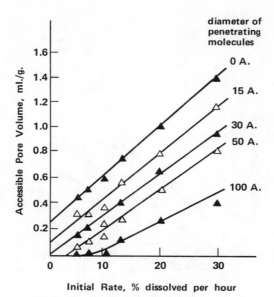

FIGURE 7. Relationship between the initial rate of cellulose digestion (by a cellulase enzyme preparation) and the volume of water accessible to molecules of different sizes. (From Stone, J. E., Scallan, A. M., Donefer, E., and Ahlgren, E., *Cellulases and Their Applications,* Advances in Chem. Ser. No. 95, Hajny, G. T. and Reese, J. T., Eds., American Chemical Society, Washington, D.C., 1969, 236. With permission.)

ester cross-links by NaOH results in additional subsequent swelling in water and that this finally may account for the increase in digestibility of hardwoods.

Wang et al.[100] showed that treatment of hardwood with liquid ammonia produced amides, and that the amide groups were derived from lactone or ester groups in the original 4-O-methyl-glucurono-xylan. Tarkow and Feist[95] found that the nitrogen content as ammonium and amide in ammonia-treated hardwood was equivalent stoichiometrically to the carboxyl content measured by calcium ion exchange for NaOH-treated wood. They concluded that the chemical action of liquid ammonia on the hardwood is consistent with that predicted from the action of dilute NaOH on wood. A clear relationship between the fiber saturation point (swelling) and the content of bound nitrogen following treatment with liquid ammonia was shown. Liquid ammonia is able to cleave the uronic ester cross-links. The rate of reaction was much slower than for NaOH.

The importance of the increased swelling capacity for the accessibility of the wood carbohydrates for the enzymes of rumen microorganisms is thoroughly discussed by Tarkow and Feist[95] and Stone et al. (1969).[94d] It may increase the possibility of proteins within enzyme systems to penetrate or diffuse through the substance, and it may provide conditions for more rapid movement of enzymes in order to come in proper stereorelationships with the vulnerable bonds in the carbohydrate substrate. Stone et al.[94d] found linear relationships between initial rate of cellulase enzyme digestion and the pore volume (Figure 7) or the surface area (Figure 8) in cotton fibers accessible to molecules of different sizes. Different swelling capacity (i.e., different accessible pore volume and surface area) of cotton was obtained by treatment with phosphoric acid. These observations indicate that the possibility for enzyme molecules to penetrate cell walls could be an important factor determining the digestibility of fibrous materials.

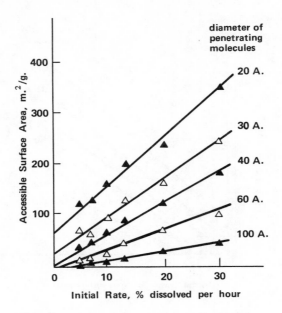

FIGURE 8.   Relationship between the initial rate of cellulose digestion and the surface area of cellulose accessible to molecules of different sizes. (From Stone, J. E., Scallan, A. M., Donefer, E., and Ahlgren, E., *Cellulases and Their Applications,* Advances in Chem. Ser. No. 95, Hajny, G. T. and Reese, E. T., Eds., American Chemical Society, Washington, D.C., 1969, 237. With permission.)

TREATMENT OF DIFFERENT PLANT MATERIALS

Ololade et al.[75] observed that alfalfa stems responded less to alkali treatment than did barley straw. Corn stover was intermediate but had a high initial digestibility. The effect of alkali treatment on both in vitro and in vivo digestibility was greater for corn stover than for alfalfa stems.[53] Similar low responses of alfalfa stems were noted by Jones and Klopfenstein[47] and Klopfenstein et al.[52] Ammonia treatment increased the enzyme digestibility of legume stems (alfalfa and bean) by only 9 and 13 percentage units, while the increase for straws from barley, fescue, oat, perennial ryegrass, rice, and wheat were from 25 to 36 units[99] (Table 3). Summers and Sherrod[94e] observed that alfalfa was not appreciably affected by alkali treatment, while the digestibility of sorghum stubble, corn stover, corn cobs, wheat straw, and bermudagrass was significantly improved. Legume stems thus seem to respond less to mild alkali treatment than straws from grasses and grains.

Grain and grass straws with digestibilities varying between 35 and 55% seem to show equal levels of responses to alkali treatment.[58,84,99]

The digestibility of rice hulls could be substantially increased only after treatment with high amounts of NaOH (15 to 20 g/100 g hulls), and the digestibility did not increase to more than approximately 50%.[20,63] Strong alkali treatment severely depressed intake in sheep. The enzyme digestibilities of hulls from peanuts, rice, and safflower were increased to only 15, 34, and 51%, respectively, by treatment with NaOH under pressure, while the digestibility of straws was increased to 70 to 80% by the same treatment.[33] Similarly, Summers and Sherrod[94e] found that the in vitro digestibility of cotton seed hulls was only slightly improved by alkali treatment, and that of peanut hulls was depressed by treatment. On the other hand Levy et al.[61] obtained a high increase in the feeding value of cotton hulls by alkali treatment.

The digestibility in vitro for dry matter in 16 different species of hardwoods (untreated) varied between 2 and 37%.[67] The response to soaking in diluted alkali varied greatly, the increase in digestibility being from 5 to 50 percentage units.[27,67] The greatest response was found for American basswood (+ 50%), while the effects on aspen, ash, birch, and maple were intermediate (+ 15 to + 30%). A dry matter digestibility of 50 to 55% could be obtained for basswood and aspen, while the digestibility of all other species after treatment were below 40%. Softwoods are practically undigestible (0 to 5%)[67] and does not respond to mild alkali treatment.[6,67] Huffman et al.[41] reported a substantial increase in digestibility in vitro of poplar wood, a slight increase for alder, and almost no effect of alkali treatment on fir.

Thus, a substantially increased but still very limited feeding value can be obtained by alkali treatment on low digestibility materials of certain kinds of hulls and hardwoods (basswood and aspen), and they may be used as feed only for animals with very low energy requirements. Most of these kinds of materials cannot be made into useful feeds for animals by mild alkali treatment.

EFFECT OF ALKALI TREATMENT ON DIGESTIBILITY IN VIVO, FEED INTAKE, FEED EFFICIENCY, AND ANIMAL PERFORMANCE

The effect on the apparent digestibility in vivo of treating straw with increasing amounts of NaOH is demonstrated by the results shown in Table 11. It is indicated that the digestibility is increased with an increasing level of NaOH up to 4 to 5%. At higher levels of NaOH the digestibility is either kept constant or increased very little.

The pattern of increase in digestibility in vivo is thus different from that in vitro. The digestibility in vitro is markedly increased up to 8 to 10% NaOH (see Section on sodium hydroxide). This deviation is also indicated by direct comparisons made by Piatkowski et al.[84] and Kristensen et al.[58] (Table 11). The reason for this difference is not evidenced. The in vitro methods measure solubility rather than digestibility, and this possibly explains the discrepancy. Some of the material, which is rendered soluble by alkali treatment (see earlier section) may be undigestible (e.g., lignin). Some other soluble and digestible parts may pass the rumen without being digested, owing to a high rate of passage. The rate of passage of solubles is assumed to be high because of a high water intake caused by the increased sodium content of NaOH-treated feeds. Such effects are indicated by the results of Hogan and Weston[38] (Table 12). Only around 40% of the cell contents (or neutral detergent solubles), which are normally very highly digestible, were digested in the stomach. The cell contents constituted approximately 25% of the organic matter. It was also indicated that the neutral detergent solubles originating from the cell wall constituents were poorly digested in the whole tract. More detailed analysis of the real composition of the various fractions would add greatly to the knowledge of the utilization of alkali-treated materials.

The data of Hogan and Weston[38] showed that 80 to 85% of the cell wall constituents were digested, and that 80 to 90% of this digestion occurred in the stomach. They also indicated that a higher percentage of the cell wall constituents were digested at the low level of feeding than at the high level. Other experiments have shown that the level of feeding has a marked influence on the apparent digestibility of alkali treated straw in vivo. An example of this is shown in Table 13. This effect must be taken into consideration, when results of digestibility experiments are compared. Differences between in vitro and in vivo digestibilities are more pronounced at high levels of intake than at low levels.

Table 11
EFFECT OF LEVEL OF NaOH ON THE APPARENT DIGESTIBILITY OF STRAW

| Experimental feed | | Treatment | Experimental animals | Level of feeding | In vitro % DMD | In vitro % OMD | In vivo % ODM | Ref. |
Kind	% of total ration							
Wheat straw	70[a]	Untreated	male calves 9 to 18 months	95[c]			55	94b
		3.3% NaOH[b]		121			70	
		6.7%		105			71	
		10.0%		89			71	
Wheat straw	70[d]	Untreated	sheep 54.2 ± 4.1 kg of live weight	1603[f]			48.5	2
		2.4% NaOH[e]		1952			55.3	
		4.1%		2139			59.5	
		5.6%		1792			60.4	
Means for rye, Wheat, oat and Barley straw	85[g]	Untreated	sheep	940[i]		39.9	49.8	84
		2% NaOH[h]				50.0	55.5	
		4%				62.4	56.5	
		5%				66.7	64.9	
		6%				71.6	65.0	
		8%				77.5	68.3	
Barley straw	65[j]	Untreated	sheep mean live weight 58.0 kg	39.7[l]			63.7 (51.6)[m]	46
		4.5% NaOH[k]		39.2			71.7 (64.2)	
		6.75%		39.5			72.5 (65.5)	
		9.0%		39.3			72.7 (65.6)	
Barley straw Means for 2 trials	75[n]	Untreated	adult sheep	800[p]	43.1		47.5[q]	58
		2.1% NaOH[o]		1070	53.8		53.6	
		2.9%			59.5		60.3	
		4.1%			66.4		64.3	
		5.8%			72.3		67.7	

a 79% chaffed straw (air dry basis) mixed with 10% groundnut cake, 10% molasses, and 1% mineral mixture and fed, together with 1 kg/animal day of green, wilted forage.

b Spray treatment, 1000ℓ water or alkali solution per t of straw; percentages on air dry basis.

c Dry matter, g/kg $W^{0.75}$ per day (*ad lib.*).

d Straw mixed with 20% soybean meal, 8% molasses, and 2% mineral mixture on dry matter basis and pelleted.

e Spray treated and mixed (33% alkali solution) and immediately pelleted; percentages on dry matter basis.

f Grams of organic matter per animal per day (*ad lib.*).

g Chaffed straw added 100 mℓ water or alkali solution per 100 g; mixed with 11% barley grain, 2% urea and, 2% mineral mixture.

h Spray treatment, neutralization with acetic acid, percentages on air dry basis.

i Grams of feed per animal per day (restricted).

j Straw and a concentrate mixture fed separately, percentages on dry matter basis.

k Spray treatment, 60 mℓ/100 g straw, neutralized with HCl; percentages on air dry basis.

l Dry matter, g/kg $W^{0.75}$ per day (excluding added ash) (restricted).

m Derived digestibility of straw assuming 85% digestibility of concentrate.

n Straw and soybean meal fed separately.

o Spray treated and mixed (33% alkali solution) and immediately pelleted, percentages on dry matter basis.

p Grams of feed per animal per day (restricted).

q Derived digestibility of straw assuming 90% digestibility of soybean meal.

Table 12
DIGESTION OF ORGANIC MATTER (OM), PLANT
CELL WALL CONSTITUENTS (CWC) AND PLANT
CELL CONTENTS (CC) IN THE STOMACH AND
INTESTINES OF SHEEP FED AT THREE LEVELS ON
A DIET BASED ON ALKALI TREATED STRAW[a]

	Level of feeding		
	High	Medium	Low
OM intake, g/day	723	444	310
OM apparently digested, % of intake	68.9	69.6	70.5
Apparent OM digestion in stomach, % of intake	45.5	55.6	50.0
Dietary OM truly digested in stomach, % of intake	62.7	69.0	67.4
CWC digested in the whole tract, % of intake	80.1	84.7	85.0
CWC digested in stomach, % of intake	65.5	77.4	71.1
% of total digested	82.7	91.9	83.6
Plant CC digested in stomach, % of intake	39	34	44

[a]   8 kg NaOH in 30 ℓ water per 100 kg ground wheat straw, mixed for 90 min and allowed to react for 16 hr. Neutralized. Diet composed of treated straw supplied with urea and minerals. The straw used at the medium level of intake was from another batch than that used at the high and low levels. The results may therefore not be directly comparable.

From Hogan, J. P. and Weston, R. H., *Aust. J. Agric. Res.,* 22, 956, 1971. With permission.

Table 13
THE EFFECT OF THE LEVEL OF FEEDING ON THE
APPARENT DIGESTIBILITY OF ALKALI TREATED[a]
BARLEY STRAW IN SHEEP[b]

	Restricted feeding	*Ad libitum* feeding
Soybean meal, kg DM per animal per day	0.17	0.17
Treated straw, kg DM per animal per day	0.50	1.67 ± 0.11
Crude protein in ration, % of DM	17.0	9.1
Digestibility of organic matter, %	77.5 ± 0.7	68.7 ± 0.6

[a]   Chaffed straw mixed with a strong alkali solution (15 ℓ 33% NaOH solution/ 100 kg straw) and stored. Temperature rise in the stack to maximum 65°C.
[b]   Adult sheep.

The digestibility of dry matter or organic matter in rations containing 70 to 98% alkali treated straw, corn cobs, or corn stover has varied between 60 and 70% dependent on the composition of the diet and the level of intake.[17,25,28,37,53,74,94a] (see also Table 11). The maximum digestibility of organic matter in NaOH-treated straw alone has been 65 to 70%, when fed at restricted levels near maintenance requirement. Carmona and Greenhalgh[17] found that the soaking and washing procedure was more sufficient in increasing the digestibility of straw than the spraying procedure. The digestibility

Table 14
EFFECT OF NH₃-TREATMENT ON THE APPARENT DIGESTIBILITY OF VARIOUS PLANT MATERIALS IN SHEEP

Experimental feed	Treatment	Untreated, %		Treated, %		Ref.
		DMD	OMD	DMD	OMD	
Wheat straw	5% anhyd. NH₃	36	—	50	—	22
Oat straw	5% anhyd. NH₃	47	—	55	—	
Barley straw	5% anhyd. NH₃	43	—	55	—	
Barley straw	3.5% anhyd. NH₃ (3 weeks)	47	—	60	—	4
	3.5% anhyd. NH₃ (8 weeks)	47	—	64	—	
Oat straw	5.3% anhyd. NH₃	—	45	—	55	91
Corn stover	3% aqueous NH₃	—	57	—	65	74
	5% aqueous NH₃	—	57	—	67	

Table 15
EFFECT ON VOLUNTARY INTAKE OF TREATMENT WITH INCREASING AMOUNTS OF NaOH IN RATIONS WITH PREDOMINANTLY TREATED FEEDS

Experimental feed	Diet	Treatment % NaOH	Experimental animals	Intake			Ref.
				DMI	OMI	DOMI	
Wheat straw	Ca. 70% straw + groundnut cake, molasses, and green forage	0[a]	Male calves 9-18 months	95[c]	—	46	94b
		3.3		121	—	75	
		6.7		105	—	64	
		10.0		89	—	53	
Wheat straw	70% straw, 20% soybean meal, 8% molasses	0[b]	Sheep 54.2 ± 4.1 kg of live weight	—	1603[d]	779	2
		2.4		—	1952	1075	
		4.1		—	2139	1269	
		5.6		—	1792	1085	
Barley straw	Intake data for straw only. Concentrates fed separately	0[a]	Sheep, mean live weight 60 kg	35.5[c]	—	—	46
		4.5		57.1	—	—	
		9.0		46.9	—	—	
Rice hulls	50% rice hulls, 50% lucerne meal	0[b]	Lambs, mean live weight 25 kg	—	756[d]	317	20
		2.5		—	719	262	
		5.0		—	1025	443	
		10.0		—	793	325	
		15.0		—	545	284	

[a]   Air dry basis.
[b]   Dry matter basis.
[c]   g/kg $W^{0.75}$ per day.
[d]   g/animal per day.

of organic matter in chopped and soaked barley straw was 71%, while that of chopped and sprayed straw was 61%. The difference may be explained partly by a 30% higher intake of the sprayed material. However, the results obtained by Hogan and Weston[38] (Table 12) would suggest a higher digestibility of washed materials, as the poorly digested solubles are removed by the washing.

Sundstøl et al.[94f] reviewed experiments with ammonia treatment and concluded that the digestibility of dry matter in straw may be increased 10 to 15 percentage units. Results from ammonia treatments are shown in Table 14. In these experiments, long term treatments are used without acceleration by heat or high pressure.

Results of experiments estimating the effect of the level of NaOH on the voluntary intake of treated roughages are shown in Table 15. A maximum intake of straw is apparently obtained at about 4 to 5% NaOH, and a marked decrease occurs at higher levels. The increase in intake at lower levels of NaOH is presumably brought about by increased digestibility and increased rate of digestion. The intake of the rice hull containing rations,[20] were also increased at 5% NaOH compared to the control and 2.5% NaOH, although the digestibility was not altered before at the 15% NaOH level. The reason for the decrease in intake at high levels of NaOH is not precisely known, but is to be found in the high intake of sodium or possibly unreacted alkali. These results are obtained in experiments, where the material has been spray treated. If the unreacted alkali is removed by washing, the intake may be increased at higher levels of NaOH. Thus, Carmona and Greenhalgh[17] noted higher intake after treatment with 18% NaOH than with 9%, when barley straw was soaked and washed.

The maximum intake of dry matter (DMI) or organic matter (OMI) was 27 to 36% higher than the intake of the control in the experiments with mixed rations. In the experiment of Jayasuriya and Owen,[46] where the intake was measured for straw only, the intake was increased 61%.

The effect of treatment on the voluntary intake is also illustrated by the results shown in Table 16. In these experiments the rations contained high proportions of roughages (50 to 100%). The intake of rations containing only straw, supplied with small amounts of nitrogen sources, was increased 60 to 80% in lambs.[17,25] The intake in lambs of rations containing 65 to 70% straw was increased 30 to 60% by treatment with NaOH.[37,93] When the ration contained 50% straw and 50% concentrate, the intake in lambs was increased 51% after NaOH-treatment.[32] NaOH-treatment raised the intake in lambs of rations containing 75%[92] and 80%[54] corn cobs. The increases were 18 and 26%, respectively.

The increase in intake after alkali treatment is possibly less in young, growing cattle than in lambs. In the experiment of Flachowsky et al.[28] with young bulls, the ration contained 60 to 61% straw, when 1.5 kg concentrates was given along with the pellets, and 71 to 73%, when 0.5 kg concentrate was given. The intake of straw was increased 17 to 18% after alkali treatment (Table 16). Similarly, the rations in the experiment by Foldager[29] contained 65 to 70% straw, and the intake of straw in the heifers increased 6 to 23% after treatment. The intake of rations containing 50% wheat straw was increased 21% in steers.[13] The intake of straw in steers increased 30% after treatment with ammonia.[78]

As the digestibility of straw is increased by alkali treatment, the intake of digestible organic matter (DOMI) is increased more than the intake of dry matter or organic matter. Thus, the maximum increase in intake of dry matter was 27%, while the increase in intake of digestible organic matter was 63%[94b] (Table 15). Similarly, in the experiment of Ali et al.,[2] the maximum increase in intake of organic matter was 33%, and the increase in intake of digestible organic matter was 63%. In both experiments the rations contained 70% straw.

The daily gain in lambs fed rations containing 65% straw was increased 115 and 72 g by alkali treatment, when these rations contained either soybean meal or urea, respectively, as nitrogen sources[93] (Table 16). Lambs fed rations containing 70% straw gained 123 and 126 g faster per day on treated than on untreated straw, when the nitrogen sources were soybean meal or urea, respectively.[37]

Young bulls fed pellets containing 82% straw gained 307 and 312 g faster after alkali treatment, when the pellets were supplied with 1.5 and 0.5 kg concentrates, respectively[28] (Table 16). The daily weight gain in steers fed rations containing 50% wheat straw was 270 g higher on treated than on untreated straw, when the diet contained no urea.[13] In experiments, where only the straw was given *ad libitum* and fixed amounts of other feeds were supplied, the daily gain in steers and heifers was 100 to 170 g higher after treatment of the straw with alkali[29,78] (Table 16).

Table 16

EFFECT OF ALKALI TREATMENT ON VOLUNTARY INTAKE AND LIVE WEIGHT GAIN

Experimental feed	Treatment[a,b]	Diet	Experimental animals	Intake (g DM/day[a] or g DM/kg W^{0.75} per day[b])	Relative	Weight gain g/day	Ref.
Oat straw	Untreated	100% straw	Lambs 8 to 10 months	43.2[b]	100		25
	8% 8% NaOH[b]	97.5% straw + 2.5% urea		48.6	113 (100)		
	8%NaOH	100% straw		24.3	56		
		97.5% straw + 2.5% urea		79.9	185 (164)		
Oat straw	Untreated	65% straw + soybean meal, corn, molasses	Lambs initial weight 20 to 28 kg	870[a]	100	62	93
	12% NaOH[a]			1,290	148	177	
	Untreated	65% straw + urea, corn molasses		820	94 (100)	53	
	12% NaOH[a]			1,110	128 (136)	125	
Barley straw	Untreated chopped(C)	Straw ad libitum + 8 g soya protein per 100 g straw	Lambs Initial weight 52 to 56 kg	26.7[b]	100		17
	Untreated milled(M)			36.2	136(100)		
	9% NaOH[a](C)			37.1	139		
	18% NaOH[a](C)			44.2	166		
	8% NaOH[b](M)			48.4	181		
	8% NaOH[b](M)			53.6	201(148)		
Barley straw	Untreated 8% NaOH[b]	50% straw + 50% concentrates	Lambs	560	100	77	32
				848	151	140	
Barley straw	Untreated	46% straw + 54% grass silage (DM-basis)	Lambs	14.7[d]	100		96
	7% NaOH[b]			24.1	164		
Barley straw	Untreated	2 kg conc., 3 to 5 kg silage, straw ad libitum	Steers initial weight 287 kg	2,800[a]	100	403	78
	3.5% NH₃ (anhyd.)			3,650	130	573	
Barley straw	Untreated	1 kg molasses, 0.6 kg soybean meal, 0.6 kg fodderbeet DM, straw ad libitum	Heifers initial weight 270 kg	18.0[d]	100(100)[c]	635	29
	3% NH₃ (anh.)			19.3	107(114)	741	
	5% NaOH[b] (chop.)			18.5	103(106)	630	
	5% NaOH[b] (pell.)			20.4	113(123)	787	

Table 16 (continued)

EFFECT OF ALKALI TREATMENT ON VOLUNTARY INTAKE AND LIVE WEIGHT GAIN

Experimental feed	Treatment[a,b]	Diet	Experimental animals	Intake (g DM/day[a] or g DM/kg W^0.75 per day[b])	Relative	Weight gain g/day	Ref.
Wheat straw	Untreated	70% straw + ground wheat and soybean meal	Lambs initial weight 30 kg	924[a]	100	37	37
	4% NaOH[b]			1,208	131	160	
	Untreated	70% straw + ground wheat and urea		627	68(100)	-46	
	4% NaOH[b]			990	107(158)	80	
Wheat straw	Untreated	82% straw, 15% wheat, 3% urea in pellets + 1.5 kg conc.	Young bulls initial weight 148 kg	25[d]	100(100)[c]	724	28
	4% NaOH[b]			27	108(117)	1031	
	Untreated	82% straw, 15% wheat, 3% urea in pellets + 0.5 kg conc.		26	100(100)	638	
	4% NaOH[b]			29	112(118)	950	
Wheat straw	Untreated	50% straw, 0 urea	Steers initial weight 265 kg	9,000[a]	100	970	13
	4% NaOH[b]	50% straw, 0 urea		10,900	121	1240	
	4% NaOH[b]	50% straw, 1% urea		9,300	103	850	
	4% NaOH[b]	50% straw, 2% urea		8,800	98	850	
Corn cobs	Untreated	80% corn cobs, (ensiled) + 20% soybean meal	Lambs	976[a]	100	68	54
	4% NaOH[b]			1,230	126	109	
Corn cobs (ens.)	Untreated	75% corn cobs (ensiled)	Lambs initial weight 23 kg	628[a]	100	30	92
	4% NaOH[b]	25% corn gluten + urea		738	118	82	
	4% NaOH[b]	75% cobs (ensiled), 25% soybean meal		976		137	
	3% NaOH + 1% Ca(OH)$_2$[b]			1,190		169	

a   Soaked.
b   Sprayed.
c   Intake data for straw only.
d   Grams per kilogram live weight.
e   Relative intake for straw only.

In an experiment of Koers et al.,[55] steers were fed rations containing 80% untreated or alkali treated (4% NaOH) and ensiled corn cobs. The weight gain was increased from 300 g/day to 720 g by the treatment.

Javed and Donefer[45] compared NaOH-treated oat straw to dehydrated alfalfa meal, both constituting 77.5 to 85% of the rations. The daily intake in lambs of the alfalfa and treated straw rations were 1.72 and 1.65 kg, respectively, and the daily weight gain 180 and 140 g. Terry et al.[96] obtained approximately equal results, when comparing a ration of highly digestible grass silage with a ration consisting of 54% silage and 46% NaOH-treated (7% NaOH) barley straw (dry matter basis). Silage and straw were mixed immediately before feeding. A similar mixture of silage and untreated straw was also fed. The digestibilities of organic matter in sheep of pure silage, silage + treated straw and silage + untreated straw were 78.4, 74.0, and 64.2%, respectively, and the voluntary intake of dry matter 23.7, 24.1, and 14.7 g/kg live weight per day. The intake in calves with an initial live weight of 100 kg was 17.5 g organic matter per kg live weight per day of pure silage and 20.1 g organic matter of silage + treated straw and the corresponding live weight gain 440 and 420 g/day.

Mowat[69] compared a ration of pure corn silage with rations consisting of 75% corn silage + 25% NaOH-treated (6% NaOH) corn stover or barley straw and rations with 50% corn silage + 50% treated corn stover or barley straw. The treated materials were ensiled and the rations mixed immediately, before they were fed to bull calves (233 kg of live weight) or heifer calves (215 kg). Rations containing 25% treated materials were comparable to corn silage in rate and efficiency of gain. At 50% treated materials the dry matter intake was slightly reduced and rate of gain significantly depressed. The digestibility of dry matter in corn silage and a mixture of 75% corn silage and 25% treated straw was approximately equal (76 and 75%).

Maeng et al.[62] fed rations consisting of NaOH-treated (6% NaOH) barley straw, alfalfa silage, and mixtures of silage and treated straw in the ratios of 50:50 and 75:25 adequately supplied with soybean meal, minerals, and vitamins. The mixing of these to feeds seemed to enhance the utilization of energy. The digestibility in sheep of energy in pure straw was 66.5%. Assuming a constant digestibility of energy in silage in all diets, the calculated digestibility of energy in the treated straw in the two mixed diets were 69.1 and 77.9%.

Piatkowski et al.[86] compared corn silage and mixtures of corn silage and NaOH-treated (5% NaOH) barley straw in the ratios of 83:17 and 66:34. The feeds were mixed at the time of ensiling. The digestibilities of organic matter in sheep of these rations were approximately equal, 66.4, 68.9, and 68.3% for silage/straw ratios of 100:0, 83:17, and 66:34, respectively. The voluntary intake in dry cows of pure corn silage and of a silage/straw mixture of 66:34 were 7.4 and 7.2 kg dry matter per cow per day, respectively.

Rounds et al.[92] in three experiments treated ground corn cobs with several different chemicals and mixtures of chemicals, ensiled the treated material and fed it to lambs. Corn cobs treated with 4% NaOH and supplied with 15% corn gluten meal plus urea resulted in approximately the same feed intake and weight gain as corn silage. The intake was 1090 and 1044 g dry matter per day and the weight gain 118 and 115 g/day for treated corn cobs and untreated corn silage, respectively. Treatment of corn cobs with 3% NaOH + 1% $Ca(OH)_2$ invariably gave higher intakes and weight gains than treatment with 4% NaOH. Results from one experiment are shown in Table 16. Ensiling with NH_4OH always resulted in lower intakes and weight gains than ensiling with NaOH or mixtures of NaOH and $Ca(OH)_2$. It was stated that the strong ammonia odor may have contributed to the reduced intake. Neutralization of the NH_4OH-treated material had only minor effects on intake and gain.

Levy et al.[61] obtained results, which were different from those given here before.

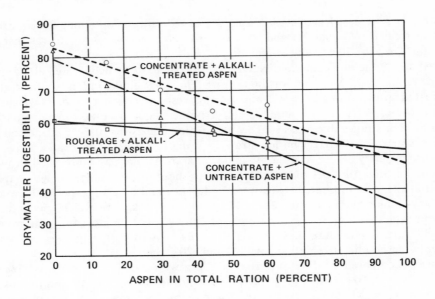

FIGURE 9.   In vivo digestibility of rations containing various proportions of untreated or alkali treated aspen sawdust with alfalfa meal or ground corn as main basal components. (From Mellenberger, R. W., Satter, L. D., Millett, M. A., and Baker, A. J., *J. Anim. Sci.*, 32, 760, 1971. With permission.)

They treated wheat straw and cottonseed hulls with 4 and 8% NaOH, and fed both untreated and treated materials to young bulls (experimental weight interval 230 to 480 kg). The animals were constantly fed 6 kg of concentrate per day, and the roughages were given *ad libitum*. The intake of roughages declined linearly with increasing rate of NaOH application. Live weight gain and carcass gain was equal on untreated and 4% NaOH treated roughages, while they were increased at 8% NaOH. The daily weight gain was 903, 908, and 967 g, and carcass gain 497, 484, and 522 g at 0, 4, and 8% NaOH, respectively. The dressing percentage and the amount of depot fat were significantly increased with increasing NaOH application. Thus, the feed efficiency and energy conversion was substantially improved by NaOH-treatment.

Young bulls (weight interval 220 to 440 kg) were fed untreated and NaOH-treated (5% NaOH) barley straw in pelleted rations containing 20 and 40% straw plus barley, beet molasses, and soybean meal.[58] One to 1.5 kg hay was given per day. The intake of rations containing untreated and treated straw were equal (7.8 kg dry matter at 20% straw in the pellets and 8.5 kg at 40% straw). The daily gain at 20% untreated and treated and at 40% untreated and treated straw were 1374, 1336, 1209, and 1322 g, respectively. Correspondingly, the daily carcass gain was 736, 707, 624, and 685 g. Thus, there were no responses to treatment at the low level (20%) of straw. With 40% straw in the ration, the daily growth rate was 113 g and the carcass gain 61 g higher on treated than on untreated straw.

In the two last mentioned experiments, rations with low proportions of straw and high proportions of concentrate were used. It is known that a high proportion of easily digestible carbohydrates in the ration may depress the digestion of cellulose, presumably because a low pH value decreases the activity of cellulolytic microorganisms. This may explain a poorer utilization of alkali treated roughages in such rations. Mellenberger et al.[66] clearly demonstrated the difference in utilization of untreated and alkali-treated aspen sawdust in rations, where the basal components were either alfalfa meal or concentrate (ground corn). The results of one of these experiments are shown in Figure 9. When the basal component was alfalfa meal, the digestibility of the rations declined linearly with increasing content of sawdust indicative of constant utilization independent of the ration composition. In the concentrate based rations, the digestibility was relatively lower at low levels of sawdust than at higher levels, and the relationship between the content of sawdust and the digestibility of the ration was curvelinear. Thus, factors affecting digestibility were apparently changed, as the ration changed from predominantly starch to higher contents of cellulose.

As pointed out earlier, the digestibility in vivo of alkali-treated straw is not increased further, when the rate of application of NaOH is increased above 4 to 5%. In the experiment of Levy et al.,[61] the rate of weight gain in young bulls was increased when the straw and cotton seed hulls were treated with 8% NaOH compared to 4% NaOH. Jackson[44] suggested that if concentrates constitute a great part of the ration, higher levels of NaOH may be essential as the high content of unreacted alkali can prevent a fall in rumen pH and thereby improve the utilization of the ration. In such rations high levels of NaOH would not depress feed intake either.

The alkaline feed may also have particular advantages, when fed in mixtures with acid feeds as silages. As shown earlier, very promising results have been obtained with such rations.[62,69,86,96]

The content of nitrogen and minerals in cellulose rich plant materials is often inadequate. In order to obtain a good utilization of these roughages, and especially of the alkali-treated roughages, it is necessary to supply them with such nutrients. The importance of an adequate supplementation with nitrogen was clearly shown by Donefer et al.[25] (Table 16). However, several studies indicate that cellulolytic microorganisms may require amino nitrogen in the diet, if maximum activity is to be obtained. Thomsen and Neimann-Sørensen[97] and Møller and Thomsen[73] noted that the efficiency of digestion of cellulose by rumen microorganisms in vitro was higher when soybean meal was added to the inoculum than if isonitrogeneous amounts of urea were added. Higher feed intakes and weight gains have been obtained in feeding experiments with lambs and steers when soybean or cotton seed meal were given as nitrogen sources instead of urea (see Table 16). The daily intake in lambs of a ration containing 70% straw was 200 to 300 g of DM higher when supplied with soybean meal than with urea, and the weight gain was approximately 80 g higher.[37] Saxena et al.[93] found only small differences in intake and weight gain comparing urea and soybean meal as nitrogen sources in a ration containing 65% untreated oat straw. When the straw was treated, the daily intake of DM and the weight gain were increased 180 and 52 g, respectively, on soybean meal compared to urea. The daily intake of dry matter in lambs from a ration containing 75% NaOH-treated straw was increased 238 g when the nitrogen source was soybean meal compared to urea, and the daily weight gain was increased 55 g.[92] Braman and Abe[13] used varying proportions of cotton seed and urea in rations with 50% NaOH-treated wheat straw for steers. The feed intake was reduced, and the weight gain was strongly reduced on urea containing rations.

High intakes of alkali-treated roughages may also be obtained in dairy cows. Cows in early lactation were constantly fed 7.7 kg of concentrate plus 5 kg of beet molasses per day. Chopped, untreated, or NaOH-treated (5% NaOH) barley straw was fed *ad libitum*[57] (Table 17). The cows consumed 5.8 kg DM in untreated and 7.6 kg in treated straw, which corresponded to 1.08 and 1.41% of the live weight. Thus, treatment increased the intake of straw by 31%. However, this difference may be too great. A change over design was used in this experiment, and in the second period, the group who changed from treated to untreated straw dropped to a very low level of intake. During the first period the difference between the intake of treated and untreated straw was 14%. This result is in accordance with the result of an experiment of Rissanen and Kossila,[91] in which untreated and NH₃-treated, long oat straw were compared

Table 17

DAILY FEED INTAKE, MILK YIELD, AND LIVE
WEIGHT CHANGE IN DAIRY COWS FED
UNTREATED AND NaOH-TREATED, CHOPPED
BARLEY STRAW[57]

	Untreated straw	Treated straw
Feed		
Concentrates, kg	7.7	7.7
Beet molasses, kg	5.0	5.0
Straw, kg DM	5.8	7.6
Straw, kg DM/100 kg live weight	1.08	1.41
Total DM, kg/100 kg live weight	3.06	3.39
Performance		
Milk, kg	24.0	24.6
Fat, %	3.34	3.64
4% FCM, kg	21.7	23.3
Weight change, g	136	414

Table 18

DAILY FEED INTAKE AND MILK YIELD AND TOTAL LIVE
WEIGHT CHANGE IN DAIRY COWS FED UNTREATED AND NH₃-
TREATED, LONG OAT STRAW IN RATIONS WITH VARYING
AMOUNTS OF GRASS SILAGE[91]

	Group					
	1 Untr. straw	2 Treat. straw	3 Untr. straw	4 Treat. straw	5 Untr. straw	6 Treat. straw
Feed						
Straw, kg DM	6.1	6.9	4.4	5.1	2.6	3.0
Silage, kg DM	—	—	3.2	3.2	6.5	6.3
Barley, kg DM	5.4	5.1	4.1	4.0	2.7	2.7
Concentrates, kg	3.9	3.7	3.4	3.3	2.6	2.5
Dry molassed beet pulp, kg	2.0	1.8	2.0	1.9	1.8	1.9
Total DM intake, kg	15.8	16.1	15.9	16.3	15.1	15.4
Performance						
4% FCM, kg	19.0	19.4	18.9	18.5	18.7	19.5
Fat, %	4.38	4.41	4.43	4.49	4.30	4.65
Weight change, kg	19	23	16	22	25	31

(Table 18). In this experiment the straw was fed *ad libitum* together with varying amounts of grass silage. The intake of treated straw was in average 16% higher than that of untreated straw.

Greenhalgh et al.[32] compared complete rations containing 50% concentrates and 50% either untreated or NaOH-treated (8% NaOH) barley straw. The cows consumed 10.8 and 13.4 kg DM of the rations containing untreated and treated straw, respectively, the difference being 24% (Table 19). Complete rations containing 40% untreated and NaOH-treated (soaked and washed) sugarcane bagasse were compared in an experiment of Randel et al.[89] The experiment lasted from day 6 to day 306 of lactation. The intake of the ration with treated bagasse was 17% higher than the intake of the ration with untreated bagasse (Table 20).

Junker[48] fed chopped and pelleted, NaOH-treated wheat straw to dairy cows from

Table 19
DAILY FEED INTAKE AND MILK
YIELD IN COWS FED COMPLETE
RATIONS CONTAINING 50%
UNTREATED OR NaOH-TREATED
BARLEY STRAW[32]

	Untreated straw	Treated straw
Intake, kg DM	10.8	13.4
Milk, kg	17.6	19.0
Fat, %	3.54	3.74

Table 20
FEED INTAKE, MILK YIELD, AND WEIGHT
CHANGE IN DAIRY COWS FED COMPLETE
RATIONS CONTAINING 40% UNTREATED
OR NaOH-TREATED SUGARCANE
BAGASSE[89]

	Untreated bagasse	Treated bagasse
Intake, kg DM/day	12.5	14.6
Milk, kg/day	12.5	17.2
Milk, kg in 300 days	3,739	5,158
Fat, %	3.38	3.13
Weight change, g/day	40	220

before calving to 3 months after. Three cows, which were given 11 to 13 kg of concentrates consumed 9.9 to 10.5 kg straw in average during day 11 to day 100 of lactation. In another experiment with cows in midlactation the amount of concentrates was less than half of the above-mentioned, but the intake of straw was approximately the same. Cows in late lactation, which were fed 4.9 kg of concentrates consumed 9.9 kg (8.7 kg DM) of pelleted, NaOH-treated barley straw.[57] When the straw was neutralized with HCl, the intake was increased to 10.4 kg DM in straw.

Rissanen and Kossila[91] found that the intake of NH_3-treated straw DM was reduced with approximately 0.6 kg for each kilogram DM given in grass silage (Table 18). However, the amounts of concentrates were also reduced with increasing amounts of silage. Kristensen[57] fed NH_3-treated long barley straw *ad libitum* together with constant amounts of concentrate and increasing amounts of grass silage. The intake of straw DM was decreased with 0.8 kg per kilogram increase in the amount of grass silage (Table 21).

The marginal effect on milk production of increased energy intake and therefore the effect of treatment of roughages on milk yield is dependent on the level of feeding. A small effect on a high level of feeding was obtained by Kristensen[57] (Table 17). The milk yield was 0.6 kg and the yield of 4% fat corrected milk (FCM) 1.6 kg higher on treated than on untreated straw. But the weight gain was appreciably increased. Piatkowski[81] fed 7 kg per cow per day of pelleted mixtures containing approximately 40% untreated or NaOH-treated wheat straw. The pellets were fed together with silages, dried grass, and grain. The milk yield was 24.8 and 25.0 kg, the fat content 4.20 and 4.34%, and the amount of 4% FCM 25.5 and 26.2 kg on untreated and treated straw, respectively. The straw constituted less than 15% of the total ration dry matter.

Table 21
DAILY FEED INTAKE, MILK YIELD, AND
WEIGHT CHANGE IN DAIRY COWS FED
NH₃-TREATED, LONG BARLEY STRAW
IN RATIONS WITH VARYING AMOUNTS
OF GRASS SILAGE[57]

	Groups		
	1	2	3
Feed			
Concentrates, kg	7.0	7.0	7.0
Fodderbeets, kg DM	2.2	2.2	2.2
Grass silage, kg DM	—	3.0	5.9
Straw, kg DM	6.0	3.5	1.2
Performance			
Milk, kg	20.9	20.6	21.4
Fat, %	3.86	3.78	3.84
Weight change, g	235	314	308

Greenhalgh et al.[32] fed complete rations with 50% straw and obtained a somewhat higher increase in milk production by treatment (Table 19). Much bigger differences were found in the total lactation experiment with treated and untreated bagasse of Randel et al.[89] (Table 20). The average daily milk yield was 4.7 kg higher on the ration with treated bagasse than on that with untreated. At the same time the weight gain was higher on the treated ration.

Soper et al.[94c] replaced 0, 12, and 23% cracked corn in complete rations with corn cobs treated with 3% NaOH + 1% Ca(OH)₂. Increasing amount of treated cobs in the ration reduced milk production and tended to decrease live weight gain.

From the feeding experiments with treated roughages (having a digestibility of 60 to 70% in sheep fed at a restricted level), it is generally concluded that the net energy value of such roughages is equal to the value of hays with equal digestibilities.

Many experiments have shown high increases in water intake and urine excretion, when animals are fed rations high in NaOH-treated roughages. These increases are caused by the high content of sodium in the treated feed. The sodium has to be excreted with the urine. Thus, the concentration of sodium in the urine of sheep was increased to nearly 1 g/100 g urine, and the amount of urine increased from 1086 mℓ/day on untreated straw to 3763 mℓ on straw treated with 5.6% NaOH.[2] This is a strain to the animal, and can have adverse consequences. Voigt and Piatkowski[98] found that cows fed NaOH-treated straw excreted increased amounts of protein and ammonia and a decreased amount of urea in the urine; and the creatinine clearance was increased. The authors stated that these observations could be an indication of overloading of the kidneys. A lot of feeding experiments with great amounts of spray treated roughages have been carried out without any visible effect on the health of the animals. Adverse effects on animal health have been reported from one experiment with young bulls.[64] The bulls were fed 2 kg concentrate per day plus chopped, NaOH-treated (4 to 4.5%) barley straw. The intake of straw was high and the average daily live weight gain 1256 g during the first 12 weeks of the experiment. After that the animals stunted, and the growth rate fell strongly. It was necessary to limit the intake of straw. More investigations on these problems are needed.

Another aspect that should apparently be studied is the possibility that certain degradation products in treated roughages could have adverse effects on digestion. It is found that the activity of a number of biological systems is more or less inhibited in the presence of certain concentrations of alkali lignins.[26] The authors suggested that the inhibiting effect of alkali lignins on the activity of various animal enzymes and extracellular microbial products such as enzymes, toxins, and antibiotics may indicate that a number of biological systems in the digestive tract may be inhibited or interfered with, if the animals fodder contains alkali lignins.

REFERENCES

1. Adeleye, I. O. A. and Kitts, W. D., The effect of nitrogen source on in vitro cellulose digestion of chemically treated oat straw and poplar wood, *J. Agric. Sci.*, 82, 571—573, 1974.
2. Ali, C. S., Mason, V. C., and Waagepetersen, J., The voluntary intake of pelleted diets containing sodium hydroxide-treated wheat straw by sheep. I. The effect of the alkali concentrations in the straw. II. The effect of the proportion of beet pulp-molasses in the diet, *Z. Tierphys., Tierernähr. u. Futtermittelkd*, 39, 173—191, 1977.
3. Anderson, D. C. and Ralston, A. T., Chemical treatment of ryegrass straw: in vitro dry matter digestibility and compositional changes, *J. Anim. Sci.*, 37, 148—152, 1973.
4. Arnason, J. and Mo, M., Ammonia Treatment of Straw, 3rd Straw Utilization Conf., Oxford, United Kingdom, February 24-25, 1977.
5. Baker, A. J., Effect of lignin on the in vitro digestibility of wood pulp, *J. Anim. Sci.*, 36, 768—771, 1973.
6. Baker, A. J., Millett, M. A., and Satter, L. D., Wood and wood-based residues in animal feeds, in *Cellulose Technology Research*, Turbak, A. F., Ed., ACS Symp. Ser. 10, 1975, 75—105.
7. Becker, K. and Pfeffer, E., Untersuchungen über den Aufschluss von Stroh mit Ammoniak, *Das Wirtschaftseigene Futter*, 23, 83—87, 1977.
8. Beckmann, E., Die Veredelung von Getreidestroh und Lupinen zu hochwertigen Futtermitteln, *Festschrift der Kaiser Wilhelm Gesellschaft zur Förderung der Wissenschaften zu ihrem zehnjährigen Jubiläum*, Verlag Julius Springer, Berlin, 1921, 18—26.
9. Bergner, H., Müller, J., Marienburg, J., Adam, K., and Zimmer, H. J., Untersuchungen Zur Charakterisierung von Strohpellets. III. Mitt. Herstellung von Strohpellets mit ^{14}C- und ^{15}N-markierten Aufschlussmitteln und ihre Chemische Untersuchung, *Arch. Tierernähr.*, 24, 567—576, 1974.
10. Bergner, H., Zimmer, H. J., and Münchow, H., Untersuchungen zur Charakterisierung von Strohpellets, VI. Mitt. Verdaulichkeitsuntersuchungen an Weizenstrohpellets, *Arch. Tierernähr.*, 24, 689—700, 1974.
11. Bolduan, G. and Piatkowski, B., Untersuchungen zum Aufschluss von Getreidestroh. I. Mitt. Behandlung mit Natronlauge und die anschliessende Neutralisierung, *Arch, Tierernähr.*, 22, 485—492, 1972.
12. Bolduan, G., Voigt, J., and Piatkowski, B., Untersuchungen zum Aufschluss von Getreidestroh. III. Mitt. Einfluss der Behandlung mit Natronlauge auf die Pansenfermentation in Versuchen an Kühen, *Arch. Tierernähr.*, 24, 149—157, 1974.
13. Braman, W. L. and Abe, R. K., Laboratory and in vivo evaluation of the nutritive value of NaOH-treated wheat straw, *J. Anim. Sci.*, 46, 496—505, 1977.
14. Breirem, K. and Homb, T., *Förmidler og Förkonservering*, Forlag Buskap og Avdratt A/S, Gjøvik, Norway, 1970.
15. Bryant, M. P., Nutritional requirements of the predominant rumen cellulolytic bacteria, *Fed. Proc.*, 32, 1809—1813, 1973.
16. Carmona, J. F., Utilizacion de paja de trigo tratada con alcali. Instituto National de Investigaciones Agrarias, *An. Prod. Anim.*, 3, 73—78, 1972.
17. Carmona, J. F. and Greenhalgh, J. F. D., The digestibility and acceptability to sheep of chopped or milled barley straw soaked or sprayed with alkali, *J. Agric. Sci.*, 78, 477—485, 1972.
18. Chandra, S. and Jackson, M. G., A study of various chemical treatments to remove lignin from coarse roughages and increase their digestibility, *J. Agric. Sci.*, 77, 11—17, 1971.
19. Chaturvedi, M. L., Singh, U. B., and Ranjhan, S. K., Effect of alkali treatment of wheat straw on feed consumption, digestibility and VFA production in cattle and buffalo calves, *Indian J. Anim. Sci.*, 43, 677—683, 1973.
20. Choung, C. C. and McManus, W. R., Studies on forage cell walls. III. Effects of feeding alkali-treated rice hulls to sheep, *J. Agric. Sci.*, 86, 517—530, 1976.

21. Clarke, S. D. and Dyer, I. A., Chemically degraded wood in finishing beef cattle rations, *J. Anim. Sci.*, 37, 1022—1026, 1973.
22. Coxworth, E., Kernan, J., Nicholson, H., Chaplin, R., and Manns, J., A Report On a Search for Economical Farm Scale Methods of Improving the Feeding Value of Straw of the Canadian Prairies — Use of Ammonia and other Bases, Proc. 12th Ann. Nutr. Conf. Feed Manufact., Toronto, Canada (cited by Sundstøl, et al.[94])
23. Cross, H. H., Smith, L. W., and DeBarth, J. V., Rates of in vitro forage fiber digestion as influenced by chemical treatment, *J. Anim. Sci.*, 39, 808—812, 1974.
24. Dekker, R. F. H. and Richards, G. N., Effect of delignification on the in vitro rumen digestion of polysaccharides of bagasse, *J. Sci. Food Agric.*, 24, 375—379, 1973.
25. Donefer, E., Adeleye, I. O. A., and Jones, T. A. O. C., Effect of urea supplementation on the nutritive value of NaOH-treated oat straw, in *Cellulases and Their Applications*, Adv. Chem. Ser. 95, Hajny, G. J. and Reese, E. T., Eds., American Chemical Society, Washington, D.C., 1969, 328—339.
26. Farstad, L. and Näss, B., Effects of alkali lignins on some biological systems, *Acta Agric. Scand.*, 27, 123—128, 1977.
27. Feist, W. C., Baker, A. J., and Tarkow, H., Alkali requirements for improving digestibility of hardwoods by rumen microorganisms, *J. Anim. Sci.*, 30, 832—835, 1970.
28. Flachowsky, G., Löhnert, H. J., Wolf, J., Müller, B., and Krämer, E., NaOH-Zusatz bei der Strohpelletierung und Prüfung der Pellets im Tierversuch, *Tierzucht*, 30, 359—362, 1976.
29. Foldager, J., Halm til opdråt, paper presented at NJF-seminar: halm; håndtering, behandling og udnyttelse, Middelfart, Denmark, March 28—31, 1978.
30. Frederiksen, J. Højland, *Fordøjelighed og optagelse af natriumhydroxydbehandlet byghalm hos udvoksede får med tilskud af hel byg plus urinstof eller sojaskrå*, Medd.nr. 247, Statens Husdyrbrugsforsøg, Copenhagen, 1978.
31. Gharib, F. H., Goodrich, R. D., Meiske, J. C., and El Serafy, A. M., Effect of grinding and sodium hydroxide treatment on poplar bark, *J. Anim. Sci.*, 40, 727—733, 1975.
32. Greenhalgh, J. F. D., Pirie, R., and Reid, G. W., Alkali-treated barley straw in complete diets for lambs and dairy cows, *Anim. Prod.*, 22 (Abstr.), 159, 1976.
33. Guggolz, J., McDonald, G. M., Walker, H. G., Garrett, W. N., and Kohler, G. O., Treatment of farm wastes for livestock feed, *J. Anim. Sci.*, 33 (Abstr.), 284, 1971.
34. Guggolz, J., Kohler, G. O., and Klopfenstein, T. J., Composition and improvement of grass straw for ruminant nutrition, *J. Anim. Sci.*, 33, 151—156, 1971.
35. Guggolz, J., Saunders, R. M., Kohler, G. O., and Klopfenstein, T. J., Enzymatic evaluation of processes for improving agricultural wastes for ruminant feeds, *J. Anim. Sci.*, 33, 167—170, 1971.
36. Görsch, R., Zimmer, H. J., Adam, K., and Bergner, H., Untersuchungen zur Charakterisierung von Strohpellets. IV. Mitt. Abbau von ^{15}N-markierten Strohpellets im Verdauungskanal des Schafes, *Arch. Tierernähr.*, 24, 577—588, 1974.
37. Hasimoglu, S., Klopfenstein, T. J., and Doane, T. H., Nitrogen source with sodium hydroxide treated wheat straw, *J. Anim. Sci.*, 29 (Abstr.), 160, 1969.
38. Hogan, J. P. and Weston, R. H., The utilization of alkali-treated straw by sheep, *Aust. J. Agric. Res.*, 22, 951—962, 1971.
39. Homb, T., Foringsforsøk Med Lutet Halm, 64th report of the Inst. Anim. Nutr., Royal Agric. Coll., Norway, 1948.
40. Homb, T., Sundstøl, F., and Arnason, J., Chemical treatment of straw at commercial and farm levels, in *New Feed Resources*, FAO Animal Production and Health paper No. 4, Food and Agriculture Organization, Rome, 1977, 25—38.
41. Huffmann, J. G., Kitts, W. D., and Krishnamurti, C. R., Effects of alkali treatment and gamma irradiation on the chemical composition and in vitro rumen digestibility of certain species of wood, *Can. J. Anim. Sci.*, 51, 457—464, 1971.
42. Hvidsten, H. and Homb, T., A survey of cellulose and Beckmann-treated straw as feed, Reprint No. 62, Inst. Anim. Nutrit., Royal Agric. Coll., Norway, 1947.
43. Jackson, M. G., The alkali treatment of straws, *Anim. Feed. Sci. Technol.*, 2, 105—130, 1977.
44. Jackson, M. G., Treating straw for animal feeding, FAO Animal Production and Health Paper No. 10, Food and Agriculture Organization, Rome, 1978, 81.
45. Javed, A. H. and Donefer, E., Alkali-treated straw rations for fattening lambs, *J. Anim. Sci.*, 31 (Abstr.), 245, 1970.
46. Jayasuriya, M. C. N. and Owen, E., Sodium hydroxide treatment of barley straw; effect of volume and concentration of solution on digestibility and intake by sheep, *Anim. Prod.*, 21, 313—322, 1975.
47. Jones, M. J. and Klopfenstein, T. J., Chemical treatments of poor quality roughages, *J. Anim. Sci.*, 26 (Abstr.), 1492, 1967.

48. Junker, T., Untersuchungen über die Verbesserung des Futterwertes von Stroh für Wiederkauer durch trockenen Aufschluss beim Brikettieren. Dissertation, Universität Göttingen, 1976, 160.

49. Kamstra, L. D., Moxon, A. L., and Bentley, O. V., The effect of stage of maturity and lignification on the digestibility of cellulose in forage plants by rumen microorganisms in vitro, *J. Anim. Sci.,* 17, 199—208, 1958.

50. Kellner, O., *Die Ernährung der landwirtschaftlichen Nutztiere,* 1 Ausg., Paul Parey, Berlin, 1905.

51. Kishan, J., Ranjhan, S. K., and Netke, S. P., Effect of alkali spray treatment and supplementation of molasses and urea on the utilization of wheat straw by buffalo calves, *Indian J. Anim. Sci.,* 43, 609—614, 1973.

52. Klopfenstein, T. J., Bartling, R. R., and Woods, W. R., Treatments for increasing roughage digestion, *J. Anim. Sci.,* 26 (Abstr.), 1402, 1967.

53. Klopfenstein, T. J., Krause, V. E., Jones, M. J., and Woods, W., Chemical treatment of low quality roughages, *J. Anim. Sci.,* 35, 418—422, 1972.

54. Koers, W. C., Klopfenstein, T. J., and Woods, W., Sodium hydroxide treatment of corn cobs, *J. Anim. Sci.,* 29 (Abstr.), 163, 1969.

55. Koers, W., Woods, W., and Klopfenstein, T. J., Sodium hydroxide treatment of corn stover and cobs, *J. Anim. Sci.,* 31 (Abstr.), 1030, 1970.

56. Krause, V., Klopfenstein, T. J., and Woods, W., Sodium hydroxide treatment of corn silage, *J. Anim. Sci.,* 27 (Abstr.), 1167, 1968.

57. Kristensen, V. F., Halm til malkekøer. Foderoptagelse, fodersammensatning og mälkeproduktion, paper presented at the NJF seminar: halm; håndtering, behandlig og udnyttelse, Middelfart, Denmark, March 28-31, 1978.

58. Kristensen, V. F., Andersen, P. E., Stigsen, P., Thomsen, K. V., Andersen, H. R., Sørensen, M., Ali, C. S., Mason, V. C., Rexen, F., Israelsen, M., and Wolstrup, J., Sodium hydroxide-treated straw as feed for cattle and sheep, 464. beretn. fra, Statens Husdyrbrugsforsøg, Copenhagen, 1977, 218.

59. Lampila, M., Experiments with alkali straw and urea, *Ann. Agric. Fenn.,* 2, 105—108, 1963.

60. Lehmann, F., Strohaufschliessung, Verlag O., Elsner, Berlin, 1917.

61. Levy, D., Holzer, Z. Neumarkt, H., and Folman, Y., Chemical processing of wheat straw and cotton byproducts for fattening cattle, *Anim. Prod.,* 25, 27—37, 1977.

62. Maeng, W. J., Mowat, D. N., and Bilanski, K., Digestibility of sodium hydroxide-treated straw fed alone or in combination with alfalfa silage, *Can. J. Anim. Sci.,* 51, 743—747, 1971.

63. McManus, W. R. and Choung, C. C., Studies on forage cell walls. II. Conditions for alkali treatment of rice straw and rice hulls, *J. Agric. Sci.,* 86, 453—470, 1976.

64. Matre, T., Halm i oppdrett og kjøtproduksjon, paper presented at NJF-seminar: halm; håndtering, behandling og udnyttelse, Middelfart, Denmark, March 28—31, 1978.

65. Mellenberger, R. W., Satter, L. D., Millett, M. A., and Baker, A. J., An in vitro technique for estimating digestibility of treated and untreated wood, *J. Anim. Sci.,* 30, 1005—1011, 1970.

66. Mellenberger, R. W., Satter, L. D., Millett, M. A., and Baker, A. J., Digestion of aspen, alkali-treated aspen and aspen bark by goats, *J. Anim. Sci.,* 32, 756—763, 1971.

67. Millett, M. A., Baker, A. J., Feist, W. C., Mellenberger, R. W., and Satter, L. D., Modifying wood to increase its in vitro digestibility, *J. Anim. Sci.,* 31, 781—788, 1970.

68. Moore, W. E., Effland, M. J., and Millett, M. A., Hydrolysis of wood and cellulose with celluloytic enzymes, *J. Agr. Food Chem.,* 20, 1173—1175, 1972.

69. Mowat, D. N., NaOH-stover or silage in growing rations, *J. Anim. Sci.,* 33 (Abstr.), 1155, 1971.

70. Mowat, D. N., Processing Forages with Sodium Hydroxide, Univ. of Guelph Nutr. Conf. Feed Manufact., Ontario, Canada, 34—39, 1973.

71. Müller, J. and Bergner, H., Untersuchungen zur Charakterisierung von Strohpellets. VIII. Mitt. Veränderungen von Strohlignin durch Ammoniakeinwirkung, *Arch. Tierernähr.,* 25, 37—45, 1975.

72. Müller, J., Bergner, H., and Marienburg, J., Untersuchungen zur Charakterisierung von Strohpellets. IX. Mitt. Herstellung und Untersuchung von ^{14}C- ^{15}N- Harnstoff-Strohpellets, ^{14}C- Saccharose-Strohpellets, ^{32}P-Phosphat-Strohpellets and ^{3}H-Strohpellets, *Arch. Tierernähr.,* 26, 221—229, 1976.

73. Møller, P. D. and Thomsen, K. V., Nitrogen utilization from forages by ruminants. Laboratorial and physiological measurements, in *Quality of Forage,* Proc. of a seminar organized by the NJF, Uppsala, Inst. för Husdjurens Utfodring och Vård, Uppsala, Sweden, Rapport nr., 54, 1977, 27—54.

74. Oji, U. I., Mowat, D. N., and Winch, J. E., Alkali treatments of corn stover to increase nutritive value, *J. Anim. Sci.,* 44, 798—802, 1977.

75. Ololade, B. G. and Mowat, D. N., Influence of whole-plant barley reconstituted with sodium hydroxide on digestibility, rumen fluid and plasma metabolism of sheep, *J. Anim. Sci.,* 40, 351—357, 1975.

76. Ololade, B. G., Mowat, D. N., and Winch, J. E., Effect of processing methods on the in vitro digestibility of sodium hydroxide treated roughages, *Can. J. Anim. Sci.,* 50, 657—662, 1970.

77. Owen, E., Perry, F. G., Burt, A. W. A., and Pearson, M. C., Variations in the digestibility of barley straws, *Anim. Prod.,* 11 (Abstr.), 272, 1969.
78. Pestalozzi, M. and Matre, T., Forsøk med ammoniakkbehandlet halm til kastrater, cited by Matre, 1978, 1977.
79. Phoenix, S. L., Bilanski, W. K., and Mowat, D. N., In vitro digestibility of barley straw treated with sodium hydroxide, *Trans. ASAE,* 15, 1091—1093, 1972.
80. Phoenix, S. L., Bilanski, W. K., and Mowat, D. N., In vitro digestibility of barley straw treated with sodium hydroxide at elevated temperatures, *Trans. ASAE,* 17, 780—782, 1974.
81. Piatkowski, B., Die Wirkung von kompaktierten Stroh-Koncentrat-Gemischen unter Verwendung von rohem und NaOH-behandeltem Getreidestroh in Versuchen an Milchkühen. I. Mitt. Verdaulichkeit und Milchproduktion, *Arch. Tierernähr.,* 26, 131—137, 1976.
82. Piatkowski, B., Alert, H. J., Voigt, J., Jacopian, V., and Philipp, B., Untersuchungen zum Aufschluss von Gretreidestroh mit Natronlauge. VII. Mitt. Aufschluss von Stroh durch Einweichen in verdünter Natronlauge bei Wiederverwendung des Washwassers sowie die Wirkung des Defibrierens, *Arch. Tierernähr.,* 27, 459—468, 1977.
83. Piatkowski, B., Bolduan, G., Zwierz, P., and Kauffold, P., Untersuchungen zum Aufschluss von Getreidestroh. II. Mitt. Einfluss der Behandlung mit Natronlauge auf Struktur und Verdaulichkeit, *Arch. Tierernähr.,* 23, 435—445, 1973.
84. Piatkowski, B., Bolduan, G., Zwierz, P., and v. Lengerken, J., Untersuchungen zum Aufschluss von Getreidestroh mit Natronlauge. IV. Mitt. Beziehungen zwischen der NaOH-Koncentration im behandelten Stroh, dem säurefällbaren Ligninanteil und der Verdaulichkeit in vivo und in vitro, *Arch. Tierernähr.,* 24, 513—522, 1974.
85. Piatkowski, B. and Nagel, S., Ergebnisse über die Kau-und Wiederkauaktivität von Kühen bei Rationen mit Getreidestroh verschiedener physikalischer Form und nach chemischer Behandlung mit Natronlauge, *Arch. Tierernähr.,* 25, 575—582, 1975.
86. Piatkowski, B., Schmidt, L., Weissbach, F., Voigt, J., Peters, G., and Prym, R., Untersuchungen zum Aufschluss von Getreidestroh mit Natronlauge. VI. Mitt. Die Wirkung einer gemeinsamen Einsilierung von NaOH-behandeltem Stroh und Grünmais auf die Silagequalität, Pansenfermentation und Verdaulichkeit, *Arch. Tierernähr.,* 24, 701—710, 1974.
87. Pigden, W. J. and Heaney, D. P., Lignocellulose in ruminant nutrition, in *Cellulases and Their Applications,* Adv. Chem. Ser. 95, Hajny, G. T. and Reese, E. T., Eds., American Chemical Society, Washington, D.C., 1969, 245—261.
88. Pigden, W. J., Prichard, G. I., and Heaney, D. P., Physical and chemical methods for increasing the available energy content of forages, *Proc. 10th Int. Grassl. Congr.,* Helsinki, 397—401, 1966.
89. Randel, P. F., Ramirez, A., Carrero, R., and Valencia, I., Alkali-treated and raw sugarcane bagasse as roughages in complete rations for lactating cows, *J. Dairy Sci.,* 55, 1492—1495, 1972.
90. Rexen, F., Stigsen, P., and Kristensen, V. F., The effect of a new alkali technique on the nutritive value of straws, in *Feed Energy Sources for Livestock,* Swan, H. and Lewis, D., Eds., Butterworths, London, 1976, 65—82.
91. Rissanen, H. and Kossila, V., Untreated and ammonized straw with or without silage to dairy cows, in *Quality of Forage,* Proceeding of a seminar organized by the NJF, Uppsala, Inst. för Husdjurens Utfodring och Vård, Uppsala, Sweden, Rapport nr. 54, 1977, 177—180.
92. Rounds, W., Klopfenstein, T., Waller, J., and Messersmith, T., Influence of alkali treatments of corn cobs on in vitro dry matter disappearance and lamb performance, *J. Anim. Sci.,* 43, 478—482, 1976.
92a. Saarinen, P., Jensen, W. J., and Alhojärvi, J., Digestibility of high yield chemical pulp and its evaluation, *Acta Agral. Fenn.,* 94, 41, 1959.
93. Saxena, K., Otterby, D. E., Donker, J. D., and Good, A. L., Effects of feeding alkali-treated oat straw supplemented with soybean meal or nonprotein nitrogen on growth of lambs and on certain blood and rumen liquor parameters, *J. Anim. Sci.,* 33, 485—490, 1971.
94. Schieman, R., Nehring, K., Hoffmann, L., Jentsch, W., and Chudy, A., *Energetische Futterbewertung und Energienormen,* VEB Deutsche Landwirtschaftsverlag, Berlin, 1971.
94a. Schultz, T. A. and Ralston, A. T., Effects of various additives on nutritive value of ryegrass straw silage. II. Animal metabolism and performance observations, *J. Anim. Sci.,* 39, 926—930, 1974.
94b. Singh, M. and Jackson, M. G., The effect of different levels of sodium hydroxide spray treatment of wheat straw on consumption and digestibility by cattle, *J. Agric. Sci.,* 77, 5—10, 1971.
94c. Soper, I. G., Owen, F. G., and Nielsen, M. K., Hydroxide treated corn cobs fed with corn silages in complete rations, *J. Dairy Sci.,* 60, 596—601, 1977.
94d. Stone, J. E., Scallan, A. M., Donefer, E., and Ahlgren, E., Digestibility as a simple function of a molecule of similar size to a cellulase enzyme, in *Cellulases and Their Applications,* Adv. Chem. Ser. 95, Hajny, G. T. and Reese, E. T., Eds., American Chemical Society, Washington, D.C., 1969, 219—238.

94e. Summers, C. B. and Sherrod, L. B., Sodium hydroxide treatment of different roughages, *J. Anim. Sci.,* 41 (Abstr.), 420, 1975.

94f. Sundstøl, F., Coxworth, E., and Mowat, D. N., Improving the nutritive value of straw and other low-quality roughages by treatment with ammonia, *World Anim. Rev.,* 26, 13—21, 1978.

95. Tarkow, H. and Feist, W. C., A mechanism for improving the digestibility of lignocellulosic materials with dilute alkali and liquid ammonia, in *Cellulases and Their Applications,* Adv. Chem. Ser. 95, Hajny, G. T. and Reese, E. T., Eds., American Chemical Society, Washington, D.C., 1969, 197—217.

96. Terry, R. A., Spooner, M. C., and Osbourn, D. F., The feeding value of mixtures of alkali-treated straw and grass silage, *J. Agric. Sci.,* 84, 373—376, 1975.

97. Thomsen, K. V. and Neimann-Sørensen, A., Fertilizer influence on the nutritional value of pasture and forage in view on proteins for ruminants, in *Fertilizer Use and Protein Production,* Int. Potash Inst., "Der Bund" AG, Berne, Switzerland, 1975.

98. Voigt, J. and Piatkowski, B., Untersuchungen zum Aufschluss von Getreidestroh. V. Mitt. Die Wirkung des Natriums im NaOH-behandelten Stroh auf die Zusammensetzung von Blut and Harn sowie auf die Exkretion verschiedener Verbindungen, *Arch. Tierernähr,* 24, 589—600, 1974.

99. Waiss, A. C., Jr., Guggolz, J., Kohler, G. O., Walker, H. V., Jr., and Garrett, W. N., Improving digestibility of straws for ruminant feed by aqueous ammonia, *J. Anim. Sci.,* 35, 109—112, 1972.

100. Wang, P. Y., Bolker, H. I., and Purves, C. B., Ammonolysis of uronic ester groups in birch xylan, *Can. J. Chem.,* 42, 2434—2439, 1964.

101. Wilson, R. K. and O'Shea, J., In vitro production of volatile fatty acids and dry matter digestibility of wheat straw as affected by alkali treatment, *Irish J. Agric. Res.,* 3, 245—246, 1964.

102. Wilson, R. K. and Pigden, W. J., Effect of sodium hydroxide treatment on the utilization of wheat straw and poplar wood by rumen microorganisms, *Can. J. Anim. Sci.,* 44, 122—123, 1964.

103. White, J., Utilization of ammoniated rice hulls by beef cattle, *J. Anim. Sci.,* 25, 25—28, 1966.

104. Waagepetersen, J. and Thomsen, K. V., Effect on digestibility and nitrogen content of barley straw of different ammonia treatments, *Anim. Feed Sci. Technol.,* 2, 131—142, 1977.

105. Zimmer, H. J., Bergner, H., Müller, J., and Marienburg, J., Untersuchungen zur Charakterisierung von Strohpellets. V. Mitt. Herstellung und chemische Untersuchung von Weizenstrohpellets, *Arch. Tierernähr.,* 24, 681—688, 1974.

EFFECT OF PROCESSING ON NUTRITIVE VALUE OF FEEDS: ACID TREATMENT

J. L. L'Estrange

INTRODUCTION

The most widespread method of acid preservation of feeds is that of ensilage. In this process a green crop or other material of high moisture content is preserved by organic acids formed by microbial fermentation of substrates in the crop. Acids, both organic and inorganic, may be added to improve fermentation or to substitute for it. Fish silage, made by direct acidification of ground fish and fish offals, after which liquefaction takes place, is a modification of this process.

A second and more direct form of acid treatment of feeds is the addition of organic acids, especially propionic, to moist cereal grains at harvesting. These can then be stored without being dried. A third form is the addition of acids to the hay crop to improve its preservation.

SILAGE ADDITIVES

Before dealing with the use of acids as silage additives, a brief account will be given of the principles of ensilage and the factors which influence the type of fermentation. The review is confined to the two major crops ensiled, grassland herbage and hybrid maize. Fish and fish byproducts may also be preserved as fish silage, a topic which will be dealt with separately.

Chemical Reactions During Ensilage

The agents involved in chemical changes in the ensiled crop are enzymes in the crop and microorganisms. The respiratory enzymes in the harvested crop remain active until the oxygen supply is used up. They break down soluble carbohydrate to CO_2, water, and heat, and cause an increase in the temperature of the ensiled material. The other plant enzymes which are active during ensilage are the proteolytic enzymes which convert plant proteins to polypeptides and amino acids.

The microorganism in silage can be categorized as follows:

1. *Heterofermentative bacteria;* which under anaerobic conditions utilize carbohydrate to form lactic acid.
2. *Heterofermentative bacteria;* which utilize carbohydrate, forming lactic acid, acetic acid, mannitol, ethanol, and carbon dioxide.
3. *Clostridial bacteria;* which utilize carbohydrate, proteins, and lactic acid, forming butyric acid, branched-chain fatty acids, ammonia, and amines.
4. *Molds and yeast;* which are aerobic and utilize carbohydrate, protein, and organic acids.

The development of the microbial population during ensilage has been reviewed.[1,2,3] In summary it appears that after ensilage the aerobic organisms die rapidly when the oxygen supply is used up. Anaerobic organisms then multiply rapidly for some days before gradually declining in numbers, the rate of growth and decline varying with the conditions. Coliforms dominate at first and are then dominated by leuconositics and streptococci. They in turn are dominated by lactobacilli and pediococci. If the pH is low enough at this stage, i.e., around 4.0, the clostridia will not develop and there is little further microbial activity. If the pH is high the clostridia may develop and result in spoilage of the silage. Fungi, being dependent on oxygen, survive only near the surface and with effective sealing can virtually be eliminated. Yeasts can multiply in wilted crops and contribute both to heating during ensilage and to deterioration after unloading.[4]

The main compounds which contribute to the formation of organic acids during ensilage are the water-soluble carbohydrates. In herbage these are mainly the simple sugars and fructosans[5,6] though hemicellulose may also contribute.[7,8] Plant enzymes are responsible for the hydrolysis of fructosans and sucrose to simple sugars, after which the homolactic and heterofermentative bacteria are involved, the former being more efficient in the production of acid and lowering of the pH than the latter.[1] Under the conditions favorable to the clostridial organism, butyric acid is produced from the saccharolytic varieties from lactic acid; two molecules of lactic forming one molecule of butyric.[1] Generally 80 to 90% of the total nitrogen in herbage at harvesting is present as protein. Even in well-preserved silage, 50 to 60% of the protein is hydrolyzed to polypeptides and amino acids,[1] primarily by plant enzymes.[3] Further breakdown of the amino acids by clostridia may occur yielding amides such as putrescine, glutamine, cadavarine, histamine, γ-aminobutyric acid, tryamine, tryptamine, and ammonia.[9-10] The production of these compounds is enhanced if the pH remains high or if it falls slowly.[10] With the formation of organic acids, the pH of the ensiled crop falls rapidly after filling the silo. Under favorable conditions it reaches about pH 4.0 within a few days and remains at this level subsequently. Under unfavorable conditions it may reach only about pH 4.5 after some days and then it gradually increases for several weeks going above 5.0. This increase is due to the formation by clostridia, of butyric acid from lactic acid, as butyric is a weaker acid than lactic. Ammonia resulting from protein breakdown also causes an increase in pH.

The buffering constituents in herbage are mainly the organic acids, citric, malic, and glyceric, while in silage they are mainly lactic, acetic, and butyric acids.[11]

Factors Which Influence the Type of Fermentation

Typical fermentation patterns for wilted and unwilted grasses obtained in laboratory silos,[12] are shown in Table 1. The main factors, apart from silage additives, which affect the type of fermentation are as follows:

Moisture content — The moisture content of ensiled materials can vary from 86% for materials harvested at an early stage of maturity to less than 50% for wilted material. In general a high moisture content results in a higher rate of fermentation and so wilting is one way of restricting fermentation. In addition, the high moisture content is less favorable to lactic acid accumulation and usually results in higher levels of butyric and acetic acid, a higher pH and a greater amount of protein breakdown.[5,6,10]

Water-soluble carbohydrate content — Being the main substrate for the microbial population the level of water-soluble carbohydrate influences the rate and type of fermentation, a high level usually resulting in a rapid accumulation of lactic acid and a sharp fall in pH. In the grass crop the content can vary considerably. It is higher in bright sunny weather,[13] and also at the early flowering stage of the plant.[14] Maize, when harvested at the doughy stage, is high in soluble carbohydrate and consequently tends to give good fermentation whereas legumes are low and often present problems.

Chopping and laceration of the material — Laceration and bruising of the herbage at harvesting leads to exudation of sap from the plant which promotes bacterial growth, the formation of lactic acid and a rapid fall in pH.[15]

Sealing of the silo — Restriction of the exposure to air, either through the use of tower silos or by the use of plastic or rubber sheets to cover the exposed material,

Table 1

FERMENTATION OF UNWILTED AND WILTED GRASS[12]

Material	Composition at ensilage		pH	Composition of silage				
	H_2O	Water-soluble carbohydrate (% in DM)		Water-soluble carbohydrate (% in DM)	Lactic acid (% in DM)	Acetic acid (% in DM)	Butyric acid (% in DM)	Ammonia-nitrogen (% of total nitrogen)
S-37 Cocksfoot (cut on 18th May)								
Unwilted	86.3	11.1	4.88	<1	1.4	7.9	0.2	11.4
Wilted	73.2	11.8	4.15	<1	7.2	1.4	<0.1	6.2
S-23 Perennial ryegrass (cut on 29th May)								
Unwilted	83.7	16.4	3.77	<1	15.4	3.4	0	8.5
Wilted	70.7	18.2	4.00	1.6	9.7	2.3	<0.1	7.6
Eynsford Lucerne (cut on 20th May)								
Unwilted	86.3	11.6	5.07	<1	2.9	11.5	0.3	14.4
Wilted	70.1	9.7	4.50	<1	7.9	2.6	0.1	9.4

improves the quality of fermentation by lowering the rate of aerobic breakdown of carbohydrate.[16] In addition it decreases losses from mold growth on the surface.

Compounds Used as Silage Additives

Where the quality of fermentation is poor, nutrient losses are generally high during ensilage. The nutritional value of the silage may be further reduced due to low intakes of such silage by animals. In addition, with well-fermented material — ensiled at a high dry matter content — such as maize or wilted herbage, losses can occur during the feeding period when the silage is exposed to air. This is known as aerobic degradation and is caused by bacteria and yeasts. It is characterized by an increase in temperature and pH and some loss of dry matter.

Silage additives have been developed accordingly to solve the above problems. They can be broadly classified into two types, those which stimulate and those which inhibit microbial activity. The main stimulatory additive is molasses which acts by providing a source of soluble carbohydrate. Most of the inhibitory additives are acids or acid forming compounds (Table 2).

Mineral Acids

Complete inhibition of microbial growth was advocated by the Fingerling process patented in Germany in 1926.[17] This process involved the addition of hydrochloric acid to the crop to lower the pH to 2.0. The process was not commercially successful because the resulting silage was unpalatable to livestock. A less severe mineral acid treatment was developed by Virtanen[18] and was named after him, the AIV process. In this process a mixture of 2 N hydrochloric and 2 N sulfuric acid are added at about 60 ℓ/t of the crop, the rate varying according to the buffering capacity of the crop to lower the pH below 4.0. The result is almost complete inhibition of the action of plant enzymes and growth of bacteria due to the low pH and hence a minimum of chemical changes in the crop during ensilage. In addition to sulfuric and hydrochloric acid, other compounds could be used in this process.[17] These are sulfuryl chloride alone or mixed with thionyl chloride (which dissolves to produce mixtures of hydrochloric, sulfuric, and sulfurous acid), phosphoric acid, and the compounds phosphorus pentachloride and phosphorus oxychloride (which forms hydrochloric acid and phosphoric acid). More recently, the salts, ammonium bisulfate and sodium bisulfate were used on the same principle of direct acidification but with the advantage regarding application of being solids.

The AIV acids and salts have generally gone out of favor in recent years, partly because of problems of handling and of corrosion and partly because of low intakes by animals of silage made with them.

Organic Acids

Organic acids have antimicrobial properties and their potential as silage additives was studied extensively in the past.[17] Formic acid, which has proved the most successful, was advocated as an additive in Germany as early as 1926. It became widely used only in the past 10 years, following the invention in Norway of a simple applicator for the acid at harvesting and reports of its beneficial effect on animal performance.[19-22] It is usually added as an 85% solution at a rate of 2.3 ℓ/t, which lowers the pH to about 4.6. An example of this is Add-F developed by British Petroleum (U.K.) Ltd. At this level it does not completely suppress bacterial growth but it supresses the undesirable ones including the putrefactive coliform and butyric acid bacteria.[23] The beneficial effect is thought to be due directly to its effect on the hydrogen-ion concentration of the silage.[24] When applied at much higher rates it can completely inhibit

Table 2
LIST OF CHEMICALS USED AS SILAGE ADDITIVES INCLUDING NORMAL RECOMMENDED RATES AND LEVEL OF ACTIVE INGREDIENTS

Name	Chemical formula of active ingredients	Usual recommended rate per tonne of herbage	Level of active ingredient per tonne of herbage
AIV acids, Sulfuric Hydrochloric Phosphoric	H_2SO_4 HCl H_3PO_4	60 l of 2 N solution (usually 1:1 mix of the acids) 9.2 kg of 68% solution of phosphoric acid	120 equivalents of acid 63 equivalents of acid
AIV salts, Sodium bisulfate Ammonium bisulfate	$NaHSO_4$ NH_4HSO_4	10 kg of solid 6.5 kg of solid	96 equivalents of acid 112 equivalents of acid
Sodium metabisulfite	$Na_2S_2O_5$	2—4 kg of solid	10.5 to 21 mol of bisulfite
Calcium formate plus sodium nitrite mixture (20:3)	$Ca(CHO_2)_2$ $NaNO_2$	1.3—1.75 kg of moisture	26.7 to 54.3 equivalents of formate[a]
Formic acid	HCOOH	2.3 l of 85% solution of l formic acid	42.5 equivalents of acid
Formaldehyde plus formic acid mixture	HCHO HCOOH	5 l of solution containing 22% formaldehyde and 26% formic acid	36.7 mol of HCHO 28.2 equivalents of HCOOH
Formaldehyde plus sulfuric acid mixture	HCHO H_2SO_4	2.5 to 5 l of solution containing 22% formaldehyde and 14.6% sulfuric acid	18.4 to 36.7 mol of HCHO 7.4 to 14.8 equivalents of H_2SO_4
Propionic acid	C_2H_5COOH	0.3 to 1 l of 100% solution of propionic acid	4.0 to 13.7 equivalents of C_2H_5COOH

[a] As mode of action is not clear, it is not possible to specify further the amount of active ingredients.

Table 3

EFFECT OF FORMIC ACID TREATMENT ON GRASS SILAGE
FERMENTATION[20]

Herbage type	Formic acid application per tonne (liters)	DM% in herbage	Silage pH	Silage composition (% of DM)	
				Lactic	Butyric
June cut	2.3	19.3	3.8	2.45	0.04
	None	19.5	4.2	0.78	0.34
September cut	2.3	16.7	3.7	3.30	0.05
	None	16.7	4.5	1.63	0.29

fermentation in the silage,[25] but this process is too expensive to be used in practice. An example of its effect on fermentation of grass silage is shown in Table 3.

Propionic acid has strong antimicrobial properties and is the most widely used of the other organic acids. There is evidence that it is inferior to formic acid when used for grass silage.[26] It may have more potential for the control of aerobic deterioration of silage. When applied at ensilage to wilted alfalfa or haylage at a level of 0.8%, it helped control heating and reduced top spoilage.[27] When added to wilted herbage at ensilage at a level of 1% it reduced dry matter losses during feeding from 8.4 to 1.1%.[28] When added to maize silage at a level of 0.33% at ensilage, it reduced dry matter losses through heating.[29] It has also been added to maize silage after opening the silo, at a level of 0.5% and found to reduce losses.[30] In a comparison of organic acids added at levels up to 2% at ensilage, propionic acid was found to be more useful than formic or acetic acid in controlling fungal growth in maize silage following exposure to air.[31]

Recent laboratory experiments with other short chain fatty acids as silage additives indicate that caproic acid is the most effective at inhibiting fermentation in silage than the shorter chain fatty acids,[32-34] but to date it has not been developed on a commercial scale.

Sodium Metabisulfite

This compound was examined as a silage additive by Cowan et al.[35] on the basis that it was a bactericide and had been used successfully for many years in the wine industry. When added at a level of 3 to 4 kg/t of chopped crop, it was found to improve the type of fermentation by reducing the production of acetic and lactic acid, virtually eliminating the production of butyric acid and inhibiting protein breakdown to a satisfactory degree.[36,37] Its mode of action was thought to be due to the action of the bisulfite ion rather than the hydrogen ion as the pH remained fairly high. Subsequent experiments showed it to be a useful additive provided it is well mixed into chopped material and provided the material is not too wet, as otherwise it may be leached away.[15]

Calcium Formate/Sodium Nitrate Mixture

Formate salts, usually mixed with sodium nitrite, have been used as silage additions for many years.[17] A mixture which includes approximately 20 parts calcium formate and 3 parts sodium nitrite along with some carbohydrate and trace elements, has been patented as Kylage by Pan Britanica Industries Ltd. It is added as a fine powder at a rate of 1.3 to 1.75 kg/t of fresh crop. Its mode of action is unclear, but it appears that the calcium formate is converted to formic acid which is the main active ingredient.[15] Oxides of nitrogen are also formed which may act as sterilizing agents and help also with the formation of formic acid. The overall effect of the mixture is rather similar to that of sodium metabisulfite, i.e., it improves fermentation and reduces proteolysis and nutrient losses.[15] Responses to it are variable and it seems essential to mix it very evenly into a chopped crop which is not too high in moisture.

Formaldehyde

Formaldehyde either alone, or in combination with an organic or a mineral acid, has recently been developed as a silage additive. Two processes are currently used commercially. One developed in Finland[38] recommends the addition of 5 ℓ of a solution of 22% formaldehyde and 26% formic acid per tonne of fresh herbage. The second, developed in England and sold as Sylade by Imperial Chemical Industries Ltd., recommends the addition of a solution of 2.5 to 5 ℓ of 20 to 25% formaldehyde, 14.6% sulfuric acid, along with antipolymerization agents and corrosive inhibitors.

The mixtures above are designed to promote good fermentation and also to protect the plant protein from degradation both in the silage and in the rumen of animals consuming the silage. Regarding fermentation, much higher levels of formaldehyde alone can completely inhibit microbial activity, but the silage is extremely unpalatable to livestock.[39] At the level used in the mixtures, formaldehyde alone may result in a clostridial-type fermentation and it is thought that the inclusion of acid with it reduces that risk.[40] The level of sulfuric acid used in the mixture is only about 5% of that used in the AIV process and has been shown by Crawshaw[41] to have only a slight effect on the crop pH. Regarding protein protection, there is evidence that formaldehyde at the recommended levels decreases considerably protein breakdown during ensilage.[42-44] Its effect on protein utilization of the animal will be dealt with in a later section.

Loss of Nutrients During Ensilage
General

The average dry matter losses during ensilage are about 16%.[17] They arise through a number of ways. The major source is through respiration and fermentation, which amounts to about 6% in laboratory silos[6,8] and 11% in practical silos.[17] Seeping from the silo can account for losses, the extent of which are directly related on the dry matter content of the ensiled crop,[45] being negligible at 30% DM and in practice rarely exceeding 6%.[17] Mold growth on the surface where oxygen is available may render material inedible, while a further loss can occur through aerobic degradation during feeding caused by yeasts and bacteria.

From studies up to 1960, Watson and Nash,[17] concluded that silage additives had little effect on dry matter losses except for acid additives which reduced losses from about 16% down to 12%. More recently, Waldo et al.[22] found that formic acid as an additive reduced energy losses from 14.4% down to 8.1%. Losses caused by aerobic degradation have received considerable attention in recent years. These losses are higher with material ensiled at a high dry matter content when they may be up to 10%.[46] The effect of additives on such losses is complex. Losses may be reduced by organic acids, especially propionic acid,[27-31] but may be increased with additives such as formaldehyde which restrict fermentation during ensilage and accordingly leave a high level of substrate available for oxidation.[16]

Energy

During ensilage there is often an increase in the gross energy concentration of the material due to the formation of high energy compounds such as ethanol. Fermentation in the rumen, as indicated by the proportions of volatile fatty acids produced, is little influenced as a result of ensilage of grass[47,48] or maize,[49,50] though fermentation activity was found to be higher after silage than after grass consumption.[48] The digestibility of the dry matter and organic matter of well-made silage is generally similar to that of the crop before ensilage,[51-53] but may be reduced where a clostridial type of fermentation occurred,[53] where wilting was carried out,[52] or where a high level of formaldehyde was used as an additive.[54]

The metabolizable energy (ME) of normal silage is very similar to that of the original crop before ensilage[54] and is about 84% of the digestible energy.[55] The efficiency of utilization of the ME of silage for maintenance and production appears to be similar to that for dried material of the same digestibility,[54,56-59] though a very low value of efficiency for fattening was obtained for one silage containing a high content of lactic acid.[55]

From the foregoing it might be expected that silage additives are unlikely to affect appreciably energy digestion by animals from silage. Data to 1960 for silages made with a wide range of additives support this conclusion.[17] Recent studies with formic acid showed that it caused a slight increase in energy digestibility[21,22,48] though this effect was only observed with prewilted silage in the latter report. There is little published data on metabolism experiments directly examining the effects of silage additives on the efficiency of utilization of digestible or metabolizable energy from silage. Data on animal performance indicate that their effect on improving performance can generally be accounted for by increased feed intake, though there is some evidence also for increased efficiency of utilization of digestible energy.[21,22]

Protein

The average loss of crude protein during ensilage in experiments up to 1960 was the same as for dry matter, i.e., about 16%, losses being reduced by acid additives and by wilting.[17] Of more importance is the change in nutritional value of the protein. During ensilage over half of the plant protein is hydrolyzed to polypeptides and amino acids. Where the pH is higher, there is further breakdown to ammonia and amines.[10] Such breakdown does not necessarily result in reduced utilization by the ruminant, as it is well established that the microorganism in the rumen can synthesize microbial protein from nonprotein nitrogen. The extent of such synthesis is determined by the amount of available energy in the rumen,[60] which in turn is influenced by the type of fermentation during ensilage and by the extent of supplementation of the diet with energy rich feeds.

In normal silage the digestibility of the nitrogen is very similar to that in the original material.[49,53] Where overheating takes place the digestion of the protein may be considerably reduced, probably as a result of the Maillard reaction.[61] The retention of nitrogen by animals is generally lower than for the same material before ensilage.[54] This may be increased by supplementation of the diet with energy sources such as cereals,[62-65] due presumably, to the increased supply of energy for the rumen microorganisms.

The effect of type of fermentation and of silage additives on nitrogen utilization from silage can be considerable and can often account for the beneficial effects of such treatments on the performance of animals. Syrjala[66] found that the growth rate of lambs was higher when fed silage treated with formic acid than when fed untreated silage of the same origin, which had been allowed to ferment to a much greater degree. However, supplementation of the lambs with 15% sucrose or starch resulted in a much greater improvement in growth rate of the lambs on the untreated than on treated silage, indicating that the degree of fermentation was of importance in affecting nitrogen utilization. In an experiment with sheep, Demarquilly and Dulphy[67] found that silage made from grass or lucerne which was treated with formic acid alone, or with a small amount of formaldehyde, resulted in a higher nitrogen retention than untreated silage. It was suggested from the results that nitrogen retention may be adversely affected by ensilage, if the water-soluble nitrogen content decreases to less than 50% of the total nitrogen and if the silage is highly digestible. In another study[68] it was observed that in grass silage, the protein nitrogen content fell progressively after a 132-day period. This was accompanied by an increase in water-soluble nonprotein nitrogen and that simultaneously there was a decline in nitrogen retention of sheep fed the silage.

It may be concluded that the restriction of fermentation by silage additives will normally improve nitrogen utilization of the silage and that this effect is due to the decrease in breakdown of both the carbohydrate and the protein fraction during ensilage.

Formaldehyde treatment of feeds may improve nitrogen utilization by the animal by protecting against protein degradation in the rumen, its effectiveness being dependent on the level used and on the protein content of the feed.[69] When used as a silage additive it may reduce the amount of protein degradation both during ensilage and in the rumen, and thereby increase the supply of digestible protein to the small intestine. With dried material increased absorption of amino acids was obtained with formaldehyde added at a level of 4 g/100 g crude protein.[70] This corresponds to about 1.2 kg/t of fresh herbage, assuming crude protein and dry matter contents of 15 and 20%, respectively, and is similar to that recommended as a silage additive, i.e., 5 ℓ of 22% solution per tonne. The beneficial effects of such levels on nitrogen retention by animals have yet to be established. In one report with cattle, improved nitrogen retention was obtained[42] whereas in another with lambs there was no effect.[66] There is also evidence that the proportion of energy digested in the rumen is increased by formaldehyde treatment of herbage at ensilage and that some improvement in the efficiency of energy absorption may result.[69]

Minerals

The losses of minerals during ensilage are accounted for by seepage or effluent from the silo. They are related very closely to the dry matter content of the ensiled material, being negligible at 30% dry matter. In an examination of research reports from 1938 to 1960, Watson and Nash[17] found average loss of total ash from silage was 9.1% while the corresponding values for calcium and phosphorus were 14.4 and 17.7%. Losses of other minerals are not well-documented. In one report the percentage loss of magnesium, potassium, and sodium tended to be higher than that of calcium and phosphorus.[71]

As mineral losses arise through seepage from the silage rather than through fermentative changes, silage additives are unlikely to affect them.

Vitamins

The important vitamin in green crops in relation to ruminant feeding is vitamin A, present as the precursor carotene. Vitamin C is also present in large amounts but is not required by the ruminant and is important only in relation to its effect on the vitamin C content of milk.

In published work up to 1960, Watson and Nash[17] found that the average loss of carotene in silage was 25% for unwilted and 48% for wilted material. These losses are much lower than where green crops are conserved as hay, where losses can be up to 100%. The effect of additives on carotene losses was small, though the minimum loss was found when acid additives were used.[17] Losses of vitamin C were found to be high during ensilage, being about 50% in the AIV process and substantially higher in other methods.[17]

Intake of Silage by Animals

The amount which sheep and cattle will consume when it is offered to them *ad libitum* is of critical importance in determining their performance, particularly if it is the sole feed. It is generally found that the dry matter intake from silage is less than that from the original crop fed fresh or in a dried form, the difference being greater with sheep than with cattle. In addition the inverse relationship between dry matter intake and organic matter digestibility which is well established for ruminants fed fresh or dry herbage,[72] does not hold so well in the case of silage.[54,67] A number of factors which may affect silage intake have been identified.

Dry Matter Content

With grass silage fed to sheep and cattle it is generally found that the higher the dry matter content of the silage, up to about 40%, the higher the dry matter intake.[73-79] With maize silage, dry matter content has less influence on intake.[80,81] It has been demonstrated that the water content of the silage is unlikely to directly affect feed intake.[82,83] Rather the effect is an indirect one, associated in some way with the effect of moisture content on fermentation in the silage.

Fineness of Chopping

Chopping of herbage prior to ensiling has generally resulted in increased dry matter intake compared with unchopped material.[84,85] This effect, unlike that of water, is probably a direct one, as chopping of long silage prior to feeding resulted in a substantial increase in intake of sheep,[86,87] and to a lesser extent, of cattle.[85,87]

Type of Fermentation

Studies on the relationship between silage intake and its chemical composition indicate that type of fermentation is important. A positive correlation was found between intake and lactic acid content,[53,79,88] while negative correlations were found between intake and the contents of acetic acid,[53,79,88,89] propionic acid,[53,89] total volatile fatty acids,[53] and ammonia nitrogen as a percentage of total nitrogen.[77,88] It may be inferred, therefore, that a clostridial type of fermentation will usually lend to reduced silage intake. It was shown by McLeod et al.[90] that neutralizations of silage with sodium bicarbonate increased intake, while addition of lactic acid to silage decreased intake, indicating that the extent of fermentation may also affect intake.

The mechanism by which silage fermentation influences intake is not clear, as many of the compounds, which in silage were found to be negatively correlated with intake, have little or no effect when supplemented at levels as high as they are likely to occur in silage. These include acetic acid,[91-93] butyric acid,[92,94] lactic acid,[93,95-96] and ammonium butyrate.[92]

Other compounds in badly fermented silage which have been implicated as reducing feed intake are the amines, following the observation by Neumark et al.,[97] that infusion of histamine depressed intake of sheep. There is, however, no evidence to date of a similar effect when amines are mixed into the feed.

The possibility remains that the influence of type of fermentation on feed intake is brought about not through the accumulation during ensilage of appetite-depressing compounds, but rather by a loss in the nutritional value of some of the nutrients. The most likely nutrient is protein, the utilization of which may be reduced either as a result of its own breakdown during ensilage or of the breakdown of soluble carbohydrate required for microbial protein synthesis in the rumen.

Silage Additives

Beneficial effects on feed intake have been found with the use of some silage additives including formic acid[20,48] while adverse effects on intake have been found with mineral acids[98-100] and with high levels of formaldehyde.[39]

Table 4
EFFECT OF DIETARY HYDROCHLORIC ACID ON VOLUNTARY FEED INTAKE OF SHEEP AND CATTLE FED A BASAL DIET OF GRASS MEAL PELLETS

Experiment	Level of HCl (mmol/kg feed DM)[a]	Dietary pH (diluted 1:5)	DM intake as % of control	Animals used	Ref.
1	320	3.8	81	Sheep	106
2	157	4.7	89	Sheep	107
	314	3.7	64		
	470	2.9	51		
	628	2.6	41		
3	300	4.4	87	Sheep	95
	500	3.8	71		
	300	4.4	79	Cattle	95
	500	3.8	66		
4	280	4.4	83	Sheep	107
	280 (IR)	5.6	97		
	560	3.2	52		
	560 (IR)	5.6	62		

[a]   Acid added to feed except for two treatments marked IR in Experiment 4 where it was given intraruminally.

Formic acid is metabolized in the rumen to methane and carbon dioxide,[101] and so is unlikely to impose any nutritional stress on ruminants. In addition, more than half of what is added to silage disappears before feeding,[21] and thus the level in silage is normally less than 1.5% of the dry matter. When added at a level of 4.1% of the feed dry matter to grass meal pellets, formic acid did not affect feed intake or acid-base balance of sheep.[92] The beneficial effect of formic acid on silage intake, when it occurs, is most likely due to its effect on silage fermentation. Other short chain fatty acids, such as propionic acid, if used as silage additives, are also unlikely to adversely affect intake as they are normal metabolites in the rumen and do not affect acid-base balance of animals when fed in the free acid form at levels of about 5% of the feed dry matter.[92]

The necessity to neutralize AIV silage with alkali salts to improve animal performance was known for many years.[18,98-100] Low intakes of silage were also obtained when made with ammonium bisulfate[102-104] and sodium bisulfate.[105] Recent studies involving the direct feeding of these acids and salts to animals have helped explain their effect on feed intake.[95,106-110]

In the case of hydrochloric acid, a linear relationship was found between the level of acid feed and the depression in feed intake of sheep (Table 4). The reduction in intake was generally associated with the induction of metabolic acidosis. There also was some evidence for oropharyngeal sensitivity to the acid in the feed. Its effect was greater when added to the feed than when infused intrarumenally (Table 4).

In the case of sulfuric acid and the sulfate salts, an additional factor involved was the high intake of sulfate-S (Table 5). It was observed as the level of the neutral salt, sodium sulfate, was progressively increased in the diet up to 1.5% sulfur, voluntary feed intake gradually decreased down to 41% of control. Similar levels of dietary sulfate did not affect intake of rats,[111] and so its effect on ruminants is associated with the unique way in which they metabolize sulfur. Dietary sulfate is rapidly reduced to sulfide in the rumen,[112] which is then absorbed into the blood stream and oxidized to sulfate in the liver,[113] before being excreted in the urine in this form. Sulfide is a more

Table 5
EFFECTS OF DIETARY SULFATE FROM DIFFERENT SOURCES
ON VOLUNTARY FEED INTAKE OF SHEEP FED A BASAL DIET
OF GRASS MEAL PELLETS

Experiment	Compound[a]	% Sulfate—Sulfur added to feed DM	DM intake as % of control	Ref.
1	Sodium sulfate	1.0	78	108
	Sodium bisulfate	1.0	79	
	Ammonium bisulfate	1.0	59	
	Ammonium sulfate	1.0	55	
	Sulfuric acid	1.0	53	
2	Sulfuric acid	0.68	70	106
3	Sodium sulfate	0.5	81	109
		1.0	52	
		1.5	41	
4	Sodium Sulfate	1.0	72	110
	Sodium sulfate (IR)	1.0	63	
	Sulfuric acid	1.0	25	
	Sulfuric acid (IR)	1.0	52	

[a]   Acid added to feed except for two treatments marked IR in Experiment 4 where it was given intraruminally.

toxic form of sulfur than sulfate and when infused at high levels into the rumen it resulted in depressed rumen mobility and caused respiratory distress and collapse.[114] In addition, Dougherty et al.[115] showed that hydrogen sulfide can exert a toxic effect on the nervous system when it was absorbed via eructation of gas from the rumen to the lungs, but not when absorption in that way was prevented by blocking the trachea. High intake of dietary sulfate was not accompanied by such clinical symptoms of toxicity,[109] and so the mechanism by which it reduces intake is not clear.

The effect of high levels of dietary sulfate on feed intake is very important with regard to silage additives as, unlike the direct mineral acid effect, it cannot be overcome by neutralization with alkali salts. These results help to explain the original observation of Virtanen,[18] that the consumption of silage was much lower when sulfuric acid was used alone than when a mixture of sulfuric and hydrochloric was used and also why very low intakes were obtained with silage made with sodium bisulfate[105] and ammonium bisulfate.[102-104] When sulfuric acid or the sulfate salts are used alone the level of sulfate-S added is about 1% of the feed dry matter whereas with the acid mixture only about half of that level is added.

It may be noted that the level of sulfuric acid used in the formaldehyde/sulfuric acid mixture supplies only about 0.12% sulfate-S to the feed dry matter and so is unlikely to affect feed intake.

In the case of sodium metabisulfite, there is also the possibility of reduced feed intake, if it is used at a very high level, as it was shown by Alhassen and Satter[116] that infusion of sodium sulfite at increasing levels progressively decreased feed intake of cows. Sulfite is also rapidly reduced to sulfide in the rumen[112] and later excreted in the urine as sulfate-sulfur,[117] following oxidation in the liver. It should therefore effect intake in the same way as sulfate. The recommended level of application of 3 to 4 kg/t of herbage, supplies about 0.6% sulfite-S to the feed dry matter and might be expected to have a small effect. The effect would be much less than with sodium bisulfate which would normally supply about 1% S to the feed dry matter, and such a difference between the two additives has been observed in practice.[105]

Formaldehyde, when used at high levels, has a very adverse effect on feed intake of sheep.[39] This effect occurs at a level of about 8 g formaldehyde per 100 g of protein,[40] which is roughly double the level at which it is normally used as a silage additive. The mechanism by which formaldehyde depresses intake is not clear. One suggestion is that it affects the viability of the microorganisms in the rumen.[40]

ACID TREATMENT OF FISH FOR MANUFACTURE OF FISH SILAGE

Fish silage is a liquid product formed from fish and fish offals which are treated with acid. The product may also be described as liquid fish protein. It is used primarily as a protein supplement for pigs.

Acid Additives and Manufacture

The original process was based on the AIV process for grass silage and was developed in Sweden in 1936 by Edin.[118] The mineral acids were added to minced fish to lower the pH to 2.0, at which satisfactory preservation was achieved. Initially a 14 N mixture of hydrochloric and sulfuric acid was used but subsequently 14 N or 50% sulfuric acid was used. The amount of acid required was calculated from the crude protein and ash content as follows:

$$\text{Liters } 14N \text{ acid per 100 kg fresh material} = 0.14 \times \text{ \% crude protein} + 0.9 \times \text{ \% ash.}$$

When mixing the fish material with the acid, wooden shovels and concrete vats coated with bitumen were used. Before feeding the fish silage was partly neutralized by the addition of chalk to raise the pH to 4.0, which required 20 to 50 kg.

The use of formic acid instead of mineral acids was developed in 1942 by Olsson.[119] It was found that with formic acid it was necessary to reduce the pH only to 4.0 to achieve satisfactory preservation. The amount of formic acid was based on the ash content and on the time of year as follows: During the winter, liters 90% formic acid per 100 kg fresh material = 0.25 × % ash. During the summer the amount of formic acid was doubled.

This process had two major advantages over the mineral acids. First, it required only about ¼ of the amount of acid because formic acts as a good preservative at pH 4.0, and second, it did not require neutralization before feeding.

Other methods for the manufacture of fish silage are reviewed by Petersen[120] and Tatterson and Windsor.[121] Most of these were based on mixtures of mineral acids with formic acid, though one process involved direct fermentation of the fish by the addition of molasses and a culture of lactic acid bacteria. Recently a process using 3.5% by weight of 85% formic acid has been described.[122] In this process the raw materials should be reduced in size to pieces less than 4 mm in diameter. The acid must be thoroughly dispersed throughout, while periodic agitation is necessary to achieve rapid liquefaction. The temperature should be above 20°C for liquefaction to take place quickly. which should be within a few days.

When the fish silage is prepared from oily fish such as herring, an additional oil separation stage is usually carried out.[122] This involves heating to 65° to 70°C after the material becomes liquid, followed by decantation and centrifugation. The oil fraction is then removed and the other fractions combined to form deoiled fish silage.

Chemical Reactions During Ensilage

It is generally assumed that liquefactions of the fish are carried out by gut enzymes which spread throughout the mass, although there is some doubt that the gut is the sole origin.[121] Fish which has been heat treated at 100°C does not liquify and it is

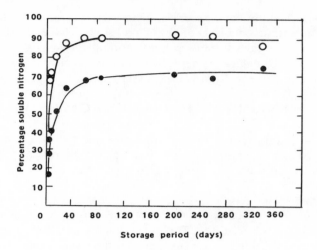

FIGURE 1.   Soluble nitrogen as a percentage of total nitrogen during storage of sprat silage. ——●——
●——, + 2°C; ——○——○——○, + 23°C. (From Tatterson, I. N. and Windsor, M. L., *J. Sci. Food Agric.,* 25, 369, 1974. With permission.)

assumed that this is because the autolytic enzymes have been destroyed by heat. Micro-organisms seem to play no role in liquefaction or preservation. During storage of fish silage made with 3.5% formic acid, the activity of aerobic bacteria was found to be very low which lactic acid bacteria, clostridia, and coliforms could not be detected.[123]

The major chemical change during storage was an increase in the amount of nitrogen soluble in 20% trichloracetic acid, from about 10% at the start to about 80% after some weeks. The rate of increase was curvelinear and was accelerated by a rise in temperature (Figure 1). Protein breakdown appeared to end at the amino acid stage, as no major losses of amino acids were found during storage,[121,123] while the content of ammonia-nitrogen as a percentage of total nitrogen remained constant at about 4%.[123]

The major change observed in the lipid fraction during storage was a marked increase in the content of free fatty acids and a decrease in the iodine value of the fat (Table 6). The change in free fatty acids was highest in the early part of the storage period while the fall in iodine value was linear with time.[121]

Effect of Ensilage on Nutritive Value

The objective with fish silage is to produce a stable high quality protein feed to be used as a protein supplement primarily for pigs and poultry. Because the produce is liquid it cannot be incorporated too readily into dry feeding systems. Its use therefore is confined mainly to pig feeding.

In the case of fish silage made from white fish, the product contains about 20% dry matter, the composition of which contains 70% crude protein, 3% ether extract, and 16% ash. The amino acid composition of the protein is very similar to that of white fish meal.[123] One of the most useful chemical tests for protein quality of fish meal is the measurement of available lysine, based on the reaction between FDNB reagent and the ε-amino group of lysine, developed by Carpenter.[124] This measurement cannot be made accurately on fish silage as hydrolysis, during storage, leads to the production of free lysine. It might be expected, however, that because heat treatment is not carried out to any great extent, little lysine would be converted into an unavailable form.

In a feeding experiment, Smith and Adamson[123] found that the performance of pigs from 30 to 90 kg live weight, was slightly lower when fed white fish silage contributing

Table 6
CHANGES IN THE OIL DURING STORAGE OF FISH SILAGE

	Type of silage									
	Whole sprat		Whole herring		Herring offal		Whole sand eels		Whole mackerel	
Analysis	+2°C	+23°C	+2°C	+23°C	+2°C	+23°C	+2°C	+23°C	+2°C	+23°C
Free fatty acid (as oleic acid)										
1. At start	6.6	6.6	5.0	5.0	2.7	2.7	4.2	4.2	6.1	6.1
2. After 12 months	17.4	20.0	18.7	21.4	7.9	12.6	21.5	25.7	12.1	17.2
Iodine value										
1. At start	135	135	147	147	147	147	175	175	163	163
2. After 12 months	123	114	124	104	110	100	94	145	142	108

From Tatterson, I. N. and Windsor, M. L., *J. Sci. Food Agric.,* 25, 369, 1974. With permission.

10.3% of the feed dry matter, compared with a similar ration in which the protein supplement was a mixture of soybean meal and white fish meal. Both rations were formulated to provide 0.9% lysine in the dry matter. No difference was detected in the flavor or odor of the meat from pigs or the two treatments. In the same experiment, a serious odor and flavor problem was found in the carcass of pigs fed herring fish silage containing 43.9% ether extract and which was not deoiled, even at a level of 5% in the feed dry matter.

Deoiled herring fish silage (17.6% dry matter which contained 75.6% crude protein and 3.9% oil) was compared with white fish meal, both supplying about 25% of the feed dry matter, by Whittemore and Taylor.[125] The results for energy and nitrogen digestibility and for nitrogen retention of the pigs, were very similar for the two sources, the fish silage giving slightly better values. In a further study by Hillyer et al.[126] deoiled herring meal, the dry matter of which contained 65.6% crude protein and 4.2% ether extract, was fed to pigs at 7.5 or 15% of the feed dry matter, in a comparison with soybean meal as the protein supplement providing the same levels of lysine. No palatability problems were observed with any of the diets. Pigs fed the rations from 30 to 60 kg live weight, performed significantly better on the fish silage treatments. Differences in the aroma of carcass fat could be detected between treatments, the importance of which has yet to be clarified.

It is concluded from the limited studies carried out that the nutritive value of fish is well maintained during ensilage. Further studies are required on the nutritional importance of the change in free fatty acids and iodine value of the lipids, which occur during storage. From studies with dry materials,[127] it is unlikely these changes will have an adverse effect on animal health or performance.

ACID TREATMENT OF CEREAL GRAINS

Cereal grains for use as animal feeds are usually harvested at a moisture content which is high enough to support microbial growth. These organisms include fungi or molds, yeasts, actinomycetes, and bacteria.[128] Animals fed badly stored moist cereals on which mold growth had occurred, perform poorly and have a very low efficiency of food conversion.[129] It is generally thought that fungi are responsible for such adverse effects. Of greatest concern to animal health are the species *Aspergillus flavus, Fusarium* and *Pennicillium,* which may provide toxins which are dangerous not only to animals but also indirectly to man consuming animal products where they may accumulate.[130] Thus, *A. flavus* produces aflatoxins, one of which can be secreted in milk as aflatoxin M. Apart from health hazards the growth of microorganisms on moist cereals results in considerable loss of nutrients.

Preservation of cereal grains can be done in the following ways:

1.   By drying to a moisture content of 14 to 15% at which microbial activity is reduced to a minimum.

2.   By ensiling in airtight containers so that the oxygen supply becomes used up and is insufficient to promote the growth of fungi. The disadvantage of this process is that when the grain is exposed to air there is rapid growth of molds and bacteria, and spoilage results. This presents serious problems regarding the feeding of such material.

3.   By treating with chemicals to control the microorganisms.

The use of organic acids, especially propionic, was examined and developed successfully for this purpose by British Petroleum Chemicals International Ltd.[131,132]

Effects of Acid Treatment on Growth of Microorganisms

A number of experiments have shown that during storage of moist cereals heating and mold growth was prevented by application of propionic acid alone or in various mixtures of propionic with other acids including formic, acetic, and butyric.[129,133-136] In barley ensiled under anaerobic conditions without additives, Livingstone et al.[136] found that the microorganisms increased per gram of DM as follows: fungi to 7.0×10^7, aerobic bacteria to 3.3×10^5, and anaerobic bacteria to 8.0×10^4. The fungi included three main types, *Endomycopsis chodatii, A. fumigatus,* and *A. flavus.* Somewhat similar results for the growth of bacteria and fungi in maize stored under anaerobic conditions were recently reported by Bothast et al.[128] These authors observed that the growth of bacteria, molds, yeasts, and actinomycetes increased to over 10^6/g DM in similar instances of maize unsealed, stored, and untreated. A mixture of propionic-acetic acids (80 to 20) applied at a level of 1.2%, controlled the growth of all organisms.

The rate of application of propionic acid required to control mold growth depends on the moisture content and on the length of storage, increasing from about 0.4% at 18% moisture to about 1.7% at 40% moisture (Table 7). Sorghum grain with a high tannin content required less propionic acid than grain with a low tannin content.[138] Propionic acid may be partly replaced by acetic acid[139] or formic acid.[140] There is evidence that formic, acetic, and propionic acid, applied at 1.5%, are equally effective at inhibiting mold growth.[141]

Other additives which were found to be fairly successful are isobutyric acid applied at 1.5%,[128] ammonium isobutyrate applied at 1.75%,[128] or 1.5%[142] and ammonia applied at 0.5%.[128] The organic acids were found to control bacteria as well as fungi, whereas ammonium isobutyrate and ammonia had little effect on bacteria.[128]

Loss of Nutrients During Storage of Acid-Treated Cereal Grains

Ware et al.[143] found that maize harvested at about 25% moisture, had a total dry matter loss from the time of harvesting to consumption by cattle, which was 3.4% when preserved with propionic, 3.1% when preserved with ammonium butyrate, and 7% when it was artificially dried to 14% moisture. Dry matter losses during anaerobic ensilage of maize at moisture levels of 33.1, 27.5, and 21.5% were 5.6, 3.7, and 2.7%, respectively.[144] In that experiment the major end products of fermentation were lactic acid and ethanol while very small amounts of acetic, propionic, and butyric acid were

Table 7
LEVEL OF PROPIONIC ACID RECOMMENDED FOR MOIST GRAIN STORAGE[137]

	Propionic acid (g) per 100 kg net weight	
Moisture % in grain	1-Month storage	6-Months storage[a]
18	350	450
20	400	500
22	450	600
24	500	700
26	550	800
28	650	950
30	800	1100
35	1150	1400
40	1400	1750
45	1650	2100

[a]   Amount should be increased by 3% for each additional month of storage over 6 months.

also found. The levels of these compounds increased with increasing moisture content, while the pH fell. In contrast Jones et al.[145] observed that with corn of 34% moisture preserved with 1.5% propionic acid, very little lactic acid was formed, indicating that fermentation was negligible.

Vitamin E appears to be the main nutrient which is appreciably reduced during the storage of moist cereals. Young et al.[146] found that α-tocopherol in moist maize which was treated with propionic acid or ensiled under anaerobic conditions, decreased from 9.3 mg/kg DM at harvest to 1.0 mg/kg DM following storage for 230 days, whereas in artificially dried maize it decreased only to 8.2 mg/kg DM during the same period. The loss of α-tocopherol in naturally dried maize was intermediate. A similar effect was observed by Lawrence,[147] who obtained a value for α-tocopherol of 0.5 mg/kg DM in moist maize preserved with propionic acid after 4 months storage compared with 7.2 mg/kg DM in maize stored after drying. Allen et al.[148] found much lower values of α-tocopherol in propionic acid-treated barley and oats compared with dried grain after storage.

The risk of vitamin E deficiency in animals fed moist cereals, particularly where selenium is also deficient, has been noted.[148] It is important therefore to ensure that rations based on moist cereals are supplemented adequately with vitamin E.

The only other notable chemical change reported in acid-treated cereals was an increase in free glucose from 0.07 to 0.7% while the starch content decreased from 63 to 58%.[149] Thus, the acid treatments result in some starch hydrolysis.

Performance of Animals on Acid-Treated Cereals

A large number of feeding trials have been carried out on acid-treated cereals compared with dried or ensiled cereals and have been reviewed by Jones et al.[150] The following are the main conclusions from those studies and from studies subsequently carried out.[151,152]

General Health

No adverse effects on the health of animals as a result of acid-treatment of cereals has been observed. The acids are well tolerated, as results of an experiment with pigs showed that up to 8% of the ration could be supplied as a mixture of acetic, propionic,

Table 8

THE EFFECT OF STORAGE AND PREPARATION OF BARLEY ON
PERFORMANCE AND CARCASS CHARACTERISTICS OF 144 PIGS FROM
25 to 90 kg LIVE WEIGHT[151]

	Treatments			
	Dry milled*a*	Acid treated and milled*b*	Dry rolled*a*	Acid treated and rolled*b*
Modulus of firmness	1.82	4.75	2.79	5.14
Feed intake (kg/day)	1.79	1.80	1.86	1.80
Growth rate (kg/day)	0.706	0.737	0.712	0.678
Feed utilization (kg DM/kg live weight gain)	2.56	2.47	2.63	2.67

a   Barley at 14% moisture.
b   Barley at 21% moisture, treated with 0.8% propionic acid.

and butyric acid (40:40:20) without adverse effect on performance.[153] This is a much higher level than ever required for grain preservation. The volatile fatty acids, acetic, propionic, and butyric are produced normally in the digestive tract of the pig,[154] and to a much greater extent in the forestomach of ruminants, and so it is not surprising that they are well tolerated. However, pigs receiving moist barley treated with 1.3% of a mixture of formic and propionic acid (70 to 30), were found to have some yellow-brown discoloration of the carcass fat.[140] Similar discoloration has not been observed with other acids and so formic acid may in some way have caused this problem. Formic acid in the ruminant is metabolized to CH_4 and CO_2 in the rumen,[101] but its fate in the monogastric animal does not appear to have been established.

Swine

In most experiments the growth rate and feed conversion efficiency has been similar for pigs fed acid-treated moist cereals and dried cereals, where the physical treatment of the cereals before feeding has been the same.[129,140,151] In one experiment the feed conversion efficiency was better with the acid-treated grain[135] and in another the opposite was found.[155]

With barley diets, any differences between the performance of pigs fed acid-treated and dried material, can be explained by the fact that while the dried material was usually ground before feeding, the acid-treated material, because of the difficulty of grinding, was usually fed rolled or in a coarsely ground form. Evidence for the superiority of ground over rolled material when both were preserved with propionic acid, was recently observed (Table 8).

The experiments, where cereals ensiled anaerobically were compared with dried or acid-treated cereals, have shown inferior performance of pigs on the ensiled material.[136,156,157] It was not established whether this effect was caused by toxins in the ensiled material or loss of nutrients during ensilage.

Poultry

A limited number of experiments carried out on the feeding of acid-treated cereals to poultry, have shown that generally, the performance of both broilers and laying hens is as good as with the feeding of dried cereals.[150] Technical problems with the handling of moist cereals for poultry has, to date, discouraged their adoption for the poultry industry.[150]

Cattle

Research work on the feeding of acid-treated cereals to beef cattle has generally shown no difference in live weight gains between cattle fed acid-treated or dried material, in the case of barley, sorghum, and maize.[150] There is some evidence that the feed conversion efficiency on a dry matter basis is better for the acid-treated material[143,150,158] and that this is related to increased energy and protein digestion.

ACID TREATMENT OF HAY

Nutrient Losses During Hay-Making

In hay-making the moisture content of the crop is lowered by natural drying to a level where plant enzymes and microorganisms are relatively inactive and so the material can be stored for winter feed. In moist temperate climates this may be difficult to achieve, particularly if the crop is harvested at an early stage of maturity, and as a result the loss of nutrients may be very high.

The loss of nutrients arise through the following ways:

1. Respiration after cutting due to plant enzymes, leads to a dry matter loss of 2 to 8% under good weather conditions and up to 16% under poor conditions.[159]
2. Mechanical fragmentation of the material as it dries, leads to average losses of 19.1% in grass crops and 38.9% in lucerne.[160,161]
3. Leaching of soluble nutrients by rain may result in very substantial losses depending on the weather conditions at harvesting.
4. Heating in the stack as a result of plant enzymes and microbial growth can cause losses.

The extent of losses through heating increases as the temperature rises. This in turn depends largely on the moisture content of the material being stacked. Watson and Nash[17] quoted figures which showed that the temperature increases almost linearly from about 28°C for material containing 18.5% moisture to 50°C for material containing 29% moisture, while dry matter losses were 2% at 48°C, 13% at 60°C, 17% at 77°C, and 29% at 80°C. In addition they reported that heating caused a reduction in the digestibility of the dry matter, especially of the protein fraction, which presumably is similar to that in overheated silage.[61]

The other problem caused by heating in the stack is the growth of molds, which are a potential health hazard to animals and man. If the temperature goes above 40°C, organisms develop, which may cause respiratory disease and mycopic abortion.[162]

Acid Additives

Acid treatment of hay has been developed to reduce nutrient losses, ease the harvesting operation, and reduce health hazards from mold growth. Acid treatment may take place at two stages, first at the cutting stage where it can be described as chemical conditioning of the crop, and second when the harvested crop is being stacked.

Chemical Conditioning of Hay

Various chemicals, including acids, have been added to hay at the cutting stage to lower the surface resistance to moisture loss and so increase the rate of drying. Laboratory studies showed that treatment of grass leaves with formic acid increased considerably the rate of evaporation of water.[163] The effect, however, was greater than that obtained with propionic acid and formic acid applied in the field for grass cut for hay, where it lasted only for a few hours after application.[164] Klinner[162] applied a commercial blend of formic/propionic acid at 0.15 and 0.29% of the crop weight at harvesting and obtained some increase in crop drying rate.

Results to date indicate that further developments are required before acid treatment of hay at the cutting stage becomes adopted commercially.

Acid Addition to the Stack

The most successful of the acids used for application to the stack is propionic acid. Results in Holland by Schukking[165] showed that the amount of propionic acid required to control heating in the stack varied according to the dry matter of the hay, being 2, 1½, and 1% for hay at 60, 65, and 70% dry matter, respectively. The loss of propionic acid varied from 35 to 55%. In Britain, the problem of application of the acid has been examined and a special window applicator developed which gave much better results than baler application.[162] The effectiveness of ammonium propionate, made by combining 4 parts of propionic acid with 3 parts of ammonia (SG = 0.88) and 1 part of water, was also measured.[162] It was found that losses of the additive were considerably less for ammonium propionate than for propionic acid (i.e., 37% vs. 59%) but that it must be applied at approximately twice the rate volumetrically to give the same amount of the acid radical. It also was less corrosive than propionic acid. Overall, the results to date show some control of heating the stack as a result of acid treatment. The beneficial effects on animal health and performance have not been reported to date.

REFERENCES

1. **Whittenbury, R.,** Microbiology of grass silage, *Proc. Biochem.,* 3, 27—31, 1968.
2. **Whittenbury, R., McDonald, P., and Bryan-Jones, D. G.,** A short review of some biochemical and microbiological aspects of ensilage, *J. Sci. Food Agric.,* 18, 441—444, 1967.
3. **Woolford, M. K.,** Some aspects of the microbiology and biochemistry of silage making, *Herbage Abstr.,* 42, 105—111, 1972.
4. **Beck, T. and Gross, F.,** Causes of the differences in the keeping properties of silage, *Wirtschaftseigene Futter,* 10, 298—312, 1964.
5. **Anderson, B. K. and Jackson, N.,** Conservation of herbage of varying dry matter content in air tight metal containers with reference to the carbohydrate fraction, *J. Sci. Food Agric.,* 21, 228—234, 1970.
6. **Anderson, B. K. and Jackson, N.,** Conservation of wilted and unwilted grass ensiled in air-tight metal containers with and without the addition of molasses, *J. Sci. Food Agric.,* 21, 235—241, 1970.
7. **McDonald, P., Stirling, A. C., Henderson, A. R., and Whittenbury, R.,** Fermentation studies on wet herbage, *J. Sci. Food Agric.,* 13, 581—590, 1962.
8. **McDonald, P., Stirling, A. C., Henderson, A. R., and Whittenbury, R.,** Fermentation studies on inoculated herbages, *J. Sci. Food Agric.,* 15, 429—436, 1964.
9. **MacPherson, H. T. and Slater, J. S.,** γ-amino-*n*-butyric, aspartic, glutamic, and pyrrolidone carboxylic acid: their determination and occurrence in grass during conservation, *Biochem. J.,* 71, 654—660, 1959.
10. **MacPherson, H. T. and Violante, P.,** The influence of pH on the metabolism of arginine and lysine in silate, *J. Sci. Food Agric.,* 17, 128—130, 1961.
11. **Playne, M. J. and McDonald, P.,** The buffering constituents of herbage and of silage, *J. Sci. Food Agric.,* 17, 264—268, 1966.
12. **Wilson, R. F. and Wilkins, R. J.,** Formic acid as a silage additive. I. Effects of formic acid on fermentation in laboratory silos, *J. Agric. Sci.,* 81, 117—124, 1973.
13. **Mackenzie, D. J. and Wylam, C. B.,** Analytical studies on the carbohydrates of grasses and clovers. VIII. Changes in carbohydrate composition during the growth of perennial ryegrass, *J. Sci. Food Agric.,* 8, 38—45, 1957.
14. **Waite, R.,** The water-soluble carbohydrates of grasses. III. First and second year growth, *J. Sci. Food Agric.,* 8, 422—428, 1957.
15. **Murdoch, J. C.,** Making and Feeding Silage, *Farming Press (Books),* Lloyds, Chambers, Ipswich, U.K., 1962, 33—34.

16. Ruxton, I. B., Clark, B. J., and McDonald, P., A review of the effects of oxygen on ensilage, *J. Br. Grassl. Soc.,* 30, 23—30, 1975.

17. Watson, S. J. and Nash, M. J., *The Conservation of Grass and Forage Crops,* Oliver and Boyd, Edinburgh, 1960.

18. Virtanen, A. I., The A.I.V. method of preserving fresh fodder, *Emp. J. Exp. Agric.,* 1, 143—155, 1933.

19. Saue, O. and Breirem, K., Comparison of formic and silage with other silages and dried grassland products in feeding experiments, in *Proc. 3rd. Gen. Meet. European Grasslands Federation,* Brunswick, 1969, 282—284.

20. Castle, M. E. and Watson, J. W., Silage and milk production, a comparison between grass silages made with and without formic acid, *J. Br. Grassl. Soc.,* 25, 65—71, 1970.

21. Waldo, D. R., Smith, L. W., Muller, R. W., and Moore, L. A., Growth, intake and digestibility from formic acid silage versus hay, *J. Dairy Sci.,* 52, 1609—1616, 1969.

22. Waldo, D. R., Keys, J. E., Jr., Smith, L. W., and Gordon, C. H., Effect of formic acid on recovery, intake, digestibility and growth from unwilted silage, *J. Dairy Sci.,* 54, 1—8, 1971.

23. Taylor, M. M. and Phillips, J. D., The bacterial flora of the four silages. (Appendix to Castle, M. E. and Watson, J. N., Silage and milk production, a comparison between grass silages made with and without formic acid), *J. Br. Grassl. Soc.,* 25, 70—71, 1970.

24. Nørdgaard Pedersen, E. J., Møller, E., and Skovborg, E. B., Experiments on the addition of formic acid and AIV acid in the ensiling of pasture crops, *Tidsskr. Planteavl,* 72, 356—366, 1968.

25. Henderson, A. R. and McDonald, P., Effect of formic acid on the fermentation of grass of low dry matter content, *J. Sci. Food and Agric.,* 22, 157—163, 1971.

26. Devuyst, A., Arnould, R., Vanbelle, M., and Moreels, A., L'acide propionique comme conservant pour ensilage, *Rev. Agric.,* 25(6/7), 891—908, 1972.

27. Yu, Y. and Thomes, J. W., Effect of propionic acid and ammonium isobutyrate on preservation and nutritive values of alfalfa haylage, *J. Dairy Sci.,* 41, 1458—1467, 1975.

28. Daniel, P., Honig, H., Weise, F., and Zimmer, E., The action of propionic acid in the ensilage of green fodder, *Wirtschaftseigene Futter,* 16, 239—252, 1970.

29. Leaver, J. D., The use of propionic acid as an additive for maize silage, *J. Br. Grassl. Soc.,* 30, 17—21, 1975.

30. Cook, J. E., The use of additives to improve the stability of maize silage in aerobic conditions, *Maize Bull. No. 54,* The Maize Development Association, Tunbridge Wells, Kent, U.K., 1973, 13—16.

31. Britt, D. G., Huber, J. T., and Rogers, A. L., Fungal growth and acid production during fermentation and refermentation of organic acid treated corn silage, *J. Dairy Sci.,* 58, 532—539, 1975.

32. McDonald, P. and Henderson, A. R., The use of fatty acids as grass silage additives, *J. Sci. Food Agric.,* 25, 791—795, 1974.

33. Woolford, M. K., Microbiological screening of the straight chain fatty acids (C_1-C_{12}) as potential silage additives, *J. Sci. Food Agric.,* 26, 219—228, 1975.

34. Ohyama, Y. and Masaki, S., Chemical composition of silages treated with fatty acids, with special references to the changes in sugars, *J. Sci. Food Agric.,* 28, 78—84, 1977.

35. Cowan, R. L., Bratzler, J. W., and Swift, R. W., Use of sodium metabisulphite as a preservative for grass silage, *Science,* 116, 154, 1952.

36. Alderman, G., Cowan, R. L., Bratzler, J. W., and Swift, R. W., Some chemical characteristics of grass and legume silage made with sodium metabisulphite, *J. Dairy Sci.,* 38, 805—810, 1954.

37. Bratzler, J. W., Cowan, R. L., and Swift, R. W., Grass silage preservation with sodium metabisulphite, *J. Anim. Sci.,* 15, 163—176, 1956.

38. Poutiainen, E. and Huida, L., The quality and digestibility of silage made with different preservatives, *Keotoiminta ja Käytanto,* 27, 2, 1970.

39. Brown, D. C. and Valentine, S. C., Formaldehyde as a silage additive. I. The chemical composition and nutritive value of frozen lucerne, lucerne silage and formaldehyde — treated lucerne silage, *Aust. J. Agric. Res.,* 23, 1093—1100, 1972.

40. Wilkinson, J. M., Wilson, R. F., and Barry, T. N., Factors affecting the nutritive value of silage, *Outlook Agric.,* 9, 3—8, 1976.

41. Crawshaw, R., An approach to the evaluation of silage additives, *A.D.A.S. Q. Rev.,* 24, 1—15, 1977.

42. Waldo, D. R., Keys, J. E., Jr., and Gordon, C. H., Formaldehyde and formic acid as a silage additive, *J. Dairy Sci.,* 56, 229—232, 1973.

43. Barry, T. N., Effect of treatment with formaldehyde, formic acid and formaldehyde-acid mixtures, on the chemical composition and nutritive value of silage. II. Mature herbage, *N.Z. J. Agric. Res.,* 19, 185—191, 1976.

44. Barry, T. N., Effect of treatment with formaldehyde, formic acid and formaldehyde-acid mixtures on the chemical composition and nutritive value of silage. I. Silage made from immature pasture compared with hay, *N.Z. J. Agric. Res.,* 18, 285—289, 1975.

45. Zimmer, E., The influence of prewilting on nutrient losses, particularly on the formation of fermentation gas, *Tagungsbar. Dtsch. Akad. Landwirtschactswiss. Berlin,* 92, 37, 1967.
46. Zimmer, E., Factors Influencing Fodder Conservation, in *Proc. Int. Meet. Animal Production from Temperate Grassland,* Dublin, 1977, 121—125.
47. Anderson, B. K. and Jackson, N., Volatile fatty acids in the rumen of sheep fed grass, unwilted and wilted silage and barn dried hay, *J. Agric. Sci.,* 77, 483—490, 1971.
48. Donaldson, E. and Edwards, R. A., Feeding value of silage: silages made from freshly cut grass, wilted grass and formic acid treated wilted grass, *J. Sci. Food Agric.,* 27, 536—544, 1976.
49. Wilkinson, J. M., Huber, J. T., and Henderson, H. E., Acidity and proteolysis as factors affecting the nutritive value of corn silage, *J. Anim. Sci.,* 42, 208—218, 1976.
50. Andrieu, J. and Demarquilly, C., Composition of rumen fluid of sheep receiving green or ensiled maize *ad libitum, Ann. Zootech.,* 23, 301—312, 1974.
51. Harris, C. E., The digestibility of fodder maize and maize silage, *Exp. Agric.,* 1, 121—123, 1965.
52. Harris, C. E., Raymond, W. F., and Wilson, R. F., The Voluntary Intake of Silage, *Proc. 10th. Int. Grasslands Conf.,* Helsinki, 1966, 564—567.
53. Demarquilly, C., Chemical composition, fermentation characteristics, digestibility and voluntary intake of forage silages. Changes compared to the initial green forage, *Ann. Zootech.,* 22, 1—35, 1973.
54. Wilkins, R. J., The nutritive value of silages, in *8th. Nutr. Conf. Feed Manufacturers University of Nottingham, England,* 1974, 167—189.
55. Kelly, N. C. and Thomas, P. C., Some Aspects of the Energy Utilization of High Quality Silages, 4th Silage Conf. Hurley, U.K., 1976.
56. El Serafy, A. M., Goodrich, R. D., and Meiske, J. C., The influence of fermentation on the utilization of energy from alfalfa-brome forage, *J. Anim. Sci.,* 39, 780—787, 1974.
57. Preston, R. L., Net energy evaluation of cattle finishing rations containing varying proportions of corn grain and corn silage, *J. Anim. Sci.,* 41, 622—624, 1975.
58. Boekholt, H. A., van der Honing, Y., and van Es, A. J. H., Results of Some Energy Balance Trials with Dairy Cows Fed Rations Containing Maize Silage, Paper 6 in 4th Silage Conference, Hurley, U.K., 1976.
59. Smith, J. S., Wainman, F. W., and Dewey, P. J. S., The energy value of sheep of three Aberdeenshire silages, Paper 4, in 4th Silage Conf., Hurley, U.K., 1976.
60. Hogan, J. P. and Weston, R. H., Quantitative aspects of microbial protein synthesis in the rumen, in *Physiology of Digestion and Metabolism in the Ruminant,* Phillipson, A. T., Ed., Oriel Press, 1970, 474—485.
61. Yu, Y. and Veira, D. M., Effect of artificial heating on alfalfa haylage on chemical composition and sheep performance, *J. Anim. Sci.,* 44, 1112—1118, 1977.
62. Thompson, D. J., The digestibility and utilization of fresh grass, hay and silage by sheep, *Anim. Prod.,* 10, 240, 1968.
63. Griffiths, T. W., Some nutritional effects on the addition of rolled barley to an all silage basal diet, *Anim. Prod.,* 11, 286, 1969.
64. Griffiths, T. W. and Spillane, T. A., Nutritional effects on cereals and protein supplementation of an all silage basal diet, *Anim. Prod.,* 12, 259—260, 1970.
65. Griffiths, T. W., Spillane, T. A., and Bath, I. H., Protein and energy interrelationships in silage-based diets for growing cattle, *Anim. Prod.,* 13, 386, 1971.
66. Syrjala, L., Effect of preservation on the utilization of silage protein, Paper 14, in 4th Silage Conf., Hurley, U.K., 1976.
67. Demarquilly, C. and Dulphy, J. P., Effect of Ensiling on Feed Intake and Animal Performance, Proc. Int. Meet. Animal Production from Temperate Grassland, Dublin, 1977, 53—61.
68. Fujita, H., Length of storage period of grass silage as a regulating factor in the utilization of silage nitrogen in ruminants, *Jpn. J. Zootech. Sci.,* 47, 224—232, 1976.
69. Barry, T. N., The effectiveness of formaldehyde treatment in protecting dietary protein from rumen microbial degradation, *Proc. Nutr. Soc.,* 35, 221—229, 1976.
70. Hemsley, J. A., Hogan, J. P., and Weston, R. H., Protection of forage protein from ruminal degradation, in *Proc. 11th. Int. Grasslands Congr.,* University of Queensland Press, Australia, 1970, 703—706.
71. Antoine, A., La réduction des parties a la conservation des fourrages, *Bull. Inst. Agron. Stn. Tech. Gemblaux,* 25, 9—37, 1957.
72. Blaxter, K. L. and Wilson, R. S., The voluntary intake of roughages by steers, *Anim. Prod.,* 4, 351—358, 1962.
73. Dodsworth, T. L. and Campbell, W. H. McK., Report on a further experiment to compare the fattening values, for beef cattle, of silage made from grass cut at different stages of growth, together with the results of some supplementary experiments, *J. Agric. Sci.,* 13, 166—177, 1953.
74. Dodsworth, T. L., Further studies on the fattening value of grass silage, *J. Agric. Sci.,* 44, 383—393, 1954.

75. Murdoch, J., The effects of prewilting herbage on the composition of silage and its intake by cows, *J. Br. Grassl. Soc.*, 15, 70—73, 1960.

76. Logan, V. S. and Haydon, P. S., The effect of moisture content of forage stored in polyvinyl silos on intake and performance of diary cows, *Can. J. Anim. Sci.*, 44, 125—131, 1964.

77. Gordon, C. H., Derbyshire, J. C., Wiseman, H. G., Kane, E. A., and Melin, C. G., Preservation and feeding value of alfalfa stored as hay, haylage and direct cut silage, *J. Dairy Sci.*, 44, 1299—1311, 1961.

78. Gordon, C. H., Derbyshire, J. C., Jacobson, W. C., and Humphrey, J. L., Effects of dry matter in low-moisture silage on preservation, acceptability and feeding value for dairy cows, *J. Dairy Sci.*, 48, 1062—1068, 1965.

79. Jackson, N. and Forbes, T. J., The voluntary intake by cattle of four silages differing in dry matter content, *Anim. Prod.*, 12, 591—599, 1970.

80. Dinius, D. A., Hill, D. L., and Noller, C. H., Influence of supplementary acetate feeding on the voluntary intake of cattle fed green corn and corn silage, *J. Dairy Sci.*, 51, 1505—1507, 1968.

81. Wilkinson, J. M., Voluntary intake and efficiency of utilization of whole crop maize silage, *Anim. Feed Sci. Technol.*, 1, 441—454, 1976.

82. Thomas, J. W., Moore, L. A., Okamoto, M., and Sykes, J. F., A study of factors affecting rate of intake of heifers fed silage, *J. Dairy Sci.*, 44, 1471—1483, 1961.

83. Mayes, R. W. and L'Estrange, J. L., Effect of Water on Voluntary Feed Intake of Sheep, Proc. Int. Meet. Animal Production from Temperate Grassland, Dublin, 1977, 160.

84. Murdoch, J. C., The effect of length of silage on its voluntary intake by cattle, *J. Br. Grassl. Soc.*, 20, 54—58, 1965.

85. Dulphy, J. P. and Demarquilly, C., The effect of type of forage harvester and chopping fineness on the feeding value of silage, *Ann. Zootech.*, 22, 199—217, 1973.

86. Thomas, P. C., Kelly, N. C., and Wait, M. K., The effect of physical form of a silage on its voluntary consumption and digestibility by sheep, *J. Br. Grass. Soc.*, 31, 19—22, 1976.

87. Deswysen, A. and Vanbelle, M., Effect of Chopping Before and After Ensiling on the Voluntary Intake of Silage by Sheep and Cattle, Proc. 4th. Silage Conf., Hurley, U.K., 1976.

88. Wilkins, R. J., Hutchinson, K. J., Wilson, R. F., and Harris, C. E., The voluntary intake of silage by sheep. I. Interrelationships between silage composition and intake, *J. Agric. Sci.*, 77, 531—537, 1971.

89. Brown, D. C. and Radcliffe, J. C., Relationship between intake of silage and its chemical composition and *in vitro* digestibility, *Aust. J. Agric. Res.*, 23, 25—33, 1972.

90. McLeod, D. S., Wilkins, R. J., and Raymond, W. F., The voluntary intake by sheep and cattle of silages differing in free acid content, *J. Agric. Sci.*, 75, 311—319, 1970.

91. Hutchinson, K. J. and Wilkins, R. J., The voluntary intake of silage by sheep. II. The effects of acetate on silage intake, *J. Agric. Sci.*, 77, 539—543, 1971.

92. L'Estrange, J. L., Mayes, R. W., and Ryan, D., Effects of Organic Acids and Ammonium Salts on Feed Intake and Metabolism of Sheep, in *Proc. Int. Meet. Animal Production from Temperate Grasslands*, Dublin, 1977, 159.

93. Senal, S. H. and Owen, F. G., Relation of acetate and lactates to dry matter intake and volatile fatty acid metabolism, *J. Dairy Sci.*, 49, 1075—1079, 1966.

94. Senal, S. H. and Owen, F. G., Relation of dietary, acetic, and butyric acids to intake, digestibility, lactation performance, and ruminal and blood levels of certain metabolites, *J. Dairy Sci.*, 50, 327—333, 1967.

95. Morgan, D. J. and L'Estrange, J. L., Effect of dietary additions of hydrochloric and lactic acid on feed intake and metabolism of sheep and cattle, *Ir. J. Agric. Sci.*, 15, 55—63, 1976.

96. Morgan, D. J. and L'Estrange, J. L., The effect of lactic acid when supplemented in the feed or intraruminally upon voluntary food intake and metabolism of sheep in relation to factors which affect silage intake, *J. Br. Grassl. Soc.*, 32, 217—224, 1977.

97. Neumark, H., Bondi, A., and Volcani, R., Amines, aldehydes and keto-acids in silages and their effects on feed intake by ruminants, *J. Sci. Food Agric.*, 15, 487—492, 1964.

98. Woodward, T. B. and Shepherd, J. B., A statistical study of the influence of moisture and acidity on the palatability and fermentation losses of ensiled hay crops, *J. Dairy Sci.*, 25, 517—523, 1942.

99. Lepard, O. L., Page, E., Maynard, L. A., Rasmussen, R. A., and Savage, E. S., The effect of phosphoric acid silage on the acid-base balance in dairy cows, *J. Dairy Sci.*, 23, 1013—1022, 1940.

100. King, W. A., Comparison of limestone and sodium bicarbonate as neutralizers for phosphoric acid oat silage, *J. Dairy Sci.*, 26, 975—981, 1943.

101. Vercoe, J. E. and Blaxter, K. L., The metabolism of formic acid in sheep, *Br. J. Nutr.*, 19, 523—530, 1965.

102. McCarrick, R. B., Keane, E., and Tobin, J., The nutritive values of ammonium bisulphate and molassed silages. I. Comparisons of acceptability and feeding value for beef and dairy cattle, *Ir. J. Agric. Sci.,* 4, 115—123, 1965.

103. McCarrick, R. B., Poole, D. B. R., and Maguire, M. F., The nutritive values of ammonium bisulphate and molassed silages. II. The effects of sulphate levels on performance of growing and mature cattle, *Ir. J. Agric. Sci.,* 4, 125—133, 1965.

104. McCarrick, R. B., Maguire, M. F., Poole, D. B. R., and Spillane, T. A., The nutritive values of ammonium bisulphate and molassed silages. III. Effects of level of ammonium bisulphate applied to herbage on silage quality and animal performance, *Ir. J. Agric. Sci.,* 4, 135—142, 1965.

105. Schmekel, J., The influence of additive on silage consumption, *Lantbrukshoegsk. Ann.,* 33, 785—804, 1967.

106. L'Estrange, J. L. and Murphy, F., Effects of dietary mineral acids on voluntary food intake, digestion, mineral metabolism, acid-base balance of sheep, *Br. J. Nutr.,* 28, 1—17, 1972.

107. L'Estrange, J. L. and McNamara, T., Effects of dietary hydrochloric acid on voluntary food intake and metabolism of sheep in relation to the use of mineral acids as silage additives, *Br. J. Nutr.,* 34, 221—231, 1975.

108. L'Estrange, J. L., Clarke, J. J., and McAleese, D. M., Studies on high intakes of various sulphate salts and sulphuric acid in sheep. I. Effects on voluntary feed intake, digestibility and acid-bas balance, *Ir. J. Agric. Res.,* 133—150, 1969.

109. L'Estrange, J. L., Upton, P. K., and McAleese, D. M., Effects of dietary sulphate on voluntary feed intake and metabolism of sheep. I. A comparison between different levels of sodium sulphate and sodium chloride, *Ir. J. Agric. Res.,* 11, 127—144, 1972.

110. Upton, P. K., L'Estrange, J. L., and McAleese, D. M., Effects of dietary sulphate on voluntary feed intake and metabolism of sheep. II. Palatability and metabolic effects of sodium sulphate and sulphuric acid, *Ir. J. Agric. Res.,* 11, 145—158, 1972.

111. Upton, P. K. and L'Estrange, J. L., Effects of high intakes of dietary sodium sulphate and sodium chloride on voluntary food intake of rats, *Proc. Nutr. Soc.,* 35, 20A, 1976.

112. Lewis, D., The reduction of sulphate in the rumen of the sheep, *Biochem. J.,* 56, 391—399, 1954.

113. Anderson, C. M., The metabolism of sulphur in the rumen of the sheep, *N.Z. J. Sci. Technol.,* 37, 379—394, 1956.

114. Bird, P. R., Sulphur metabolism and excretion studies in ruminants. X. Sulphide toxicity in sheep, *Aust. J. Biol. Sci.,* 25, 1087—1098, 1972.

115. Dougherty, R. W., Mullenax, C. H., and Allison, M. J., Physiological phenomena associated with eructation in ruminants, in *Physiology of Digestion in the Ruminant,* Dougherty, R. W., Ed., Butterworths, Washington, D.C., 1965, 159—170.

116. Alhassen, W. S. and Satter, L. D., Observations on sodium sulphite administration to the ruminant, *J. Dairy Sci.,* 57, 981, 1968.

117. Keener, H. A., Teeri, A. E., Harrington, R. V., and Baldwin, R. R., Metabolic fate of S^{35} in the lactating cow when fed $S350_2$ preserved silage, *J. Dairy Sci.,* 36, 1—7, 1953.

118. Edin, H., Studies on the protein problem in consequence of the closed import. Methods for wet preservation of animal offal, *Nord. Jordbrugsforsk,* 22, 142, 1940.

119. Olsson, N., Experiments Concerning the Preservation and Use of Fish Products as Feedingstuff for Hens and Chicks, Report 7, Domestic Animal Research Institute of the Agricultural College, Lantbrokshögskolans Husdjursförsöksanstalt Experimentalfälet, Göteborg, Sweden, 1942.

120. Petersen, H., Acid preservation of fish and fish offal, *FAO Bull.,* 6, 18—26, 1953.

121. Tatterson, I. N. and Windsor, M. L., Fish Silage, *J. Sci. Food Agric.,* 25, 369—379, 1974.

122. Tatterson, I. N., The Preparation and Storage of Fish Silage, Proc. Torry Research Station Symp. Fish Silage, Aberdeen, 1976.

123. Smith, P. and Adamson, A. H., Pig Feeding Trials with White Fish and Herring Liquid Protein (Fish Silage), Proc. Torry Research Station Symp. Fish Silage, Aberdeen, 1976.

124. Carpenter, K. J., The estimation of the available lysine in animal-protein foods, *Biochem. J.,* 77, 604—610, 1960.

125. Whittemore, C. T. and Taylor, A. G., Nutritive value to the growing pig of deoiled liquefied herring offal preserved with formic acid (fish silage), *J. Sci. Food Agric.,* 27, 239—243, 1976.

126. Hillyer, G. M., Peers, D. G., Morrison, R., Parry, D. A., and Woods, M. P., Evaluation for on-farm use of deoiled herring silage as a protein feed for growing pigs, Proc. Torry Research Station Symp. Fish Silage, Aberdeen, 1976.

127. Lea, C. H., Parr, L. T., L'Estrange, J. L., and Carpenter, K. J., Nutritional effects of autoxidized fats in animal diets. III. The growth of turkeys on diets containing oxidized fish oil, *Br. J. Nutr.,* 20, 123—133, 1966.

128. Bothast, R. J., Adams, G. H., Hatfield, E. E., and Lancaster, E. B., Preservation of high-moisture corn: a microbiological evaluation, *J. Dariy Sci.,* 58, 386—391, 1975.

129. Cole, D. J. A., Dean, G. W., and Luscombe, J. R., Single cereal diets for bacon pigs. II. The effect of method of storage and preparation of barley on performance and carcass quality, *Anim. Prod.*, 12, 1—6, 1970.

130. Christensen, C. M. and Kaufmann, H., *Grain Storage: The Role of Fungi in Quality Loss,* University of Minnesota Press, Minneapolis, 1969.

131. Bee, R., Chemical conservation, *Agriculture (London),* 75, 114—118, 1968.

132. Huitson, J. J., Cereal preservation with propionic acid, *Process Biochem.,* 3, 31—32, 1968.

133. Jones, G. M., Preservation of high moisture corn with volatile fatty acids, *Can. J. Anim. Sci.,* 50, 739—741, 1970.

134. Young, L. G., Moisture content and processing of corn for pigs, *Can. J. Anim. Sci.,* 50, 705—709, 1970.

135. Young, L. G., Brown, R. G., and Sharp, B. A., Propionic acid preservation of corn for pigs, *Can. J. Anim. Sci.,* 50, 711—715, 1970.

136. Livingstone, R. M., Denerley, H., Steward, C. S., and Elsley, F. W. H., *Anim. Prod.,* 13, 547—556, 1971.

137. Fink, F., Preserving grain and shelled corn with propionic acid, *Landtechnik,* 13, 334—336, 1971.

138. Nelson, L. R. and Cummins, D. G., Effects of tannin content and temperature on storage of propionic acid treated grain sorghum, *Agron. J.,* 67, 71—73, 1975.

139. Larsen, H. J., Jorgensen, N. A., Barrington, G. P., and Niedermeier, R. P., Effect of organic acids on preservation and acceptability of high moisture corn, *J. Dairy Sci.,* 55, 685, 1972.

140. Perez-Aleman, S., Dempster, D. G., English, P. R., and Topps, J. H., Moist barley preserved with acid in the diet of the growing pig, *Anim. Prod.,* 13, 271—277, 1971.

141. Thomke, S., Feeding value of barley for fattening pigs. II. Effect of difficulties in protein content due to locality, *Z. Tierphysiol. Tierernaehr. Futtermittelkd.,* 28, 132—147, 1972.

142. Fontenot, J. P., Lucas, D. M., and Webb, K. E., Ammonium isobutyrate as a preservative for high moisture grain, *J. Anim. Sci.,* 36, 221, 1973.

143. Ware, D. R., Self, H. L., and Hoffman, M. P., Comparison of chemically preserved and artificially dried corn for finishing yearling steers, *J. Anim. Sci.,* 44, 722—728, 1977.

144. Goodrich, R. D., Byers, F. M., and Meiske, J. C., Influence of moisture content, processing and reconstitution on the fermentation of corn grain, *J. Anim. Sci.,* 41, 876—881, 1975.

145. Jones, G. M., Performance of dairy cows fed propionic acid-treated high-moisture shelled concentrations for complete rations, *J. Dairy Sci.,* 56, 207—211, 1973.

146. Young, L. G., Lun, A., and Forshaw, R. P., Selenium and α-tocopherol in stored corn, *J. Anim. Sci.,* 35, 1112, 1972.

147. Lawrence, T. L. J., Some effects of physical form and tocopherol supplementation on the acceptability and utilization of high moisture maize grain by the growing pig, in *Proc. Br. Soc. Anim. Prod.,* 4, 118, 1975.

148. Allen, W. M., Parr, W. H., Bradley, R., Swannock, K., Barton, C. R. Q., and Tyler, R., Loss of vitamin E in stored cereals in relation to a myopathy of yearling cattle, *Vet. Rec.,* 94, 373—375, 1974.

149. Holmes, J. H. G., Bayley, H. S., and Horney, F. D., Digestion and absorption of dry and high-moisture maize diets in the small and large intestine of the pig, *Br. J. Nutr.,* 30, 401—410, 1973.

150. Jones, G. M., Mowat, D. N., Elliot, J. I., and Moran, E. T., Jr., Organic acid preservation of high moisture corn and other grains and the nutritional value: a review, *Can. J. Anim. Sci.,* 54, 499—517, 1974.

151. Cole, D. J. A., Brooks, P. H., English, P. R., Livingstone, R. M., and Luscombe, J. R., Propionic acid-treated barley in the diets of bacon pigs, *Anim. Prod.,* 21, 295—302, 1975.

152. Ware, D. R., Self, H. L., and Hoffman, M. P., Comparison of chemically preserved and artificially dried corn for finishing yearling steers, *J. Anim. Sci.,* 44, 722—728, 1977.

153. Bowlands, J. P., Young, B. A., and Milligan, L. P., Influence of dietary volatile fatty acid mixtures on performance and on fat composition of growing pigs, *Can. J. Anim. Sci.,* 51, 89—94, 1971.

154. Friend, D. W., Cunningham, H. M., and Nicholson, J., The production of organic acids in the pig. II. The effect of diet on levels of volatile fatty acids and lactic acid in sections of the alimentary tract, *Can. J. Anim. Sci.,* 43, 156—168, 1963.

155. English, P. R., Topps, J. H., and Dempster, D. G., Moist barley preserved with propionic acid in the diet of the growing pig, *Anim. Prod.,* 17, 75—83, 1972.

156. Livingstone, R. M. and Livingstone, D. M. S., The use of moist barley in diets for growing pigs, *Anim. Prod.,* 12, 561—568, 1970.

157. Jones, G. M., Donefer, E., and Elliot, J. I., Feeding value for dairy cattle and pigs of high moisture corn preserved with propionic acid, *Can. J. Anim. Sci.,* 50, 483—489, 1970.

158. McKnight, D. R., MacLeod, G. K., Buchanan-Smith, J. G., and Mowat, D. N., Utilization of ensiled or acid-treated high-moisture shelled corn by cattle, *Can. J. Anim. Sci.,* 53, 491—496, 1973.

159. Klinner, W. E. and Shepperson, G., The state of haymaking technology — a review, *J. Br. Grassl. Soc.*, 30, 259—266, 1975.
160. Klinner, W. E., Design and performance characteristics of an experimental crop conditioning system for difficult climates, *J. Agric. Eng. Res.*, 20, 149—165, 1975.
161. Wood, G. M., Biggar, G. W., and Klinner, W. E., Performance of Commerical Mowing and Crop Conditioning Machines, Report 5, Paper 2, Nat. Inst. Agric. Eng., Silsoe, U.K., 1972.
162. Klinner, W. E., Mechanical and Chemical Field Treatment of Grass for Conservation. Rep. No. 21, National Institute of Agricultural Engineers, U.K., 1976.
163. Thaine, R. and Harris, C. E., Formic acid as a desiccant for grass leaves, *J. Agric. Sci.*, 80, 349—351, 1973.
164. Weineke, F. and Hartman, D. The use of propionic acid for the chemical conditioning of forage crops, *Landtech. Forsch.*, 19, 23—25, 1971.
165. Schukking, S. K., Treatment of silage and hay with organic acids, 2nd. Silage Conf., Hurley, U.K., 1972.

EFFECT OF PROCESSING ON NUTRIENT CONTENT OF FOODS AND FEEDS: WATER TREATMENT

J. O. L. King

INTRODUCTION

The water treatment of animal feeds can improve the nutritional value in several ways, and these are set out below. However, there are two general precautions which apply to foods fed soaked: (a) forcing animals to consume an excessive quantity of water may limit the intake of other nutrients because of the limited capacity of the digestive system, (b) soaked foods are liable to turn sour when kept for more than 1 or 2 days, especially during warm weather, and, although foods in such a state are not necessarily toxic, they might constitute a health risk and should not be fed.

IMPROVED DIGESTIBILITY

The soaking of hard grains, such as maize, wheat, and barley, improves digestibility largely by ensuring more complete mastication. For horses and farm animals this process has now largely been superseded by crushing and grinding, but the soaking of the food fed to old dogs, which have lost most of their teeth, is still practiced.

INCREASED SAFETY

By increasing the water content certain foods are made safer so that their nutrient contents can be utilized.

1. Dried sugar beet pulp readily absorbs water and swells and must be soaked in two or three times its weight of water for some time before it is fed, especially when used for horses. Failure to do this can result in impaction of some part of the alimentary canal because the dry pulp absorbs water from the canal wall and swells rapidly causing a blockage.
2. Dusty foods, particularly when dry and powdery, can cause respiratory distress. This is observed most frequently in horses. Products such as grassmeal tend to produce an excessively dusty ration and this effect can be reduced by damping the food before it is fed or feeding it as a wet mash.
3. Parakeratosis, a disease of pigs in which there is a check in growth and the development of typical skin lesions, caused by a zinc deficiency is seen when certain rations are fed dry to fattening pigs but is not usually observed when these rations are fed wet. The reason is not clearly understood.

PRODUCTION OF A LAXATIVE EFFECT

Bran has a very mild laxative effect when fed dry, but when soaked in water the laxative effect is more pronounced.[1] This special merit of bran mash is its capacity to produce bulky moist feces which are regularly and completely evacuated without causing purging. Bran mashes are often fed to animals of all species at the time of parturition, and are used to prevent disorders in working horses when idle and confined to a stable, for instance over weekends.

INCREASED RATE OF FOOD INTAKE

When high yielding cows are fed in milking parlors during milking it is difficult for a satisfactory amount of concentrate food to be consumed in the time available. If yields are not to be reduced either extra feeding outside the parlor is needed, involving additional labor, or the cows must be allowed to remain in the parlor after they have been milked, which delays the milking operation. As cows can drink faster than they can eat, they can take in much more nutrient in liquid form than in a solid state in a given time.

In one trial with four different water-to-concentrate ratios (air-dry, 0.5, 1.0, and 1.5 units of water per unit of concentrate) the data showed that as the ratio of water to concentrate increased there was a decrease in eating time and that the method of eating changed.[2] The tongue was placed in the air-dry concentrate and the amount that adhered to the tongue was drawn into the mouth. The cows used their tongues to pull the sticky concentrate into their mouths when 0.5 of a unit of water was added per unit of concentrate. Concentrates with a water-to-concentrate ratio of 1:1 were completely wetted, and the cows put their mouths into the feed and swallowed the amount taken into their mouths. The cows drank the slurry concentrates (water-to-concentrate ratio of 1.5:1) and used their tongues to clean the manger. In another experiment water was added to a concentrate mix at the rate of 1.5 units of water per unit of feed and the water addition increased the eating rate significantly.[3] On the dry mix the cows ate 0.28 kg/min, whereas the rate increased to 0.40 kg with the addition of water.

More recent work has shown that a concentration of 1 unit of meal mixed in 2.5 units of water produced a wet slurry suitable for drinking.[4] Once the amount of concentrate approaches one unit per two units of water there is a tendency for cows to lick the material and the rate of consumption is lowered. Farmers with rotary parlors, which generally leave cows less time to take in their ration than static ones, have found liquid feeding particularly advantageous.

IMPROVED PRODUCTIVITY

Beef Cattle

Workers determined the effect of three moisture concentrations (10, 50, and 75%) on the nutritive value of two complete fattening diets containing 25 and 45% of poor roughage fed to young cattle. They concluded that the soaking of such diets for any length of time tended to improve their digestibility and energy value.[5] The concentration of propionic acid in rumen liquor was increased significantly. In a further experiment intact male Israeli-Friesian cattle were grown from an initial weight of about 300 kg to a slaughter weight of about 520 kg. Live weight gains were 956, 1080, and 1025 g/day and carcass gains were 516, 584, and 563 g/day for diet moisture contents of 10, 50, and 75%, respectively.[6] This improvement in performance obtained by increasing the water content was attributed to the accumulated effect of three factors, increased dry matter intake, improved digestibility, and an increased concentration of propionic acid. Similar experiments conducted with Zebu cattle and buffalo showed that presoaking of wheat straw for 1 to 2 hr with ordinary tap water may give an overall beneficial effect. The feed intake was improved significantly and the availability of digestible energy and metabolizable energy was increased.[7] It was also found that the production rates of total volatile fatty acids and acetic acid were higher in the animals fed on water-soaked wheat straw.

Pigs

In the early part of this century the meal given to pigs was commonly soaked in water before feeding and, although this practice lost popularity it is now being used again because labor can be saved by soaking meal in a tank and circulating it to the feeding troughs via a pipeline. For this reason a number of workers have compared wet and dry feeding in pigs, most indicating that water treatment has a beneficial effect. In one experiment a diet was fed dry, wetted (1:1 feed and water) and soaked (12 hr, 1:1 feed and water).[8] Wetting or soaking the diet increased the rate and efficiency of weight gain by limited-fed pigs, but only the rate of weight gain of those full-fed.

Other investigators reporting the results of a coordinated trial carried out at 18 centers showed that pigs fed a restricted amount of food dry (with water available *ad libitum*) took on an average about 10 days longer to reach bacon weight and required nearly 0.09 kg more food per 0.45 kg live weight gain than did pigs fed the same amount of food mixed with water.[9]

Opposing results were obtained by other workers who compared pigs fed on dry meal with pigs fed meal soaked for 24 hr and recorded no significant differences in daily live weight gain and food conversion efficiency.[10,11] However, in one of these experiments these two parameters were improved by the use of a mixture of skim milk and water,[10] and in the other small improvements in feed conversion efficiency were recorded when soaked meal was fed through a pipeline system.[11] One of these investigations demonstrated that pig feed left to soak in water soon became acidic and that the production of lactic acid by *lactobacilli* prevented multiplication of potential pathogens in the meal.[11] These organisms, which include the *Escherichia coli* group, require a less acid medium for their development.

The amount of water added to meal when wet feeding is practiced appears to have only slight effects on the performance of the pigs and the ratio of four parts of water to one part of food required for satisfactory pumping in mechanized liquid feeding systems does not result in a substantially worse performance than a ratio of 2.5 parts to 1. However, a ratio of water to dry feed of 6:1 may reduce live weight gain and food intake,[12] while a ratio of 1.25:1 will decrease the rate of weight gain. A comparison between a basal ration fed wetted, just before feeding, or soaked for 6 hr during the day and 16 hr during the night in a warm environment showed that pigs receiving the soaked diet grew more rapidly and used their food more efficiently than pigs fed the ration wetted.[14]

Poultry

Eight wheat samples were improved significantly (13 to 19%) in feeding value for chicks by water treatment and chick feeding experiments with wheat milling fractions (flour, bran, germ, shorts, middlings, and wheat millfeed) showed that water treatment improved the utilization of all milling fractions, except flour.[15] When nitrogen retention of chicks fed basal rations was compared with that of chicks which grew significantly better on rations containing water treated grain it was found that water treatment increased energy utilization without affecting protein utilization.[16] In another experiment the increased metabolizable energy of water treated barley was found to be due to greater utilization of the fat, nitrogen, and nitrogen-free extract in the barley.[17] A higher percentage of body fat was found in chicks and poults receiving water treated barley diets, although the greater average weight was not due solely to fat deposition.[18] As the improvement in nutritive value achieved by water treatment of wheat and its components resided largely in the starchy portion of the grain the water treatment of starch was evaluated.[19] The soaking of potato starch improved growth, food conversion, and metabolizable energy value in chick feeding trials.

Several suggestions have been made to account for these improvements noted in poultry feeding experiments. One is that the soaking improves the nutritive value because of better availability due to a breakdown of cellular structure.[17] This view is supported by the fact that potato starch, which is improved by soaking, has large granules.[19] Another is that enzymes or other factors produced by microorganisms bring about the nutritional improvement of barley by water treatment.[20,21] Enzyme supplementation has been found to produce similar effects.[21,22] However, the addition of enzymes to water-treated barley does not improve its nutritive value above that of the water-treated barley alone, and an assay of water-treated barley showed it to contain approximately 33 g of bacitracin-like activity per 1016 kg.[22,23] This result indicated that the superiority of water-treated barley over enzyme supplements was due to the presence of an effective antibiotic in the treated barley for chick growth.

INCREASED EFFICACY OF GROWTH ADDITIVES

The volume of water added to meal diets fed to pigs can influence the efficacy of growth promoting additives. In one experiment using a ration supplemented with 20 mg of oxytetracycline hydrochloride per 0.45 kg pigs which received the meal mixed with 2.5 parts of water per 1 part of meal grew significantly more rapidly between the 9th and 15th weeks of age than did pigs which were given the same meal with 3 parts of water per 1 part of meal.[24] In an experiment of similar design using a diet containing 0.1% of copper sulfate pigs between weaning and the attainment of a bodyweight of 45 kg grew significantly more rapidly and utilized their food significantly more efficiently when 3 parts of water were used per part of meal than did others given 2.5 parts of water per 1 part of meal.[25]

The length of time which water is left in contact with a meal containing a growth additive may also influence the response. In one experiment a basal diet was compared with this diet fortified with 20 mg of oxytetracycline hydrochloride or with 0.1% of copper sulfate.[14] The findings indicate that between weaning and bacon weight (90 kg) the pigs on the antibiotic supplemented diet soaked in 3 units of water per unit of meal for 6 hr during the day and 16 hr during the night gained in average weight slightly, but not significantly, more rapidly than did others on this diet wetted with water in the same proportion before feeding. Pigs receiving the copper sulfate fortified ration soaked grew significantly more rapidly than those fed this ration wetted. Up to 45 kg live weight food conversion efficiency was also significantly improved, but not subsequently or over the whole period. It is possible that the advantageous effect of soaking copper containing rations was due to an increase in the rate of absorption of copper sulfate as it went into solution in the soaking water.

IMPROVED FOOD UTILIZATION BY DRYING AFTER SOAKING

There is some evidence that the soaking of a meal in water followed by drying improves food utilization. One of the first papers on this subject showed that barley soaked in tap water for 8 hr and then dried at 70°C improved both growth rate and food utilization in young chickens compared with diets containing untreated barleys.[26] Further work has confirmed that the nutritional improvement of feed for chicks obtained by treatment with water can be carried through when the feed is subsequently dried.[27] When dried soaked barley replaced one eighth, one quarter, one half, and all the barley in a pelleted diet containing 3% of animal fat, improved production efficiency was obtained as the level increased, with optimal results at the half replacement level.[28]

IMPROVED FEEDING VALUE BY GERMINATION AND DRYING

The feeding value of grains can be markedly improved by soaking in water and germination, possibly followed by drying. The nutritive value of corn was significantly improved when the grain was partially germinated, dried, and then included in chick diets.[29] The response from wheat approached significance but little or no response was obtained from germinated barley.

REFERENCES

1. **Sheehy, E. J.**, *Animal Nutrition*, Macmillan, New York, 1955, 226.
2. **Dalton, H. L., Huffman, C. F., and Ralston, N. P.**, The effect of feeding concentrates with different degrees of fineness and water contents on the eating and milking times in dairy cattle, *J. Dairy Sci.*, 36, 1279—1284, 1953.
3. **Hupp, E. W. and Lewis, R. C.**, Effect of adding water to the concentrate mix on maximum milking rate, average milking rate and eating rate, *J. Dairy Sci.*, 41, 724, 1958.
4. **Clough, P. A.**, Feeding 'porridge' in the parlor, *Dairy Farmer*, 19, 18-21, 1972.
5. **Holzer, Z., Levy, D., Tagari, H., and Volcani, R.**, Soaking of complete fattening rations high in poor roughage. I. The effect of moisture content and spontaneous fermentation on nutritional value, *Anim. Prod.*, 21, 323—335, 1975.
6. **Holzer, Z., Tagari, H., Levy, D., and Volcani, R.**, Soaking of complete fattening rations high in poor roughage. II. The effect of moisture content and of particle size of the roughage component on the performance of male cattle, *Anim. Prod.*, 22, 41—53, 1976.
7. **Chaturvedi, M. L., Singh, V. B., and Ranjhan, S. K.**, Effect of feeding water-soaked and dry wheat straw on feed intake, digestibility of nutrients and VFA production in growing Zebu and buffalo calves, *J. Agric. Sci.*, 80, 393—397, 1973.
8. **Becker, D. E., Jensen, A. H., Harmon, B. G., Norton, H. W., and Breidenstein, B. C.**, Effect of restricted diet on the performance of finishing pigs, *J. Anim. Sci.*, 22, 1116, 1963.
9. **Braude, R. and Rowell, J. G.**, Comparison of dry and wet feeding of growing pigs, *J. Agric. Sci.*, 68, 325—330, 1967.
10. **Barber, R. S., Braude, R., and Mitchell, K. G.**, Effect of soaking the meal ration of growing pigs in water or skim milk, *Anim. Prod.*, 4. 313—318, 1962.
11. **Smith, P. A.**, A comparison of dry, wet and soaked meal for fattening bacon pigs, *Exp. Husb.*, 30, 87—94, 1976.
12. **Rerat, A. and Ferrier, C.**, Influence du taux d'hydretation du regime sur la crossance et la composition corporella du porc, *Ann. Zootech.*, 14, 39, 1965.
13. **Cunningham, H. M. and Friend, D. W.**, Studies of water restriction on nitrogen retention and carcass composition of pigs, *J. Anim. Sci.*, 25, 663—667, 1966.
14. **King, J. O. L.**, Influence of the water treatment of rations supplemented with copper sulphate and an antibiotic on the growth rate of pigs, *Exp. Husb.*, 18, 25-31, 1969.
15. **Leong, K. C., Jensen, L. S., and McGinnis, J.**, Improvement of the feeding value of wheat fractions for poultry, *Poult. Sci.*, 39, 1269, 1960.
16. **Adams, O. L. and Naber, E. C.**, Studies on the mechanism of the chick growth promoting effect achieved by water treatment of grains and their components, *Poult. Sci.*, 40, 1369, 1961.
17. **Stutz, M. W., Matterson, L. D., and Potter, L. M.**, Metabolizable energy of barley for chicks as influenced by water treatment or by presence of fungal enzyme, *Poult. Sci.*, 40, 1462, 1961.
18. **Willingham, H. E.**, Effect of enzymes and water treated barley on carcass composition, water and feed consumption and faeces moisture of chicks and poults, *Poult. Sci.*, 43, 137, 1964.
19. **Naber, E. C., Saini, S. S., and Touchburn, S. P.**, The effect of hydration and gelatinization of starch on growth and energy utilization by the chick, *Poult. Sci.*, 41, 1669, 1962.
20. **Thomas, J. M., Jensen, L. S., Leong, K. C., and McGinnis, J.**, Role of microbial fermentation in improvement of barley by water treatment, *Proc. Soc. Exp. Biol. Med.*, 103, 198-200, 1960.
21. **Willingham, H. E., Jensen, L. S., and McGinnis, J.**, Studies on the role of enzyme supplements and water treatment for improving the nutritional value of barley, *Poult. Sci.*, 38, 539-544, 1959.
22. **Jensen, L. S., Fry, R. E., Allred, J. B., and McGinnis, J.**, Improvement in the nutritional value of barley for chicks by enzyme supplementation, *Poult. Sci.*, 36, 919-921, 1957.
23. **Willingham, H. E., McGinnis, J., Nelson, F., and Jensen, L. S.**, Relation of superiority of water treated barley over enzyme supplements to antibiotics, *Poult. Sci.*, 39, 1307, 1960.
24. **King, J. O. L.**, The effect of water intake on the response of growing pigs to an antibiotic supplement in the diet, *Vet. Rec.*, 72, 1090—1092, 1960.

25. King, J. O. L., The effect of water intake on the efficacy of copper sulphate as a growth stimulant for pigs, *Vet. Rec.,* 75, 651-653, 1963.

26. Fry, R. E., Allred, J. B., Jensen, L. S., and McGinnis, J., Influence of water treatment on nutritional value of barley, *Proc. Soc. Exp. Biol. Med.,* 95, 249-251, 1957.

27. Lepkovsky, S. and Furuta, F., The effect of water treatment of feeds upon the nutritional values of feeds, *Poult. Sci.,* 39, 394-398, 1960.

28. Rose, R. J. and Arscott, G. H., Further studies on the use of enzymes, soaking and pelleting barley for chicks, *Poult. Sci.,* 39, 1288, 1960.

29. Adams, O. L. and Naber, E. C., Effect of physical and chemical treatment of grains on growth of and feed utilization by the chick, *Poult. Sci.,* 48, 853-858, 1969.

EFFECT OF PROCESSING ON NUTRITIVE VALUE OF FEEDS: ENZYME TREATMENT

Olav Herstad

INTRODUCTION

Enzymatic digestion of ingested food is necessary before nutrients can be absorbed in the intestine. Usually, digestive enzymes may be secreted in sufficient amounts in the alimentary canal. Additionally, there are enzymes in certain feeds which may facilitate the digestive processes. Ungerminated cereals, for instance, contain an abundance of β-amylase. During germination of grains an appreciable supply of α-amylase is produced. Both of these enzymes are important in digestion of starch.

Certain molds and bacteria are capable of producing large quantities of enzymes. Most of the commercially available microbial enzymes used in food processing have been derived from strains of the *Aspergillus flavous-oryzae* group, the *Aspergillus niger* group and of *Bacillus subtilis*. However, many other types of nonpathogenic and nontoxic organisms could well be used as sources of useful enzymes. Commercial enzymes are usually standardized with respect to functional enzyme activity for particular purposes.[1]

Supplementary enzyme preparates are widely used by bakeries and breweries to convert grain starch into sugars for yeast fermentation. In processing of animal feed, enzyme treatment has up to now been used to only a limited extent.

ENZYME SUPPLEMENTATION OF FEEDS

In animal nutrition, enzyme supplementation of feeds has been used to counteract insufficiency in the animal's own enzyme capacity and thereby to increase the digestibility and feed efficiency. During the 1960s, numerous experiments on enzyme-supplemented feeds were carried out. In Table 1 a survey of experiments on enzyme treatments of feed performed since the mid-1950s is shown.

The data in Table 1 do not comprise a complete bibliography, but show clearly that the interest has been concentrated on barley fed to chicks. This will be dealt with in a separate chapter. Besides, the positive effect of phytase treatment of some feeds for *chicks* should be stressed. Most vegetable feeds are rich in phosphorus but 50 to 80% of it is present as phytin phosphorus, which is known to be poorly digested by chicks. It has been found that supplementing cottonseed meal, soybean meal, and wheat bran rations with a phytase (50 mℓ/kg of feed) obtained from *Aspergillus ficcum* hydrolyzed the phytin in the meals almost completely.[26,27] Chicks are able to utilize hydrolyzed phytate phosphorus as efficiently as inorganic phosphorus.[46] Consequently, the need to supply poultry rations with inorganic phosphorus may be reduced by enzyme treatment of vegetable feeds. Phytase hydrolysis has been found to cause a reduction in the gossypol toxicity of glanded cottonseed meal and further to reduce the zinc requirement of chicks. Phytase treatment also may increase the metabolizable energy (ME) content in feeds, though conflicting results with soybeans are reported.[26,27]

For *pigs* there seem to be only very limited data available on enzyme-supplemented feeds. It has been found that baby pigs may be deficient in proteolytic and amylolytic digestive enzymes.[30,47] However, supplementation of starter diets containing soybean protein, fish protein, and starch with protease and amylase has failed to influence the digestibility in baby pigs.[31,32] Considerable increase in weight after protease supplementation of soybean protein and casein diets was found in an experiment reported by

Table 1
SURVEY OF EXPERIMENTS WITH ENZYME-SUPPLEMENTED FEED

Species	Diets	Enzyme	Performance	Ref. Positive effect	Ref. No effect
Chicks	Barley	Amylase	Weight gain	2—9	9,10
			Digestibility	11,17	
		M.U.E.[a]	Weight gain	12—19	20—22
			Digestibility	23	
		Glucanase	Weight gain	24 (high viscose barley)	24 (low viscous barley)
	Corn	Amylase	Weight gain		10,16
		M.U.E.	Weight gain		17,25
		M.U.E.	Digestibility		25
	Rye	M.U.E.	Weight/Digest.	25	
	Wheat	M.U.E.	Weight gain		10,16
	Wheat bran	Phytase	Digestibility	26	
	Soybean meal	Phytase	Digestibility	27	26 (Neg. on ME)
	Cottonseed meal	Phytase	Digestibility	26,27	
Laying hens	Barley	M.U.E.	Laying		10,18
	Corn/Milo	M.U.E.	Laying	28,29	
Baby pigs	Soybean	Protease	Weight gain	30	31
	Casein	Protease	Digestibility		31
	Fish-protein	Protease	Digestibility		
	Starch	Amylase	Digestibility		32
Calves	Milk replacer	M.U.E.	Weight gain		33,34
		Protease	Weight gain	35	
		Amylase	Weight gain		35
Cattle	Alfalfa	M.U.E.	Weight gain	36,37,38	39
	Grass hay	M.U.E.	Digestibility		36
	Silage	Cellulase	Weight gain		40,41
	Grain	Protease	Digestibility	42	
Sheep	Cottonseed hulls	M.U.E.	Digestibility	43	44
		Protease	Digestibility	45	
	Roughages	Amylase	Digestibility		39
	Alfalfa	Cellulase	Digestibility		36
	Grain				

[a] M.U.E. = Multiple or unspecified enzyme preparates.

Lewis et al.[30] With older pigs, Burnett and Neil[48] obtained results suggesting that enzyme supplementation may have little value for practical purposes. Similarly, enzyme supplementation of milk-replacers for *young calves* has been tried without convincing results. In *cattle* and *sheep,* the addition of enzymes to alfalfa-corn diets,[36,37] silage,[38] cottonseed hulls,[43] and roughage[45] has revealed a positive effect on growth rate and digestibility. Other experiments, however, have failed to yield a positive response.[39-41,44]

In most of the studies reporting beneficial effect of enzyme-supplemented feeds, growth rate has been the trait most affected. It is not clear how much this increased growth rate can be associated with increased digestibility. Probably, the enzyme preparates which affect weight gain, may contain growth promoting factors other than the enzyme per se.

In recent years, very few investigations on enzyme treatment of animal feed have been published in western languages. However, according to available abstracts, several reports from eastern Europe indicate positive response from enzyme-supplemented feed for broilers,[50] laying hens,[51,52] pigs,[53,54] and cattle.[55-57]

ENZYME TREATMENT OF BARLEY FOR CHICKS

From the literature, it seems that enzyme supplementation of barley for use in chick rations has been studied with special interests. The reports in Table 1 show that enzyme supplementation in most cases has increased the nutritional value of barley, while other grains, i.e., corn and wheat, have been unaffected. The responses have been shown to be dependent on the geographical area and weather conditions where the barley is grown.[9] Generally, barley grown under arid conditions, as on the Pacific coast of the U.S., has responded positively to enzyme supplementation, while the effect on barley grown in more humid areas has been more inconsistent. The response has also been more pronounced for younger than for older birds.[14,18]

The most successful enzyme preparates used have been those containing the starch-splitting α-amylase given in amounts of 0.1 to 0.3% of the diet. Results similar to those found for enzyme supplementation have also been obtained for water-treated or soaked barley.[2,9,10,12,16,19] Water treatment improves the enzyme content in the grain and this has been considered, at least partly, as explaining the positive effect of water treatments, though other factors may also be involved.

It has been generally observed that chicks and hens fed barley produce, to a greater or less extent, sticky, wet droppings which adhere to wire floors or lead to poor litter conditions. This problem is often found to be reduced with enzyme supplementation and water treatment.[2-5,8]

Burnett[58] found that production of sticky droppings was connected to a glucan and β-glucan in the hemicellulose fraction of barley as well as the content of endo-β-glucanase. In low-enzyme barley the glucans gave rise to fairly stable highly viscous conditions in the small intestine. Supplementing such barley with enzyme preparates containing β-glucanase improved the litter condition.

Also, Gohl et al.[59] found that the viscosity of barley was mainly caused by a water-soluble glucan but that a water-soluble arabinoxylan might be a contributing factor. The factors causing the viscosity were concentrated in the layers between the bran and the center of the kernels.

Besides the influence of weather conditions, the viscosity is found to be dependent on stage of ripeness. Early harvested barley is generally more viscous than late harvested. A relationship appears to exist between viscosity and nutritional value of barley.[24,60]

The viscosity of barley decreases with water treatment or when an enzyme capable of hydrolyzing β-glucan is added. Gohl[24] showed that weight gain and feed conversion by broilers increased when high viscosity barley was water-treated or supplemented with β-glucanase. These treatments were found to have no effect with medium-viscosity barley.

A relationship between enzyme content in barley and viscosity was reported by Munck,[61] who also showed that sprouting in the field during the harvesting season, one of the obvious results of humid climate, reflects a high α-amylase content in the seed. The enzyme content was strongly linked to harvest time, and large varietal differences in viscosity and enzyme content occurred. As the author pointed out, it is unlikely that α-amylase as such contributes much to reduction in viscosity. An increase in the α-amylase activity is, however, likely to be correlated with enhancement of other enzymes such as proteases and peptidases as well as β-glucanases and other carbohydrases.

The nutritive value of barley for chicken is, as mentioned above, dependent on several factors, such as growing conditions, harvesting time and variety. For commercially available barley, information about these factors may be difficult to obtain. A simple viscosity test should, however, give a good indication as to whether the barley is suitable for chicken feed without any treatment or if enzyme supplementation is needed.

The viscosity can be measured by different methods. The method used by Gohl et al.[62] was described as follows: "Viscosity was measured in an Eprecht Rheomat 15, using system "B" at pH 3.0 and 35°C. The viscosity was measured in a suspension of 20 g of the abraded material mixed with 70 mℓ 0.1 *M* phosphate-citrate buffer. Care was taken to measure the viscosity at the same interval (5 min) after mixing the meal with the buffer."

A barley with viscosity of 1.35 cp was characterized as "high viscosity barley", which responded positively with regard to growth rate of broilers upon supplementation with 0.2 g β-glucanase per kilogram of feed.[24]

Viscosity problems in barley are certainly more serious for chicks than for other animals, because the birds may have a low water content in the intestine and also low enzyme activity in the upper part of their digestive tract.[63] Munck[61] stresses that while viscous barley causes problems when fed to chicks it may be a positive value in pigs by preventing diarrhea.

REFERENCES

1. **Beckhorn, E. J., Labbee, M. P., and Underkofler, L. A.,** Production and use of microbial enzymes for food processing, *J. Agric. Food Chem.,* 13, 30—34, 1965.
2. **Arscott, G. H., Rose, R. J., and Harper, J. A.,** An apparent inhibitor in barley influencing efficiency of utilization by chicks, *Poult. Sci.,* 39, 268-270, 1960.
3. **Arscott, G. H. and Rose, R. J.,** Use of barley in high-efficiency broiler rations. IV. Influence of amylolytie enzymes on efficiency of utilization, water consumption and litter condition, *Poult. Sci.,* 39, 93-95, 1960.
4. **Arscott, G. H.,** Use of barley in high-efficiency broiler rations. VI. Influence of small amounts of corn on improvement of barley, *Poult. Sci.,* 42, 301-304, 1963.
5. **Arscott, G. H., Hutto, D. C., and Rachapaetayakom, P.,** Use of barley in high-efficiency broiler rations. VII. Pancreatic enlargement in chicks fed barley containing diets, *Poult. Sci.,* 44, 432-434, 1965.
6. **Baelum, J. and Petersen, V. E.,** II. Forsøg med slagtekyllinger, *Forsøgslaboratoriets Årbog,* Copenhagen, 191-214, 1971.
7. **Herstad, O. and McNab, J. M.,** The effect of heat treatment and enzyme supplementation on the nutritive value of barley for broiler chicks, *Br. Poult. Sci.,* 16, 1-8, 1975.
8. **Rose, R. J. and Arscott, G. H.,** Use of barley in high-efficiency broiler rations. V. Further studies on the use of enzymes, soaking and pelleting barley for chicks, *Poult. Sci.,* 41, 125-130, 1962.
9. **Willingham, H. E., Leong, K. C., Jensen, L. S., and McGinnis, J.,** Influence of geographical area of production on response of different barley samples to enzyme supplements or water treatment, *Poult. Sci.,* 39, 103-108, 1960.
10. **Adams, O. L. and Naber, E.,** Effect of physical and chemical treatment of grains on growth of and feed utilization by the chick, *Poult. Sci.,* 48, 853—858, 1969.
11. **Leong, K. C., Jensen, L. S., and McGinnis, J.,** Effect of water treatment and enzyme supplementation on the metabolisable energy of barley, *Poult. Sci.,* 41, 36-39, 1962.
12. **Anderson, J. O., Dobson, D. C., and Wagstaff, R. K.,** Studies on the value of hulless barley in chick diets and means of increasing this value, *Poult. Sci.,* 40, 1571—1583, 1961.
13. **Berg, L. R.,** Enzyme supplementation of barley diets for laying hens, *Poult. Sci.,* 38, 1132—1139, 1959.
14. **Berg, L. R.,** Effect of adding enzymes to barley diets at different ages of pullets on laying house performance, *Poult. Sci.,* 40, 34—39, 1961.
15. **Daghir, N. J. and Rottensten, K.,** The influence of variety and enzyme supplementation on the nutritional value of barley for chicks, *Br. Poult. Sci.,* 7, 159—163, 1966.

16. **Fry, R. E., Allred, J. B., Jensen, L. S., and McGinnis, J.**, Influence of enzyme supplementation and water treatment on the nutritional value of different grains for poults, *Poult. Sci.,* 37, 372—375, 1958.

17. **Leong, K. C., Jensen, L. S., and McGinnis, J.**, Influence of fibre content of diet on chick growth response to enzyme supplements, *Poult. Sci.,* 40, 615—619, 1961.

18. **Petersen, C. F. and Sauter, E. A.**, Enzyme sources and their value in barley rations for chick growth and egg production, *Poult. Sci.,* 47, 1219—1224, 1968.

19. **Willingham, H. E., Jensen, L. S., and McGinnis, J.**, Studies on the role of enzyme supplements and water treatment for improving the nutritional value of barley, *Poult. Sci.,* 38, 539—544, 1959.

20. **Baelum, J.**, Forsøg med slagtekyllinger, *Forsøgslaboratoriets Årbog,* Copenhagen, 300—301, 1962.

21. **Frölich, A.**, Some nutritional effects on chick growth, *K. Lantbrukshögsk. Annaler,* 28, 43—47, 1962.

22. **Herstad, O., Raastad, N., and Höie, J.**, Enzymtilskott til kraftforblandinger for kyllinger, *Meld. Nor. Landbrukschöegsk.,* 45(4), 1—11, 1966.

23. **Potter, L. M., Stutz, M. W., and Matterson, L. D.**, Metabolizable energy and digestibility coefficients of barley for chicks as influenced by water treatment or by presence of fungal enzyme, *Poult. Sci.,* 44, 565—573, 1965.

24. **Gohl, B., Aldén, S., Elwinger, K., and Thomke, S.**, Influence of beta-glucanase on feeding value of barley for poultry and moisture content of excreta, *Br. Poult. Sci.,* 19, 41—47, 1978.

25. **Moran, E. T., Jr., Lall, S. P., and Summers, J. D.**, The feeding value of rye for the growing chick: effect of enzyme supplements, antibiotics, autoclaving and geographical area of production, *Poult. Sci.,* 48, 939—949, 1969.

26. **Miles, R. D., Jr. and Nelson, T. S.**, The effect of the enzymatic hydrolysis of phytate on the available energy content of feed ingredients for chicks and rats, *Poult. Sci.,* 53, 1714—1717, 1974.

27. **Rojas, S. W. and Scott, M. L.**, Factors affecting the nutritive value of cottonseed meal as a protein source in chick diets, *Poult. Sci.,* 48, 819—835, 1969.

28. **Ely, C. M.**, Factors influencing laying hen response to enzyme supplements, *Poult. Sci.,* 42, 1266, 1963.

29. **Gleaves, E. W. and Dewan, S.**, Influence of a fungal-enzyme in corn and milo layer rations, *Poult. Sci.,* 49, 596-598, 1970.

30. **Lewis, C. J., Catron, D. V., Liu, C. H., Speer, V. C., and Ashton, G. C.**, Enzyme supplementation of baby pig diets, *J. Agric. Food Chem.,* 3, 1047—1050, 1955.

31. **Cunningham, H. M. and Brisson, G. J.**, The effect of proteolytic enzymes on the utilization of animal and plant proteins by newborn pigs and the response to predigested protein, *J. Anim. Sci.,* 16, 568—573, 1957.

32. **Cunningham, H. M. and Brisson, G. J.**, The effect of amylases on the digestibility of starch by baby pigs, *J. Anim. Sci.,* 16, 370—376, 1957.

33. **Fries, G. F., Lassiter, C. A., and Huffman, C. F.**, The effect of enzyme supplementation of milk replacers on the growth rates of calves, *J. Dairy Sci.,* 41, 1081—1087, 1958.

34. **Yang, M. G., Bush, L. J., and Odell, G. V.**, Enzyme supplementation of rations for dairy calves, *J. Agric. Food Chem.,* 10, 322—324, 1961.

35. **Rust, J. W., Jacobson, N. L., and McGilliard, A. D.**, Supplementation of dairy calf diets with enzymes. I. Effect on rate of growth, *J. Anim. Sci.,* 22, 1104—1108, 1963.

36. **Burroughs, W., Woods, W., Ewing, S. A., Greig, J., and Theurer, B.**, Enzyme addition to fattening cattle rations, *J. Anim. Sci.,* 19, 458—464, 1960.

37. **Nelson, L. F. and Catron, D. V.**, Comparison of different supplemental enzymes with and without diethylstilbestrol for fattening cattle, *J. Anim. Sci.,* 19, 1279, 1960.

38. **Rovics, J. J. and Ely, C. M.**, Response of beef cattle to enzyme supplements, *J. Anim. Sci.,* 21, 1012, 1962.

39. **Ward, J. K., Richardson, D., and Tsien, W. S.**, Value of added enzyme preparations in beef cattle rations, *J. Anim. Sci.,* 19, 1298, 1960.

40. **Heinemann, W. W.**, Grazing and drylot cattle fed grain and a cellulolytic enzyme, *J. Anim. Sci.,* 23, 451—453, 1964.

41. **Leatherwood, J. M., Mochrie, R. D., and Thomas, W. F.**, Some effects of a supplementary cellulase preparation on feed utilization by ruminants, *J. Dairy Sci.,* 43, 1460—1464, 1960.

42. **Rust, J. W., Jacobson, N. L., McGilliard, A. D., and Hatchkiss, D. K.**, Supplementation of dairy calf diets with enzymes. II. Effect on nutrient utilization and on composition of rumen fluid, *J. Anim. Sci.,* 24, 156—160, 1965.

43. **Grainger, R. B. and Stroud, J. W.**, Effect of enzymes on nutrient digestion by wethers, *J. Anim. Sci.,* 19, 1263—1264, 1960.

44. **Theurer, B., Woods, W., and Burroughs, W.**, Influence of enzyme supplements in lamb fattening rations, *J. Anim. Sci.,* 22, 150—154, 1963.

45. Ralston, A. T., Church, D. C., and Oldfield, J. E., Effect of enzymes on digestibility of low quality roughages, *J. Anim. Sci.,* 21, 306—308, 1962.

46. Nelson, T. S., Shier, T. R., Wodzinski, R. J., and Ware, J. H., The availability of pythate phosphorus in soybean meal before and after treatment with a mold phytase, *Poult. Sci.,* 47, 1842-1848, 1968.

47. Hudman, D. B., Douglas, W. F., Hartman, P. A., Ashton, G. C., and Catron, D. V., Digestive enzymes of the baby pig. Pancreatic and salivary amylase, *J. Agric. Food Chem.,* 5, 691—693, 1957.

48. Burnett, G. S. and Neil, E. L., The influence of processing and of certain crude enzyme preparations on the utilization of cereals by pigs, *Anim. Prod.,* 6, 237—244, 1964.

49. Theurer, B., Woods, W., and Burroughs, W., In vitro studies on the proteolytic activity of enzyme preparations, *J. Anim. Sci.,* 22, 146—149, 1963.

50. Anderson, P. and Mitrevits, E., Cytorosemin in broiler rations, *Nutr. Abstr. Rev.,* 46, 811, 1966.

51. Tolokonnikov, Yu and Berezhnova, L. M., Effect of enzyme preparations on production of eggs and meat by laying hens, *Nutr. Abstr. Rev.,* 46, 705, 1976.

52. Zakirov, M. Z., Nazarenko, P. P., Akhmedzhanov, Sh. A., and Mirzarakhimova, M., Effect of the enzyme preparation oriyzin on egg production of hens, *Nutr. Abstr. Rev.,* 47(B), 353, 1977.

53. Grigorov, V., Orekhova, T., Malysheva, L., and Burdyugova, V., Premixes with enzymes in diets for pigs, *Nutr. Abstr. Rev.,* 47(B), 488, 1977.

54. Poleacu, J., Vintila, M., and Mihai, I., Enzyme preparations in the feed of early weaned and intensively fattened piglets, *Nutr. Abstr. Rev.,* 47B, 61, 1977.

55. Maslov, D., Lysenko, N., Umnov, A., Kokovich, N., and Yurchenko, L., Biologically active substances in diets for fattening cattle, *Nutr. Abstr. Rev.,* 47B, 327, 1977.

56. Tolokonnikov, Yu. A. and Torzhkov, N. J., Effect of bacterial enzyme preparations on the productivity of cattle and the quality of beef, *Nutr. Abstr. Rev.,* 47B, 257, 1977.

57. Skvortsov, V. A. and Kudrin, A. G., Activity of oxidizing enzymes in feeds and milk yield of cows, *Nutr. Abstr. Rev.,* 47B, 113, 1977.

58. Burnett, G. S., Studies of viscosity as the probable factor involved in the improvement of certain barleys for chickens by enzyme supplementation, *Br. Poult. Sci.,* 7, 55—75, 1966.

59. Gohl, B., Nilsson, M., and Thomke, S., Distribution of soluble carbohydrates in barley grain at late stage of maturity and relation to viscosity, *Cereal Chem.,* 55, 341—347, 1978.

60. Thomke, S., On the influence of different stage of ripeness on the production value of barley fed to chickens, laying hens, rats, and mice, *Acta Agric. Scand.,* 22, 107—120, 1972.

61. Munck, L., Improvement of nutritional value in cereals, *Hereditas,* 72, 1—128, 1972.

62. Gohl, E., Larsson, K., Nilsson, M., Theander, O., and Thomke, S., Distribution of carbohydrates in early harvested barley grain, *Cereal Chem.,* 54, 690—698, 1977.

63. Sturkie, P. D., Secretion of gastric and pancreatic juice, pH of tract, digestion in alimentary canal, liver and bile, and absorption, in *Avian Physiology,* 3rd ed., Sturkie, P. D., Ed., Springer-Verlag, New York, 1975, 196—209.

EFFECT OF PROCESSING ON NUTRIENT CONTENT OF FEEDS: IRRADIATION

W. R. McManus

INTRODUCTION

Application of ionizing radiation to foods and feeds is beneficial in their preservation. Benefit arises from the ability of such radiation to destroy many life forms which otherwise would spoil food.

Irradiation leads to little, if any, rise in temperature of food during treatment. Irradiation of water has only a small effect in raising the temperature. A dose of 10^5 rads is sufficient to raise tissue temperature by 0.25°C and is thus of advantage in the preservation treatment of raw or lightly cooked foods.[1] The use of irradiation technology promises a wider distribution of perishable foodstuffs in the fresh or near-fresh state. The fact that penetrating radiation passes through appreciable thicknesses of material makes irradiation attractive and suitable as a sterilizing process.

The major adverse features of irradiation include the induction of unpleasant flavors and odors, color changes, structural alterations of molecular configurations of some nutrient substances, the production of active radicals, and the difficulty of satisfactorily demonstrating that irradiation products in the food are not toxic or hazardous in either the short or the long terms. Viruses, which are readily destroyed by heat, are more resistant to radiation than are bacteria, and it cannot be guaranteed that viruses will be completely killed by radiation.

The approximate doses of radiation to kill various organisms are outlined below:

Higher animals, including mammals	400—1,000 rads
Insects	1,000—100,000 rads
Nonsporulating bacteria	50,000—1,000,000 rads
Sporulating bacteria	1,000,000—5,000,000 rads
Viruses	1,000,000—20,000,000 rads

Irradiation of foods is used for a number of purposes other than microbiological control. These include inhibition of sprouting of some vegetables, e.g., potatoes and onions, retardation of the ripening of fruits and vegetables, disinfestation from various insects, devitalization of helminthic parasites, and the facilitation of the rehydration of various dried foods. Much higher doses of radiation have been applied to wood and to lignocellulosic roughage feedstuffs in attempts to increase their digestibility to ruminants.[2-10]

Processing Treatments

Units

Radiation power from electron accelerators is generally specified in watts of electron beam power, whereas that from a radioisotope source is more often expressed in curies and electron volts. One kW of radiation is equivalent to the radiation output from 68,000 Ci of cobalt-60. Watts and curies are related by the expression:

$$P = 5.92 \times 10^{-3} \ C \ E \qquad (1)$$

where P is in watts, C in curies, and E in MeV (million electron volts).

The unit of absorbed dose is the rad. One rad is equal to the absorption of 100 ergs of the energy carried by the radiation per gram of material. This is equivalent to 1.26 × 10⁻⁶ Wh/lb (0.456 kg). If all the emitted radiation is absorbed by a material then 1 kWh will process 800 lb (362.88 kg) of material to a dose of 1,000,000 rad (1 Mrad). This is 1 kWh (whether in X-rays, gamma rays, or electrons) is equivalent to 800 Mrad-lb/hr or 362.88 Mrad-kg/hr.[11]

Only certain types of ionizing radiations possess properties which make them suitable for the treatment of food. These are electromagnetic radiations in the form of gamma (γ) or X-rays and beams of electrons or negative beta (β) particles within a certain energy range. β- and γ-radiations (including X-rays) are most suited for food preservation as they are highly penetrating, the effective depth depending on their energy. Gamma radiation derived from radioisotopes has the greater penetrating power. For a particular purpose the most suitable radiation source is dictated by such considerations as those of dosage required, food thickness, radiation penetration, machine efficiency, and cost.[12,13] Some of these factors are interrelated.

Dose

Application of ionizing rays to food processing has been classified according to dose.[14]

1.   Low dose treatments (to achieve such objectives as sprout inhibition of vegetables, insect disinfestation, and destruction of parasites in fresh meat), 0.01 to 0.1 Mrad.
2.   Medium dose treatments (to achieve destruction of Salmonellae, "pasteurization" of meats, destruction of superficial mold spores on fruit), 0.1 to 1.0 Mrad.
3.   High dose treatments (required to achieve commercial sterility in foods), 2 to 5 Mrads.

More recently a newer classification of terminology has been proposed.[15,16]

Radappertization (Type I) — The application to foods of doses of ionizing radiation (2 to 5 Mrad) sufficient to reduce the number and/or activity of viable organisms to such an extent that very few, if any, are detectable in the treated food by any recognized method (viruses being excepted). In the absence of postprocessing contamination no microbial spoilage or toxicity should become detectable with presently available methods, no matter how long or under what conditions the food is stored.

Radicidation (Type II) — The application to foods of doses of ionizing radiation (0.1 to 1.0 Mrad) sufficient to reduce the number of viable specific nonspore-forming pathogenic microorganisms (other than viruses) so that none is detectable in the treated food by any standard method.

Radurization (Type III) — The application to foods of doses of ionizing radiation < 1.0 Mrad and usually < 0.1 Mrad) sufficient to enhance keeping quality by causing substantial reduction in the numbers of viable specific spoilage microorganisms.

As examples of irradiation treatments, Table 1 lists a few foods categorized in accordance with the processes which are applicable to each. The pertinent microbial considerations would differ for each of the three categories of irradiation treatment, because the surviving microflora would not normally be the same for each treatment. Also, for each category the significance of the surviving organisms would differ depending on whether the physiocochemical nature of the food would allow, or would not allow, proliferation of organisms that survive the irradiation treatment, when the food was held under recommended conditions of storage.[16] For detailed consideration of microbiological specifications and testing methods for irradiated feed the reader is referred to a report by the International Atomic Energy Agency.[16]

Irradiation of foods at levels of energy greater than 10 million electron volts causes

Table 1
CHARACTERISTIC APPLICATIONS OF SPECIFIC IRRADIATION TREATMENTS

Perishable foods	Semipreserved foods with salt and/or sugar added, variable water activity	Semipreserved canned foods (some baby foods, some seafoods)	Foods already preserved by other means	Fully preserved canned foods
1. *Raw* non-acid high water activity, held in chilled or frozen storage (liquid egg, poultry, meat, fish, and other sea foods)	1. Nonacid, and cured and/or smoked (vacuum packed fish, meat products)		1. Low water activity (powdered eggs, meringue powder, all animal feed ingredients, spices)	
Radappertization not applicable Radicidation applicable	Radappertization not practical Radicidation applicable	Radappertization not practical Radicidation not recommended *Clostridium Botulinum* might develop	Radappertization applicable Radicidation is extremely important	Radappertization applicable
Radurization applicable	Radurization applicable	Radurization not recommended *C. Botulinum* might develop	Radurization not required	
2. As 1, but *cooked* (Complete dinners, cooked meats, and some delicatessen items) The cooking process required to render foods in this category organoleptically acceptable should be adequate to eliminate food-borne pathogens *except* spore forming bacteria; In view of this bactericidal effect good commercial practice should produce safe food without additional radiation treatment	2. Acid and cured and/or smoked, (some meat and fish products)		2. Low water activity and acidity (salami); On the whole	

Table 1 (continued)

CHARACTERISTIC APPLICATIONS OF SPECIFIC IRRADIATION TREATMENTS[16]

Perishable foods	Semipreserved foods with salt and/or sugar added, variable water activity	Semipreserved canned foods (some baby foods, some seafoods)	Foods already preserved by other means	Fully preserved canned foods
	Radappertization not practical Radicidation applicable Radurization applicable		Radappertization Radicidation Radurization not applicable, but there may be marginal products in which radicidation and radurization may find application	
	3. Acid alone (herrings and some other fish products; mayonnaise and salads) Radappertization not practical Radicidation not recommended *Clostridium Botulinum* might develop Radurization not recommended *C. Botulinum* might develop		3. High water activity chemically preserved Radappertization NOT practical Radicidation applicable Radurization not recommended *Clostridium Botulinum* might develop	

From IAEA, Tech. Rep. Ser. No. 104, International Atomic Energy Agency, Vienna, 1970, 1-121. With permission.

many elements to become activated, notably nitrogen (threshold level, 10.5 MeV), carbon (threshold level 18.8 MeV), and oxygen (threshold level, 15.8 MeV).[12] Ionizing rays form chemically active radicals or ions which attack and damage radiation sensitive substances. As a general indication, 1 Mrad produces between 1 to 10 mM of active radicals per kilogram of irradiated meat tissue.[14] Thus, it is probable that structural alterations in molecular arrangement can occur in foods with applications of energy greater than 10 MeV. The practical significance of such induced changes represent a major area of concerned inquiry in the irradiation of foodstuffs basically to establish the wholesomeness, nontoxicity, and safety as well as the acceptability of the treated product.

The tests which must be applied to an irradiated food to establish its safety for consumption are broadly similar to those generally applied to ensure the safety of food additives.[12] The reader is referred to recommended procedures for the testing of intentional food additives to establish their safety in use as formulated by the Joint FAO/WHO Expert Committee on Food Additives.[17,18]

Protocols for Animal Feeding Studies

While extrapolation of data from animal tests to man is sometimes unsatisfactory, animal tests generally represent a sound biological approach in the assessment of possible adverse effects consequent upon the consumption of irradiated food, and are a necessary investigative step. However, species and strains do differ in radiation sensitivity. To assist and guide the reader, protocols for animal feeding studies on irradiated food are shown below.[12]

AN EXAMPLE OF PROTOCOLS FOR ANIMAL FEEDING STUDIES ON IRRADIATED FOOD[12]

These protocols indicate the type of animal feeding studies which may be used to assess the safety for consumption of irradiated food. The methods and procedures outlined are based on those currently being used by the U.S. Atomic Energy Commission after discussion with the U.S. Food and Drug Administration.

Long-Term Tests
Experimental Design

Three species of experimental animals, e.g., rats, dogs, chickens, are used in the investigation of each irradiated food item. Duration of studies is two years. The following groups are used:

Control group — Fed basal diet (well-standardized "open formula" commercial ration).

Experimental control group — Fed unirradiated test food, preferably up to 35% of the total dry solids in the diet (otherwise the maximum amount of total diet commensurate with providing a well-balanced, nutritionally adequate and acceptable diet for each species).

Low-level test group — Fed irradiated test food which constitutes an identical portion of the total diet to that specified for Group 2. Test food irradiated at 1 × dose level.

High-level test group — Fed irradiated test food which constitutes an identical portion of the total diet to that specified in Group 3. Test food irradiated at 2 × dose level.

The diets fed to the four groups of animals are to be nutritionally balanced, isocaloric, and adjusted, if necessary to contain equal amounts of total protein.

Each of the above treatment groups consists of the following numbers of animals. At least two replications in each test group are desirable:

1.   Rats — 80 (40 males and 40 females).
2.   Dogs — 8 (4 males and 4 females).
3.   Chickens — 20 (10 males and 10 females).

Experimental Observations

Detailed observations which are to be made on experimental animals include studies concerning growth, food consumption, reproduction and lactational performance, hematology, mortality, gross pathology, and histopathology. Supplementary studies include blood and tissue enzyme determinations, urine analysis and estimations of the nutritional value of the irradiated food.

Studies on the rat are designed to provide data on both chronic toxicity and carcinogenicity over the 2-year period. Rat and chicken studies are carried out through three generations subsequent to the parent generation, whereas dog studies are carried out on one generation subsequent to the parent stock.

Standard and acceptable statistical procedures are used for evaluation of the results and randomness is stressed as an essential feature throughout the entire experimental design.

Short-Term (Subacute) Studies

These experiments are considered adequate for investigations on certain foods when extensive studies have previously been conducted on similar or closely related irradiated foods.

Experimental Design

Studies of 90 days duration are conducted using two species of animals, rats and dogs. Treatment groups consist of control (nonirradiated) and irradiated (specified irradiation dose) groups similar to those outlined for long-term tests. A minimum of 50 rats and 8 dogs or 8 monkeys is required in each treatment group. A minimum of two replications is used, with equal numbers of male and female animals in each replicate.

Experimental Observations

General observations include behavior, general appearance, growth, and mortality. Clinical studies include hematology, urine analysis, gross pathology, histopathology, and digestive enzyme analyses. Results are evaluated by standard and acceptable statistical procedures.

PARTICULAR

Radiation Products and Effects of Irradiation on Food Components

Ultraviolet radiation affects only those compounds which absorb energy at the particular wavelength of that radiation. In contrast X-rays, gamma, and electron rays having energies between 100 kV and 10 MeV may affect all food components, these energy levels being much higher than the ionization energies of the most stable compounds.[19]

The quantitative effect of a radiation chemical reaction is expressed by the G-value, which indicates the number of the transformed molecules per 100 eV of absorbed radiation. The amounts in mg/100 g can be calculated from the G-values using the following relationship:[19]

$$C = 1.04 \times G \times M \times D \times 10^{-4} \qquad (2)$$

where C = concentration of the transformed molecules in mg/100g, G = G-value, the number of transformed molecules/100 eV of absorbed radiation, also called the radicals value, M = molecular weight of substance irradiated, and D = radiation dose, krad.

Water

As water is the major constituent of moist foods most of the primary reactions caused by irradiation will take place in the water phase, with the formation of OH· radicals (G = 2.7), H atoms (G = 0.6), and solvated electrons (G = 2.7). These unstable primary water radicals react with other and with substances dissolved in the water.

Carbohydrates

In aqueous solutions carbohydrates are attacked by OH· radicals at a relatively high rate (rate constant number $k \sim 1 \times 10^9 \ M^{-1} \ sec^{-1}$).[20] H atoms and solvated electrons are much less active ($k \sim 1 \times 10^6 - \sim 1 \times 10^7 \ M^{-1} \ sec^{-1}$). OH· radicals abstract preferably the hydrogen of the C—H bond, forming water. The resulting C· radicals react further by disproportionation, dimerization, dehydration, and radiation induced β-splitting to form stable compounds.[21-24] Among radiolytic products of carbohydrates aldonic acid, alduronic acid, ozones, and arabinose have been found. C_2- and C_3-splitting products (glyoxal, glycolaldehyde, glyceraldehyde, dihydroxyacetone) have been found as well, with early determinations of G-values being of the order of 0.8 while more recent determinations indicate G-values of the order of 0.1.[22,24,25] Some 25 radiolysis products of glucose have been determined, but G-values of product formation have yet to be determined for most of them.[19]

Irradiation causes chain degradation and other chemical changes in cellulose which lead to decrease in molecular strength. Irradiation with doses in excess of 1.0 Mrad results in carbonyl group formation, carboxyl group formation, and chain cleavage in the ratio of 20:1:1. For cellulose irradiated at 10^6 rads (vacuum), the G-value determined by electron paramagnetic resonance has been reported to be 1.2, which would indicate a G-value of 0.6 at half that dose (500 krad).[6]

Table 2 shows some G-values of radiolytic products of glucose.

Proteins and Amino Acids

Solvated electrons and hydrogen atoms arising from ionizing radiation are largely nonreactive toward carbohydrates and lipids, but together with OH· radicals they react with amino acids and proteins.[19] Aliphatic amino acids react mainly with OH· radicals ($k \sim 1 \times 10^7 - 1 \times 10^9 \ M^{-1} \ sec^{-1}$). Aromatic amino acids like phenylalanine, tyrosine, and tyrotophan show higher reaction rates ($k \sim 1 \times 10^9 - 7 \times 10^9 M^{-1} \ sec^{-1}$).[20] Solvated electrons have a high reactivity with sulfur-containing amino acids (cysteine, cystine; $k \sim 1 \times 10^{10} \ M^{-1} \ sec^{-1}$).[20] Dissolved proteins are attacked by OH· radicals as well as by solvated electrons with high reaction rates.[19] OH· radicals react with hydrogens attached to the C atoms neighboring the COOH groups ($\propto$ −C atoms). The resulting radicals are stabilized by disproportionation or dimerization. Aromatic amino acids are hydroxylized by OH· radicals on the aromatic groups; with heterocyclic amino acids a splitting of the heterocyclic ring is observed. Solvated electrons split off NH_2 and SH groups.[19]

G-values of the decomposition rates of amino acids in pure aqueous solution are of the order one to ten. The chemical constitution of the main radiolytic products of all

Table 2
G-VALUES OF THE RADIOLYTIC PRODUCTS OF 0.05 M (1%) AQUEOUS SOLUTIONS OF GLUCOSE AND YIELD AT 500 krad (PRESENCE AND ABSENCE OF OXYGEN)

Compound	G-values		Yield at 500 krad (mg/ 100 g)		
	O_2 present	O_2 absent	O_2 present	O_2 absent	Ref.
Glucose (decomposition)	−3.50	−3.50	−32.5	−32.5	25
Gluconic acid	0.35	0.40	3.5	4.1	25
Glucuronic acid	0.90	—	9.0	—	25
Glucosone	—	0.40	—	4.1	25
Erythrose	0.25	—	1.6	—	25
Deoxycarbonyls and other deoxy compounds	—	0.30	—	2.7	23
2-Deoxygluconic	—	1.00	—	9.4	26, 27
C_2-fragments[a]	0.85	0.80	2.6	2.4	25
C_3-fragments[a]	0.80	0.80	3.7	3.7	25

Note: The dash means that either the concentration has not been determined or the substance is not present.

[a]   The presence of C_2 and C_3 fragments may be doubtful according to newer studies.[24]

Reprinted with permission from *Int. J. Appl. Radiat. Isot.,* 26, Diehl, J. F. and Scherz, H., Estimation of radiolytic products as a basis for evaluating the wholesomeness of irradiated foods, ©1975, Pergamon Press, Ltd.

Table 3
DECOMPOSITION OF SOME AMINO ACIDS IRRADIATED IN AQUEOUS SOLUTION. G-VALUES AND CALCULATED LOSS AT 500 Krad

Compound	Concentration (mol ℓ)	Atmosphere	G-values	Loss at 500 krad (mg/100 g)	Ref.
Glycine	1.0	N_2	−4.4	−17.2	30
Alanine	1.0	N_2	−5.0	−23.1	31
Serine	0.1	O_2	−5.5	−30.0	32
	0.1	N_2	−2.5	−13.1	32
Threonine	0.1	O_2	−9.0	−55.6	32
	0.1	N_2	−6.8	−42.0	32
Methionine	0.01	O_2	−6.5	−50.1	33
	0.01	N_2	−3.8	−29.5	33
Cysteine	0.01	Vacuum	−9.3	−58.5	34
Phenylalanine	0.015	O_2	−2.9	−24.8	35
Tyrosine	0.003	O_2	−0.62	−5.8	35
Tryptophan	0.02	Argon	−0.7	−7.4	36

Reprinted with permission from *Int. J. Appl. Radiat. Isot.,* 26, Diehl, J. F. and Scherz, H., Estimation of radiolytic products as a basis for evaluating the wholesomeness of irradiated foods, ©1975, Pergamon Press, Ltd.

amino acids is known and the reader is referred to excellent review articles by Liebster and Kopoldova[28] and Garrison.[29]

Table 3 indicates the G-values and the amounts of total destruction for several amino acids.[19] Table 4 shows the main radiolytic products for three amino acids when irradiated in aqueous solution.

Experimental evidence indicates that the level of protein destruction of irradiated

Table 4
RADIOLYSIS PRODUCTS OF ALANINE (1.0 M AQUEOUS SOLUTION, ABSENCE OF OXYGEN),[31] CYSTEINE (0.01 M AQUEOUS SOLUTION, ABSENCE OF OXYGEN),[34] AND METHIONINE (0.001 M AQUEOUS SOLUTION, ABSENCE OF OXYGEN)[19,37]

Alanine		Cysteine		Methionine	
Compound	G-value	Compound	G-value	Compound	G-value
Alanine (decomposition)	−5.0	Cysteine (decomposition)	−9.3	Methionine (decomposition)	−3.0
Ammonia	4.48	Alanine	2.6	Methionine sulfoxide	0.30
Acetaldehyde	0.59	Cystine	3.4	α-Aminobutyric acid	0.46
Pyruvic acid	1.92	Hydrogen sulfide	2.5	3-Methylthiopropylamine	0.59
Propionic acid	1.04	Hydrogen	1.1	Methional	0.08
Ethylamine	0.17			Carboxylic acid	0.24
				Mercaptan + disulfide	0.61
				Ammonia	1.48
				Carbon dioxide	1.45

Reprinted with permission from *Int. J. Appl. Radiat. Isot.*, 26, Diehl, J. F. and Scherz, H., Estimation of radiolytic products as a basis for evaluating the wholesomeness of irradiated foods, ©1975, Pergamon Press, Ltd.

foods is not as high as the above G-values for dilute solutions of pure amino acids would indicate. Amino acids bound in proteins are more radiation-resistant than free amino acids, and proteins irradiated in the complex matrix of foodstuffs are more radiation resistant than are isolated proteins.[19] Proteins and peptides do not cleave at the peptide bond but at the side chains giving rise to hydrocarbons such as *n*-alkanes, benzene, and toluene. Sulfur containing proteins yield sulfides, disulfides, and mercaptans.

Table 5 indicates some experimental data as to changes in protein and amino acid status of foods following irradiation.

Lipids

Direct bond cleavage of the various compounds making up meat accounts for many compounds that have been isolated. Lipids give rise to *n*-alkanes, *n*-alkenes, and *n*-alkynes. Radiolysis of lipids causes a splitting of ester bonds by various mechanisms.[55,56] Lipids containing saturated fatty acids yield alkanes and *l*-alkenes; unsaturated fatty acids yield *l*-alkenes, alkadienes, and alkatrienes.[55] The main alkanes found after irradiation of tripalmitin are $C_{15}H_{32}$ (G = 0.14); $C_{14}H_{30}$ (G = 0.03) and $C_{17}H_{36}$ (G = 0.01). The main component of the *l*-alkenes was $C_{14}H_{28}$ (G = 0.02).[57] Aldehydes, ketones, and esters were also found.[57] It has been calculated that 500 krad radiation applied to 100 g tripalmitin would yield 2.4 mg of alkanes and 0.3 mg of *l*-alkenes.[19,57]

Analysis of hydrocarbons in irradiated pork (30% fat) has shown linear increase of heptadecane and hexadecadien, the two main volatile products, with radiation dose. Application of 500 krad (0.5 Mrad) resulted in formation of about 10 μg of each of these hydrocarbons per gram of fat giving an estimated product yield of radiolytic factors of 0.3 mg/100 g meat.[19,58]

The full implications of the effects of irradiation upon food and feed substances have yet to be established by research. However, for many useful applications of irradiation using 0.5 Mrad or less it has been suggested that levels of potentially toxic radiation-induced compounds in experimental diets are probably well below 0.01 mg/100 g.[19]

Table 5

CHANGES IN PROTEIN AND AMINO ACID STATUS OF FOODS FOLLOWING IRRADIATION

Irradiated item	Radiation dose (Mrad)	Detected consequence	Ref.
Beef	5.0 and 20	No significant change in amino acid composition using chromatography; some ammonia	38
	6.0	Volatile compounds in mg/100 g: alkanes 1.2, alkenes 1.4, aldehydes 0.15, S-compounds 0.10, alcohols 0.10, ketones <0.05, alkylbenzenes <0.01, esters <0.01	39
	2.0	Carbonyls, 0.5 mg/100 g	40
	3.75	Increased volatile bases, amines (as ammonia) by 2.5 mg/100 g	41
Cod fillets	10.0	No significant change in amino acid composition	42
Codfish	4.5	29% loss of cysteine; also found H_2 (G = 0.2), CO_2 (G = 0.2), CH_4 (G = 0.04)	43
	0.6	Protein nutritive value unaffected (microbiological assay)	44
Fish meal (dried)	1.0	No reduction in protein nutritive value	45
Fish meal	1.0	No loss of available lysine	46
Blood meal	1.0	No loss of available lysine	46
Haddock fillets	2.5	No significant change in amino acid composition	47
Clams	4.5	No significant change in amino acid composition	48
Shrimp	0.25	No change in protein efficiency ratio (PER) when fed to rats; no change in growth response of the microorganism *Tetrahymena pyriformis*	49
Wheat flour	0.2 to 1.0	No change in growth response of the microorganism *T. pyriformis*	49
Wheat	0.2 to 2.0	No change in protein efficiency ratio (PER) when fed to rats	49
	5.0	Gluten lost 60% of its methionine	45
Wheat bran	5.0	No significant change in amino acid composition	50
Kidney beans (*Phaseolus vulgaris* L.)	5.0	No change in amino acid composition (amino acid analyzer); however, low food intakes, and weight loss occur in conventional rats fed irradiated diets; gnotobiotic rats did not show these effects when fed irradiated diets	51
Animal diet	10.0	No significant change in amino acid content	52
	7.0	No significant change in amino acid content	53
	3.5	No significant change in amino acid content	54
Soybean protein solutions	0.1—2.5	Free SH groups increased following irradiation with 0.5 Mrad and the solution became more insoluble by freezing and thawing than did unirradiated protein; disc electrophoretic pattern of the 11 S component showed drastic change at 0.5 Mrad, with an increase of slow-moving components	73

Nucleic Acids

Radiation also affects other compounds in food. Nucleic acids can have phosphate-diester bonds of the polynucleotide chains split, as well as deamination and oxidation of pyramidines and splitting of the purine ring system. The extent of damage is likely to be small and only measurable at high doses of radiation. Irradiation of calf thymus DNA (5 mg/mℓ) with 1 Mrad in 0.01 M phosphate buffer resulted in only 10% of purine and pyrimidine bases being found to be altered.[59,60] Excepting some organs (thymus) average nucleic acid content of foods tends to be below 100 mg/100 g and purine and pyramidine bases constitute less than one third of this. It is probable that an irradiation dose of 0.5 Mrad would result in less than 0.1 mg of a particular reaction product per 100 g food.[19]

Vitamins

Some vitamins seem to be sensitive to irradiation. Among fat-soluble vitamins, vitamin E (α-tocopherol, $C_{29}H_{50}O_2$; β-tocopherol, $C_{28}H_{48}O_2$; and its isomer γ-tocopherol ($C_{28}H_{48}O_2$); and among the water-soluble vitamins, vitamin B_1 (thiamin) are most affected by irradiation. Vitamin B_1 is present in foods in concentrations of less than 1 mg/100 g, thus its radiolysis products are below this concentration.[19] Radiation doses up to 1 Mrad have been applied to crystalline thiamin dichloride, thiamin in solution, and thiamin in dried whole egg.[61] Both in the dry state and in aqueous solutions, radiation-induced reactions continue to affect thiamin over a considerable period of time after irradiation. Thiamin losses decreased in the following order: dry thiamin dichloride, aqueous solution, hydrochloric acid solution, dried whole egg.[61]

Excepting nuts which may contain about 50 mg/100 g of vitamin E, most foods contain less than 10 mg vitamin E per 100 g.[19] A radiation dose of 0.5 Mrad may destroy some 50% of vitamin E. The nature of the radiolysis products is not known, but chromatographic analysis indicates formation of at least five reaction products.[19] Nuts could contain around 5 mg of individual vitamin E decomposition products per 100 g when irradiated at 0.5 Mrad.[19] In other foods the maximum could be closer to 1 mg/100 g.[19]

Vitamin C (ascorbic acid, $C_6H_8O_6$) is one of the most radiosensitive vitamins. Irradiated citrus fruit and juice have been studied.[62,63] Application of 0.4 to 1.0 Mrad radiation reduced vitamin C content of orange juice by 21.2% to 70.2%. Caratenoid losses in orange juice ranged between 10.5% and 23.7%.[63] Keeping lemons irradiated with 0.2 Mrad in a controlled atmosphere reduced postirradiation losses of vitamin C.[62] It is to be noted that irradiation to control various citrus molds can affect oils in the skin of the fruit.

Steroids

Studies of the radiolysis of pure solutions of various steroids have been conducted and many of the reaction products so far identified are identical with various steroids isolated from natural sources.[64]

Sterols (any saturated or unsaturated alcohol derived from cyclopentanoperhydro-phenanthrene) yield normal and isoalkanes.

Sensory Compounds

In many foods aroma compounds are present in low concentrations, and radiation-induced decomposition products of the aroma compounds can only be formed in even lower concentrations. Relatively high levels of taste- and odor-active compounds are present in spices and irradiation effects on these have been studied in some detail.[19]

Capsaicin, the pungent principle of paprika has been found to remain at a level of 34 mg% after irradiation at 1.6 Mrad.[65]

Gas chromatography of various spices irradiated with a dose of 4.5 Mrad revealed radiation-induced compounds at a level of less than 0.01% of the total volatile constituents.[66]

When high doses of radiation are applied to meats a typical foreign flavor develops which is similar but not identical with that associated with heat scorching. It affects consumer acceptance. Of the common commercial meats, beef develops the greater flavor intensity, followed by lamb and veal while pork and chicken meats are less sensitive. The best method so far devised for suppressing this effect has been to irradiate at subfreezing temperatures (−30°C to −80°C) which suggests the flavor to arise from the action of free radicals, probably originating in the water present.[67] Methional, 1-nonanal and phenylacetaldehyde and mercaptans are probably major sources of flavor alteration.[67]

Tables 6 and 7 show data indicating the suppression of flavor alteration when irradiation of meats is conducted at different temperatures. Part B of Table 6 shows that irradiation flavor increases and overall acceptance decreases with an increase in radiation dose.[74] Table 8 shows some data as to maximum radiation dose which will not incur a sensory change, as well as data on product life extension under conditions of good refrigeration.[67]

Irradiation of rice, with doses up to 0.2 Mrad, to successfully disinfect it from the rice weevil (*Situphilus oryzea* L.) did not alter the organoleptic (appearance, odor, flavor, texture) properties of the rice when evaluated on the hedonic scale by 12 panelists. The thiamin content of this rice was stable up to 0.001 Mrad, and doses up to 0.007 Mrad gave 100% mortality of the rice weevil larva.[71]

Other Structural Effects

Sterilizing doses (2 to 5 Mrad) tenderize meat by degradation of collagen and, with storage, irradiated meats may suffer complete loss of texture due to autolysis. This may be prevented by precooking.[69,70] Studies of irradiation of beef meat at levels up to 5 Mrad either before or after freezing have shown irradiation sequence and freezing rate to have marked effects on total water uptake. Highest total water uptake (rate and extent of rehydration) was found for meat irradiated when fresh and frozen slowly at −22°C.[75] These effects indicate that changes in tissue structure and membrane permeability have occurred.

The softening effect of radiation has been considered as a means of reducing the cooking time for dehydrated vegetables in soups. Table 9 shows the irradiation doses required for selected dehydrated fruits and vegetables to reduce the cooking time during rehydration from 10 to 20 min down to 1 or 2 min. Any deleterious effects on flavor or appearance are considered minor.[67]

A major chemical consequence of the irradiation of cellulose is reduction in chain length.[68] Table 10 shows the effect of radiation dose upon solubility of raw cotton linters and low quality forage roughages such as are fed to ruminants.[39] The crude fiber and cellulose content of cotton linters, peanut hulls, corn cobs, and sugar cane bagasse irradiated at doses of 10, 20, or 40 Mrad decreased with each increase in level of radiation.[2]

In vitro dry matter digestibility (IVDM) of forage roughage feedstuffs is increased by irradiation.[3,9] When fed to sheep irradiated wheat straw and rice straw (25, 50, 75 Mrad) does not lead to increased in vivo digestibility because feed particle size is so decreased the ingesta passes too quickly through the gastrointentinal tract to allow complete utilization of the fermentation potential in the rumen.[10]

Wholesomeness

Irradiated foods are not yet general commodities of the market place. Table 11

Table 6
EFFECT OF IRRADIATION TEMPERATURE AND DOSE ON SENSORY AND HEDONISTIC CHARACTERISTICS OF BEEF MEAT

Effect of irradiation temperature on beef steak flavor (Dose 6 Mrad)[68]

Irradiation temperature °C	Flavor score[a]
25	5.07
−25	4.37
−80	4.08
−140	3.52
−196	2.35

Reproduced from "Radiation Preservation of Foods," 1965, with the permission of the National Academy of Sciences, Washington, D.C.

Effect of radiation dose on odor, flavor, and preference of beef — 2 to 3% fat (U.S. choice beef roast irradiated at −185°C)[74]

Radiation dose (Mrad)	Off[b] odor	Irradiation[b] flavor	Preference[b]
0	1.00[c]	1.77[c]	7.64[c]
3.0	2.05	2.62	6.40[d]
4.5	2.20	3.18	6.18
6.0	2.41	3.36	5.54

[a] Flavor score recorded on a six-point scale; 1 = no irradiated flavor and 6 = very much. 95% confidence level ± 0.35.
[b] Sensory and hedonistic evaluation using a nine-point scale; 1 = none, 9 = extreme.
[c] Significantly different from other samples.
[d] Significantly different from 6.0 Mrad sample. N = 23, significant at 5% level.

Reprinted with permission from *J. Agric. Food Chem.*, 23, 1037. © 1975 American Chemical Society.

shows various applications of irradiation to foods which are in use and the doses administered.[72]

Wholesomeness of irradiated foods is difficult to prove. Table 12 shows a summary of some toxicity trials with various feedstuffs. Major aspects of concern are the identity and identification of toxic factors.

Recent reports of induction of lethal mutations in mice after consuming extracts from irradiated potatoes, and the discovery that polyploidy results in children and in rats from their being fed irradiated wheat grain constitute findings of major significance.[78,80,81] These reports emphasize the possible dangers inherent in consumption of some irradiated foods and feedstuffs and indicate the need for further critical research. Research should center upon the establishment of causal relationships in this area. Attention should also be paid to assessing the incidence and significance of radiolysis products in food-chain sequences of relevance to agriculture.

Table 7
EFFECT OF IRRADIATION TEMPERATURE ON PREFERENCE RATINGS[a] OF IRRADIATED HAM

Storage (months)	n[b]	3.5 to 4.4 Mrad at °C				Unirradiated control
		+ 5	−18	−40[c]	−80	
1	30	—	5.9	5.9	6.8	7.5
1	30	5.6	6.1	6.4	7.1	6.9
4	30	5.5	5.8	5.6	6.6	6.1
12	32	5.4	—	—	6.2	6.9
12	32	6.1	—	—	6.8	6.4
Overall average		5.65	5.93	5.97	6.7	6.76

[a]   Nine-point hedonic scale; 9 = "like extremely"; 1 = "dislike extremely"; 5 = "neither like nor dislike". Samples scoring above 5.0 are considered acceptable.

[b]   n = Number of taste test panelists.

[c]   In this study irradiation with 3.5 to 4.4 Mrad at −40°C is adequate to control the flavor change.

From FAO/IAEA, Tech. Rep. Ser. No. 114, International Atomic Energy Agency, Vienna, 1970, 60-65. With permission.

Table 8
MINIMUM RADIATION DOSE WITHOUT DETECTABLE SENSORY CHANGE AND LIFE EXTENSION UNDER GOOD REFRIGERATION (MARINE AND FRESH-WATER PRODUCTS)

Product	Radiation dose (krad)	Life extension (days)
Fish — sea		
Haddock	200 to 250	18
Perch	350	18
Atlantic mackerel	350	30
Cod	150	18
Petrale sole	400	25 to 38
Grey sole	100	20
Halibut	400	12 to 23
Pollock	150	18
Fish — fresh water		
Channel cat	100	—[a]
Yellow perch	200	13 to 18
White fish	300	16 to 20
Mollusks		
Clams	800	—
Oysters	200	—
Crustaceans		
Shrimp	200	5 to 14
King crab	200	14 to 37
Blue crab	250	28
Lobster	250	10 to 18

[a]   Not determined.

From FAO/IAEA, Tech. Rep. Ser. No. 114, International Atomic Energy Agency, Vienna, 1970, 60-65. With permission.

Table 9
RADIATION DOSE FOR SELECTED DEHYDRATED VEGETABLES AND FRUITS TO REDUCE COOKING TIME ON REHYDRATION[a]

Product	Approximate preferred dose (Mrad)
White onion flakes	0.3
Tomato flakes	0.6
Potato dice	1.0
Carrot dice	2.0
Dried fresh peas	2.0
Leek	2.4
Bell pepper dice (green or red)	2.5
Cabbage flakes	3.0
Green lima beans	3.0
Celery flakes	3.3
Cut green beans	4.0
Okra pieces[b]	4.0
Beet cubes	>4.0
Apples	5.0
Prunes	9.0

[a] Doses to reduce cooking time during rehydration from 10 to 20 min to 1 to 2 min; any adverse effects on flavor or appearance are considered minor.

[b] The gum of okra, which accounts for a major characteristic of this vegetable, is destroyed by irradiation.

From FAO/IAEA, Tech. Rep. Ser. No. 114, International Atomic Energy Agency, Vienna, 1970, 60-65. With permission.

Table 10
EFFECT OF IRRADIATION ON SOLUBILITY IN WATER OF CELLULOSE AND LOW QUALITY FEEDSTUFF ROUGHAGE

	% Solubility			
Radiation dose (Mrad)	Raw[a] cotton lint	Wheat[b] straw (*Triticum aestivum*) (41.5% CF)[c]	Rice[a] straw (*Oryza sativa*) (78.2% CWC)[d]	Nassella[a] grass (*Nassella trichotoma*) (86.0% CWC)[d]
0	2.6	5.0	9.0	4.3
0.1	—	7.5	—	—
1.0	—	7.5	—	—
5.0	2.5	—	8.5	4.6
10.0	3.9	7.4	7.6	5.0
25.0	6.4	—	10.5	5.9
100.0	17.2	24.5	22.4	14.2
200.0	31.2	—	34.2	24.3
250.0	—	72.0	—	—
500.0	—	77.0	—	—
750.0	—	85.0	—	—
1000.0	—	87.0	—	—

[a] 14-hr mechanical shaking of 1-g sample per 100 ml water, filtered, dried.[9]

[b] 2-hr incubation of 250 mg sample in 25 ml buffered (pH 6.9), centrifuged, washed, dried.[3]

[c] CF = crude fiber, A.O.A.C.

[d] CWC = cell wall content, neutral detergent fiber, A.O.A.C.

Table 11
DOSES OF RADIATION USED FOR FOOD ITEMS

Product	Purpose	Radiation dose (krad)
Fruit and vegetables		
Potatoes	Sprout inhibition	10
Onions	Sprout inhibition	6
Dried fruits	Disinfestation	100
Fresh fruits and vegetables	Radurization	200—400
Mushrooms	Radurization	250
Asparagus	Radurization	200
Strawberries	Radurization	200
Cocoa beans	Disinfestation	20—50
Spices and condiments	Radicidation	800—1000
Grain and grain products	Disinfestation	30
Meat and fish		
Combined treatment	Radurization	600—800
Kitchen ready	Radurization	800
Poultry	Radurization	600
Shrimps	Radurization	50—100
Other products		
Dried fruit concentrates	Disinfestation	70
Deep frozen meals	Radappertization	2500

From Ullman, R. M., *Peaceful Uses of Atomic Energy,* Vol. 12, International Atomic Energy Agency, Vienna, 337-346. With permission.

Table 12
OUTLINE OF FINDINGS OF SOME STUDIES AS TO WHOLESOMENESS OF IRRADIATED FOODS

Food	Radiation dose (krad)	Type of study	Detected effects	Ref.
Potatoes	60	Short-term toxicity to monkeys	None significant	73
		Long-term toxicity to mice and rats	Changes in ovary weight of rats	73
		Reproduction studies on mice for generations	None significant	73
Potatoes	20	Gamma irradiated, fed to male and female rats (after 3 months storage) 10 g potato/rat/day (400 rats); mutagenic study	None significant	76
Potatoes	30	Fast electron irradiated, fed to male and female rats (after 3 months storage) 10 g potato/rat/day (400 rats); mutagenic study	None significant	76
		Fast electron, 5 generations of rats (60, parent generation)	None significant	76
Potato extract	10	Gamma irradiated, fed to male mice as gavage of steam-boiled extract (after potatoes being stored 40 to 90 days following irradiation); (intake equivalent to 3 g raw potato/head/day) (800 mice); mutagenic study	None significant	76
		Gamma irradiated, alcoholic extract given by gavage to mice. Short-term studies 7 to 35 days; mutagenic study	Dominant lethal mutations induced	78

Table 12 (continued)
OUTLINE OF FINDINGS OF SOME STUDIES AS TO WHOLESOMENESS OF IRRADIATED FOODS

Food	Radiation dose (krad)	Type of study	Detected effects	Ref.
Potato extract	12	Gamma irradiated, alcoholic extract given by gavage to mice; short-term studies 7 to 35 days; mutagenic study	None significant	77
Wheat	20 to 200	Rats and mice fed for 5 generations; mutagenic study	None detected	79
Wheat	75	Cytotoxicity study; fifteen children (2 to 5 years) initially malnourished (kwashiorkor) were rehabilitated over 42 days in hospital under 3 dietary regimes in which the diet had 20 g wheat/ kg bodyweight and provided 2 g wheat protein/kg bodyweight/day		80
		1. Unirradiated wheat (5 children)	Other than 'normal' chromosomal aberrations none detected	
		2. Freshly gamma-irradiated wheat (5 children)	4 out of 5 children showed polyploid cells 28 days after start of treatment; other abnormal cells present	
		3. Wheat stored (84 days) after gamma irradiation and prior to feeding; (5 children)	Reduced incidence of polyploid cells detected	
		Cytotoxicity study; 30 weanling rats (male, female); wheat comprised 70% of diet and was fed for 84 days		81
		1. Unirradiated wheat (10 rats)	0.04% incidence of polyploid cells	
		2. Freshly gamma irradiated wheat (10 rats)	0.58% incidence of polyploid cells, significantly different from unirradiated treatment	
		3. As above (42 rats)	Significant increase in polyploid cells	
		4. Wheat stored (90 days) after gamma irradiation at 4°C and prior to feeding (10 rats)	0.10% incidence of polyploid cells (not different from unirradiated treatment)	
Rat feed	2500 to 4500	3 generations of rats	None detected	82

REFERENCES

1. Lea D. E., *Actions of Radiations on Living Cells,* 2nd ed., Cambridge University Press, 1955, 2.
2. Ammerman, C. B., Evans, J. L., Tomlin, D. C., Arrington, L. R., and Davis, G. K., Gamma irradiation and subsequent digestion in the rumen of cellulose and various roughages, *J. Anim. Sci.,* 18 (Abstr.), 1518, 1959.
3. Pritchard, G. I., Pigden, W. J., and Minson, D. J., Effect of gamma irradiation on the utilization of wheat straw by rumen microorganisms, *Can. J. Anim. Sci.,* 42, 215—217, 1962.
4. Garnett, J. L. and Merewether, J. W. T., Chemical effects from the irradiation of wood, in *Proc. Conf. Tech. Use of Radiation,* Australian Atomic Energy Commission, Sydney, 1960, 76—81.
5. Smith, D. M. and Mixer, R. Y., The effects of lignin on the degradation of wood by gamma irradiation, *Radiat. Res.,* 11, 776—780, 1959.
6. Dilli, S., Ernest, I. T., and Garnett, J. L., Radiation-induced reactions with cellulose. IV. Electron paramagnetic resonance studies of radical formation, *Aust. J. Chem.,* 20, 911—927, 1967.
7. Pigden, W. J., Pritchard, G. I., and Heaney, D. P., Physical and chemical methods for increasing the available energy content of forages, *Proc. 10th Int. Grassl. Congr. Helsinki,* 1966, 397—401.
8. Teszler, O., Kiser, L. H., Campbell, H. A., and Rutherford, H. A., Effect of nuclear radiation on fibrous material. III. Relative order of stability of cellulosic fibers, *Text. Res. J.,* 28, 456—462, 1958.
9. McManus, W. R., Manta, L., McFarlane, J. D., and Gray, C. H., The effects of diet supplements and gamma irradiation on dissimilation of low-quality roughages by ruminants. II. Effects of gamma irradiation and urea supplementation on dissimilation in the rumen, *J. Agric. Sci.,* 79, 41—53, 1972.
10. McManus, W. R., Manta, L., McFarlane, J. D., and Gray, C. H., The effects of diet supplements and gamma irradiation on dissimilation of low quality roughages by ruminants. III. Effects of feeding gamma-irradiated base diets of wheaten straw and rice straw to sheep, *J. Agric. Sci.,* 79, 55—66, 1972.
11. Clouston, J. G., Irradiation plants for industrial processing, *Food Technol. Aust.,* 16—21, 1964.
12. WHO, *Technical Basis for Legislation on Irradiated Foods,* World Health Organization, Geneva, 1965, 1—56.
13. Goldblith, S. A., in *Radiation Preservation of Foods,* Publ. No. 1273, National Academy Science — National Research Council, Washington, D.C., 1965.
14. Rolfe, E., Processing and preservation of animal products for human consumption, in *Nutrition of Animals of Agricultural Importance,* Part 1, The science of nutrition of farm livestock, Cuthbertson, Sir David, Ed., Pergamon Press, Oxford, 1969, 521—592.
15. Goresline, H. E., Ingram, M., Macuch, P., Mocquot, G., Mossel, D. A. A., Niven, C. F., and Thatcher, F. S., Tentative classification of food irradiation processes with microbiological objectives, *Nature (London),* 204, 237—238, 1964.
16. IAEA, *Microbiological Specifications and Testing Methods for Irradiated Food,* Tech. Rep. Ser., International Atomic Energy Agency, No. 104, Vienna, 1970, 1—121.
17. WHO, *Procedures for the Testing of Intentional Food Additives to Establish Their Safety for Use,* Food and Agriculture Organization, Nutrition Meetings Report Series No. 17, World Health Organization Tech. Rep. Ser. No. 144, Rome, 1958.
18. WHO, *Evaluation of the Carcinogenic Hazards of Food Additives,* Fifth Report of the Joint FAO/ WHO Expert Committee on Food Additives, Food and Agriculture Organization, Nutrition Meetings Report Series No. 29, World Health Organization Tech. Rep. Ser. No. 220, Rome, 1961.
19. Diehl, J. F. and Scherz, H., Estimation of radiolytic products as a basis for evaluating the wholesomeness of irradiated foods, *Int. J. Appl. Radiat. Isot.,* 26, 499—507, 1975.
20. Anbar, M. and Neta, N., A compilation of specific bimolecular rate constants for the reactions of hydrated electrons, hydrogen atoms, and hydroxyl radicals with inorganic and organic compounds in aqueous solution, *Int. J. Appl. Radiat. Isot.,* 18, 493—523, 1967.
21. Norman, R. O. C. and Prichet, R. J., Electron-spin resonance studies. XIII. The oxidation of mesoinositol and some monosaccharides by the hydroxy radical, *J. Chem. Soc., B,* No. 12, 1329—1332, 1967.
22. Hartmann, V., Von Sonntag, C., and Schulte-Frohlinde, D. Z., γ-Radiolysis of 2-deoxy-D-ribose in aqueous solution, *Z. Naturforsch. Teil B,* 25-, 1394—1404, 1970.
23. Scherz, H., Formation of deoxycompounds and malondialdehyde in irradiated aqueous solutions of carbohydrates and related compounds, *Radiat. Res.,* 43, 12—24, 1970.
24. Dizdaroglu, M., Scherz, H., and von Sonntag, C. Z., Radiation chemistry of alcohols. XVI. γ-Radiolysis of meso-erythritol in aqueous solution, *Z. Naturforsch. Teil B.,* 27, 29—42, 1972.
25. Phillips, G. O., Chemical effects of ionizing radiations on aqueous solutions of aldohexoses, *Radiat. Res.,* 18, 446—460, 1963.
26. Kawakishi, S. H. and Namaki, N., *Shokuhin Shosha,* 4, 29, 1969.
27. Schubert, J., *Improvement of Food Quality by Irradiation,* Proc. FAO/IAEA Panel Meeting, International Atomic Energy Agency, Vienna, 1974, 1.

28. **Liebster, J. and Kopoldova, J.**, The radiation chemistry of amino acids, *Adv. Radiat. Biol.,* 1, 157—226, 1964.
29. **Garrison, W. M.**, Radiation-induced reactions of amino acids and peptides, *Radiat. Res. Rev.,* 3, 305—326, 1972.
30. **Weeks, B. W. and Garrison, W. M.**, Radiolysis of aqueous solutions of glycine, *Radiat. Res.,* 9, 291—304, 1958.
31. **Sharpless, N., Blair, A. E., and Maxwell, C. R.**, The effect of ionizing radiation on amino acids. IV. pH effects on the radiation decomposition of alanine, *Radiat. Res.,* 3, 417—422, 1955.
32. **Pageau, R. and Mehran, A. R.**, Radiochemical degradation of threonine and serine, *Nature (London),* 212, 98—99, 1966.
33. **Kopoldova, J., Liebster, J., and Gross, L.**, Radiation chemical reactions in aqueous solutions of methionine and its peptides, *Radiat. Res.,* 30, 261—273, 1967.
34. **Wilkening, V. G., Lal, M. Arends, M., and Armstrong, D. A.**, The cobalt-60 γ radiolysis of cysteine in deaerated aqueous solutions at pH values between 5 and 6, *J. Phys. Chem.,* 72, 185—190, 1968.
35. **Wheeler, O. H. and Montalvo, R.**, Radiolysis of phenylalanine and tyrosine in aqueous solution, *Radiat. Res.,* 40, 1—10, 1969.
36. **Armstrong, R. C. and Swallow, A. J.**, Pulse and γ-radiolysis of aqueous solutions of tryptophan, *Radiat. Res.,* 40, 563-579, 1969.
37. **Tajima, M., Morita, M., and Fujimaki, M.**, Radiation chemistry of foods. V. Material balance and additional effect of second argeric solutes in the radiolysis of aqueous methionine solution, *Agric. Biol. Chem.,* 36(7), 1129—1134, 1972.
38. **Rhodes, D. N.**, The treatment of meats with ionizing radiations. XII. Effects of ionizing radiation on the amino acids of meat protein, *J. Sci. Food Agric.,* 17, 180—182, 1966.
39. **Merritt, C., Jr.**, Qualitative and quantitative aspects of trace volatile components in irradiated foods and food substances, *Radiat. Res. Rev.,* 3, 353—368, 1972.
40. **Batzer, O. F., Sribney, M., Doty, D. M., and Schweigert, B. S.**, Production of carbonyl compounds during irradiation of meat and meat fats, *J. Agric. Food Chem.,* 5, 700—703, 1957.
41. **Burks, R. E., Jr., Baker, E. B., Clark, P., Esslinger, J., and Lacey, J. C. J. J.**, Detection of amines produced on irradiation of beef, *J. Agric. Food Chem.,* 7, 778—782, 1959.
42. **Maslennikova, N. V.**, Effect of γ-radiation on the amino acid composition of the muscular tissue of fish, *Rybn. Khoz.,* 45, 72—74, 1970; *Food Sci. Technol. Abstr.,* 2, 6R213, 1970.
43. **Underdal, B., Nordal, J., Lunde, G., and Eggum, B.**, Effect of ionizing radiation on the nutritional value of fish (cod) protein, *Lebensm. Wiss. Technol.,* 6, 90—93, 1973.
44. **Kennedy, T. S. and Ley, F. J.**, Studies on the combined effect of gamma radiation and cooking on the nutritional value of fish, *J. Sci. Food Agric.,* 22, 146—148, 1971.
45. **Kennedy, T. S.**, Nutritional value of foods treated with γ-radiation. II. Effects on the protein in some animal feeds, egg, and wheat, *J. Sci. Food Agric.,* 16, 433—437, 1965.
46. **Van Der Schaaf, A. and Mossel, D. A. A.**, Gamma radiation sanitation of fish and blood meals, *Int. J. Appl. Radiat. Isot.,* 14, 557—562, 1963.
47. **Brooks, R. O., Ravesi, E. M. Gadbois, D. F., and Steinberg, M. A.**, Preservation of fresh unfrozen fishery products by low-level radiation, *Food Technol.,* 20(11), 99—102, 1966.
48. **Brooke, R. O., Ravesi, E. M., Gadbois, D. F., and Steinberg, M. A.**, Preservation of fresh unfrozen fishery products by low level radiation, *Food Technol.,* 18(7), 116—120, 1964.
49. **Srinivas, H., Vakil, U. K., and Sreenivasan, A.**, Evaluation of protein quality of irradiated foods, *J. Food Sci.,* 40, 65—69, 1975.
50. **Moran, E. T., Jr., Summer, J. D., and Bayley, H. S.**, Effect of cobalt-60 γ-irradiation on the utilization of energy, protein, and phosphorus from wheat bran by the chicken, *Cereal Chem.,* 45, 469—479, 1968.
51. **Rattray, A. S., Palmer, R., and Pusztai, A.**, Toxicity of kidney beans (*Phaseolus vulgaris,* L.) to conventional and gnotobiotic rats, *J. Sci. Food Agric.,* 25, 1035—1040, 1974.
52. **Udes, H., Hiller, H. H., and Juhr, N. C. Z.**, Changes in the quality and quantity of crude protein in a rat and mouse diet due to various sterilization methods, *Z. Versuchstierkd.,* 13, 160—166, 1971.
53. **Eggum, B. O. Z.**, Effect of sterilization on the protein quality of feed mixtures, *Z. Tierphysiol. Tierernaehr. Futtermittelkd.,* 25, 204—210, 1969.
54. **Sickel, E., Diehl, J. F., and Grünewald, T. Z.**, Comparison of the suitability of heat- and irradiation-sterilized prestarter feed for the adaptation of gnotobiotic piglets to the specific pathogen-free status, *Z. Tierphysiol. Tierernaehr. Futtermittelkd.,* 25, 258—269, 1969.
55. **Dubravcic, M. F. and Nawar, W. W.**, Radiolysis of lipids: mode of cleavage in simple triglycerides, *J. Am. Oil Chem. Soc.,* 45, 656—660, 1968.
56. **Dubravcic, M. F. and Nawar, W. W.**, Effects of high energy radiation on the lipids of fish, *J. Agric. Food Chem.,* 17, 639—644, 1969.
57. **Nawar, W. W.**, *Progress in the Chemistry of Fats and Other Lipids,* Vol. 22, Part 2, Pergamon Press, Oxford, 1972, 91.

58. Nawar, W. W. and Balboni, J. J., Detection of irradiation treatments in foods, *J. Assoc. Off. Anal. Chem.*, 53, 726—729, 1970.

59. Hems, G., Chemical effects of ionizing radiation on deoxyribonucleic acid in dilute aqueous solution, *Nature (London)*, 186, 710—712, 1960.

60. Scholes, G., Ward, J. F., and Weiss, J., Mechanism of the radiation-induced degradation of nucleic acids, *J. Mol. Biol.*, 2, 379—391, 1960.

61. Diehl, J. F., Thiamin in bestrahlten Lebensmitteln. I. Einfluss verschiedener Bestrahlungsbedingungen und des Zeitablaufs nach der Bestrahlung, *Z. Lebensm. Unters. Forsch.*, 157, 317—321, 1975.

62. Shrikhande, A. J. and Kaewubon, N., Effect of controlled atmosphere on irradiated lemon fruits. *Radiat. Bot.*, 14, 315—321, 1974.

63. Hussain, A. and Maxie, E. C., Effect of gamma rays on shelf life and quality of orange juice, *Int. Biodetlor. Bull.*, 10, 81—86, 1974.

64. Swallow, A. J., *Radiation Chemistry of Organic Compounds,* Pergamon Press, Oxford, 1960, 186.

65. Farkas, J., Beczner, J., and Incze, K., *Radiation Preservation of Food,* Proc. International Atomic Energy Agency, Vienna, 1973.

66. Tjaberg, T. B., Underdal, B., and Lunde, G., The effect of ionizing radiation on the microbiological content and volatile constituents of spices, *J. Appl. Bacteriol.*, 35, 473—478, 1972.

67. FAO/ IAEA, *Training Manual on Food Irradiation Technology and Techniques,* Tech. Rep. Ser. No. 114, International Atomic Energy Agency, Vienna, 1970, 60—65.

68. Kaffman, F. L., Harlan, J. W., and Rasmussen, C. E., 1964 U.S. Army Natick Lab. Contract No. DA-19-129-QM-2000, Swift and Company, final report, cited by, Urbain, W. M. in *Radiation Preservation of Foods,* Publ. 1273, National Academy of Science, National Research Council, Washington, D.C., 1965.

69. Coleby, B., Ingram, M., and Shepherd, H. J., Treatment of meats with ionizing radiations. VI. Changes in quality during storage of sterilized raw beef and pork, *J. Sci. Food Agric.*, 12, 417—424, 1961.

70. Rhodes, D. N. and Meegungwan, C., Treatment of meats with ionizing radiations. IX. Inactivation of liver autolytic enzymes, *J. Sci. Food Agric.*, 13, 279—282, 1962.

71. Abdullah, N., Siagian, E. G., Isnaeni, N., and Ismachin, M., Laboratory work on food irradiation in Indonesia, in *Peaceful Uses of Atomic Energy,* Vol. 12, International Atomic Energy Agency, Vienna, 1972, 337—346.

72. Ulmann, R. M., Introducing irradiated foods to the producer and consumer, in *Peaceful Uses of Atomic Energy,* Vol. 12, International Atomic Energy Agency, Vienna, 1972, 299—308.

73. Sato, T., Present status of food irradiation in Japan, in *Peaceful Uses of Atomic Energy,* Vol. 12, International Atomic Energy Agency, Vienna, 1972, 325—336.

74. Merritt, C., Angelini, G., Wierbicki, E., and Shults, G. W., Chemical changes associated with flavour in irradiated meat, *J. Agric. Food Chem.*, 23, 1037—1041, 1975.

75. Watson, E. L., Ni, Y. W., and Richards, J. F., Effect of irradiation and freezing rate on rehydration of freeze dried beef, *J. Can. Inst. Food Sci. Technol.*, 7, 232—235, 1974.

76. Zajcev, A. N., Shillinger, J. I., Kamaldinova, Z. M., and Osipova, I. N., Toxicologic and hygienic investigation of potatoes irradiated with a beam of fast electrons and gamma-rays to control sprouting, *Toxicology*, 4, 267—274, 1975.

77. Levinsky, H. V. and Wilson, M. A., Mutagenic evaluation of an alcoholic extract from gamma-irradiated potatoes, *Food Cosmet. Toxicol.*, 13, 243—246, 1975.

78. Kopylov, V. A., Osipova, I. N., and Kuzin, A. M., Mutagenic effects of extracts from gamma-irradiated potato tubers on the sex cells of male mice, *Radiobiologiya*, 12, 58—63, 1972.

79. Aravindakshan, M., Vakil, U. K., and Sreenivasan, A., *Studies on Wholesomeness Testing of Gamma-Irradiated Wheat,* BARC Rep. No. 455, Bhabha Atomic Research Centre, Trombay, India, 1970.

80. Bhaskaram, C. and Sadasivan, G., Effects of feeding irradiated wheat to malnourished children, *Am. J. Clin. Nutr.*, 28, 130—135, 1975.

81. Vijayalaxmi, C., Cytogenetic studies in rats fed irradiated wheat, *Int. J. Radiat. Biol.*, 27, 283—285, 1975.

82. Vas, K., Hungarian research on the use of atomic energy in food and agriculture, in *Peaceful Uses of Atomic Energy,* Vol. 12, International Atomic Energy Agency, Vienna, 1972, 85—96.

CHANGES IN THE NUTRITIVE CONTENT AND VALUE OF FEED CONCENTRATES DURING STORAGE*

B.J. Francis and J.F. Wood

ABSTRACT

The chemical changes which can take place in feed concentrates during storage and their effect upon nutritional quality are discussed. Important factors initiating these changes are length of storage, temperature, and atmospheric oxygen. More emphasis is given to likely biochemical and chemical reactivity than to the causal agents (insects and molds) which are often responsible for generating heat and high moisture in stored bulk material. Depending upon the nature of the product and the conditions of storage, chemical reactions can proceed within the macronutrient fractions which then reduce the digestibility, destroy micronutrients or render the material unacceptable to animals. Quantitative information related to any reduction in the nutritive value which concentrates have suffered in storage is limited and is mostly derived from studies with food and from the knowledge of chemical reactions which occur when food or feed processing involves heat.

INTRODUCTION

Feed concentrates are nonfibrous starchy or proteinaceous materials suitable for feeding all types of animals. They are relatively dry materials (85 to 95% dry matter) and have a low lipid content (10 to 14%). Together with roughages and supplements they make up the bulk of material used in animal feeds. Concentrates also include some materials that are neither strictly concentrates nor roughages, for example, molasses, dried sugar beet pulp, and lucerne (alfalfa) meal. Concentrates may be classified under the following headings:[15]

1. Cereals, e.g., maize, barley, wheat, sorghum and millet, and byproducts from their processing, e.g., milling offals and brewery residues
2. Oilseed expeller cakes and extracted meals, e.g., those from cottonseed, groundnuts, soybeans, sesame, and coconut
3. Grain legumes, e.g., various types of beans, peas, grams, and lentils
4. Animal products, e.g., fish meal, meat meal, dried blood, bone meal, feather meal and poultry byproducts, and dried milk products
5. Dried roots and tubers, e.g., cassava and potatoes
6. Leaf meals, e.g., those from dried grass and alfalfa
7. Byproducts from sugar processing, e.g., molasses and dried beet pulp

All these are commonly used in the production of compounded feedingstuffs which inevitably entails storage of both raw materials and finished feeds. The proximate composition of the more important feed concentrates is given in Table 1.

In this article two aspects of nutritive value are considered. The first is related to the level of specific nutrient components and the second to changes which, while not necessarily reducing a nutrient level, cause the animal to reject or limit its intake of

* From Francis, B. J. and Wood, J. F., *Trop. Sci.*, 22(3), 1980, in press, British Crown Copyright. With permission of the Controller of Her Britannic Majesty's Stationery Office.

Table 1
PROXIMATE COMPOSITION OF CONCENTRATED FEED
MATERIALS (%)

Materials	Dry matter	Crude protein	Oil	Crude fiber	Mineral matter	Nitrogen free extract
Maize	87.0	9.9	4.4	2.2	1.3	69.2
Barley	85.0	9.0	1.5	4.5	2.6	67.4
Wheat	87.0	12.2	1.9	1.9	1.7	69.3
Sorghum	89.6	10.8	2.8	2.3	2.0	71.7
Wheat bran	90.1	16.4	4.5	10.0	6.1	53.1
Rice bran	90.8	12.4	13.6	11.6	13.3	39.9
Cottonseed cake (undecorticated)	92.4	28.0	5.2	21.4	4.6	33.2
Cottonseed cake (decorticated)	90.0	41.1	8.0	7.8	6.7	26.4
Groundnut cake (undecorticated)	90.0	30.3	9.1	23.0	5.7	21.9
Groundnut cake (decorticated)	90.0	45.4	6.0	6.5	5.7	26.4
Soy meal (extracted)	89.0	44.8	1.5	5.1	5.5	32.1
Sesame cake	91.0	44.7	11.9	4.5	8.9	21.0
Coconut cake	90.0	21.2	7.3	11.4	5.9	44.2
Broad beans (*Vicia faba*)	89.0	23.4	2.0	7.8	3.4	52.4
Bengal gram (*Cicer arietinum*)	89.0	20.1	4.5	4.9	2.9	56.6
Lentils (*Lens esculenta*)	89.0	24.2	1.8	3.1	2.2	57.7
Dried cassava roots	94.4	2.8	0.5	5.0	2.0	84.0
Dried potatoes	91.4	9.7	0.3	2.1	4.3	75.0
Dried grass	90.0	15.0	2.6	20.9	10.8	40.7
Alfalfa meal	92.7	21.1	3.3	17.5	11.5	39.3
Beet molasses	80.5	8.4	0	0	10.1	62.0
Cane molasses	73.4	3.0	0	0	8.6	61.7
Dried beet pulp	91.2	8.8	0.6	19.6	3.5	58.7
Fish meal (white)	87.0	61.0	3.5	—	21.0	1.5
Fish meal (oily)	93.5	62.1	8.3	0.7	18.2	4.2
Pure meat meal	89.2	72.2	13.2	—	3.8	—
Dried blood	68.0	81.0	0.8	—	2.7	1.5
Bone meal (cooked)	93.6	26.0	5.0	1.0	59.1	2.5
Feather meal	94.6	87.4	2.9	0.6	3.7	0
Poultry byproducts meal	93.4	55.4	13.1	1.6	18.7	4.6
Dried skimmed milk	89.7	32.8	1.5	—	7.5	47.9

From Cockerell, I., Francis, B., and Halliday, D., *Proc. CENTO Conf. Dev. Feed Resources,* Tropical Products Institute, London, 1971, 182. With permission.

feed. The scheme of this paper is based on an appreciation of the function of feed concentrates which, briefly stated, assumes that cereals normally provide most of the energy content of feeds and may often supply most of the protein requirements of mature animals. However, because growing monogastric animals require larger proportions of protein in their rations than would be supplied by cereals alone, oilseed residues from soybeans, groundnuts and cottonseed are widely used as protein supplements together with fish and meat products. The feed concentrates under discussion are not generally considered good vitamin sources per se except perhaps some of the cereals which are rich in the water-soluble B complex and leafmeals which are used as sources of both xanthophyll and carotenoid pigments. Losses in vitamin activity can occur in storage but, almost without exception, residual activity is supplemented in feeds with standard mineral/vitamin mixes.

Many of the changes which are discussed here and which occur in the nutritive value of concentrates during storage are the result of chemical modification and interaction of components of the energy and protein fractions which either reduce or destroy the availability of nutrients to animals.

Table 2
ANTICIPATED BIOLOGICAL AND CHEMICAL ACTIVITY AT VARIOUS RH LEVELS FOR FEED CONCENTRATES

Moisture (%)	RH at 20—30°C (%)	Biological activity	Chemical activity
Up to 8	30	Not significant	Lipid oxidation — increase in peroxide compounds
8—14	30—70	Possible insect infestation	Increase in uric acid content
	>60	Mite infestation	Maillard-type reaction
14—20	70—90	Insect infestation	Production of mycotoxins
		Growth of storage fungi	Lipolysis — increase in FFA and off-flavors
20—25	90—95	Insect infestation	Increased microbial production of toxins
		Mold growth including field fungi Bacterial growth	
>25	—	Bacterial growth Germination	Loss of physical form Gross spoilage Depolymerization of starches and proteins

This deterioration occurs when the moisture level of products increases or during heating. Dry seeds such as cereals and legumes are largely biochemically inactive and most chemical changes that may take place are thought to be nonenzymic.[87] When stored at low moisture content, viable seeds respire at a level which cannot be measured and hence utilize extremely low levels of substrate; mass is therefore preserved.[14] Other changes induced by tissue enzymes (e.g., lipases and lipoxidases) are arrested during the processing of vegetable and animal products in the production of some concentrates; such processing involves heat (precooking, expelling, rendering) which destroys enzymes because of thermal denaturation of protein.[85] However, in these cases deleterious nonenzymic reactions occur which may continue at a much slower rate during storage, e.g., the Maillard reaction.

Loss in nutritive value can occur not only through chemical deterioration but also through gross weight loss due to rodent and insect attack. It cannot be overemphasized that economic loss in the value of stored concentrates caused by rodents and insects is normally much greater than that brought about by chemical reactions taking place in nutrient fractions.[1,30,34,65]

FACTORS AFFECTING NUTRITIVE CHANGES

Ambient Conditions

Ambient conditions which effect stored concentrates are mainly temperature and relative humidity and to a lesser extent atmospheric oxygen and actinic light. Temperature and humidity not only influence the rate at which chemical changes may take place but also the growth of insect pests and fungi. Certain chemical changes are closely related to the activity of fungi and insects but in practical terms of storage it is not always possible to distinguish these from those related to ambient conditions. It should be noted that although moisture content can influence chemical changes which are not biologically induced, its greatest effect is on the biological changes shown in the Table 2.

Microorganisms

The most important microorganisms which attack stored feed materials are the fungi; bacteria being of less importance due to the high relative humidity (RH) necessary for their growth. Bacteria are of importance particularly in feed materials of animal origin mainly as pathogenic contaminants such as *Salmonella* spp.

Mold growth reduces the value of feed materials by adversely affecting flavor, enhancing chemical reactions which reduce nutritive value and producing toxic metabolites. Animals are influenced by the organoleptic properties of their feed and the mustiness caused by molds that affects acceptability. Enzymes secreted by fungi, the most common being lipases, enhance the lipolysis of glycerides. Fungi also assist in the development of ketonic rancidity of oil. The heating effect of fungi (particularly by *Aspergillus candidus* and *A. flavus*) may also assist in the development of nonenzymic browning.[14] The most serious aspect of fungal contamination of feeds concerns the possible production of toxic metabolites, by far the most important being aflatoxin.[38]

Insect Infestation

Consumption of feed by insects may lead to powdering and weight loss while palatability may be reduced by insect frass and webbing, etc. In the case of wheat grains there may be an additional loss in nutritive value due to preferential consumption of certain nutritious and anatomically differentiated parts (endosperm, germ).

In serious insect infestations the heat produced by insect respiration can cause the temperature of a grain bulk to rise appreciably. If the moisture of stored grain is below the advised safe level this heating is described as "dry heating". In the absence of fungal growth, temperatures may rise to 38 to 40°C but thereafter insects are unable to tolerate higher temperatures and move to cooler areas. The danger of heat production in such instances can readily be appreciated as it would appear that many chemical reactions leading to nutritive deterioration show a normal relation to temperature.[12]

Use of moldy and infested feed has generally resulted in both poor food intake and weight gain in laboratory and farm animals; off-flavors, the presence of insects and frass, specific nutrient deficiencies, and the presence of excreta, uric acid and mycotoxins have been put forward as possible causative factors.[34,38,66,75,79] From an analytical point of view, the proximate analysis of infested grain is similar to that of low test weight or damaged grain, i.e., higher in protein and fiber but reduced in nitrogen free extract.[5] Use of damaged grain has proved feasible although metabolizable energy values may be lower than for sound grain. In insect damaged feed the changes in the distribution of proximate analysis parameters is brought about by a loss of part or parts of a seed (i.e., germ, endosperm, or both). In wheat infested with *Sitophilus oryzae* and stored for 14 weeks, the specific nutrient loss was calculated as 42% of nitrogen free extract, and a 19% loss of protein.[24]

CHANGES ASSOCIATED WITH ENERGY FRACTIONS

Carbohydrates

Carbohydrates may be grouped into two classes: sugars and nonsugars (Table 3). Starch is the energy reserve of cereals and certain tuberous roots both of which are used as concentrated energy sources in nonruminant animal feeding and for fattening cattle. Cereal starch is concentrated in the endosperm of the grain: 70% of wheat endosperm and 98% of maize endosperm is starch. The levels of nitrogen free extract shown in Table 1 for cereals approximate the sum of soluble carbohydrate and starch content.

The present knowledge of changes in carbohydrates during storage has been obtained from studies with cereals, particularly wheat, because of its importance in human nutrition.[71] Changes in the free sugars (about 2.5% in wheat) occur at moisture contents above 14% with an increase in reducing sugars and a decrease in nonreducing

Table 3
CLASSIFICATION OF CARBOHYDRATES

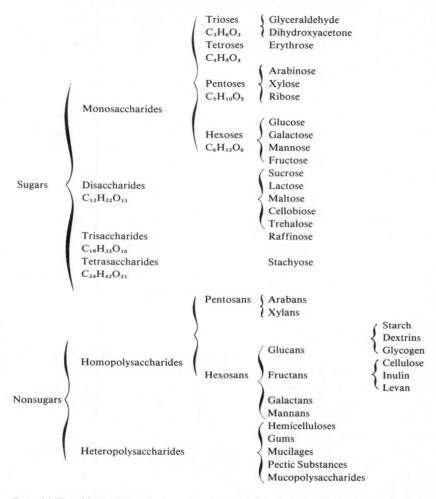

From McDonald, P., Edwards, R. A., and Greenhalgh, J. F. D., *Animal Nutrition,* 2nd ed., Longman, London, 1973, 8. With permission.

sugars (Table 4); with increasing respiration, sugars are converted to carbon dioxide and water. Below 14% moisture and at 20°C cereal respiration is slow; respiration of wheat grains free of storage fungi, and even with moisture contents of 14 to 18% and kept at 35°C, remains so low as not to be detectable.[36] The respiration rate of grains in actual storage has not been measured at moisture contents in equilibrium with relative humidities at which deterioration commences (70 to 80%) whether grains are free from or invaded by storage fungi. By the time the moisture content of the mass becomes high enough for respiration to be measured the seeds are dead and decayed through fungal growth and activity.[14] While the seeds are respiring and viable and in the absence of fungal contamination, high temperature, and humidity, loss of starch substrate by enzymic activity is, therefore, subliminal. Respiration, however, increases with temperature and moisture; amylases become active and convert starches to dextrins and maltose and glycosidases convert sucrose to glucose and fructose. Such enzyme activity may increase the digestibility of carbohydrate by making sugars avail-

Table 4
SUGAR ANALYSIS OF STORED BARLEY

	Moisture (%)	Storage time (weeks)	Reducing sugars (% DM[a])	Nonreducing sugars (% DM)
Starting material	12.6	0	0.25	2.55
	12	30	0.25	2.44
	14	30	0.26	2.39
	16	30	0.33	2.01
	17	30	0.39	1.75
	18	30	0.48	1.33
	20	23	0.40	1.35
	22	11	0.39	1.43
	23	7	0.31	1.49
	26	3	0.30	1.88

[a]   DM = dry matter.

Adapted from Lund, A., Pedersen, H., and Sigsgaard, P., *J. Sci. Food Agric.*, 22, 460, 1971. With permission.

able for assimilation. With increasing relative humidities and water production, gross spoilage is likely to be more important than loss of carbohydrate.

Lipids

With few exceptions concentrate feedstuffs are low in lipids (Table 1). One of the major components of compound feeds is barley which has only about 2% oil; with the exception of oats (5% oil) most other cereals contain similar levels to barley. Nearly all protein concentrates originate from material high in lipids which has been processed to remove the oil; oilseeds are crushed or solvent-extracted, animal fat is separated during rendering, and fish oil is removed during the manufacture of fish meal. Nevertheless, up to 10% residual lipids can be present in oilcake.

The larger part of feed concentrate lipids are essentially glycerides, triesters of glycerol and fatty acids with smaller quantities of phospholipids, carotenoids and other fat-soluble vitamins, and sterols. The fatty acid composition of a number of fats and oils of importance to the feedstuffs industry and of the fats contained in some widely used feeds are shown in Table 5 and the fatty acid composition of cereals in Table 6. Those of vegetable origin are characterized by a high degree of unsaturation. The fatty acids of coconut oil and palm kernel oil contain a preponderance of the shorter chain saturated fatty acids and very little unsaturated fatty acid. It should be noted that palm oil from the kernel mesocarp is very different in composition from palm kernel oil (Table 5). In animal fats the major saturated fatty acids are palmitic (CI6:0) and stearic (CI8:0) and the major unsaturated acid is oleic (CI8:1). Unlike the unsaturated oils in plants (e.g., linseed), fish oils are characterized by a considerable quantity of highly unsaturated long chain fatty acids.

The fatty acids in cereal grains comprise large amounts of palmitic (CI6:0) and linoleic (CI8:2) (Table 6). Pasture lipids account for some 5 to 10% of the dry matter of herbage. The major lipid fractions in such material comprise mono- and digalactosyl derivatives of diglycerides and the fatty acids present are mostly unsaturated, with linolenic (CI8:3) predominating.

All lipids of biological origin can undergo hydrolysis and/or oxidation in storage (Figure 1). When either or both reactions have proceeded to any great extent the product is described as rancid.

Table 5
FATTY ACID COMPOSITION OF FATS AND OILS (% BY WEIGHT)

	Vegetable									Marine		Animal	
	Linseed	Sunflower	Maize	Cotton seed	Ground nut	Soy bean	Coconut	Palm kernel	Palm	Codliver	Whale	Beef tallow	Lard
Saturated													
Caprylic $C_{8:0}$	—	—	—	—	—	—	5—9	2—5	—	—	—	—	—
Capric $C_{10:0}$	—	—	—	—	—	—	4—10	3—7	—	—	—	—	—
Lauric $C_{12:0}$	—	—	—	—	—	—	44—52	44—52	—	—	—	—	—
Myristic $C_{14:0}$	—	—	T-2	T-2	T	—	7—19	14—19	1—6	2—6	4—9	2—5	T-1
Palmitic $C_{16:0}$	6—16	9—16	7—10	13—25	6—13	7—12	—	7—10	32—47	6—14	10—22	25—32	26—30
Stearic $C_{18:0}$	—	—	T-4	1—3	2—6	3—6	1—4	1—4	1—6	T-1	2—4	21—34	12—18
Arachidic $C_{20:0}$	T	—	T-2	T-1	T-2	—	T	T	T	—	—	T	T
Unsaturated													
$C_{14:1}$	—	—	—	—	—	—	—	—	—	T-2	1—4	T-1	T
Palmitoleic $C_{16:1}$	—	—	—	T	T	—	T	T	T	10—20	12—18	1—5	2—5
Oleic $C_{18:1}$	13—36	14—70	23—50	18—44	39—65	22—34	5—8	11—19	40—52	25—29 }	33—45 }	20—40	41—51
Linoleic $C_{18:2}$	10—25	20—72	34—61	34—55	16—39	49—60	1—3	T-3	5—11			T-4	3—8
Linolenic $C_{18:3}$	30—60	T	—	—	T	2—10	—	—	T	—	—	T-1	—
C_{20}'s	—	—	—	—	T	—	—	—	—	25—32	8—19	T-1	T-1
C_{22}'s	—	—	—	—	—	—	—	—	—	10—20	1—11	T-1 }	T-1
C_{24}'s	—	—	—	—	—	—	—	—	—	T	—	— }	—
M. pt. °C.	~20	—	—	15	5—7	—	21—24	23—28	27—42	—	—	45—50	36—40

From Armstrong, D. G. and Ross, I. P., *Proc. Univ. Nottingham School Agric. 2nd Nutr. Conf. Feed Manuf.*, Swan, H. and Lewis, D., Eds., J & A Churchill, London, 1968, 2. With permission.

Table 6
COMPOSITION OF THE FATTY ACIDS OF CEREAL LIPIDS

Fatty acids	Wheat (%)	Barley (%)	Oats (%)	Rice (%)	Maize (%)	Sorghum (%)	Millet (%)
Saturated							
$C_{14:0}$ Myristic	0·1	1·0	—	—	—	—	—
$C_{16:0}$ Palmitic	24·5	11·5 }	15·9 }	17·6	12·4—15·6	12·3	16·7—25·0
$C_{18:0}$ Stearic	1·0	3·1 }	— }		1·7—2·4	0·8	1·8—8·0
Unsaturated							
$C_{16:1}$ Palmitoleic	0·8	—	—	—	—	—	—
$C_{18:1}$ Oleic	11·5	28·0	40·4	47·6	29·3—37·5	34·3	20·2—30·6
$C_{18:2}$ Linoleic	56·3	52·3	43·7	34·0	43·4—55·4	49·9	40·3—51·7
$C_{18:3}$ Linolenic	3·7	4·1	—	0·8	1·0—1·2	2·7	2·3—5·0
Others and unsapon.	1·9	—	—	—	—	—	0·3—1·0

From Kent, N. L., *Technology of Cereals,* 2nd ed., Pergamon Press, Oxford, 1975, 59. With permission.

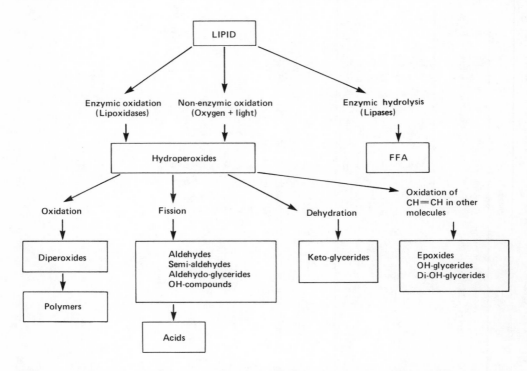

FIGURE 1.   Simplified diagramatic outline of changes in lipids during storage. (From Carpenter, K. J., in *Proc. Univ. Nottingham School of Agric. 2nd Nutr. Conf. Feed Manut.,* Swan, H. and Lewis, D., Eds., J & A. Churchill, London, 1968, 54. With permission.)

Oxidative Rancidity

Oxidative rancidity is a complex phenomenon and the most common and most important type of rancidity encountered in animal feeds. It is likely to occur when there is a high degree of unsaturation in the oils and fats present, as in fish meal. The mechanism of lipid oxidation involves, primarily, autoxidation reactions accompanied by secondary reactions of the lipid hydroperoxides produced, which may or may not themselves be oxidative. Autoxidation, which is by no means confined to the lipid fraction of feeds, can be defined as the direct reaction of the substances with molecular oxygen. It is important to realize that it is not necessary for feed materials to possess a high proportion of lipids for autoxidation to be a serious storage problem.

The main products of lipid autoxidation are hydroperoxides. These are invariably tasteless and odorless but the secondary reactions which they may undergo may lead to off flavors, destruction of essential fatty acids, reactions with protein, and the formation of toxic compounds. The possible paths of these secondary reactions are both diverse and complex and have not yet been fully investigated, but a number of reactions have been established and are discussed fully elsewhere.[52,53]

The oxidation of unsaturated fats apart from producing off flavors can cause the degradation of the natural aroma, flavor, and coloring substances present and also destroy those fat-soluble vitamins which are susceptible to oxidation, e.g., vitamins, A, D, and E. Another important consideration is that autoxidation of highly unsaturated fats is exothermic and so contributes to heat being generated by other systems in stored material. This has been demonstrated in stacked fish meal and oil seeds. The generation of heat in this way together with the products of oxidation can enhance the course of further degradative changes particularly of the nonenzymic Maillard-type (itself an exothermic reaction) as well as destroying the more heat-labile vitamins (e.g., thiamin). In some cases the generation of heat may be sufficient to cause spontaneous combustion.

Ketonic rancidity is a type of oxidative rancidity which is caused by the action of some fungi of the *Aspergillus* and *Penicillium* spp. and characteristic methylalkyl-ketones are produced in oils and fats.[52] The effect is particularly marked when the lipid fraction is rich in oleic acid since a most objectionable taste and odor result.

Hydrolytic Rancidity

The hydrolysis of oils and fats can be brought about by lipases naturally occurring within the tissues of the commodity or by lipases originating from insects and molds.

The hydrolysis of oils and fats, which involves the cleavage of triglycerides to yield free fatty acids, di- and monoglycerides, and glycerol, does not necessarily lead to the development of organoleptically detectable rancidity. It is most marked when fatty acids of low molecular weight are present, for example, those found in coconut and palm kernel. Oils containing mainly fatty acids with more than 14 carbon atoms and only a moderate degree of unsaturation do not develop off flavors when slightly hydrolyzed. In general it may be stated that hydrolytic rancidity is of much greater significance for extracted edible oils than for animal feedingstuffs.

It has been demonstrated that completely hydrolyzed oils can be nutritious and harmless even if they have an extremely rancid and offensive smell.[13] Nevertheless it is generally accepted that oxidation of fat with the introduction of off-odors and off-flavors tends to reduce the palatability of a diet and reduce food intake and consequently the gross intake of all major and minor nutrients.

CHANGES ASSOCIATED WITH PROTEIN FRACTIONS

In the absence of heat, the nutritive value of the protein fraction in stored concentrates should undergo little change. Heat, however, damages protein and reduces both its nutritive value and digestibility.[12,78] There is sufficient evidence to show that such protein damage and duration of heating are normally related so that prolonged storage below 100°C may be as damaging as a much shorter treatment above 100°C. Excessive heating leads to the production of melanoidins (brown nitrogenous polymers and co-polymers) and complete destruction of organic matter with the elimination of ammonia, carbon dioxide, and hydrogen sulfide (as may happen in the spontaneous combustion of oilseeds and oil press cake during storage).

Some of the complex changes involved in the heating of protein are generally and collectively described as the nonenzymic browning or Maillard reaction, of which recent reviews have been published.[21,33,76,77] As is well-known, the relevance of this reaction is primarily the reduction in the availability of lysine in monogastric nutrition. Changes to protein because of heat damage are not necessarily limited to protein-nonprotein reactions (i.e., sugar/amine or oxidized lipid carbonyl/amine bonding) as similar damage occurs in protein preparations which are essentially carbohydrate-free (fish and meat products) and in pure protein preparations.[9] It is generally accepted that severe heat damage involves other amino acids (in particular methionine and cystine) as well as lysine, all of which become nutritionally unavailable, a condition which coexists with a general reduction in the digestibility of the protein. The degradation of protein then not only involves loss of essential amino acids but also an increase in inter and intra molecular bonding which decreases the solubility of protein and renders it resistant to normal protease activity. Poor protein quality in feed concentrates has in fact been correlated with low values in the availability or loss of certain essential amino acids notably lysine, cystine, and methionine.

While the effect of heat on rendering lysine unavailable in such products as oilseed meal and in milk drying is well-known, the mechanisms by which methionine can follow a similar fate and become unavailable is not understood although this condition has been demonstrated.[58] Destruction of the sulfur amino acids generally takes place at much higher temperatures than are experienced during the storage of feed concentrates.[88] It is relevant to note that a distinction can be drawn between the low temperature protein-sugar reaction which occurs under mild conditions, and the damage to protein which occurs at temperatures near or beyond 100°C. The former can take place at normal storage temperatures but the latter normally only occurs during industrial or commercial processes such as oilseed crushing and the drying of fish meal.

Experiments describing the change in the nutritive value of some stored concentrates which occur purely as a single or multiple effect of time, temperature, or humidity and in the absence of insect or fungal activity are few. A particular example of detailed and controlled work in this field is described in the next section.[89]

STORAGE EFFECT ON VITAMINS

Vitamins present in feed concentrates are distributed between the various macronutrient fractions. Those that are loosely classified as water-soluble are associated with the carbohydrate fraction while the fat-soluble class are present in lipid tissue and as the plant carotenoid precursors of vitamin A. Specific examples of loss of vitamin activity are discussed in the next section.

CHANGES IN CONCENTRATES DURING STORAGE

Cereals and Milling Products

Changes which may affect the nutritive value of cereals during storage were briefly indicated in the preceeding sections; these include a decrease in available protein and nonreducing sugars and an increase in reducing sugars and FFA. With good storage practice, these changes can be insignificant; wheat stored in the UK for 16 years underwent little quality change.[69] In fact cereal which remained sound during prolonged storage for some years maintained its feeding value when fed as a source of energy in mixed diets to laboratory animals. No significant decrease in protein efficiency (PE) was noted after 5 years (except for one sample from a different location where the PE decreased from 2.72 to 1.81).[11] Changes during the storage of wheat, wheat germ, barley, rice, and corn have been reported.[51,54,71,72,83] However, under good commercial

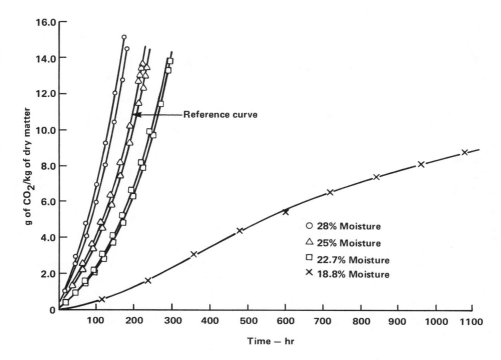

FIGURE 2.   Carbon dioxide production of maize held at 65°F (18.3°C). (From Saul, R. A., *Iowa Farm Sci.*, 22(1), 22, 1976. With permission.)

storage conditions, little enzyme activity occurs in the sugar or polysaccharide fractions. Regarding the levels of free sugar mentioned above with barley, it has been shown that as the level of grain moisture increases beyond 14% there is a corresponding increase in the activity of glycosidic enzymes, the level of reducing sugars (glucose and fructose) increases and nonreducing sugars (sucrose) decreases, an activity common to other cereals (Table 4).[51] At the higher moisture levels the grain temperature increased and feeding quality was impaired. A correlation was found between moisture content, increase in mold growth, carbon dioxide production, and changes in the sugar fractions.

Respiration and Germination Effects

If high moisture and temperature levels are tolerated in storage, germination and then starch hydrolysis will occur. The germination of barley can be initiated at 25% moisture and at temperatures above 22°C when loss of dry matter by respiration occurs.[39] It has also been shown in stored maize that the production of carbon dioxide by oxidation of carboyydrates during respiration results in a loss of dry matter. The evolution of 14.7 g of CO_2 per kilogram of dry matter is equivalent to the loss of 1% dry matter. Figures 2, 3, and 4 indicate the reduction in rate of deterioration of maize as measured by CO_2 production against an increase in moisture and temperature during storage.[83]

Simultaneously with the activity of respiratory enzymes, other enzyme systems (e.g., cytases, hemicellulases, gumases, gluconases) bring about the breakdown of polysaccharides to smaller molecules. Proteolytic enzymes also become active and are important in the release of the starch-hydrolyzing enzymes, the α- and β-amylases. β-amylase is released from a bound inactive state, α-amylase is formed and starch breakdown follows.

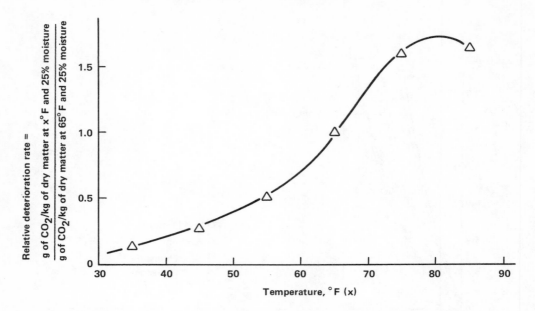

FIGURE 3.   Relative grain deterioration rate (g CO_2/kg DM at ×°F and 25% M.C.) ÷ (g CO_2/kg DM at 65°F and 25% M.C.) with respect to temperature. Maize at 25% moisture content. (From Saul, R. A., *Iowa Farm Sci.*, 22(1), 22, 1976. With permission.)

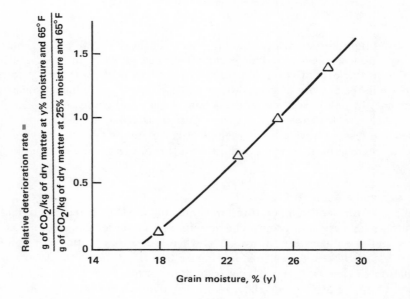

FIGURE 4.   Relative grain deterioration rate (g CO_2/kg CM at y% M.C. and 65°F) ÷ (g CO_2/kg DM at 25% M.C. and 65°F) with respect to moisture content. Maize at 65°F (18.3°C). (From Saul, R. A., *Iowa Farm Sci.*, 22(1), 22, 1976. With permission.)

Enzymes have optimum pH and temperatures at which their activity is highest and the activity of most enzymes doubles with every 10°C increase in temperature as long as the enzyme is not denatured. For barley amylase, little conversion occurs at or below 20°C, an average storage temperature for feed concentrates.[16]

Table 7
DECREASE IN AVAILABLE
LYSINE CONTENT (AS %
PROTEIN) OF FOODS
STORED UNDER VARIOUS
CONDITIONS

	Initial (%)	Maximum decrease (%)
Nonfat dry milk	6.45	87.6
Wheat	3.00	43.3
Rice	4.05	18.5
Soybeans	5.50	41.9
Chickpeas	4.70	30.0
Soybean meal	4.00	5.0
Cottonseed meal	4.20	35.7
Peanut meal	4.50	33.3

From Ben-Gera, I. and Zimmerman, G.,
J. Food Sci. Technol., 9, 113, 1972. With
permission.

The activity of starch-splitting enzymes can increase the digestibility of cereals by making sugars available for assimilation but ambient conditions under which this activity is increased are favorable to microbial growth. The warm moist conditions required for seed germination are also ideal for the growth of various microorganisms which produce starch-splitting enzymes which will also hydrolyze the starch in the exposed endosperm of broken or damaged grains. Such microorganisms are *Bacillus subtilis, B. mesentericus, Aspergillus niger,* and *A. orysae.*[16]

Changes in the carbohydrate fraction of cereals during storage is important because of the reaction between the carbonyl group of reducing sugars and ε-amino lysyl protein residues.

Cereal Protein

Cereal-based compounded feeds for nonruminants contain up to 80% cereal with a protein fraction limiting in lysine. This underlies the importance of lysine supplementation by use of protein concentrates rich in this amino acid. Table 7 shows changes in the available lysine content of wheat and rice (and a number of other feedstuffs) during storage.[8] Changes in the protein nutritive value of the wheat and rice, in which the carbonyl-amino browning reaction was assumed to be responsible, were confirmed by a controlled experiment with the exclusion of microorganisms and other pests.[89] Wheat and rice were stored in cloth bags for up to 24 months and deterioration in protein quality was measured by Net Protein Ratio (NPR), Protein Retention (PR), Net Protein Utilization (NPU), and Protein Efficiency Ratio (PER). Storage conditions of 40 and 60% relative humidity and under vacuum (in metal cans) were used. The wheat and rice underwent only slight physical change over the storage period but the decrease in nutritive value was significant.

The results for wheat and rice are given in Table 8. The values of all four parameters indicated that time had the greatest influence in decreasing protein nutritive value of the canned wheat samples. PER reflected significant changes by time and temperature, but time, temperature, and relative humidity were significant as reflected by NPR, PR, and NPU parameters. The PR was also influenced significantly by all three storage

Table 8
COMPARISON OF NUTRITIONAL PARAMETERS
AGAINST LEAST AND HIGHEST TEMPERATURE
AND RELATIVE HUMIDITY FOR STORED WHEAT
AND RICE

	Storage time (months)	RH (%)	Temp. (°C)	PER	NPR	PR	NPU
Wheat	0	—	—	3.09	4.08	42.0	50.7
	6	40	20	3.23	4.16	43.4	51.9
	24	60	40	2.69	3.55	39.2	47.5
Rice	0	—	—	3.22	4.09	50.7	58.0
	6	40	20	3.37	4.27	48.0	56.3
	24	60	40	3.23	3.93	36.1	44.1

Adapted from Yanni, S. and Zimmerman, G., *J. Food Sci. Tech.,* 7(4), 183, 1970. With permission.

factors. Rice samples were less affected by storage conditions than those of wheat. Only time had a significant influence on all nutritional parameters; temperature showed a significant influence but in no case could any influence of the relative humidity be proved. No measurable color change had taken place with either cereal. These experiments indicate that if external biological influences are completely absent, at below 60% RH and 30°C the protein nutritive change measured by biological testing of cereals stored over 2 years, although statistically significant, is extremely low and such deterioration would not be significant in practical terms of concentrate usage. Germ discoloration (associated with protein damage) increases with heavy mold infestation but can also occur in the absence of mold growth at high storage temperatures (20-30°C).[56] Fat acidity also increases with germ damage.[26]

Cereal Lipids

Changes in the lipid fraction of stored cereals can be summarized as: oxidation of unsaturated fatty acids, decrease in lipids due to oxidation, decrease in nonpolar lipids, and the rapid disappearance of glycolipids and phospholipids. In terms of feed value for animals, commercial storage of unground cereal in temperate regions and in the absence of mold attack results in relatively little change in lipids. In wheat stored at an initial moisture content of 12.6% for 8 years the FFA increased but was well below the level which would indicate deterioration.[68] In the absence of any factors which may limit feed intake the feed value of the free fatty acids which may be released by hydrolysis during storage should be similar to the feed value of the original cereal glycerides.

The enzyme lipase in raw oats is located almost entirely in the pericarp of the kernel or groat. Oat fat is distributed throughout the endosperm, germ, and aleurone layer. Raw sound oats although having an appreciable lipolytic activity remain low in FFA during storage because the fat and lipases which are found in different parts of the grain remain separated. However, unless precautions are taken to inactivate lipases during milling, problems of rancidity arise. Lipases are normally destroyed by steam treatment prior to milling, a process known as stabilization.

Cereal Vitamins

Cereal grains and their byproducts may constitute 60% or more of a compounded feed and are important sources of the water-soluble B vitamins and fat-soluble vitamin E. With the exception of yellow maize, cereals are poor sources of vitamin A activity.

Thiamin (vitamin B_1) is known to be very soluble in water and fairly stable to heat and oxidation, and among the cereals the richest sources are rice bran, rice polishings, and rice germ. During the storage of wheat of 12% moisture, thiamin losses of up to 12% in 5 months were measured under good storage conditions, but little loss was noted in stored maize.[7] The rate of loss of thiamin appears to increase with increase in grain moisture content as illustrated in Table 9 for wheat and rice. Losses of approximately 50% of thiamin in hulled rice over 3 years at 10% moisture was comparable with a similar loss in 2 months at 16% moisture.[44] Very little definite information appears to be available concerning losses of other B vitamins but it is generally believed that these vitamins, with the possible exception of pantothenic acid, are rather stable and not readily destroyed in whole grains under normal conditions of storage.[71] Riboflavin and pyridoxine are rather sensitive to light and may therefore be unstable in milled products exposed to strong light during storage.

Vitamin E activity in plants is shown by four tocopherols and four related compounds known as tocotrienols. The principal contributor to biological vitamin E activity is α-tocopherol but it is unstable in the presence of oxygen, ultraviolet light and especially to peroxidizing unsaturated fats.[4] In cereal grains α-tocopherol is concentrated in the bran and germ with wheat germ oil being a particularly rich source. On average, barley and oats contain the least α-tocopherol (0.6 mg/100g) and U.S. corn the highest (1.7 mg/100g), though wide variations are known depending on the country of origin.[4] Losses of α-tocopherol during storage may be very high; at room temperature losses of 60% from maize stored for 12 weeks and 6% per month from ground maize have been reported.[31,43] The level in wheat meal was reduced by 90% after 63 days storage and the level in wheat germ was reduced by 24 to 30% after 80 days in storage.[61,81] Another study showed a loss of only 8.6% α-tocopherol from maize artificially dried directly after harvest compared to nearly 90% loss for high moisture (20%) ensiled and propionic acid (perservative) treated grain. Atmospherically dried corn (maize) lost about 40% α-tocopherol during processing but subsequent losses over 4½ months were minimal (Table 10).[90] Similar results have been obtained for the rapid loss of α-tocopherol in high moisture propionic acid-treated barley.[2,55] In general, the decline in α-tocopherol would appear to be directly related to moisture content.

Increases in the peroxide values of fat from propionic acid-treated or ensiled barley and corn are also consistent with the loss of naturally occurring antioxidants such as tocopherols.[2,90]

Vitamin A per se is not found in plants, but various levels of provitamin activity are shown by certain carotenoid compounds which like vitamin A are readily oxidized by oxygen and light.[60] Although more than 80 naturally occurring carotenoids are known only 10 have provitamin A activity and of these the greatest activity is from β-carotene.[80] Among the cereals yellow maize is a particularly important source of vitamin A activity because of its cryptoxanthin (a carotene with 50% of the activity of β-carotene) and β-carotene contents, but xanthophyll (the carotenoid giving the yellow color to maize) has no vitamin A activity only good pigmentation properties, e.g., egg yolk coloration when used for poultry feed.[28,74] Other cereals have very little provitamin activity.

The stability of maize carotenes during storage in the dark and in closed containers is illustrated in Table 11. Carotene losses were most rapid during the early stages of storage and were approximately a logarithmic function of time. Losses in cryptoxanthin followed a similar pattern. These values are comparable with the loss of 30 to 40% of dehydrated alfalfa carotenes during 6 to 9 months storage.[10] Under cold storage carotene losses in yellow maize of 6 to 7% have been noted.[25] Storage trials with dehydrated alfalfa indicated that the rate of loss of total carotene increases with in-

Table 9
LOSS OF THIAMIN ON STORAGE

Cereal	Treatment % moisture content	Initial thiamin content (μg/g)	% Loss in storage	Ref.
Wheat	17	3.4	30% in 5 months	7
	12	3.1	12% in 5 months	
	10	—	Little change after 25—33 years on comparing stored with new crop	22
Rice				
Raw husked	10.5	3.3	4 months: 9.1 8 months: 15.1 12 months: 21.2	62
Raw undermilled	10.5	1.8	4 months: 5.5 8 months: 11.1 12 months: 22.2	
Raw milled	10.5	1.1	4 months: 9.1 8 months: 18.2 12 months: 27.3	
Hulled and unhulled	—	—	Thiamin content similar to newly harvested rice after 5 years in concrete silo	45
Hulled rice in straw bags	10	—	1 year: 8 2 years: 18 3 years: 44 4 years: 77	44
Hulled rice	2 months air storage	—	Moisture content %: 12.6 / 13.4 / 14.7 / 16.0 / 16.1 Relative loss of B_1: 0 / 28 / 36 / 51 / 82	

Table 10
LOSS OF VITAMIN E IN STORAGE

Cereal	Treatment	Initial vitamin E content (μg/g)	Loss in storage			Ref.
Corn	Room temperature	8.0	40% α tocopherol in 12 weeks			43
Wheat meal	Room temperature	—	90% α tocopherol in 63 days			61
Wheat germ	Room temperature	—	24-30% α tocopherol in 80 days			81
			7 Weeks	20 Weeks	33 Weeks	
Shelled corn	Artificially dried at 110°C 12% moisture content	9.3	5.4%	4.3%	8.6%	90
	Air-dried 9% moisture content	9.2	40.9%	36.6%	37.8%	
	Ex-field 25% moisture content	9.3	36.6%	67.7%	89.2%	
	Ex-field + propionic acid 25% moisture content	9.3	45.2%	68.8%	88.2%	
Barley	<15% moisture content	8.9	20.2% α tocopherol in 6 months			55
	<15% moisture content	9.7	18.6% α tocopherol in 6 months			
	<15% moisture content + propionic acid	8.9	47.0% α tocopherol in 6 months			
	<15% moisture content + propionic acid	9.7	36.0% α tocopherol in 6 months			

crease in moisture content; e.g., losses of 84% carotenes at 12% moisture compared to losses of 66% at 2.3% moisture for the same storage period.[42]

Rice Bran

Rice bran and polishings contain 14 to 18% oil. Naturally-occurring enzymes liberated during milling cause rapid hydrolysis of this oil as measured by FFA increase.[29] FFA may increase from about 3% in fresh bran at the rate of 1%/hr during the first 4 hours after milling reaching about 60 to 70% within a month. This lipolysis causes untreated bran to develop an unpleasant rancid odor and flavor during quite short storage periods. Drying bran to less than 4% moisture will retard but not stop this reaction. Lipolysis can be limited commercially by solvent extraction and/or the inactivation of lipases by steam treatment.[46] Steaming at just over 100°C inactivates lipases within 5 to 10 min. Dry heating at 110°C for 2 hr effectively inhibits lipolysis by lipase inactivation and the reduction of moisture content. It has been reported that bran from parboiled rice may be more stable than that from milled rice. The compositional data of bran produced from rice in Sri Lanka is given in Table 12, where the oil content of bran from parboiled rice is higher than that from raw rice.

Oilseed Products

Oilseed cake or meal is a byproduct of the oils and fats industry and is used as a source of protein in animal feeding. During storage, rancidity can develop in oilseed

Table 11
LOSS OF VITAMIN A ACTIVITY IN STORAGE

Cereal	Treatment (% moisture content/°C)	Initial carotene content (µg/g)	% Loss in storage					Ref.
			4 months	8 months	12 months	24 months	36 months	
Yellow dent corn	11/25°	4.8	16.7	47.9	62.5	64.6	79.2	73
	11/7°	4.8	35.4	27.1	29.2	29.2	45.8	
	3/25°	3.9	15.4	33.3	30.8	64.1	71.8	
	3/7°	3.9	20.5	10.3	10.3	23.1	43.6	
	(% moisture content)	Total (carotenes) (mg/kg)	6 weeks	12 weeks				
Dehydrated alfalfa	12.2	353	72	84				42
	9.2	342	64	78				
	7.8	269	55	68				
	7.1	342	71	84				
	2.5	293	59	76				
	2.3	291	51	66				

Table 12
COMPOSITION OF RICE BRAN FROM SRI LANKA[29]

Analysis (%)	Bran from raw rice (%)		Bran from parboiled rice (%)	
	1st quality	2nd quality	1st quality	2nd quality
Dry matter	89.1	88.9	89.9	90.1
Crude protein	12.5	8.9	12.8	9.1
Oil	14.6	6.5	20.3	10.4
Crude fiber	8.9	19.1	10.0	17.1
Ash	10.7	19.7	10.7	21.2
Nitrogen free extract	42.9	34.7	36.1	32.9

cake and meal and introduce problems of acceptability in finished feeds.[17] In general solvent-extracted meals contain less oil (less than 2%) than that remaining in press cake (5 to 10% oil). Although storage at high humidity can accelerate the development of rancidity, of greater importance is the risk of the development of toxin-producing molds, particularly *Aspergillus flavus*. Aflatoxin contamination has been particularly common in groundnut and cottonseed products. Complete internal carbonization of stored groundnut cake has been described due to high humidity and microbiological fermentation. Changes occurring during the 13-month storage of groundnut cake (0.86% oil) were an increase in temperature as a function of the initial moisture content, a decrease in oil content, an increase in FFA value, an inversion of nonreducing sugars, and the disappearance of others, and a slight decrease in the percentage of protein. Storage in sacks was less deleterious than in bulk, and storage in the dark better than in the light. It was recommended that the moisture level of the cake must not exceed 10%. The best method of storage was in sacks in ventilated piles at 18 to 20°C at 60 to 70% R.H.[17]

Over a period of 8 to 11 months in India some 60% of the original oil present in groundnut oil cake was lost, and the recoverable oil was darker and had a high FFA and unsaponifiable matter; these changes indicate lipolysis, oxidation, and polymerization.[82] The effect on the feeding value was not reported although evidence of the last mentioned change could point to a lowering of digestibility and nutrient availability.

Nonenzymic changes affecting the protein fraction of soybean and groundnut meal result in a decrease in protein nutritive value and antitryptic activity. In storage studies it was reported that the decrease in nutritive value was quite small and confined to samples stored at high temperature over 2 years.[89] The decrease in antitryptic activity did not result in the expected increase in nutritive value. Experimental evidence indicated that toasting soybeans before storage weakened the susceptibility of the protein to browning reactions. The browning of soybean meal was significantly influenced by temperature but not by relative humidity. With groundnut meal, analysis of variance proved that time and temperature factors had a significant influence on all parameters (Table 13).

In the same series of experiments cottonseed meal showed a marked increase in nutritive value expecially at high temperatures and relative humidity during storage and also a decrease in free gossypol content. The increase in nutritive value was attributed to the lower free gossypol content. It was not established that the disappearance of gossypol during storage was entirely due to phenolic-lysine binding. The meal in these studies contained sufficient free gossypol to bind all the epsilon-amino groups of lysine but with the sample in which the greatest reduction of free gossypol had occurred the nutritive value did not reflect a complete dietary unavailability of lysine (Table 14).[89]

Table 13
COMPARISON OF NUTRITIONAL PARAMETER AGAINST LEAST AND HIGHEST TEMPERATURE/RELATIVE HUMIDITY FOR STORED GROUNDNUT MEAL, TOASTED SOYBEAN MEAL, AND UNTOASTED SOYBEAN MEAL

	Storage time (months)	RH (%)	Temp (°C)	PER	NPR	PR	NPU
Groundnut meal	0	—	—	2.84	3.58	43.1	49.9
	6	40	20	2.71	3.60	41.8	46.3
	24	60	40	2.58	3.40	36.3	41.2
Toasted soybean meal	0	—	—	2.80	3.74	39.3	47.6
	6	40	20	2.83	3.78	33.9	43.4
	24	60	40	2.60	3.46	33.7	41.9
Untoasted soybean meal	24	60	40	1.70	2.81	26.8	38.3

Adapted from Yannai, S. and Zimmerman, G., *J. Food Sci. Technol.*, 7(4), 192—193, 1970. With permission.

Table 14
COMPARISON OF NUTRITIONAL PARAMETERS AGAINST LEAST AND HIGHEST TEMPERATURE/RELATIVE HUMIDITY FOR COTTONSEED MEAL STORED OVER 2 YEARS

Storage time (months)	RH (%)	Temp. (°C)	PER	NPR	PR	NPU	Free gossypol (%)	Food intake of rats[a]
0	—	—	0.42	2.58	18.4	36.8	—	—
12	40	20	1.20	2.98	19.3	37.3	0.53	67.3[b]
24	60	40	2.01	2.93	23.7	31.5	0.09	94.7

[a]   Grams dry diet per rat per 10 days.
[b]   T°C = 40.

Adapted from Yannai, S. and Zimmerman, G., *J. Food Sci. Technol.*, 7(4), 195, 1970. With permission.

Grain Legumes

Grain legumes are good sources of protein for livestock; some fifteen types have been listed as suitable for inclusion in animal feeds.[27] In common with other foodstuffs the fats of legumes break down during storage; this is accelerated by insect attack and leads to oxidative rancidity and an increase in free fatty acids. Such an increase was found in field beans (*Dolichos lablab*) and green gram (*Phasaolus aureus*) stored for 6 months in which the KOH equivalent in milligrams per 100 g rose from 29 to 48 in the former and from 40 to 52 in the latter. Bruchid attack raised the figures to 229 and 95 mg/100 g, respectively.[34] Both high temperature (30°C) and relative humidity (80%) affected a fourfold increase in fat acidity in *P. vulgaris* after storage for a year.[35]

Viable grain legumes can, like cereals, generate heat through respiration especially if this is accelerated through storage under unfavorable moisture conditions. Mold growth follows which can raise the temperature of stored soybeans to a maximum of 55°C. Heating can then be carried forward to higher temperatures by nonbiological oxidation with progressive browning of the seeds and eventually complete spoilage. The high content of unsaturated oil in soybeans may be responsible for a component of the spontaneous heating process which occurs more readily in these seeds than in cereal grains.[59] The same comment would apply to other oilseeds containing high levels of unsaturated oil. Soybeans are particularly sensitive to chemical deterioration and do not store well in the tropics. Discoloration of seeds occurs after prolonged exposure at temperatures as low as 38°C. Advanced physical deterioration and darkening can occur from prolonged exposure to temperatures below that at which proteins are heat denatured as indicated by the retention of urease activity which has been noted in commercial samples.

Although any sequence of biological or nonbiological heating can reduce the nutritive value of protein it can also have some beneficial effect by reducing the antitryptic activity (ATA) of legumes. However, although denaturation of proteins at 20 and 40°C has been reported in stored soybeans as evidenced by a reduction of ATA, this did not improve the overall nutritive value of the product. In soybeans stored under vacuum at 40°C the ATA after 2 years was practically zero and, in contrast to published evidence concerning heat treatment, was not accompanied by the expected increase in nutritive value; the initial PER and NPU were 2.34 and 3.35, respectively, and after 2 years these decreased to 2.17 and 3.11, respectively. The influence of storage conditions (time and temperature) was similar in fact to that of overheating.[89] Only small changes were observed in the nutritive value of chick peas stored for 2 years and only storage time appeared to significantly affect all the nutritional parameters measured.[89]

Animal Products

There are more storage problems with fish meal than with other animal products. The difficulties with meat meals occur during processing when the high temperatures used for eliminating the potential danger of *Salmonella* infection can reduce the important available lysine content. However, in common with most concentrates there are no problems with protein deterioration when stored well under dry conditions.

Dried Milk

One animal product, skim milk powder, however, is very susceptible to deterioration.[67] This is due to its unique chemical and physical structure; dried milk powder has a protein-plus-lactose content of from 60 to 90% depending upon the original material. In this high concentration of loosely associated protein and reducing sugar the Maillard carbonyl-amino reaction can proceed at ambient temperatures with the formation of melanoidins.[41] With whole milk powder oxidation of fat during storage is the most serious deteriorative change.[84] Milk powder of moisture content not above 5% will store reasonably well at moderate ambient temperatures if kept dry and in an environment of low relative humidity. The only ideal packing method is gas packing in airtight metal cans which can preserve milk powder for 7 years or more in tropical conditions. Packed in cloth bags even at 20°C and at not more than 60% RH, the NPU value of defatted milk powder dropped from an initial 51.4 to 32.1 after 2 years. After 2 years at 50% RH and 40°C the original nutritional value could not be achieved by the addition of synthetic lysine to the rat diet in sufficient quantity to replace that lost.[89] This would indicate that one or more of the other essential amino acids also underwent considerable damage. Changes in the nutritional parameters of stored defatted milk powder are given in Table 15.

Fish and Fish Meal

Fish oils readily autooxidize due to the polyethnoid acids of the C_{18}, C_{20}, C_{22}, and C_{24} series, those of the C_{18} and C_{20} groups being by far the most abundant (Table 16). Fish meals made from lean white fish with a low fat content of low reactivity are therefore more stable than meal from oily fish with highly reactive fat. During storage, oxidation and polymerization of unsaturated fish oil can limit the digestibility of pro-

Table 15
COMPARISON OF NUTRITIONAL PARAMETERS AGAINST LEAST AND HIGHEST TEMPERATURE/RELATIVE HUMIDITY FOR STORED DEFATTED MILK POWDER

Storage time (months)	RH (%)	Temp. (°C)	PER	NPR	PR	NPU
0	—	—	2.91	3.70	39.8	51.4
6	40	20	2.69	3.80	37.1	47.9
24	60	40	−0.23	1.38	−0.2	16.9

Adapted from Yannai, S. and Zimmerman, G., *J. Food Sci. Technol.,* 7(4), 182, 1970. With permission.

Table 16
COMPONENT ACIDS(% WT) IN ANIMAL DEPOT FATS

	Saturated	Unsaturated			
	C_{16} (palmitic)	C_{16} (hexadecenoic)	C_{18}	C_{20}	C_{22}
Marine fish	12—15	15—18	27—30	20—25	8—12
Domestic fowl	25—26	6—7	60	0.5—1	
Ox	27—30	2—3	40—50	0.2—0.5	
Sheep	23—28	1—2	40—50	0—6	
Pig	25—29	2—3	50—65	0.3—1	

After Hilditch, T. P. and Williams, P. N., *The Chemical Constitution of Natural Fats,* 4th ed., Chapman and Hall, London, © 1964 by T. P. Hilditch and P. N. Williams.

teins by either physically preventing the digestive enzymes from reaching peptide bonds or reacting with the amino acid residues of protein to form indigestible compounds.[3,20,48,49,64] The exothermic oxidation of fresh fish meal which occurs during storage can also enhance the browning reaction and so reduce the nutritive value of protein.

Reactive meals are "cured" before stacking and shipping, that is the bags of meal are stood apart or built up in single or double layers in order to allow the heat of oxidation to dissipate. Cured fish meal is never completely inactive and if bulk stocks are large enough there is still a danger of over-heating. Table 17 gives some indication of the relationship between reactivity in terms of the rate of heat generation (RHG), size of stack, and maximum temperature reached.[23]

Spontaneous heating of fish meal through oxidation may be prevented by stacking or packing in airtight conditions such as silos or heat-sealed plastic bags; the small amounts of oxygen present being quickly used up. The disadvantage of such storage is that the meal remains reactive and is liable to oxidation if repacked in paper bags and stacked. Alternatively antioxidants have been effectively used in preventing spontaneous heating of fish meal. Table 17 also demonstrates the use of two antioxidants, ethoxyquin (EQ) and butylated hydroxytoluene (BHT), in reducing heat generation in stacked fish meal. The RHG was reduced to a low but not insignificant value and showed a similar relationship against time to that of untreated meals. It should be

Table 17

PRODUCTION OF SPONTANEOUS HEATING IN FISH MEAL STACKS[23]

	Stack details			Stack predictions		
Size of stacks (feet)	Mass of meal in stack (tons)	RHG at 1 hr at 50°C at time of stacking (cal.×10⁻⁵/g/sec)	History of fish meal	Max. center temp. rise above ambient (°C)	Time to reach max. center temp. (hours)	Center temp. rise after 1000 hr (°C)
9×9×9	13	1.5	0.012% ethoxyquin in meal	7.4	1,165	—
63×21×6	46.2	1.5		8.45	1,037	—
21×21×21	164	1.5		30.0	3,940	12.7
21×42×21	1000	1.5		56.2	6,400	13.3
9×9×9	13	2.3	0.1% BHT in meal	12.0	1,089	—
21×21×6	46.2	2.3		13.95	1,233	—
21×21×21	164	2.3		54.0	3,460	24.4
63×42×21	1000	2.3		131.5	5,130	25.4
9×9×9	13	8.5	14 days natural curing	66.0	685	—
21×21×6	46.2	8.5		83.6	737	—
9×9×9	13	5.0	28 days natural curing	44.8	813	—
21×21×6	46.2	5.0		51.6	915	—
9×9×9	13	0.7	156 days natural curing	3.4	1,242	—
21×21×6	46.2	0.7		4.23	1,398	—
21×21×21	164	0.7		13.1	4,330	6.2
63×42×21	1000	0.7		22.0	7,520	7.2

Note: Stacking temperature was 20°C in all cases.

Table 18
STORAGE TEST OF FULL PITCHARD MEAL[19]

Ethoxyquin (ppm)	Storage time (months)	Available lysine (g/16gN)	Available methionine (g/16gN)	Corrected for constant casein values (NPU = 65.5, D = 97.5, BV = 67.2)		
				NPU	D	BV
0	1 week	7.4	2.2	64	94	68
400		7.5	2.5	74	95	78
800		7.6	2.5	74	96	77
0	4	7.5	2.6	67	91	74
400		7.6	3.0	74	95	78
800		7.8	2.8	78	95	82
0	8	6.9	2.6	69	91	76
400		7.6	2.9	74	94	79
800		7.6	3.1	79	94	84
0	12	6.9	2.5	69	93	74
400		7.5	2.9	77	96	80
800		7.3	2.7	81	96	84
0	6			63	93	68
1000(BHT)				67	93	72

Note: D = digestibility. BV = biological value. NPU, D, and BV were determined about 1½ to 2 months after the times shown in the table.

noted, however, that although antioxidants inhibit the rate of oxidation so as to retard spontaneous heating of fish meal they do not actually prevent oxidation.[50]

The nutritive value of fish meal can be reduced by the binding effect of oxidizing unsaturated lipid with protein, a 9% reduction of nutritionally available lysine was found in herring meal stored for a year at 25°C when measuring this parameter as an index of protein damage. No such fall was found in a similar sample of meal stored after the extraction of lipid.[48] It is not always possible to predict the processing conditions likely to affect denaturation of protein in stored fish meal, however, an exception was found in a fat-extracted product whose nutritive value remained unchanged after 12 years.[47] It was also reported at that time that no deterioration of quality was established in samples of whole meal and pressed meal stored from 7 to 12 years when fed as a supplement in a chick diet. However, it is generally accepted that treatment of fish meals with antioxidant effectively prevents the denaturation of both protein and lipid during storage although meals protected in this way are more likely to cause off-flavor in the flesh of animals. It is therefore advisable that inclusion levels of fish meal are carefully controlled in compounded feeds.[86] Table 18 shows the beneficial effect of adding ethoxyquin pilchard meal in order to protect nutritive value.[19]

Metabolizable energy of antioxidant-treated herring meals has been maintained at a higher level than those stored without antioxidant and significantly better growth was achieved with chick starter diets supplemented with fish meal which had been treated with anitoxidant prior to several months storage. In this example the difference in protein supplementary value between untreated and treated antioxidant meal was not related to the level of chemically determined available lysine present after storage. A number of other studies have confirmed that antioxidant stabilized fish meal has shown a higher energy value than similarly stored unstabilized meal.[18,63] Vitamin E activity is rapidly lost during the storage of fish meal due to oxidation. This loss can

Table 19
LOSS OF TOCOPHEROL IN FISH MEALS DURING STORAGE IN AIR[19]

Type of fish meal

EQ (%)	Days	Mackerel (mg/lb)	(%)	Anchovy (mg/lb)	(%)	Pilchard (mg/lb)	(%)
0	0	13.33	100.0	10.63	100.0	19.28	100.0
0	10	9.40	70.5	7.70	72.4	7.30	37.9
+0.04		11.50	86.3	8.60	80.9	13.76	71.4
+0.08		—	—	10.40	97.8	17.70	91.8
0	20	6.20	46.6	4.80	45.2	6.20	32.2
+0.04		8.30	62.3	6.60	62.1	10.0	51.9
+0.08		—	—	10.0	94.1	16.30	84.5
0	60	5.65	42.4	3.40	32.0	3.30	17.1
+0.04		6.70	50.3	4.80	45.2	5.50	28.5
+0.08		—	—	5.40	50.8	6.60	34.2
0	90	5.40	40.5	2.20	20.7	3.10	16.1
+0.04		5.20	39.0	4.60	43.3	5.50	28.5
+0.08		—	—	4.00	37.6	6.10	31.6

be reduced with antioxidants but even with the use of ethoxyquin half the tocopherol disappeared during a 3-month storage period (Table 19).[19]

Dried Roots and Tubers
Processed dry starch products such as roots and tubers are essentially inert and storage losses of these generally result from the activities of external biodeteriogens and insects.[37]

CONCLUSIONS

The most important causal agents in the loss of stored feed concentrates are rodents, insects, and molds. In terms of loss of nutritive value, physical losses in weight, and hence loss of nutritive fractions, are far in excess of any lesser damage associated with the chemical reactions which result in various degrees of impairment of nutritive value.

Although changes to protein in feeds which result in loss of digestibility are more often caused by heat pretreatment than by long storage, such nonenzymic denaturation which involves nonprotein bonding with protein can occur at ambient temperatures; when it does so, duration of storage is the most critical factor influencing change. With heating, from whatever source, protein denaturation and the browning reaction will be accelerated. Carbonyl groups originating from both sugars and oxidized fat as a result of contemporaneous chemical activity will also be involved in this type of reaction.

Lipases and lipoxidases in plant products, insects and molds increase the FFA content and the oxidation of fats in stored concentrates and so indirectly reduce the availability of nutrients to animals because of poor acceptability.

Loss of energy substrate in storage other than through removal by insects is generally small. Feeding trials with stored grains indicate little change in ME values although digestibility may be affected with infested grain due to some of the more digestible NFE fraction being consumed by insects at the expense of the fiber fraction.

Although the analysis of a moldy feed concentrate may appear to be satisfactory it is now considered particularly hazardous to use such feed since the discovery of the hepato-carcinogen aflatoxin in 1960; consideration of residual nutritional value of products spoiled by fungi are overridden by the possible presence of dangerous myco-toxins. The use of such material even at reduced inclusion rates is not recommended unless satisfactorily screened for the absence of aflatoxin.

Apart from encouraging mold growth excessive moisture has a particularly deleteri-ous effect on dried milk products. The loss of vitamins in feed concentrates is variable during storage as is the original low level of such compounds in the source material. Loss of vitamins also varies with pre-treatment and the fractionation which occurs during the production of secondary products. Losses are accepted by compounders and made good with supplementation, at recognized rates of inclusion, with standard vitamin mixes.

Dry storage conditions of insect-free material, good aeration, and moderate temper-atures with as short a storage time as is economically and practically possible are of paramount importance with feed concentrates. Any relaxation of these considerations is likely to increase both biochemical and chemical reactions and reduce the effective-ness of nutrient utilization.

ACKNOWLEDGMENTS

The publishers permission to publish Tables 3-8 and 13-16 is gratefully acknowl-edged.

REFERENCES

1. Adams, J. M., A review of the literature concerning losses in stored cereals and pulses published since 1964, *Trop. Sci.*, 19(1), 1—28, 1977.
2. Allen, W. M., Parr, W. H., Bradley, R., Swannack, K., Barton, C. R. O. and Tyler, R., Losses of vitamin E in stored cereals in relation to a myopathy of yearling cattle, *Vet. Rec.*, 94, 373-375, 1974.
3. Almquist, H. J., Changes in fat extractability and protein digestibility in fish meal during storage, *J. Agric. Food Chem.*, 4, 638-639, 1956.
4. Ames, S. R., Tocopherols, occurrence in foods, in *Vitamins: Chemistry, Physiology, Pathology, Methods*, Vol. 5, 2nd ed., Sebrell, W. H., Jr. and Harris, R. S., Eds., Academic Press, New York, 1972, 233-248.
5. Anon., Feed value of damaged grains shown, *Feedstuffs*, 49(8), 32, 1977.
6. Armstrong, D. G. and Ross, I. P., Principles of fat utilization, in *Proc. Univ. Nottingham School Agric. 2nd Nutr. Conf. Feed Manuf.*, Swan, H. and Lewis, D., Eds., J. & A. Churchill, London, 1968, 2—21.
7. Bayfield, E. G. and O'Donnell, W. W., Observations on the thiamine content of stored wheat, *Food Res.*, 10, 485—488, 1945.
8. Ben-Gera, I. and Zimmerman, G., Changes in the nitrogenous constituents of staplefoods and feeds during storage. I. Decrease in the chemical availability of lysine, *J. Food Sci. Technol.*, 9, 113—118, 1972.
9. Bjarnson, J. and Carpenter, K. J., Mechanisms of heat damage in proteins. II. Chemical changes in pure proteins, *Br. J. Nutr.*, 24, 313—329, 1970.
10. Bolton, J. L., *Alfalfa*, Leonard Hill, New York, 1962, 381—384.
11. Cabell, C. A. and Ellis, N. R., Feeding value of stored corn, *J. Anim. Sci.*, 14, 1167—1173, 1955.
12. Carpenter, K., and Booth, V. H., Damage to lysine in food processing; its measurement and its significance, *Nutr. Abstr. Rev.*, 43, 423—451, 1973.
13. Carpenter, K. J., Possible adverse effect of oxidised fat in feeds, in *Proc. Univ. Nottingham School Agric. 2nd Nutr. Conf. Feed Manuf.*, Swan, H. and Lewis, D., Eds., J & A. Churchill, London, 1968, 54—71.
14. Christensen, C. M., Microflora and seed deterioration, in *Viability of Seeds*, Roberts, E. H., Ed., Chapman & Hall, London, 1972, 59-63.

15. **Cockerell, I., Francis, B., and Halliday, D.,** Changes in the nutritive value of concentrate feeding-stuffs during storage, *Proc. CENTO Conf. Dev. Feed Resources,* Tropical Products Institute, London, 1971, 181—192.

16. **De Becze, G. I.,** Enzymes — industrial, in *Kirk-Othmer Encyclopedia of Chemical Technology,* Vol. 8, Mark, H. F., McKetta, J. J., and Othmer, D. F., Eds., Interscience, New York, 1965, 173—230.

17. **Defromont, C. and Delahaye, D.,** Etude du comportement du tourteau au cours du stockage, *Rev. Fr. Corps Gras,* 8(6), 359—375, 1961.

18. **De Groote, G.,** Energetic evaluation of unstabilized and stabilized fish meals in terms of metabolizable energy and heat energy for maintenance and growth, *Feedstuffs,* 40(51), 26-27, 5 4-61, 1968.

19. **Dreosti, G. M.,** New developments: antioxidants in fish meal, in *Symp. Compounders and Concentrate Manuf.,* International Association of Fish Meal Manufacturers, London, 1970, 60-102.

20. **Einarsson, H., Sinnhuber, R. O., and Worthington, O. J.,** Expression of oil from dried fish meal, *J. Agric. Food Chem.,* 2(18), 946—950, 1954.

21. **Feeney, R. E., Blankenhorn, G., and Dixon, H. B. F.,** Carbonyl-amine reactions in protein chemistry, *Adv. Protein Chem.,* 29, 135-203, 1975.

22. **Fifield, C. C. and Robertson, D. W.,** Milling, baking and chemical properties of Marquis and Kanred wheat grown in Colorado and stored 25 to 33 years, *Cereal Sci. Today,* 4, 179-183, 1959.

23. **Fishing Industry Research Institute,** *19th Annu. Rep.,* University of Cape Town, Cape Town, South Africa, 1965.

24. **Francis, B. J. and Adams, J. M.,** Loss of dry matter and nutritive value in experimentally infested wheat, *Trop. Sci.,* 22(1), 55-68, 1980.

25. **Fraps, G. S. and Kremmerer, A. R.,** Losses of vitamin A and carotene from feeds during storage, *Texas Agric. Exp. Stn. Bull.,* No. 557, 1937.

26. **Geddes, W. F.,** The chemistry, microbiology and physics of grain storage, *Food Technol. (Chicago),* 12(11), 7-14, 1958.

27. **Göhl, B.,** *Tropical Feeds,* Food and Agriculture Organization, Rome, 1975.

28. **Goodwin, T. W.,** *The Comparative Biochemistry of the Carotenoids,* Chapman & Hall, London, 1952.

29. **Halliday, D.,** Utilisation of Rice Bran, paper presented at the UNIDO Conf., Madras, India, October 11-16, 1971.

30. **Harris, K. L.,** Rodents, in *Storage of Cereal Grains and their Products,* Christensen, S. M., Ed., American Association of Cereal Chemists, St. Paul, Minn., 1974, 292—332.

31. **Herting, D. C. and Drury, E. J. E.,** Vitamin E content of vegetable oils and fats, *J. Nutr.,* 81, 335-342, 1963.

32. **Hilditch, T. P. and Williams, P. N.,** *The Chemical Constitution of Natural Fats,* 4th ed., Chapman & Hall, London, 1964.

33. **Hodge, J. E.,** Chemistry of browning reactions in model systems, *J. Agric. Food Chem.,* 1, 928-943, 1953.

34. **Howe, R. W.,** Losses caused by insects and mites in stored foods and feedingstuffs, *Nutr. Abstr. Rev.,* 35(2), 285-303, 1965.

35. **Hughes, P. A. and Sandsted, R. F.,** Effect of temperature relative humidity and light on the color of 'California Light Red Kidney' bean seed during storage, *Hort. Sci.,* 10(4), 421-423, 1975; from *Abstracts on Field Beans (Phaseolus vulgaris L.),* Vol. 1, Centro Internacional de Agricultura Tropical, Cali, Columbia, 1976.

36. **Hummell, B. C. W., Cuendet, L. S., Christensen, C. M. and Geddes, W. F.,** Grain storage studies. XIII. Comparative changes in respiration, viability and chemical composition of mold-free and mold-contaminated wheat upon storage, *Cereal Chem.,* 31, 143-150, 1954.

37. **Ingram, J. S. and Humphries, J. R. O.,** Cassava storage — a review, *Trop. Sci.,* 14(2), 131—148, 1972.

38. **Jones, B. D.,** Aflatoxin in feedingstuffs — its incidence, significance and control, in *Proc. Conf. Anim. Feeds Trop. Subtrop. Origin,* Tropical Products Institute, London, 1975, 273-290.

39. **Keen, E. and Dickson, A. D.,** Malts and malting, in *Kirk-Othmer Encyclopedia of Chemical Technology,* Vol. 12, Mark, H. F., McKetta, J. J., and Othmer, D. F., Eds., Interscience, New York, 1967, 861-886.

40. **Kent, N. L.,** *Technology of cereals,* 2nd ed., Pergamon Press, Oxford, 1975.

41. **King, N.,** The physical structure of dried milk, *Dairy Sci. Abstr.,* 27(3), 91-104, 1965.

42. **Knowles, R. E., Livingston, A. L., Nelson, J. W., and Kohler, G. O.,** Xanthophyll and carotene storage stability in commercially dehydrated and freeze-dried alfalfa, *J. Agric. Food Chem.,* 16, 654-658, 1968.

43. **Kodicek, E., Braude, R., Kon, S. K., and Mitchell, K. C.,** The availability to pigs of nicotinic acid in tortilla baked from maize treated with lime water, *Br. J. Nutr.,* 13, 363-384, 1959.

44. Kondo, M. and Okamaru, T., Storage of rice. VI. Physical and biochemical studies of hulled rice stored in straw bags, *Ber. Ohara. Inst. Landwirtsch. Forsch. Kurashiki,* 5, 395-406; *Chem. Abstr.,* 27, 4284, 1933.

45. Kondo, M. and Okamaru, T., Storage of rice. XVI. Storage of rice in concrete silos for five years, *Ber. Ohara Inst. Landwirtsch. Forsch. Kurashiki,* 7, 471-481, 1937; *Chem. Abstr.,* 32, 5087, 1938.

46. Kratzer, F. H. and Payne, C. G., Effect of autoclaving, hot-water treating, parboiling and addition of ethoxyquin on the value of rice bran as a dietary ingredient for chickens, *Br. Poult. Sci.,* 18, 475—482, 1977.

47. Lakesvela, B. and Aga, A. T., Nutritive value and analytical characteristics of new and up-to-12 year old herring meals, *J. Sci. Food Agric.,* 16, 743-749, 1965.

48. Lea, C. H., Parr, L. J., and Carpenter, K. J., Chemical and nutritional changes in stored herring meal, *Br. J. Nutr.,* 12, 297-312, 1958.

49. Lea, C. H., Parr, L. J., and Carpenter, K. J., Chemical and nutritional changes in stored herring meal. II. *Br. J. Nutr.,* 14, 91-113, 1960.

50. Le Roux, B. D. C., Butylated hydroxytoluene as antioxidant in fish meal, *Fish Ind. Res. Inst., 16th Annu. Rep.,* University of Cape Town, Rondebosch, South Africa, 1962, 58-60.

51. Lund, A., Pedersen, H., and Sigsgaard, P., Storage experiments with barley at different moisture contents, *J. Sci. Food Agric.,* 22, 458-463, 1971.

52. Lundberg, W. O., Oxidative rancidity in food fats and its prevention, in *Autooxidation and Antioxidants,* Vol. 2, Interscience, New York, 1962, 451-476.

53. Lundberg, W. O. and Järvi, P., Peroxidation of polyunsaturated fatty compounds, *Prog. Chem. Fats,* 9(3), 379-406, 1968.

54. Lynch, B. T., Glass, R. L., and Geddes, W. F., Grain storage studies. XXXII. Quantitative changes occurring in the sugars of wheat deteriorating in the presence and absence of molds, *Cereal Chem.,* 39, 256-262, 1962.

55. Madsen, A., Mortensen, H. P., Hjarde, W., Leerbeck, E., and Leeth, T., Vitamin E in barley treated with propionic acid with special reference to the feeding of bacon pigs, *Acta Agric. Scand. Suppl.,* 19, 169-173, 1973.

56. McDonald, C. E. and Milner, M., The browning reaction in wheat germ in relation to 'sick' wheat, *Cereal Chem.,* 31(4), 279—295, 1954.

57. McDonald, P., *Animal Nutrition,* 2nd ed., Longmans, London, 1973.

58. Miller, E. L., Carpenter, K. J., and Milner, C. K., Availability of sulphur amino acids in protein foods. 3. Chemical and nutritional changes in heated cod muscle, *Br. J. Nutr.,* 19, 547-564, 1965.

59. Milner, M. and Thompson, John B., Physical and chemical consequences of advanced spontaneous heating in stored soybeans, *J. Agric. Food Chem.,* 2(6), 303-309, 1954.

60. Moore, T., *Vitamin A.,* Elsevier, Amsterdam, 1957, 74-75.

61. Mühlefluh, J., New ways of improving the nutritive value of cereal products, *Med. Ernährung,* 4(1), 20-21, 1963.

62. Narayana, R. M., Viswanatha, T., Mathur, P. B., Swaminathan, M., and Subrahmayan, V., Effect of storage on the chemical composition of husked, undermilled and milled rice, *J. Sci. Food Agric.,* 5, 405-409, 1954.

63. Opstvedt, J., Influence of residual lipids on the nutritive value of fish meal, *Acta Agric. Scand.,* 23, 200-208, 1973.

64. Ousterhout, L. E. and Snyder, D. G., Effects of processing on the nutritive value of fish products in animal nutrition, in *Fish in Nutrition,* Heen, E. and Kreutzer, R., Eds., Fishing News (Books), London, 1962, 303-310.

65. Parpia, H. A. B., Post-harvest losses — impact of their prevention on food supplies, nutrition and development, in *Nutritional and Agricultural Development,* Scrimshaw, N. S. and Behar, M., Eds., Plenum Press, New York, 1976, 195-213.

66. Pawlak, M. and Pion, R., Effect of storage on the feed value of wheat proteins, *Ann. Biol. Anim. Biochem. Biophys.,* 10(1), 171-174, 1970; *Chem. Abstr.,* 73, 637949, 1970.

67. Patton, S., Browning and associated changes in milk and its products: a review, *J. Dairy Sci.,* 38(5), 457-478, 1955.

68. Pixton, S. W. and Hill, S. T., Long term storage of wheat. II, *J. Sci. Food Agric.,* 18(3), 94-98, 1967.

69. Pixton, S. W., Warburton, S., and Hill, S. T., Long term storage of wheat. III. Some changes in the quality of wheat observed during 16 years of storage, *J. Stored Prod. Res.,* 11 (3/4), 177-185, 1975.

70. Pomeranz, Y., Biochemical and functional changes in stored cereal grains, in *Critical Reviews in Food Technology,* Vol. 2, Issue 1, Furia, T. E., Ed., CRC Press, Boca Raton, Fla., 1971, 45-80.

71. Pomeranz, Y., Biochemical, functional and nutritive changes during storage, in *Storage of Cereal Grains and their Products,* 2nd ed., Christensen, S. M., Ed., American Association of Cereal Chemists, St. Paul, Minn., 1974, 56-114.

72. Prabhakara, B. B., Chakrabarty, T. K., Bhatia, B. S., and Nath, H., Storage changes in rice, *Indian Food Packer*, 19(5), 29-37, 1975.
73. Quackenbush, F. W., Corn carotenoids: effects of temperature and moisture on losses during storage, *Cereal Chem.*, 40, 266-269, 1963.
74. Quackenbush, F. W., Firch, J. G., Rabourn, J., McQuistan, M., Petzold, E. M., and Kargl, T. E., Analysis of carotenoids in corn grain, *J. Agric. Food Chem.*, 9, 132-135, 1961.
75. Rajan, P., Sanjeevarayappa, K. V., Daniel, V. A., Jayaraj, A. P. and Swaminathan, M., Effect of insect infestation on the chemical composition and nutritive value of maize and cowpea, *Ind. J. Nutr. Diet.*, 12, 325—332, 1975.
76. Reynolds, T. M., Chemistry of non-enzymic browning. I. The reaction between aldoses and amines, *Adv. Food Res.*, 12, 1-52, 1963.
77. Reynolds, T. M., Chemistry of non-enzymic browning. II. *Adv. Food Res.*, 14, 167-283, 1965.
78. Rice, E. E. and Beuk, J. F., The effects of heat upon the nutritive value of protein, *Adv. Food Res.*, 4, 233-279, 1953.
79. Richardson, L. R., Wilkes, S., Godwin, J., and Pierce, K. R., Effect of moldy diet and moldy soybean meal on the growth of chicks and poult, *J. Nutr.*, 78(3), 301-306, 1962.
80. Roels, O. A., Vitamins A and carotene. IV. Occurrence in foods, in *Vitamins: Chemistry, Physiology, Pathology, Methods*, Vol. 1, 2nd ed., Sebrell, W. H., Jr. and Harris, R. S., Academic Press, New York, 1967, 113-121.
81. Rothe, M., Feldheim, W. and Thomas, B., Vitamin E and cereals. III. Losses of vitamin E and carotenoids due to heat and storage, *Ernährungsforschung*, 111, 386-399, 1958.
82. Sabale, S. R., Rao, V. R., Subrahmanyam, V. V. R., and Kane, J. G., Changes occurring during the storage of groundnut oil-cake, *J. Oil Technol. Assoc. India*, 7(1), 33-35, 1975; *Nutr. Abstr. Rev.*, Ser. B., 47(6), 2999, 1977.
83. Saul, R. A., Rate of deterioration of shelled corn, *Iowa Farm Sci.*, 22(1), 21-23, 1967.
84. Shipstead, H. and Tarassuk, N. P., Chemical changes in dehydrated milk during storage, *J. Agric. Food Chem.*, 1(9), 613-616, 1953.
85. Sunde, M. L., Chemical and nutritional changes in feeds brought about by heat treatment, in *Effect of Processing on the Nutritional Value of Feeds*, National Academy of Sciences, Washington, D.C., 1973, 23—33.
86. Tarr, H. L. A. and Biely, J., Effect of processing on the nutritive value of fish meal and related products, in *Effect of Processing on the Nutritional Value of Feeds*, National Academy of Sciences, Washington, D.C., 1973, 252-281.
87. Thomas, H., Control mechanisms in the resting seed, in *Viability of Seeds*, Roberts, E. H., Ed., Chapman and Hall, London, 1972, 360-396.
88. Woodham, A. A., Food processing and nutrient availability, in *Nutr. Conf. Feed Manuf. Univ. Nottingham School Agric.*, University of Nottingham, 1967, 6-14.
89. Yannai, S. and Zimmerman, G., Influence of controlled storage of some staple foods on their protein nutritive value in lysine limited diets. I. Protein nutritive value of defatted milk powder, wheat and rice, *J. Food Sci. Technol.*, 7(4), 179-184, 1970; II. Protein nutritive value and antitryptic activity of soybeans and chickpeas, *J. Food Sci. Technol.*, 7(4), 185-189, 1970; III. Protein nutritive value and antitryptic activity of soybean meal and peanut meal, and protein nutritive value and free gossypol content of cottonseed meal, *J. Food Sci. Technol.*, 7(4), 190-196, 1970.
90. Young, L. G., Lun, A., Pos, J., Forshaw, R. P., and Edmeades, D., Vitamin E stability in corn and mixed feed, *J. Anim. Sci.*, 40(3), 495-499, 1975.

Specific Feedstuffs

EFFECT OF PROCESSING ON NUTRITIVE VALUE OF FEEDS: CEREAL GRAINS

W. M. Beeson and T. W. Perry

INTRODUCTION

The term "processing" refers to any treatment to which an ingredient — or part of an ingredient — has been subjected. Furthermore, an ingredient may be subjected to a combination of treatments, e.g., "steamed-flaked" or "dehydrated and ground." In this section an attempt will be made to differentiate among treatment effects and also combinations of treatment effects of grains on the performance of various types of animals, wherein such information is available.

Occasionally a processing technique may develop and be in common usage long before its value has been researched. A good example is that of grinding. When equipment such as the horse-operated burr mill became available, it was a status symbol to grind almost every feedstuff available for livestock. Many years later it was demonstrated that grinding did not improve the feeding value of dry corn for swine; many decades later it was shown there was no improvement in the nutritional value of dry corn for beef cattle from grinding it.

BEEF CATTLE

Several methods have been devised for improving the nutritional value of grain by processing. Since grain often makes up from 75 to 90% of the diet for finishing cattle, any process which will improve the utilization of grain should reduce the cost of producing beef.

This section will evaluate the following methods of processing grain for beef cattle: (1) steam flaking, (2) roasting, (3) micronizing, (4) popping, (5) extrusion, and (6) gelatinization.

Steam Flaking

Steam flaking resembles steam rolling, but the flaking method is more specific in that a longer time is given to the cooking or steaming process. In the flaking process, the additional amount of moisture required in the grain prior to rolling will vary with the type of grain and amount of moisture contained in the original grain. Cooking or steaming time will be approximately 12 min and by elevating the temperature to 200°F, corn which contained 15% moisture originally will contain 18% moisture after treatment. In the case of milo, the cooking time under atmospheric pressure may have to be increased to 14 min in order to elevate the moisture content to 18 or 20%.

Following cooking the grain is passed through rollers set to produce flakes 1/32 in. in thickness. As soon as the grain is rolled, it should be dried to approximately 15% moisture.

Colorado researchers have conducted a number of trials to learn more about the value of flaking corn or milo for beef cattle. Matsushima and Montgomery[1] compared thicknesses of flakes (1/32 and 1/12 in.) with finely ground (1/4 in. screen) corn. The results in Table 1 show that cattle fed ground raw corn did not gain as rapidly (2.65 lb/day) as those fed the 1/32 in. flake (2.82 lb/day) or as those fed the 1/12 in. flake (2.70 lb/day). Efficiency of feed conversion reflected the differences in rates of gain.

In subsequent research, Chapman and Matsushima[2] compared the digestibility and

Table 1
PERFORMANCE OF CATTLE FED FLAKED CORN
OF DIFFERENT THICKNESS OR FINELY
GROUND CORN — 165 DAYS

	Thin 1/32 in.	Thick 1/12 in.	Finely ground 1/4 in. screen
Number of animals	14	14	14
Initial weight, lb	485	483	490
Daily gain, lb	2.82	2.70	2.65
Daily corn consumption, lb[a]	12.4	12.7	12.8
Feed per lb gain	6.1	6.7	6.9

[a]   In addition to corn, the ration contained 0.8 lb supplement, 1.7 lb beet pulp, 0.9 lb alfalfa hay, and 5.6 lb corn silage.

From Matsushima, J. K. and Montgomery, R. L., *Colo. Farm Home Res.,* 17, 4, 1967. With permission.

Table 2
AVERAGE DIGESTION COEFFICIENTS
(%) FOR CORN AS AFFECTED BY
PROCESSING

	Thin flake	Dry extruded	Whole	Wet extruded
Dry matter	74.2	71.5	64.8	68.5
Gross energy	72.8	70.1	62.1	67.3
Crude protein	54.8	51.8	41.2	48.2
Crude fiber	22.9	21.0	16.7	20.6

From Chapman, R. J. and Matsushima, J. K., *Colo. State Univ. Anim. Sci. Highlights,* 4, 1970. With permission.

feedlot performance of cattle fed extruded, flaked, and whole shelled corn for fattening cattle (Table 2). Thin flaked corn was more highly digestible.

Oklahoma researchers Totsuek et al.[4] reported that steam process flaking of milo resulted in an increase in intake (11.3 vs. 10.6 lb/day and a subsequent increase in rate of gain (2.63 vs. 2.43 lb/day). However, because the increase in feed consumption closely paralleled the increase in rate of gain, no improvement in efficiency of conversion was obtained. Newsom et al.[5] confirmed the earlier Oklahoma research by demonstrating fattening beef cattle fed steam processed-flaked milo gained more rapidly (2.83 vs. 2.52 lb/day); but because such cattle ate less feed (16.4 vs. 17.0 lb/day), efficiency of feed conversion favored the cattle fed steam processed-flaked milo (5.90 vs. 6.77 lb feed per lb gain). The length of steaming time prior to flaking milo is important.[6] Cattle fed flaked corn which had been steamed 45 min gained 11% more rapidly (2.83 vs. 2.52 lb/day) than those whose corn had been steamed only 15 min. Efficiencies of feed conversion were comparable (Table 3).

Steam processing and flaking of milo resulted in an increased rate of gain (3.10 vs. 2.83 lb/day) and an improvement in efficiency of feed conversion (764 vs. 802 lb feed per 100 lb gain).[7] Steam processing and flaking of barley resulted in an increased rate of gain (3.10 vs. 2.88 lb/day) but no improvement in efficiency of feed conversion (732 vs. 722 lb of feed per 100 lb gain).

Table 3
EFFECT OF GRINDING, CRACKING, FLAKING OR PELLETING OF CORN FOR BEEF CATTLE — 126 DAYS

	Ground	Cracked	Flaked	Pelleted
Number of animals	20	20	20	20
Initial weight, lb	558	585	577	588
Average daily gain, lb	3.30	3.40	3.60	3.50
Daily concentrate consumption, lb[a]	18.1	19.5	18.3	18.5
Feed per lb gain	6.7	6.7	5.8	6.3

[a]   Limited hay was fed in addition.

From Hentges, J. F., Jr., Cabezas, M. T., Moore, J. E., Carpenter, J. W., and Palmer, A. Z., *Fla. Agric. Exp. Stn. Mimeo Series* AN67-4, 1966. With permission.

Roasting Corn

Roasting corn is a relatively new process of treating dry corn for feeding cattle. The roasting process consists of heating corn with dry heat to an exit temperature of about 300°F. The roasted corn has a pleasant "nutty" aroma with a puffed carmelized appearance. Very few of the kernels are actually popped. While raw corn weighs 45 lb/ft^3, the roasted corn weighs only 39 lb/ft^3, indicating expansion during the roasting process. The moisture content of the corn is decreased to 5 to 9%.

Perry[8] presented data from 6 years' research conducted at Purdue University in which roasted corn was compared to raw corn for finishing beef cattle (Table 4). Cattle fed roasted corn gained an average of 8.2% faster and required 9.7% less feed per unit gain than cattle fed unroasted corn.

Burroughs and Saul[9] reported that steers fed roasted corn gained 9% faster and 8% more efficiently than steers fed raw corn (Table 5). In this study, yearling steers with an average initial weight of 612 lb were fed essentially an all-corn diet with 1.0 lb of ground corn cobs and 1.8 lb of supplement daily. Steers on the roasted corn had essentially the same daily gain and feed efficiency as the steers fed 24% high-moisture corn. Roasted corn was superior to refrigerated corn.

In a comparison of whole, ground and roasted corn in a low-mositure alfalfa silage ration, the steers fed roasted corn gained more rapidly (2.94 vs. 2.69 lb daily) and required less feed per 100 lb of gain (732 vs 830 lb) than the steers fed on ground corn.[10] In this test, the steers fed whole corn gained slightly more (2.79 vs. 2.69 lb daily) with a better feed conversion (785 vs. 830 lb) than those fed ground corn. This experiment indicated there was a greater advantage for roasted corn during the early part of the feeding period than during the latter part.

Vetter and Burroughs[11] conducted three trials under different environmental conditions to compare the performance of heavy feeder steers (750 to 830 lb) on whole, cracked, and roasted corn. The steers were fed on a ration composed of 76% corn, 9% ground alfalfa hay, 7.75% dried beet pulp, 3.75% dry premix, and 3.30% liquid supplement. The only variable in the ration was the three types of corn (whole, cracked, and roasted). In two out of three experiments, the steers fed roasted corn gained more rapidly and required less feed per unit of gain. In a 112-day feeding trial with about 360 cattle per treatment, the steers fed roasted corn gained 3.10 lb daily while those on cracked corn gained 2.87 lb, and the feed required per 100 lb of gain was 730 and 850 lb, respectively.

Micronizing

Micronizing consists of heating the grain to 300°F by gas-fired infared generators.

Table 4

COMPARATIVE PERFORMANCE OF CATTLE FED RAW AND ROASTED CORN, 1970-1975, 6 TRIALS[8]

	1970		1971		1972		1973		1974		1975	
	Raw	300°F	Raw	300°F	Raw	300°F	Raw	320°F	Raw	300°F	Raw	375°F
Number of cattle	91	91	75	75	61	61	25	25	25	25	27	28
Initial weight, lb	509	513	550	551	507	507	516	518	550	552	524	524
Length of study, days	112	112	127	127	106	106	191	191	170	170	189	189
Daily gain, lb	2.33	2.66	2.33	2.47	2.23	2.37	2.24	2.42	2.60	2.73	2.19	2.40
% Change	—	+14%	—	+6%	—	+6%	—	+8%	—	+5%	—	+10%
Daily corn, lb	11.9	12.4	15.1	13.5	13.8	12.9	15.3	15.4	10.4	9.2	15.2	13.8
Dry feed per lb gain	6.8	6.2	7.5	6.4	6.7	6.1	7.6	7.0	7.9	7.4	8.4	7.3
% Change	—	-7.4%	—	-14.7%	—	-9.0%	—	-8%	—	-6%	—	-13%

Note. Average increase in gain = 8.2%. Average decrease in feed required per lb gain = 9.7%.

Table 5
PROCESSING CORN — HIGH ENERGY RATION

Treatment	Daily gain lb	% Change	Feed efficiency lb	% Change
Dried	2.56	0	711	0
Roasted	2.79	+ 9	658	+ 8
High-moisture	2.79	+ 9	656	+ 8
Refrigerated	2.59	+ 1	691	+ 3

From Burroughs, W. and Saul, R., *Iowa State University AS Leaflet*, R-145, 1971. With permission.

The term micronization was coined to describe this dry heat treatment since microwaves were emitted from the infrared burners used in the processing. Texas research[12] compared micronized sorghum grain with steam flaked sorghum grain in a field trial. Two lots of 100 steers each were assigned to each type of processed sorghum grain. Feedlot performance was comparable on the two treatments. Those fed micronized sorghum grain gained 4.8% more rapidly (2.62 vs. 2.50 lb/day) but required 2.4% more feed per pound gain (8.60 vs. 8.40 lb) than cattle fed steam flaked milo. However, a 1970 summary of the research indicated the cost of processing favored the micronizing technique over the steam flaking method.

Popping

Whole sorghum grain has been popped by using six gas-fired infrared generators, rated about 50,000 Btu/hr each, suspended about 6 in. above a table to heat the table and also the grain as it was conveyed through the machine. The popped grain had a temperature of 300 to 310°F. The percentage of grain which was actually popped varied from 13 to 45% and appeared to be influenced by moisture content, temperature, and the rate of flow through the machine. As the moisture content of the original grain increased from 11.3 to 14.7% the percentage of popped material increased from 27.4 to 43.2%. Further increase in moisture content to 16.8% did not noticeably enhance popping. The sorghum grain weighed 779 g/ℓ; this decreased to 393 g with 13% popped, 170 g with 45% popped, and 98 g with 100% popped.

Research studies on popped sorghum have given variable results. Ellis and Carpenter[13] reported slightly slower gains but 17% less feed per unit of gain when 40% of cracked milo was replaced with popped milo in an all-concentrate ration for yearling steers. Later Durham et al.[14] obtained daily gains of 2.79, 2.46, and 2.70 lb with almost identical feed conversion for cattle fed cracked, flaked, and popped milo. Cattle fed flaked milo consumed significantly less feed daily than those fed either cracked or popped milo.

Riggs et al.[15] have made an in-depth study of the effect of popping on the nutritional value of milo. Four types of milo were tested, rolled milo, normal run 13 to 45% popped milo, 100% popped fraction and partially popped — left over from screening out 100% popped fraction. The processed milos were self-fed in a mixture composed of 92% milo or popped milo, 7% cottonseed meal, and minerals. Aureomycin and vitamin A were added to all the mixtures to supply 22 mg aureomycin and 4840 IU vitamin A per pound of finished feed. The results of the feedlot test are given in Table 6.

The most striking feature of these data is the highly significant reduction in feed intake of the groups fed on the various fractions of popped milo. Steers fed rolled milo consumed 19 to 37% more dry matter per day than those fed popped milo. The

Table 6
ROLLED VS. POPPED MILO

Treatment	Daily gain lb	Daily feed lb	Feed efficiency lb
Rolled	3.10	21.2	690
Normal popped	2.73	14.9	550
100% popped	2.55	15.1	590
Partially popped	2.75	17.5	640

From Riggs, J. K., Sorenson, J. W., Jr., Adame, J. L., and Schake, L. M., *J. Anim. Sci.*, 30, 634, 1970. With permission.

Table 7
DIGESTIBILITY OF POPPED MILO

Treatment	Dry matter %	Crude protein %	Nitrogen free extract %
Rolled	57	39	61
Normal popped	75	39	75
100% popped	79	38	80
Partially popped	76	41	77

From Riggs, J. K., Sorenson, J. W., Jr., Adame, J. L., and Schake, L. M., *J. Anim. Sci.*, 30, 634, 1970. With permission.

reduced feed intake was accompanied by improved feed utilization but a marked reduction in daily gain.

Digestibility studies explained some of the reasons for the improved utilization of popped milo. Cattle fed the dry heat-treated grains showed significantly higher digestibility of dry matter, organic matter, nonprotein organic matter, and nitrogen-free extract, but not of ether extract, fiber, or protein as compared to unheated rolled milo. From these data it was not possible to determine whether roasting or actual popping was involved in the change in digestibility. A part of the digestibility data is given in Table 7. Significant differences were found in the volatile fatty acids in rumen samples from cattle fed different types of processed milo (Table 8).

Acetic and isovaleric acids both showed significantly higher values, but propionic acid was less in rumen samples from cattle fed rolled milo as compared to popped milo. The acetic to propionic ratio was insignificantly wider for rolled milo than in heat-treated milo. Rolled milo had a ratio of 1.82:1; normal popped, 0.88:1; 100% popped, 1.01:1; and partially popped, 1.06:1. The resulting narrower acetic to propionic ratio in the popped grains was in keeping with greater feed efficiency.

Extrusion

Matsushima et al.[16] have conducted several experiments determining the effect of an extrusion process on the nutritional value of corn and milo. Previous research by these workers has shown that steam flaking milo or corn will improve its feed value from 5 to 15%. Usually milo is improved more from steam flaking than corn. Even though flaking of grains improves their value, this method is essentially adapted to large feedlots due to the cost of the steaming and flaking equipment. Equipment for extrusion is less expensive than for steam flaking.

The new experimental machine developed for this method involves somewhat of an extrusion process. This machine has two working parts — a stator and a rotor. Dry

Table 8
MEAN VALUES FOR VOLATILE FATTY
ACIDS IN RUMEN FLUID

Fatty acids	Rolled milo	Normal popped	100% popped	Partially popped
		Mol %		
Acetic	54.90[a]	41.92[b]	44.62[b]	45.54[a,b]
Propionic	30.20[a]	47.61[b]	44.27[b]	43.01[b]
Butyric	8.78	7.81	8.29	8.91
Isovaleric	4.32[a]	0.99[b]	0.52[b]	0.76[b]
Valeric	1.20[a]	1.62[a]	2.37[b]	1.58[a]
Acetic to propionic ratio	1.80[a]	0.88[b]	1.01[b]	1.06[b]

[a,b] Values with different superscripts on the same line differed significantly — acetic, propionic, and isovaleric at the 0.01 level and valeric at the 0.05 level. Values for butyric acid did not differ significantly.

From Riggs, J. K., Sorenson, J. W., Jr. Adame, J. L., and Schake, L. M., *J. Anim. Sci.,* 30, 634, 1970. With permission.

Table 9
FEED PROCESSING EXPERIMENT — 144 DAYS

Treatment	Daily gain lb	Daily feed lb	Feed efficiency lb
Flaked corn	2.85	22.7	796
Extruded corn	2.69	23.0	857
Extruded milo	2.74	23.6	862
High-moisture corn	2.73	21.9	800

From Matsushima, J. K., McLaren, R. J., McCann, C. P., and Kellog, G. E., *Colo. State Univ. Beef Nutr. Res.,* 1969. With permission.

whole grain, without any moisture addition, gravitates into a hopper, then into a housing which contains the stator and rotor. The auger-like rotor crushes the grain and forces it through an orifice, thus producing flakes about 1 mm in thickness. The ribbon-like product will break into flakes of different sizes and shapes.

Cattle fed the extruded corn gained about 5% slower and required 8% more feed to produce a pound of gain than the cattle fed flaked corn (Table 9). The performance of the steers fed high-moisture corn was very similar in gain and feed efficiency to the steers on flaked corn. Extruded milo was the least efficient of the treatments compared.

Gelatinization

Mudd and Perry[17] conducted a series of studies on the effect of gelatinization of corn on its feeding value for cattle. Gelatinized corn was prepared by processing corn through a Wenger grain expander. The process involved grinding the corn followed by heating with steam to soften the grain. This material was then forced through a steel tube by an auger. Heat (an average of 300°F) and pressure increased as a result of force by the auger which extruded the softened material through cone-shaped holes in the expander head. The holes were smaller where the feed entered and gradually enlarged until the feed was expelled. This caused a sudden release of pressure and the escaping steam expanded the grain. The product was then dried and reground into a fine meal.

Three metabolism experiments and one feeding experiment were conducted to study the effect of gelatinization of corn upon its utilization by beef cattle. In two of the three metabolism studies, the substitution of gelatinized corn for raw corn as a major constituent of the diet resulted in a significant and linear depression in the digestibility of the nutrients. Gelatinized corn decreased feed consumption and cattle gains.

Nebraska researchers[18] reported to the contrary that substitution of 15, 30, or 45% galatinized corn for raw corn tended to increase rates of gain over that of cattle fed raw corn. Also, the efficiency of feed conversion tended to improve with the incorporation of gelatinized corn.

High Moisture

The earliest reported data on the feeding value of ensiled high moisture ground ear corn was that from Purdue University in 1958[19] in which it was reported such feed had from 12 to 15% greater feeding value, per unit of dry matter, based on comparable gains and decreased dry matter intake. The literature is in fairly close agreement on the subject. Unfortunately, the ear picker was pretty much replaced at about that time with the picker-sheller combine which dropped the cob on the ground behind the combine.

Since the above paper was published, a number of papers have appeared on the subject of high moisture grains. Merrill,[20] and subsequently Perry,[21] summarized data from 50 articles on high moisture corn, 55 on high moisture sorghum, and a lesser amount on high moisture barley. Therefore, it is the purpose of this section to make summary statements relative to the value of high moisture grains for beef cattle.

High Moisture Shelled Corn

If one were to pull out isolated research reports on high moisture shelled corn and make conclusions therefrom, almost any conclusion desired could be made. However, under rather ideal conditions of feeding, daily gain is comparable to that of cattle fed lower moisture shelled corn; feed efficiency is almost always improved at least 5% and oftentimes, as much as 10% with a range of from 6 to 10% improvement being one which cattle feeders might anticipate routinely. Therefore, on this basis, a ton of dry matter from high moisture shelled corn (25 to 28% H_2O) is worth nearly as much as a metric ton (2200 lb) of dry matter from "air dry" corn (15% H_2O).

The feeding results from "reconstituted" (dry corn treated with water to bring the moisture level to 25 to 30%) corn has been quite erratic. Generally, in the case of corn, it is recommended that original moisture high moisture corn will give more consistent benefits than "reconstituted" high moisture corn.

It has been concluded generally by research data on the subject that cattle respond to high moisture shelled corn treated with organic acids as they do ensiled high moisture shelled corn.

High Moisture Sorghum Grain

The benefits from high moisture sorghum grain over comparable dry sorghum grain is much more dramatic than it is for corn. Their feeding value in the dry form is lower than their chemical composition would predict. Basic research into this discrepancy indicates the starch of dry sorghum is less available and the protein is not utilized as well as that of corn or barley. Therefore, almost anything that can be done to sorghum grain probably will improve its nutritive value for livestock.

Cattle fed ground moist sorghum grain have required less grain dry matter per pound of gain than did those fed ground dry grain, ranging from 15 to 20% with an overall average of 20%. In practically all comparisons high-moisture sorghum grains have resulted in a lowered feed intake compared to other processing methods such as steam flaked, ground, or rolled dry grain. With similar weight gains, then, feed efficiency is increased.

Further processing of high moisture sorghum grain is important. For example, beef cattle fed ground high moisture sorghum grain gained 11% faster and required 37% less grain dry matter per unit of gain than cattle fed the same high moisture grain in whole form. Rolling either high moisture grain sorghum or dry sorghum is superior to fine grinding, for increasing efficiency of feed conversion.

The change that takes place in reconstituting grain sorghum, some have felt, is similar to that which occurs during germination in which the starch of the endosperm is liquified to an extent for use by the growing seedling.

For sorghum grain ensiled with its original high moisture, ensiling either whole or crushed is equally effective in improving its feeding value over that of dry grain. Reconstituting whole grain, which was then processed prior to feeding has given equal or greater improved daily gains and improvement in feed efficiency, compared to ground dry sorghum grain. However, grinding grain sorghum prior to reconstituting has not been satisfactory, resulting in slightly lower rates of gain and poorer feed efficiency.

The ideal average moisture content for high moisture sorghum grain is 30% with a range of 25 to 35%. Similarly, the ideal reconstituted level is 30% moisture. However, it's most difficult to add more than ten points of water from the starting point.

It is critical in the reconstituting of the sorghum grain that the grain remain in the reconstituting process a minimum of time. That minimum appears to be approximately 21 days.

High Moisture Barley

Barley kernels are physiologically mature when the moisture content drops below 40%. The ideal average moisture content for high moisture barley is 30%, similar to that for high moisture sorghum grain. All the physical advantages related to earlier harvesting for barley add to the increased feeding value of high moisture barley. Research indicates that high moisture barley has a place in cattle feeding — but not because of increased gain or because of improved feed efficiency. The chief advantage of high moisture barley appears to be its high acceptability, with cattle going on feed faster, resulting in better early gains. Cattle stay on feed easier on high moisture barley than do those on dry rolled barley.

High moisture barley should be rolled for beef cattle. In comparative studies, cattle fed whole high moisture barley gained 0.3 lb less per day and required 63 lb more feed per 100 lb gain than cattle whose high moisture barley was rolled prior to feeding.

DAIRY CATTLE

Grain processing has had very little acceptance with dairy cattle. Two problems arise in formulating grain mixes for dairy cattle — that of palatability and that of effect on butterfat production. It is a problem to keep the high producing cows consuming quantities of grain that are commensurate with their level of milk production. In general, no processing technique for grain improves palatability or acceptability appreciably. In fact, very fine grinding of grains — thus producing extra dust — may lower the acceptability of such feedstuffs by dairy cows.

Grinding

Some of the earliest research on the effect of grain processing has been that of grinding. Some have reported that finely ground grain supported normal milk production but Williamson[22] obtained lowered levels of butterfat production when finely ground grain was fed along with limited roughage to dairy cows. In the Williamson study, Holstein cows were fed alfalfa hay and corn silage along with a 16% protein grain mix which had been ground to 3 degrees of fineness. Grinding has been reported to improve the digestibility of grain by dairy cows.[23,24] Pulverizing of the grain component has been shown to reduce milk production.[25]

Pelleting

Variable results from the pelleting of the grain portion of dairy formulations have been reported. It is suggested the variability in results reported may be due in part to the physical form of the roughage — as well as the amount thereof — fed along with the grain in such comparisons. In other words, when a minimum of more coarse roughage is fed, whether or not the grain portion has been ground and pelleted is not nearly as criticial. Waldern and Cedeno[26] compared grain mixes containing 98% barley or wheat against a control mixture of barley, wheat, peas, and cottonseed meal, fed in the meal and pelleted forms to lactating cows. When fed as a meal, wheat-mixed feed was consumed at a lower level than all other grain diets in an acceptability trial. Cows fed pelleted grain produced more milk protein, solids-not-fat, fat, and fat corrected milk than those fed meal diets. Pelleting resulted in an increased molar percentage of rumen acetate and a decrease in butyrate.

In contrast to the above, Adams and Ward[27] observed cows fed hay and silage roughages produced less fat corrected milk and butterfat as a result of pelleting the grain portion of the diet. In acceptability comparisons, most cows showed no preference for pellets over meal, or vice versa. Thomas et al.[28] reported fat production was not depressed as the result of pelleting the grain portion of lactating cattle diets.

Steam-Flaking

Earlier research from England demonstrated fat level in milk could be depressed by flaking the grain, fed in conjunction with low-roughage diets.[29,30] Other English workers have reported similar results. Research by Brown et al.[31] indicated comparable digestibility of feed nutrients and of milk production for dairy cattle fed either steam-processed or rolled grain rations. Oklahoma data[23] showed little difference in performance between milo heated by various techniques, but expanded grain sorghum was shown to reduce milk production in one comparison.

In a commercial study,[22] cows fed steam-rolled corn produced more milk with a lower percentage fat than cows fed cracked corn. However, fat corrected milk production was comparable. The roughage fed was alfalfa hay *ad libitum* and corn silage limit-fed. In another test in which limit-fed alfalfa hay (10 lb/day) and *ad libitum* corn silage were fed, dairy cows fed steam-rolled corn produced comparable amounts of milk and fat, but cows fed cracked corn produced less milk with a higher fat test.

High Moisture Grain

Increased cost of energy for drying corn at harvest time has resulted in renewed interest in high moisture (25 to 30% water) grains for livestock. Generally, performance on high moisture grain has been slightly better, in terms of efficiency of feed conversion, for high moisture grains, for growing and finishing beef cattle. Changler et al.[32] found that lactating Holstein cows fed high moisture corn produced milk of higher fat content and greater quantities of fat while consuming 5% less concentrate

Table 10
FREE CHOICE VS. COMPLETE MIXED
RATIONS FOR FINISHING PIGS

	Free choice	Complete mix
Number of pigs	285	285
Average daily gain, lb	1.48	1.54
Feed per 100 lb gain	330	339
Feed cost per 100 lb gain, $	9.46	10.05

From Conrad, J. H., *National Hog Farmer,* Grundy Center,
Iowa, August 1959. With permission.

resulting in crude fiber content of dry matter consumed being less for the cows on the low moisture corn. Clark et al.[33] compared high moisture corn containing from 24 to 26% moisture, stored either in an oxygen limiting structure or treated with 1.3% propionic acid and stored with oxygen limitation. A third type of corn was that which had been dried with artificial heat. They concluded any of the three methods of providing corn was equally satisfactory for lactating dairy cows. Later research by the same station[34] reaffirmed that high moisture corn and dry corn had comparable feeding values for lactating dairy cows.

SWINE

More hogs are finished out on corn than on all the rest of the grains combined in the U.S. The total digestible nutrients (TDN) of corn for swine is listed usually as 80%, indicating hogs cannot utilize 20% of the energy contained in spite of the fact corn contains 4% of highly digestible oil and only 2% fiber.

It is the purpose of this section to review the research which has been conducted to study the effect of grain processing (1) on the nutritive value for swine and (2) on the effect on swine ulcers.

Free Choice vs. Complete Mixed Rations

Conrad[35] presented results of comparisons of free choice vs. complete mixed ration feeding from eight experiment stations and which involved 570 pigs. The basic ration was corn plus protein supplement and the pigs were confined to concrete floors (Table 10).

The above results show that pigs fed complete mixed rations on concrete grew 4% more rapidly but the free choice fed pigs rquired 3% less feed per lb of gain. Due to the saving in feed and the saving of grinding and mixing costs, the feeding of free choice rations on concrete, weaning to market, resulted in a saving of 60c for every 100 lb of pork produced, based on 1959 feed prices.

Even though free choice feeding of corn, protein supplement, and minerals usually results in more economical gains, a very low percentage of the pigs fed out today are fed on the free choice basis. There are several logical reasons for this. Swine feeders and nutritionists tend to calculate rations on the basis of percent of nutrients — especially protein — in the total mixture. Therefore, complete mixed rations lend themselves to such calculations better than does the free choice technique. Furthermore, on supplements built and formulated around soybean meal, swine may tend to overeat due to the high palatability of soybean meal. Complete mixed rations lend themselves to automation; thus, this method of feeding is attractive for large swine feeders.

Table 11
INFLUENCE OF GRINDING CORN ON
EFFICIENCY OF FEED UTILIZATION[36]

Station or feed company	Degree of grinding			
	Fine	Medium	Coarse	Whole
Illinois	3.03	3.12	3.29	—
	—	3.38	3.59	3.57
North Carolina	3.04	—	3.24	—
Purdue	3.06	3.34	3.51	—
Wisconsin	3.38	3.54	4.01	—
	3.90	4.49	4.08	—
	3.63	3.70	4.09	—
	5.25	—	5.35	—
	3.84	—	—	4.08
Allied Mills	3.39	—	3.61	—
	3.17	—	3.40	—
	3.21	—	3.44	3.35
Golden Sun Mills	3.45	—	3.73	—
Kent Feeds	3.12	3.42	3.74	—
Yoder Feeds	3.13	3.16	3.25	—

Table 12
VALUE OF PELLETING BARLEY
RATIONS FOR SWINE

	Meal	Pellet	Pellet advantages
Number of pigs	218	218	
Daily gain, lb	1.33	1.52	+ 14.3%
Feed per 100 lb gain	441	373	+ 15.4%

From Conrad, J. H., *National Hog Farmer*, Grundy Center, Iowa, August 1959. With permission.

Degree of Grinding Corn

Speer[36] summarized research results from four experiment stations and from four feed companies in which corn was compared in various forms (Table 11).

Based on the above efficiency comparisions, corn is utilized more efficiently by swine when it is ground finely, but in the second portion of this review, data are presented which indicate fine grinding of corn may cause it to be ulcer predisposing for swine.

Pelleting of Diets

Perhaps this aspect of feed processing for swine has attracted more interest than has any other processing technique. In 1959, Conrad[35] summarized the results from ten experiments in which barley was compared in a meal ration and in a pelleted meal ration. The research was conducted at four experiment stations and utilized 436 pigs (Table 12).

Pelleting of barley-containing rations increased grains 14% and resulted in a 15% feed savings, or 68 lb of feed per 100 lb of gain.

Comparisons of meal and pellets for corn-containing rations have not provided quite as great differences as those obtained for the pelleting of barley-containing rations.

Table 13
VALUE OF PELLETING CORN RATIONS FOR SWINE

	Growing pigs			Finishing pigs		
	Meal	Pellet	Pellet advantage	Meal	Pellet	Pellet advantage
Daily gain, lb	1.56	1.62	+ 3.8%	1.85	2.01	+ 8.6%
Feed per lb gain	2.81	2.49	+11.3%	3.80	3.40	+10.5%

From Becker, D. E., *Merck Agric. Memo.,* Merck and Co., Rahway, N.J., 2(1), 1966. With permission.

Table 14
SUMMARY OF 30 COMPARISONS OF
MEAL VS. PELLET RATIONS

	Meal	Pellet	Pellet advantage
Daily gain, lb	1.61	1.71	+ 6.2%
Feed per lb gain	3.52	3.36	+ 4.9%

From Jensen, A. H., *Feedstuffs,* 38(31), 24, 1966. With permission.

Table 15
SUMMARY OF PELLETING EFFECT ON
RATION UTILIZATION

	Meal	Pellet	Pellet advantage
	Average of 3 trials		
Growing pigs (to 110 lb)			
Daily gain, lb	1.63	1.62	—
Feed per lb gain	2.74	2.50	+ 9.6%
Finishing pigs (110 lb to finish)			
Daily gain, lb	1.86	1.98	+ 6.4%
Feed per lb gain	3.56	3.24	+10.0%

From Jensen, A. H., *Feedstuffs,* 38(31), 24, 1966. With permission.

Becker[37] presented Illinois data which showed approximately a 10% feed efficiency advantage for pelleting corn-based swine rations, both for growing pigs and finishing pigs (Table 13).

Jensen[38] summarized the effects of pelleting rations for swine under a wide variety of conditions and from numerous experiment stations. A summary of 30 comparisons at 7 different agricultural experiment stations covering 12 years (1953 to 1964) showed that on the average pigs consuming pelleted feed gained 6% faster and required 5% less feed per lb of gain (Table 14).

Table 15 is a summary of tests conducted in the middle 1960s by the University of Illinois. Corn-soybean meal rations were utilized in all comparisions. For the growing pig, average daily gain was not affected by the form of ration fed, but pelleting decreased feed intake by 9.4% and increased efficiency of feed conversion by 9.6%.

Table 16
EFFECT OF FORM OF DIET ON SWINE PERFORMANCE

	Meal	Pellets
Daily gain, lb	1.69	1.72
Feed per lb gain	3.22	3.12

From NCR-42, *J. Anim. Sci.*, 29, 927, 1969. With permission.

Table 17
RESPONSE OF PIGS FED LIQUID COMPARED TO DRY DIETS

Feeding method	Level of feeding	Number of observations	Percent response of liquid-fed pigs	
			Range	Average
			Daily gain	
Individual	Restricted	2	+ 2.3 to + 13.2%	+ 7.8%
Individual	Full	1	—	+ 8.8%
Group	Restricted	10	−5.0 to + 6.5%	+ 2.9%
Group	Full	5	+ 4.4 to + 7.9%	+ 6.5%
			Feed requirement	
			(− = less, + = more feed)	
Individual	Restricted	2	−2.0 to −11.9%	−7.0%
Individual	Full	1	—	+ 1.0%
Group	Restricted	10	−6.4 to + 7.0%	−1.0%
Group	Full	5	−7.6 to + 10.1%	+ 2.9%

From Speer, V. C., *Feedstuffs,* 41(8), 30, 1969. With permission.

During the finishing phase, rate of gain was increased 6.4%, feed intake was decreased 3.3%, and efficiency of feed conversion was increased 10% due to pelleting of the rations.

Nine experiment stations[39] in the north-central region of the U.S. participated in conducting cooperative research involving 99 pens (556 pigs) at 10 sites to study the effect of pelleting of diets on the performance of growing-finishing swine (Table 16). There was no significant effect of form of diet on rate of gain.

Since the cost of pelleting is one of the question marks relative to this method of feeding pigs, then one must ask, "How much does one need to save to cover the cost of pelleting?" At least a 6% feed saving due to pelleting must be obtained to break even on the cost of pelleting.

Liquid Feeding

Speer[40] summarized research which compared liquid and dry feeding swine rations. When feed intake has been restricted to less than *ad libitum*, pigs fed liquid diets have gained more rapidly than pigs fed dry diets. Although feed efficiency responses have been variable at feed levels below *ad libitum* intake, the average response would favor liquid feedings. Pigs self-fed dry diets have gained essentially the same as pigs fed liquid diets. Furthermore, in each of nine trials reviewed by Speer, pigs self-fed meal were more efficient than liquid-fed pigs.

Table 18
PERFORMANCE OF PIGS FED DRY
AND WET FEED

Item	Unit	Dry	Wet[a]
Live data			
Number of pigs		30	29
Average initial weight	lb	117	117
Average final weight	lb	203	203
Average daily gain	lb	1.39	1.36
Feed per lb gain	lb	3.72	4.03
Carcass data			
Dressing %	%	74.3	73.9
Average backfat	in.	1.30	1.24
Four lean cuts	%	51.8	52.6

[a]   Prepared by mixing equal weights of feed and water.

' From Thrasher, D. M., Roberts, H. P., and Mullins, A. M., La. State Univ. Livestock Prod. Day, Baton Rouge, January 21, 1968. With permission.

Table 19
PERFORMANCE OF PIGS FED DRY AND
LIQUID RATIONS

	Full-fed dry	Limit-fed dry	Limit-fed liquid[a] + water	Limit-fed liquid
Live data				
Number of pigs	40	40	40	40
Initial weight, lb	44	42	43	43
Final weight, lb	212	207	208	208
Daily gain, lb	1.55	1.35	1.38	1.39
Feed per lb gain	3.15	3.39	3.25	3.29

[a]   Liquid diet was prepared by mixing water with feed at the ratio of 3:1, four replications of ten pigs each per treatment.

Modified from Ellis, I. J. and Clawson, A. J., North Carolina State Univ. Pork Producers Conf., 1966. With permission.

Table 17 summarizes average weight gain and feed efficiency response of pigs fed liquid and dry diets in the U.S. These responses have been categorized by feeding method and feeding level.

Thrasher et al.[41] presented data which demonstrated a 1:1 mixture of feed and water did not improve rate of gain but resulted in 8% less efficient feed conversion when compared to feeding the diet dry (Table 18).

Ellis and Clawson[42] compared limit-feeding in both the dry and wet forms for growing-finishing pigs. There was no obvious advantage to mixing water with the diet in a ratio of 3:1 (Table 19).

Paste Feeding

Ohio researchers[43] have conducted a number of research trials to compare the value of paste feeding with that of the diet in the dry form. Teague et al.[43] presented two summaries involving three trials each in which pigs fed their diet in the paste form (1.3

Table 20

PASTE FEED VS. DRY MEAL FOR GROWING
PIGS[43]

Trial	Feed type	Daily gain (lb)	Daily feed	Feed per gain[a]
1	Dry	1.58	4.58	2.90
	Paste	1.83	5.07	2.77
2	Dry	1.54	4.47	2.90
	Paste	1.75	4.96	2.84
3	Dry	1.35	3.42	2.55
	Paste	1.57	4.12	2.63
Means[b]	Dry	1.46	4.00	2.73
	Paste	1.68	4.59	2.72

[a]   Air-dry basis.
[b]   255 pigs in three trials. Initial wt 40 to 65 lb to 120 lb.

Table 21

PASTE FEED VS. DRY MEAL FOR FINISHING
PIGS[43]

Trial	Feed type	Daily gain (lb)	Daily feed	Feed per gain[a]
1	Dry	1.88	6.72	3.58
	Paste	2.09	7.00	3.35
2	Dry	1.79	6.61	3.69
	Paste	1.92	6.86	3.58
3	Dry	1.63	5.64	3.45
	Paste	1.85	6.25	3.39
Means[b]	Dry	1.74	6.22	3.56
	Paste	1.92	6.62	3.45

[a]   Air-dry basis.
[b]   203 pigs, initial wt 110 to 120 lb to finish wt 210 lb.

to 1.5 parts water to 1.0 part air-dry mixed feed and free access to water) gained from 10 to 15% more rapidly than those fed their diet in the air-dry form. There was no difference in efficiency of feed conversion. (Tables 20 and 21.)

Roasting Corn

Purdue University[44] and the University of Illinois[45] stations have conducted research with roasted corn for swine which indicates some improvement in feed utilization by that processing technique. In the Purdue[44] trial (Table 22), finishing swine fed roasted corn gained comparably to those fed raw corn. However, those fed roasted corn required an average of 7% less feed per lb of gain than was required for those fed raw corn. Illinois[45] data covered three trials and the summary stated, "With group-fed finishing pigs there was a trend toward higher gain per unit of feed from the cooked corn diets."

Influence of Grains Processing on the Occurrence of Ulcers in Swine

Although the occurrence of stomach ulcers in swine has been reported for several years, it was not until 1963 that Perry et al.[46] reported that esophagogastric ulcers could be produced experimentally by type of ration. Since that time, a number of investigations have been conducted to elucidate nutritional factors involved in the production of ulcers in swine.

Table 22

THE VALUE OF ROASTED CORN AND OF ROASTED SOYBEANS FOR FINISHING SWINE[44]

Type soy Type corn	SOM		Roasted beans	
	Raw	Roasted	Raw	Roasted
Growth and feed data				
Number of pigs	24	24	24	24
Initial weight, lb	40	39	40	40
Daily gain, lb	1.50	1.43	1.47	1.52
Feed per lb gain	3.26	3.12	3.10	2.78

Table 23

EFFECT OF GELATINIZATION OF CORN ON ITS ULCER-PRODUCING EFFECT ON SWINE

	Raw corn	Gelatinized corn
Number of pigs	18	18
Initial weight, lb	35	35
Mortality	0	7
Ulcers	0	10

From Perry, T. W., Jimenez, A. A., Shively, J. R., Curtin, T. M., Pickett, R. A., and Beeson, W. M., *Science,* 139, 349. Copyright 1963 by the American Association for the Advancement of Science.

Gelatinized corn produces ulcers in swine — A 55% incidence of esophagogastric ulcers in swine was found by Perry et al.[46] (Table 23) when a basal ration containing gelatinized corn was fed. Gelatinized corn was prepared by processing through a Wenger grain expander. The process involves grinding the corn followed by heating with steam to soften the grain. This material was then forced through a steel tube by an auger. Heat (an average of 300°F) and pressure increased as a result of force by the auger which extruded the softened material through cone-shaped holes in the expander head. The holes were smaller where the feed entered and gradually enlarged until the feed was expelled. This caused a sudden release of pressure and the escaping steam expanded the grain. The produce was then dried and reground into a fine meal.

Starch portion of corn is responsible — Nuwer et al.[47] fractionated corn into bran, germ, and endosperm (starch). Then corn was reconstituted back into its normal ratios of bran, germ, and starch except one fraction was gelatinized whereas the other two fractions were in the raw form (Table 24). Interpretation of the data would indicate it was the starchy portion, which, when gelatinized, caused esophagogastric ulcers in swine.

Finely ground feed increases the incidence — Pickett et al.[48] found that increase in the modulus of particle size resulted in a significant ($P<0.05$) decrease in rate of gain. Also, analysis of the data on stomach ulcer index scores showed that each increase in modulus of fineness resulted in a significant ($P<0.01$) increase in stomach ulcer score

Table 24
EFFECT OF GELATINIZATION OF VARIOUS FRACTIONS
OF CORN ON ULCER PRODUCTION IN SWINE

	Number of pigs	Number of ulcers	Number of cornifications	Number of erosions
Raw corn	6	1	1	3
Gelatinized corn	6	5	1	0
Gelatinized endosperm	6	5	1	1
Gelatinized germ	6	0	1	1
Gelatinized bran	6	0	2	2

From Nuwer, A. J., Perry, T. W., Pickett, R. A., and Cutin, T. M., *J. Anim. Sci.,* 26, 518, 1967. With permission.

(decrease in severity). These data agreed with previous observations of Mahan et al.[49] that the incidence of esophagogastric lesions increases as more finely ground diets are fed. Physical form of the diet had a significant ($P < 0.05$) effect on the incidence of ulcers.

Gelatinization of all grains is not consistent — Barley, corn, milo, and wheat were compared in either the raw or gelatinized form by Riker et al.[50] Esophagogastric ulcers were found to occur in swine fed barley, corn, milo, or wheat. The gelatinization of the grains increased the incidence of lesions, expressed on the basis of ulcer index, significantly ($P < 0.01$), only in the case of corn or milo. There was no significant difference in severity of the lesions between pigs fed raw or expanded wheat or barley indicating a difference in the effect of gelatinization on different grains.

30% Oats gives protection — Maxson et al.[51] observed that the inclusion of 30% of oats reduced the severity of ulcer lesions significantly ($P < 0.01$). In the Purdue comparisons of 70% or 30% additions of either crushed oats or wheat to gelatinized corn diets were compared for swine. The addition of either 70% or 30% of wheat had no effect on ulcers caused by gelatinized corn. However, either 70% or 30% of added oats reduced the incidence of ulcers significantly ($P < 0.05$). Furthermore, it was shown the hull fraction of the oat kernel contained the protective agent. When 9% of oat hulls (the equivalent amount of hulls contained in 30% of oats) were included in the ration, a comparable protective effect was observed. The addition of 9% cellufloc to simulate the fiber content of oat hulls was without benefit, as was the case with 10% sand.

REFERENCES

1. **Matsushima, J. K. and Montgomery, R. L.,** The thick and thin of flaked corn, *Color. Farm Home Res.,* 17, 4, 1967.
2. **Chapman, R. J. and Matsushima, J. K.,** Digestibility and feedlot performance of cattle fed extruded, flaked, and whole shelled corn, *Color. State Univ. Anim. Sci. Highlights,* 4, 1970.
3. **Hentges, J. F. Jr., Cabezas, M. T., Moore, J. E., Carpenter, J. W., and Palmer, A. Z.,** The effect of method of processing on nutritive value of corn for fattening cattle, *Fla. Agric. Exp. Stn. Mimeo Series,* AN67-4, 1966.
4. **Totusek, R., Franks, L., Basler, W., and Renbarger, R.,** Methods of processing milo for fattening cattle, *Okla. Agric. Exp. Stn. Misc. Publ.,* 79, 79, 1967.
5. **Newsom, J. R., Totusek, R., Renbarger, R., Nelson, E. C., Franks, L., Neuhaus, V., and Basler, W.,** Methods of processing milo for fattening cattle, *Okla. Agric. Exp. Sta. Misc. Publ.,* 80, 47, 1968.

6. Wagner, D. G., Schneider, W., and Renbarger, R., Influence of steaming time on the nutritive value of steam flaked milo, *Okla. Agric. Exp. Stn. Misc. Publ.*, 84, 33, 1970.

7. Hale, W. H., Cuitun, L., Saba, W. J., Taylor, B., and Theurer, B., Effect of steam processing and flaking milo and barley on performance and digestion by steers, *J. Anim. Sci.*, 25, 392—396, 1966.

8. Perry, T. W., Roasted corn for finishing beef cattle, Talk presented at Western Livestock Show, Denver, January 1976.

9. Burroughs, W. and Saul, R., Roasted vs. high moisture vs. refrigerated vs. artificially dried whole shelled corn, *Iowa State University AS Leaflet*, Ames, R-145, 1971.

10. Vetter, R. L., Burroughs, W., and Wedin, W. F., Whole, ground, and roasted shelled corn in low-moisture alfalfa silage rations for yearling steers, *Iowa State University AS Leaflet*, Ames, R-147, 1971.

11. Vetter, R. L. and Burroughs, W., Effects of feeding whole shelled, cracked, and roasted corn on rate and cost of gain in short-fed steers, *Iowa State University AS Leaflet*, Ames, R-146, 1971.

12. Schake, L. M., Garnett, E. T., Riggs, J. K., and Butler, O.C., Micronized and steam flaked grain sorghum rations evaluated in a commercial feedlot, *Tex. Agric. Exp. Stn. Anim. Sci. Tech. Rep.*, No. 23, 1970.

13. Ellis, C. F., Jr. and Carpenter, J. A., Jr., Popped milo in fattening rations for beef cattle, *J. Anim. Sci.*, 25, (Abstr.), 594, 1966.

14. Durham, R. M., Ellis, C. F., and Cude, B., A comparison of flaked, popped, and cracked milo in all concentrate rations, *J. Anim. Sci.*, 26 (Abstr.), 220, 1967.

15. Riggs, J. K., Sorenson, J. W., Jr., Adame, J. L., and Schake, L. M., Popped sorghum grain for finishing beef cattle, *J. Anim. Sci.*, 30, 634—638, 1970.

16. Matsushima, J. K., McLaren, R. J., McCann, C. P., and Kellog, G. E., Processing grains for feedlot cattle — extrusion, flaking, reconstituted, and high moisture ensiled, Colo. State Univ. Beef Nutr. Res., Ft. Collins, February 28, 1969.

17. Mudd, C. A. and Perry, T. W., Raw cracked vs. expanded gelatinized corn for beef cattle, *J. Anim. Sci.*, 28, 822—826, 1969.

18. Wilson, B. and Woods, W., Influence of gelatinized corn on beef animal performance, *Nebr. Agric. Exp. Stn. Beef Cattle Prog. Rep.*, 1966, 22.

19. Beeson, W. M. and Perry, T. W., The comparative feeding value of high moisture corn and low moisture corn with different feed additives for fattening beef cattle, *J. Anim. Sci.*, 17, 368, 1958.

20. Merrill, W. G., The place of silage in production rations — feeding high moisture grain silages, *Proc. 1971 Int. Silage Conf.*, Washington, D.C., 1971, 156.

21. Perry, T. W., The feeding value of high moisture grains for beef cattle, Proc. High Moisture Grain Symp., Oklahoma State University, Stillwater, 1976.

22. Williamson, J. L., Grain processing for dairy cattle and other animals, *Effect of Processing on the Nutritional Value of Feeds*, National Academy of Science, Washington, D.C.,1972, 349.

23. Colovis, N. F., Kenner, H. A., Davis, H. A., Morrow, K. S., and Gibson, K. S., The effect of texture on the nutritive value of concentrates for dairy cattle, *Univ. New Hamp. Agric. Exp. Stn. Bull.*, 419, 11, 1955.

24. Oklahoma Feed Manufacturers Association, *Fifteenth Ann. Okla. Feed Industry Conf. and Workshop*, Oklahoma State University, Stillwater, 1964, 18.

25. Wilbur, J. W., Grinding grain for dairy cows, *Indiana Agric. Exp. Stn. Bull.*, 372, 8, 1933.

26. Waldern, D. E. and Cedeno, G., Comparative acceptability and nutritive value of barley, wheat mixed feed and a mixed concentrate ration in meal and pelleted forms for lactating cows, *J. Dairy Sci.*, 53, 317, 1970.

27. Adams, H. P. and Ward, R. E., The value of pelleting the concentrate part of the ration for lactating cattle, *J. Dairy Sci.*, 39, 1448, 1956.

28. Thomas, G. D., Bartley, E. E., Pfost, H. B., and Meyer, R. M., Feed processing. III. Effects of ground, steam flaked and pelleted hay, with and without pelleted grain, on milk composition and rumen volatile fatty acid ratios, *J. Dairy Sci.*, 51, 869, 1968.

29. Balch, C. C., Balch, D. A., Bartlett, S., Cox, C. P., and Rowland, S. J., Studies of the secretion of milk of low fat content by cows on diets low in hay and high in concentrates. I. The effect of variations in the amount of hay, *J. Dairy Res.*, 19, 39, 1952.

30. Balch, C. C., Balch, D. A., Bartlett, S., Hosking, Z. D., Johnson, V. W., Rowland, S. J., and Turner, J., Studies of the secretion of milk and low fat content by cows on diets low in hay and high in concentrates. V. The importance of the type of starch in the concentrates, *J. Dairy Res.*, 22, 10, 1955.

31. Brown, W. H., Sullivan, L. M., Cheatham, L. F., Jr., Halbach, K. J., and Stull, J. W., Steam processing versus pelleting of two ratios of milo and barley for lactating cows, *J. Dairy Sci.*, 53, 1448, 1970.

32. Chandler, P. T., Miller, C. N., and Jahn, E., Feeding value and nutrient preservation of high moisture corn ensiled in conventional silos for lactating dairy cows, *J. Dairy Sci.,* 58, 682, 1975.
33. Clark, J. H., Frobish, R. A., Harshbarger, K. E., and Gregg, R. E., Feeding value of dry corn, ensiled high moisture corn, and propionic acid treated high moisture corn feed with hay or haylage for lactating dairy cows, *J. Dairy Sci.,* 56, 1531, 1973.
34. Clark, J. H., Croom, W. J., and Harshbarger, K. E., Feeding value of dry, ensiled, and acid treated high moisture corn fed whole or rolled to lactating cows, *J. Dairy Sci.,* 58, 907, 1975.
35. Conrad, J. H., Summary of confinement nutrition research, *National Hog Farmer,* Grundy Center, Iowa, August 1959.
36. Speer, V. C., Personal communication, Iowa State University, Ames, 1971.
37. Becker, D. E., Pelleted hog feeds, *Merck Agric. Memo.,* Merck and Co., Rahway, N. J., 2(1), 1966.
38. Jensen, A. H., Pelleting rations for swine, *Feedstuffs,* 38(31), 24, 1966.
39. NCR-42, Cooperative regional studies with growing swine: effects of source of ingredients, form of diet, and location on rate and efficiency of gain of growing swine, *J. Anim. Sci.,* 29, 927, 1969.
40. Speer, V. C., Liquid feeding pigs: present status, *Feedstuffs,* 41(8), 30, 1969.
41. Thrasher, D. M., Roberts, H. P., and Mullins, A. M., Effect of restricted feeding wet and dry feeds and frequency of feeding on performance and carcass merit of pigs, La. State Univ. Livestock Prod. Day, Baton Rouge, January 21, 1968.
42. Ellis, I. J. and Clawson, A. J., A Comparison of Limited Feeding a Dry or Wetted Ration with Self-Feeding a Dry Ration, North Carolina State Univ. Pork Producers Conf., Raleigh, 1966.
43. Teague, H. S., personal communication, Ohio State University, Wooster, Ohio, 1970.
44. Perry, T. W., Conrad, J. H., Foster, J. R., Yake, W., and Peterson, R. C., Roasted Corn and Roasted Soybeans for Swine, Unpublished, Purdue University, W. Lafayette, Ind., 1971.
45. Jensen, A. H., Baker, D. H., Brown, H. W., Costa, P. M. A., and Harmon, B. G., Effects of roasting on nutritional value of corn in diets for growing-finishing swine, *Univ. Ill. Pork Industry Day,* 1971, 3.
46. Perry, T. W., Jimenez, A. A., Shively, J. R., Curtin, T. M., Pickett, R. A., and Beeson, W. M., Incidence of gastric ulcers in swine, *Science,* 139, 349, 1963.
47. Nuwer, A. J., Perry, T. W., Pickett, R. A., and Curtin, T. M., Expanded or heat processed fractions of corn and their relative ability to elicit esophagograstric ulcers in swine, *J. Anim. Sci.,* 26, 518, 1967.
48. Pickett, R. A., Fugate, W. H., Harrington, R. B., Perry, T. W., and Curtin, T. M., Influence of feed preparation and number of pigs per pen on performance and occurrence of esophagogastric ulcers in swine, *J. Anim. Sci.,* 28, 837, 1969.
49. Mahan, D. C., Picket, R. A., Perry, T. W., Curtin, T. M., Beeson, W. M., and Featherston, W. R., Influence of ration particle size on the incidence of ulcers in swine, *J. Anim. Sci.,* 25, 1019, 1965.
50. Riker, J. T., III, Perry, T. W., Pickett, R. A., and Curtin, T. M., Influence of various grains on the esophagogastric ulcers in swine, *J. Anim. Sci.,* 26, 731, 1967.
51. Maxson, D. W., Stanley, G. R., Perry, T. W., Pickett, R. A., and Curtin, T. M., Influence of various ratios of raw and gelatinized corn, oats, oat components, and sand on the incidence of esophagogastric lesions in swine, *J. Anim. Sci.,* 27, 1006, 1968.

EFFECTS OF PROCESSING ON NUTRITIVE VALUE OF FEEDS: OILSEEDS AND OILSEED MEALS

L. G. Young

INTRODUCTION

The major processes which oilseeds undergo are particle size reduction, oil extraction, heating, pressure, and mechanical separation. In general particle size reduction allows greater digestibility of the nutrients contained in whole seeds, the degree of effect depending upon the nature of the seed, the specie, and age of the animal involved and the degree of particle size reduction.

The extraction of oil from oilseeds normally involves a combination of the processes indicated above.[1,2] The removal of oil from oilseeds reduces the digestible energy value of the resulting meal, the degree of reduction being influenced by the efficiency of oil extraction (Tables 1, 2). The amount of heat involved in processing oilseeds may influence the quality of the resulting oilmeals, excess heating resulting in reduction of available nitrogen, nitrogen digestibility, and net protein utilization (Table 3).

Many oilseeds and oilseed meals contain naturally occurring factors which reduce their nutritional value unless the factors are removed, destroyed, or inactivated.[3,4] In most cases these toxic factors have their greatest effect in monogastric animals and appear to have little effect in mature ruminants.

Since these toxic factors as well as protein quality appear to be more critical in diets of monogastric animals, the remaining discussion will concentrate on oilseed and oilseed meals for monogastrics.

A general indication of the comparative nutrient content of various oilseeds and oilseed meals may be obtained by examining Table 1. The nutrients of greatest importance from a practical standpoint are content of protein, digestible energy, and lysine.

SOYBEANS AND SOYBEAN MEAL

The comparative value of unextracted soybeans vs. soybean meal for animal feeding has been recently reviewed.[5] Proper heat processing of raw soybeans either by moist heat[6-10] or by dry heat[11-18] results in performance similar to that of soybean meal and superior to raw soybeans when fed to growing-finishing pigs. Frequently the heat processed soybeans result in slightly increased feed efficiency. The specific reasons for the improved performance due to heat treating are not known but may include improved palatability as well as destruction or inactivation of toxic factors such as trypsin inhibitors, hemagglutinins, saponins, and isoflavones.[19] The conditions necessary for proper heat treatment of raw soybeans appear to involve a combination of time, temperature, and moisture.[20] Olsen et al.[20] suggests an optimum discharge temperature range of 130 to 150°C and a time of 1 min for processing raw soybeans using a gas-fired roaster.

Extruded soybeans were found to have a digestible energy value of 4.31 to 4.56 kcal/g (as fed basis) or approximately 1.12 times the value of soybean meal even though the extruded soybeans contain approximately 16 percentage units more of fat.[21] This difference is somewhat greater than that observed by Baird[22] when comparing soybean meal (44%) and extruded soybeans on a dry matter basis.

Most of the soybean meal available commercially in North America has had the oil extracted with organic solvents and with heat being used to remove residual solvent.[1] Additional heating or toasting is required to inactivate toxic substances present in the

Table 1

NUTRITIVE COMPOSITION OF SOME OILSEEDS AND OILSEED MEALS

Dry Basis

Name	Ref. no.[a]	Ash %	Fiber %	Ether extract %	Protein %	DE[b] kcal/g	Arginine %	Lysine %	Methionine %	Threonine %	Tryptophan %
Coconut meats, mech. extd.[c] grnd.	5-01-572	7.2	12.2	7.2	22.3	4.16	2.60	0.76	0.36	—	0.22
Coconut meats, solv. extd. grnd.	5-01-573	6.9	15.4	2.3	23.4	3.67	2.65	0.77	0.36	—	0.22
Cotton, seeds, grnd.	5-01-608	3.8	18.2	24.7	24.9	5.22	—	—	—	—	—
Cotton, seeds, mech. extd. grnd.	5-01-617	6.6	11.8	6.0	44.7	3.62	4.49	1.59	0.56	1.31	0.52
Cotton, seeds, prepress solv. extd. grnd.	5-07-872	7.0	14.2	1.0	45.5	—	4.61	1.75	0.53	1.32	0.48
Flax, seeds	5-02-052	5.2	6.6	37.5	24.8	5.26	—	—	—	—	—
Flax, seeds, mech. extd. grnd.	5-02-045	6.2	9.7	5.6	39.4	3.37	3.51	1.32	0.71	—	0.61
Flax, seeds, solv. extd. grnd.	5-02-048	6.3	9.7	1.9	40.0	3.55	3.54	1.33	0.66	—	0.62
Hemp, seeds, extn.[d] unspecified grnd.	5-02-367	9.5	26.3	6.4	33.5	2.25	—	—	—	—	—
Mustard, seeds, extn. unspecified grnd.	5-03-154	7.6	11.6	5.5	34.5	3.78	—	—	—	—	—
Palm, kernels, solv. extd. grnd.	5-03-486	4.3	14.4	3.9	20.2	3.47	—	—	—	—	—
Peanut, kernels, mech. extd. grnd.	5-03-649	5.9	7.6	8.2	50.6	4.04	—	1.41	0.65	—	—
Peanut, kernels, solv. extd. grnd.	5-03-650	4.9	14.3	1.3	51.8	3.11	6.45	2.51	0.44	1.64	0.55
Poppy, seeds, extn. unspecified grnd.	5-03-751	13.9	13.0	8.8	41.0	3.53	—	—	—	—	—
Rape, seeds	5-08-109	4.6	7.3	48.2	22.5	6.65	—	—	—	—	—
Rape, seeds, solv. extd. grnd.	5-03-871	7.8	12.9	1.7	41.0	3.31	2.51	2.42	0.85	1.89	0.62

Safflower, seeds	4-07-958	3.1	28.6	32.0	17.5	3.81	—	—	—	—	—
Safflower, seeds mech. extd. grnd.	5-04-109	4.1	35.0	7.2	23.5	2.65	1.57	0.71	0.40	—	0.30
Safflower, seeds wo. hulls mech. extd. grnd.	5-08-499	7.2	9.6	7.6	45.7	3.79	3.47	1.50	0.81	—	0.58
Sesame, seeds	5-08-509	6.1	11.2	46.6	24.2	5.09	—	—	—	—	—
Sesame, seeds, mech. extd. grnd.	5-04-220	11.2	5.8	9.4	48.0	3.86	5.12	1.39	1.49	—	0.83
Soybean, seeds	5-04-610	5.4	5.8	19.2	41.7	4.44	—	—	—	—	—
Soybean, seeds, mech. extd. grnd.	5-04-600	6.7	6.8	5.2	46.7	3.75	3.20	3.09	0.79	1.90	0.65
Soybean, seeds, wo. hulls solv. extd. grnd.	5-04-612	6.5	3.2	1.2	55.6	4.39	3.93	3.49	0.81	—	0.70
Sunflower, seeds	4-08-530	3.3	31.0	27.7	17.9	3.59	—	—	—	—	—
Sunflower, seeds, extn. unspecified grnd.	5-04-737	6.7	26.5	2.1	36.0	2.89	4.15	1.86	1.75	—	0.66
Sunflower, seeds wo. hulls, solv. extd. grnd.	5-04-739	7.3	12.7	2.6	49.8	3.21	10.2	3.36	1.68	3.59	1.12

[a] International reference number.
[b] Digestible energy for swine.
[c] Extruded.
[d] Extraction.

Adapted from the Atlas of Nutritional Data on United States and Canadian Feeds, National Academy of Sciences, Washington, D.C., 1971.

Table 2
EFFECT OF FAT REMOVAL ON ENERGY OF OIL MEAL FEEDS

Feed and preparation	Percent fat in meal	Energy	
		D.E. (kcal/kg)	T.D.N. %
Soybean seed			
None	18.0	3690	84
Mech-extd. grnd.	4.7	3294	75
Solv-extd. grnd.	0.9	3139	71
Flax, seed			
None	36.0	4320	98
Mech-extd. grnd.	5.2	3210	73
Solv-extd. grnd.	1.7	3129	71
Peanut, kernels			
Mech-extd. grnd.	5.9	3570	81
Solv-extd. grnd.	1.2	3367	76

From *Applied Animal Nutrition,* 2nd ed., by E. W. Crampton and L. E. Harris, W. H. Freeman, Copyright ©, 1969, 38.

Table 3
EFFECT OF VARIOUS DEGREES OF HEAT TREATMENT ON THE COMPOSITION AND NUTRITIVE VALUE OF COCONUT MEAL

Temperature of treatment °C	Protein content %	Available lysine (g/16g N)	Nitrogen digestibility %	Net protein utilization %
40	25.3	3.29	77.7	45.9
90	25.8	3.09	78.3	41.0
105	26.3	2.81	74.6	36.1
120	25.6	2.34	73.3	35.8
135	26.4	1.63	68.2	33.9
150	26.3	1.12	56.1	17.1

From Butterworth, M. H. and Feed, H. C., *Br. J. Nutr.,* 17, 445, 1963. With permission.

residue and may also increase the availability of some amino acids, e.g., cystine and methionine, in the meal. Excess heating may cause decreased protein quality by destroying or binding amino acids especially lysine.

COTTONSEED AND COTTONSEED MEAL

Although cottonseed is relatively high in protein (25%) and fat (25%) (Table 1) it contains a naturally occurring compound, gossypol, which reduces growth and results in high mortality when fed in large amounts to young monogastrics.[23,24] Steam cooking of cottonseed meats results in rupture of the resin glands releasing gossypol. The free gossypol binds to protein, especially the epsilon amino groups of the amino acid lysine, thereby reducing the availability of the lysine and lowering protein quality but reducing toxicity.[23] The free or unbound gossypol causes the toxic effects.

Whole ground cottonseed fed as either raw or "heated" meal reduced feed intake, daily gains, feed efficiency, and resulted in death of pigs.[23] Extensive reviews of gossypol and its influence on livestock have been published.[24,25] In general, cottonseed meal containing more than 0.04% free gossypol should be used cautiously in swine or poultry diets.[24,25]

Iron salts, particularly ferrous sulfate are effective in reducing gossypol toxicity by complexing with free gossypol. A weight ratio of 1:1 of iron from ferrous sulfate to free gossypol is suggested.[24-26]

Boiling of cottonseed kernels improves their nutritional value, however, pretreatment of cottonseed meats with ferrous sulfate solution followed by oven or sun drying and incorporating them into swine diets resulted in performance similar to that of pigs fed a corn soybean meal diet.[27]

There are three main processes used in oil extraction and the commercial production of cottonseed meal: screw press, prepress solvent extraction, or direct solvent extraction.[24,25,28] The trend is toward the use of solvent extraction resulting in a product of high protein quality (availability of amino acids) but also with a high content of free gossypol due to the low amount of heat involved. Screw press extraction of oil from cottonseed results in a meal of lower protein quality due to the heat causing binding of free gossypol to lysine and also a lower free gossypol content. Thus, the nutritional value of cottonseed meal will depend, to a large extent, on the amount of heat involved in the process of oil extraction. Rate of gain of pigs fed cottonseed meal, with ferrous sulfate added to give a weight ratio of 1:1 with free gossypol, was similar to that of pigs fed soybean meal. Supplementation with lysine or reduction of the fiber content of cottonseed meal resulted in improved performance.[29]

RAPESEED

Rapeseed is processed to obtain the oil. Both the seed and the meal contain goitrogenic and possibly other deleterious substances and the oil contains a high level of erucic acid.[30] The goitrogenicity of rapeseed appears to be primarily the effect of oxazolidinethione derived from the glucosinolates by enzymatic hydrolysis.[31] The inclusion of 10% ground unprocessed rapeseed depresses rate of gain and efficiency of energy utilization in diets of growing and finishing pigs[32] and broiler chicks.[33] A greater depression was observed with Span rapeseed (*Brassica campestris*) than Bronowski (*Brassica napus*), possibly due to the higher glucosinolate content of the former.[33] Autoclaving either types of ground rapeseed resulted in improved chick performance possibly through destruction of the enzyme myrosinase[33] which hydrolyzes the glucosinolates. Extruding of raw rapeseed may also destroy myrosinase activity.[21] Substituting extruded rapeseed in a corn-soybean meal diet for swine resulted in reduced digestibility of dry matter, nitrogen, and gross energy.[21]

The effect of processing on the chemical composition of rapeseed meal has been reviewed.[34] As with cottonseed meal, the higher the temperature used during processing, the lower the quality of protein in the meal for monogastric animals. Various procedures have been used to remove goitrogenic substances from rapeseed meals,[34] however, most of them have not been adopted on a commercial basis. Plant breeders are developing rapeseed varieties which have lower content of glucosinolates and are less goitrogenic.

The use of rapeseed meal in diets of various livestock has been reviewed.[31,35-37] Rapeseed meal may be used in diets of ruminants, however, feed intake may be reduced when high levels of rapeseed meal are included.[37] The deleterious factors in rapeseed meal appear to affect performance of monogastrics to a greater extent than ruminants. Besides the goitrogenic properties rapeseed meal appears to have a lower lysine content and availability of lysine as well as lower digestible or metabolizable energy values than soybean meal.[31] Because of these factors it is usually recommended that rapeseed meal should not replace all of the high quality protein, e.g., soybean meal in diets for monogastrics. The development of newer varieties of rapeseed containing lower levels of toxic factors may allow greater use of rapeseed meal in diets of animals.

PEANUT MEAL

Peanut meal tends to be deficient in lysine relative to the other amino acids although it is fairly high in total protein (45 to 55%).[25] Swine performance is reduced when peanut meal replaces more than 50% of the supplemental protein from soybean meal even with the addition of lysine.[38,39] It was suggested that factors in peanut meal other than lysine may limit performance when over 50% of the unsupplemental protein is derived from peanut meal.[39] Peanut meal may be contaminated by aflatoxins produced by *Aspergillus flavus*.[40]

SUNFLOWER MEAL

Sunflower seeds contain a high level of fiber (Table 1). Following extraction of oil, the sunflower meal contains variable but usually relatively high levels of crude fiber. The level of fiber, crude protein, and lysine in the sunflower meal can be influenced by the method of processing used. Increasing the levels of sunflower seed in corn-soybean meal diets for swine resulted in decreased feed intake and efficiency of feed utilization.[41] Controlled heating of sunflower seed before oil extraction may improve the nutritional value of the meal.[42] Even with lysine supplementation, sunflower seed meal resulted in poorer performance when fed to rats than a diet based on soybean meal.[42] The high fiber content and lower lysine content appear to be major factors limiting the use of sunflower meal in diets of monogastrics.[43]

SAFFLOWER MEAL

Safflower meal contains a high level of crude fiber and is deficient in lysine (Table 1). Processing to remove a high proportion of hulls will result in a meal (decorticated safflower meal) which contains a level of protein similar to 44% protein soybean meal. However, the high fiber levels and low lysine levels limit its use in diets for monogastrics.

LINSEED MEAL

Linseed meal is used mainly in the diet of ruminants. It contains a relatively low level of lysine when compared with soybean meal.

COCONUT MEAL

Coconut meal is low in protein and lysine relative to most of the other oilseed meals (Table 1),[44,45] and the protein is of low digestibility for swine.[45] When fed to swine, coconut meal results in decreased gain and feed efficiency, an affect which does not appear to be completely explained by low lysine content and low protein digestibility.[46]

REFERENCES

1. Crampton, E. W. and Harris, L. E., Protein Supplements, in *Applied Animal Nutrition,* W. H. Freeman, San Francisco, 1969, 261—262.
2. Pond, W. G. and Maner, J. H., Processing and its effect on nutritive value, in *Swine Production in Temperate and Tropical Environments,* W. H. Freeman, San Francisco, 1974, 426—428.
3. Liener, E., Protease inhibitors, in *Toxic Constituents of Plant Foodstuffs,* Academic Press, New York, 1969, 8—53.
4. Liener, E., in Problems with endogenous toxic factors in oilseed residues, *Proc. Georgia Nutrition Conf. for the Feed Industry,* American Feed Manufacturers Association, Atlanta, Ga., 1976, 3—21.
5. Young, L. G., *Effect of Processing on the Nutritional Value of Feeds,* National Academy of Sciences, Washington, D.C., 1973, 201—210.
6. Jimenez, A. A., Perry, T. W., Pickett, R. A., and Beeson, W. M., Raw and heat treated soybeans for growing-finishing swine and their effect on fat firmness, *J. Anim. Sci.,* 22, 471—475, 1963.
7. Robinson, W. L., *Ohio Agric. Exp. Res. Sta. Bull.,* 452, 42, 1930.
8. Hooks, R. D., Hays, V. W., Speer, V. C., and McCall, J. T., Effect of raw soybeans on pancreatic enzyme concentrations and performance of pigs, *J. Anim. Sci.,* 24, 894, 1965.
9. Coombs, G. E., Conness, R. G., Berry, T. G., and Wallace, H. D., Effect of raw and heated soybeans on gain, nutrient digestibility, plasma amino acids, and their blood constituents of growing swine, *J. Anim. Sci.,* 26, 1067—1071, 1967.
10. Young, L. G., Brown, R. G., Ashton, G. C., and Smith, G. C., Effect of copper on the utilization of raw soybeans by market pigs, *Can. J. Anim. Sci.,* 50, 717—726, 1970.
11. Jensen, A. H., Brown, H. W., Harmon, B. G., and Baker, D. C., Effects of roasting corn and soybeans fed to swine, *J. Anim. Sci.,* 31 (Abstr.), 1023, 1970.
12. Faber, J. L. and Zimmerman, D. R., Infrared roasted soybeans for baby pigs, *J. Anim. Sci.,* 31 (Abstr.), 1020, 1970.
13. Villegas, F. J., Veum, T. L., Hedrick, H. B., and McFate, K. L., Processed soybeans for swine, *J. Anim. Sci.,* 31 (Abstr.), 213, 1970.
14. Lafferty, D. T. and Hines, R. H., Full-fat soybeans for growing-finishing swine, *J. Anim. Sci.,* 33 (Abstr.), 233, 1971.
15. Ruffin, B. G., Powell, W. E., and Brown, V. L., Effect of roasted soybeans on pig growth and carcass quality, *J. Anim. Sci.,* 32 (Abstr.), 391, 1971.
16. Meade, R. J., Rust, J. W., and Hanson, L. E., Effects of level of dietary protein and source of soybean protein on performance of young pigs, *J. Anim. Sci.,* 33 (Abstr.), 236, 1971.
17. Wahlstrom, R. C., Libal, G. L., and Berns, R. J., Effect of cooked soybeans on performance, fatty acid composition, and pork carcass characteristics, *J. Anim. Sci.,* 32, 891—894, 1971.
18. McConnell, J. C., Shelley, G. C., Handlin, D. L., and Johnston, W. E., Corn, wheat, milo, and barley with soybean meal or roasted soybeans and their effect on feedlot performance, carcass traits, and pork acceptability, *J. Anim. Sci.,* 41, 1021—1030, 1975.
19. Pond, W. G. and Maner, J. H., Protein sources for swine, in *Swine Production in Temperate and Tropical Environments,* W. H. Freeman, San Francisco, 1974, 370.
20. Olsen, E. M., Young, L. G., Ashton, G. C., and Smith, G. C., Effects of roasting and particle size on the utilization of soybeans by pigs and rats, *Can. J. Anim. Sci.,* 55, 431—440, 1975.
21. Bayley, H. S. and Summers, J. D., Nutritional evaluat ion of extruded full-fat soybeans and rapeseeds using pigs and chickens, *Can. J. Anim. Sci.,* 55, 441—450, 1975.
22. Baird, D. M., Heat treated soybeans, corn, and BR grain sorghum for feeder pigs, *J. Anim. Sci.,* 42 (Abstr.), 256, 1976.
23. Clawson, A. J., Maner, J. H., Gomez, G., Mijia, O., Flores, Z., and Buitrago, J., Unextracted cotton seeds in diets for monogastric animals, I. The effect of ferrous sulfate and calcium hydroxide in reducing gassypol toxicity, *J. Anim. Sci.,* 40, 640—649, 1975.
24. Berardi, C. and Goldblatt, A., *Gassypol Toxic Constituents of Plant Foodstuffs,* Liener, E., Ed., Academic Press, New York, 1969, 211—266.
25. Pond, W. G. and Maner, J. H., Protein sources for swine, in *Swine Production in Temperate and Tropical Environments,* W. H. Freeman, San Francisco, 1974, 329—336.
26. Smith, K., Conf. on Inactivation of Gossypol with Mineral Salts, National Cottonseed Products Association, Memphis, 1966.
27. Clawson, A. J., Maner, J. H., Gomez, G., Flores, Z., and Buitrago, J., Unextracted cottonseed in diets for monogastric animals, II. The effect of boiling and oven vs. sun drying following pretreatment with a ferrous sulfate solution, *J. Anim. Sci.,* 40, 648—654, 1975.
28. Smith, K. J., Advances in oilseed protein utilization, in *Alternative Sources of Protein for Animal Production,* National Academy of Sciences, Washington, D.C., 1973, 73—83.
29. Knabe, D. A., Tanskley, T. D. Jr., Cater, C. M., and Hesby, J. H., Effect of lysine, fiber and free gossypol in CSM — supplemented growing swine diets, *J. Anim. Sci.,* 41 (Abstr.), 317, 1975.

30. VanEtten, C. H., Goitrogens, in *Toxic Constituents of Plant Foodstuffs,* Liener, E. Ed., Academic Press, New York, 1969, 103—142.
31. Bowlands, J. P., The use of rapeseed meal in pig and poultry rations, in *Feed Energy Sources for Livestock,* Swan and Lewis, Butterworths, London, 1976, 129—142.
32. Bowland, J. P. and Newell, J. A., Ground rapeseed from low erucic acid (lear) cultivars span and zephyr with or without organic acid treatment as a dietary ingredient for growing-finishing pigs, *Can. J. Anim. Sci.,* 54, 455—464, 1974.
33. Olomu, J. M., Robblee, A. R., Clandinin, D. R., and Mardin, R. T., Utilization of full-fat rapeseed and rapeseed meals in rations for broiler chicks, *Can. J. Anim. Sci.,* 55, 461—469, 1975.
34. Ruthowski, A., Effect of processing on the chemical composition of rapeseed meal, *Proc. Int. Conf. on the Science, Technology, and Marketing of Rapeseed and Rapeseed Products,* Ste. Adele, Quebec, 1970, 496—515.
35. Clandinin, D. R. and Robblee, A. R., Canadian experience with the use of rapeseed meal in rations for poultry, *Proc. Int. Conf. on the Science, Technology, and Marketing of Rapeseed and Products,* Ste. Adele, Quebec, 1970, 267—273.
36. Bowland, J. P., Rapeseed meal in swine feeding, *Proc. Int. Conf. on the Science, Technology, and Marketing of Rapeseed and Rapeseed Products,* Ste. Adele, Quebec, 1970, 274—280.
37. Ingalls, J. R., Waldern, D. E., and Stone, J. B., Rapeseed meal in ruminant rations: dairy cattle, *Proc. Int. Conf. on the Science, Technology, and Marketing of Rapeseed and Rapeseed Products,* Ste. Adele, Quebec, 1970, 281—296.
38. Orok, E. J., Bowland, J. P., and Briggs, C. W., Rapeseed, peanut, and soybean meals as protein supplements with or without added lysine: biological performance and carcass characteristics of pigs and rats, *Can. J. Anim. Sci.,* 55, 135—146, 1975.
39. Tanksley, T. D. Jr., Knabe, D. A., Hesby, J. H., Gregg, E. J., Purser, K. W., and Corley, J. R., Substituting peanut meal and lysine for soybean meal in sorghum-based swine G-F diets, *J. Anim. Sci.,* 41 (Abstr.), 329, 1975.
40. Pond, W. G. and Maner, J. H., Protein sources for swine, in *Swine Production in Temperate and Tropical Environments,* W. H. Freeman, San Francisco, 1974, 352—353.
41. Laudert, S. B. and Allee, G. L., Nutritive value of sunflower seed for swine, *J. Anim. Sci.,* 41 (Abstr.), 318, 1975.
42. Amos, H. E., Burdick, D., and Seerley, R. W., Effect of processing temperature and L-lysine supplementation on utilization of sunflower meal by the growing rat, *J. Anim. Sci.,* 40, 90—95, 1975.
43. Burdick, D., Potential for nonsoy plant proteins, *Proc. Georgia Nutrition Conf. for the Feed Industry,* American Feed Manufacturers Association, Atlanta, Ga., 1976, 35—49.
44. Pond, W. G. and Maner, J. H., Protein sources for swine, in *Swine Production in Temperate and Tropical Environments,* W. H. Freeman, San Francisco, 1974, 343—346.
45. Creswell, D. C. and Brooks, C. C., Composition, apparent digestibility, and energy evaluation of coconut oil and coconut meal, *J. Anim. Sci.,* 33, 366—369, 1971.
46. Creswell, D. C. and Brooks, C. C., Effect of coconut meal on coturnix quail and of coconut meal and coconut oil on performance, carcass measurements, and fat composition in swine, *J. Anim. Sci.,* 33, 370—375, 1971.

EFFECT OF PROCESSING ON NUTRIENT CONTENT OF FEEDS: ROOT CROPS

Guillermo G. Gómez and Julián A. Buitrago

INTRODUCTION

Many rhizomatous and tuberous crops are grown in different regions, especially in the tropics, where they have been estimated to provide the staple food for more than 200 million people.[1] Comprehensive basic information on root crops has already been compiled and published.[2] This chapter will review only some of the more important root crops, which include cassava, potatoes, sweet potatoes, yams, and taro. Table 1 gives the common and scientific names of these crops.

Root crops are used for both human and animal feeding. They are prepared in varied forms according to the dietary habits of the regions where they are produced. Roots are normally subjected to some type of cooking before being consumed by humans. Since fresh roots cannot be stored for long periods of time, they are often processed into dried meal or starch for later use in industry or as an animal feed.

The effects of postharvest technology and processing on the chemical composition of roots and their nutritive value are reviewed with emphasis on animal feeds.

FRESH ROOT CROPS

Chemical Composition

Fresh roots and tubers are characterized by a high moisture or water content making them highly perishable after harvesting. Table 2 shows the basic chemical composition of the fresh roots and tubers being reviewed. Roots and tubers contain small percentages of ether extract (0.1 to 0.4), crude fiber (0.6 to 1.5), ash (1.1 to 1.3), and relatively insignificant amounts of crude protein (N × 6.25:1.2 to 2.2). Their basic nutritional value consists in their high-quality starch represented by the nitrogen-free extract fraction, which ranges from 20 to 30% on a fresh weight basis (Table 2). Their feed value is about 1000 kcal/kg of metabolizable energy.

Postharvest Deterioration

Postharvest deterioration renders cassava roots unacceptable for human consumption, reduces their acceptability for animal feed, and lowers the quality of the starch obtained from them. Postharvest deterioration is therefore related to reduction in root quantity (loss of weight) as well as quality. Five factors have been described as the principal causes of postharvest losses.[3]

Physical damage — This includes mechanical injuries, breakage or crushing during preharvest, harvesting, handling operations (grading, packing, and transporting), and damage at the market or at the consumer level.

Temperature extremes — Yams, sweet potatoes, and cassava have been shown to suffer damage from chilling at 12°C or below. Excessive high temperatures during storage also result in serious damage in most cases. High temperature may induce black heart in potatoes, a disorder caused by asphyxiation of the central cells.[4]

Dehydration and respiration processes — Natural endogenous respiratory losses, together with transpiratory losses of water, always occur. The rate of respiration increases with temperature; therefore, respiratory losses of these roots can be expected to be higher under tropical conditions.

Sprouting — With the exception of cassava roots, which are organs of perennation and not propagation, sprouting is an additional factor of postharvest losses. High temperatures also increase losses due to sprouting.

Table 1
ROOT CROPS OF IMPORTANCE IN HUMAN AND ANIMAL FEEDING[2]

Common and scientific names	Family	Vernacular names in other languages
Cassava (*Manihot esculenta*)	Euphorbiaceae	Aipi, mandioca (Brazil) Guacamote (Mexico) Yuca (Latin America) Kamoteng, kahoy (Philippines) Kaspe (Indonesia) Kelala (India)
Potatoes (*Solanum tuberosum*)	Solanaceae	Alu (India) Papas, patatas (Latin America) Pomme de terre (France) Watalu (Pakistan) Viazi (East Africa) Jaga-imo (Japan)
Sweet potatoes (*Ipomoea batatas*)	Convolvulaceae	Batata, camote, chaco (Latin America) Dankoli, doukali (West Africa) Getica (Brazil) Imo (Japan) Kamote (Philippines) Ubi, mita-alu (Indonesia)
Yams (*Dioscorea* spp.)	Dioscoreaceae	Cará (Brazil) Ñame (Latin America) Igname (France, Italy) Ignamekolle (Germany) Inhame (Portugal)
Taro (*Colocasia esculenta*)	Araceae	Abalong, gabi (Philippines) Bari, koko (West Africa) Chonque, malangay, quiquisque (Latin America) Eddoe (West Indies) Taioba (Brazil) Dasheen (West Indies)

Table 2
CHEMICAL COMPOSITION OF FRESH ROOTS AND TUBERS

Roots and tubers	Dry matter %	Ether extract %	Crude fiber %	N-free extract %	Protein (N × 6.25)	Ash	Energy[a] kcal/kg
Cassava[b]	32.4	0.3	1.5	28.1	1.2	1.3	1054
Potato[b]	23.1	0.1	0.6	19.1	2.2	1.1	806
Sweet potatoes[b]	30.6	0.4	1.3	26.2	1.7	1.1	1026
Yams[c]	31.0	0.2	0.8	27.8	1.9	1.1	—
Taro[c]	26.9	0.1	1.0	23.8	1.8	1.2	—

[a] Metabolizable energy for swine.
[b] National Academy of Sciences, Atlas of Nutritional Data on U.S. and Canadian Feeds, Washington, D.C., 1971, 772.
[c] Food and Agricultural Organization and U.S. Department of Health, Education, and Welfare, food composition table for use in Africa, 1968, 306.

Pathological factors — Microorganisms (bacteria, fungi, viruses) and probably the most serious cause of postharvest deterioration of roots, reducing both the quantity of sound produce through massive infections and the quality as a result of blemish or surface diseases. Of minor importance are losses due to attack by insects, rodents, and nematodes.

Postharvest losses of perishable roots have been estimated at 25% of production;[5] however, not all roots and tubers show the same degree of susceptibility to postharvest deterioration. While cassava cannot be used for human consumption 2 to 5 days after harvesting, yams can normally be stored for several months and still be edible. However, yams lose from 10 to 20% weight after 3 months of storage and 30 to 60% after 6 months.[3] It was also found that there was a 50% loss after 7 months storage at room temperature.[6]

STORAGE

Changes in Root Quality

During storage of root crops such as cassava, potatoes, and sweet potatoes, several metabolic changes take place. The content of total sugars in potatoes may vary from traces to as much as 10% of the tuber dry weight. Sucrose, fructose, and glucose are the most important sugars in potatoes, and their concentrations vary according to the storage conditions. Sucrose predominates in mature potatoes, approaching values of 40% of the total sugars in freshly harvested tubers, but decreases progressively throughout the storage period.[7] Apparently fructose is the predominant sugar after 6 weeks storage at 10°C.[8] Storage of potatoes at low temperatures results in an increase in total sugars, notably reducing sugars. At storage temperatures below 10°C in an air atmosphere, the concentration of reducing sugars increases; but at similar temperatures in a nitrogen atmosphere, there is no accumulation of sugars.[9] The decrease in starch content associated with increments in sugars results in potatoes with a sugary taste but with poor texture after cooking. Levels of reducing sugars higher than 2% on a dry weight basis cause darkening of sliced potatoes[10] and the tubers are considered to be unacceptable for processing.[9]

A rapid accumulation of total sugars and a decline in starch content have been observed during the storage of cassava roots in both field clamps and storage boxes.[11] Cassava roots kept at ambient temperature exhibited similar carbohydrate changes as those stored in boxes and clamps, but the sucrose content declined very drastically as the symptoms of primary deterioration were observed. These changes were accompanied by alterations in the texture as well as in the cooking time of the stored roots. The stored root samples were of lower eating quality than freshly harvested roots.[11] Despite all chemical, biochemical, and texture changes that occurred during storage, the nutritional value of cassava meal produced from stored roots was not significantly affected. These observations suggest that the preparation of cassava meal to be used in diets for domestic animals would eliminate the limitations found in the texture and eating quality of stored roots.

There is a little information on the changes of other chemical constituents of roots and tubers during storage. Most of the available data refer to the starch and sugar contents because of their quantitative importance in the overall chemical composition. Total and soluble nitrogen appears to vary little with storage time in potatoes; however, insoluble nitrogen tends to decrease during the first months of storage, with an apparent recovery as storage continues.[12] Losses of some other chemical constituents depend on storage conditions; for instance, some vitamin C is lost in potatoes during storage. These losses are even greater as lower temperatures are used; but at temperatures of around 10 to 16°C, reduction of vitamin C is minimal.[12]

The effects of storage systems (field clamps vs. storage boxes) and the length of the storage period (2 vs. 4 weeks) on the proximate chemical composition of sweet and bitter cassava flours prepared from stored roots were reported.[11] The ether extract fraction appeared to be the one that consistently increased as the length of the storage period did. In general, during the relatively short period of storage, chemical composition was not appreciably affected by either storage system.

Methods of Storage

Due to their highly perishable nature, roots and tubers are normally stored under different conditions in order to preserve their nutritive quality especially for human consumption and also to minimize postharvest losses. Several methods for storing fresh roots have been experimented with, but all these methods have at least six main purposes:

1.   To reduce postharvest deterioration
2.   To improve palatability and acceptability
3.   To improve the physical appearance of the product, mainly as regards texture and consistency
4.   To improve digestibility
5.   To reduce or eliminate the toxic factor(s) present in the fresh product
6.   To extend the shelf-life or to preserve the product

The main methods for storing fresh roots and tubers include physical and chemical practices which may be described briefly as follows:

Refrigeration — Metabolic processes slow down with lower temperatures, reducing water, and respiratory losses during storage. Refrigeration also slows down the metabolism of pathogens and rotting is reduced. Most products with a high moisture content are damaged by being frozen, and temperatures around 12 to 16°C may be the optimum range for storage, preventing chilling, and condensation damage after return to ambient conditions. The following temperatures have been recommended to store roots: 3°C for cassava, 12°C for sweet potatoes, and ambient temperature for yams.[13]

High temperature — Storage diseases may be controlled by storing roots at high temperatures that will kill the pathogens. The control of black rot of sweet potatoes by holding roots at temperatures of 38°C has been reported.[14]

Curing — This is a very effective, simple method for controlling postharvest water and pathological losses. The process requires relatively high temperatures and humidity and involves suberization followed by the development of a wound periderm, which is effective in retarding water loss and acts as a barrier against infections.[3] In suberization, the cells and intercellular spaces just back of the wound become filled with sap, some of the starch changes to unsaturated fatty acids, and these combine with oxygen from the air to form suberin, a compound that has the property of retarding the escape of water from the wound in addition to preventing the entrance of soft rot organisms into the storage tissues beneath the wound.[15] In wound periderm formation, cells just back of the suberized layer lose their large vacuoles and function as meristematic cells, collectively called wound cambium or wound phellogen. The more rapidly the wound and ordinary periderm are developed, the sooner the fleshy roots will be protected against the attacks of rot-producing organisms, as well as the drying effects of the air. In Tables 3 and 4, the conditions used by several authors for curing and the advantages of the process in preventing weight losses in root crops are summarized.

Sprout suppressants — The simplest way to handle the problem of sprouts is by breaking them off at regular intervals as they appear. The use of chemical retardants has been tested; however, they have an inhibitory effect of periderm formation, delaying the wound healing process. Gamma radiation can also inhibit sprouting.

Chemicals — Several chemicals have been used to prevent postharvest deterioration at dosage rates that are not phytotoxic. These are generally bacteriostatic (fungistatic) and bacteriocidal (fungicidal) and include borax, sodium orthophenyl phenate, captan,

Table 3
CONDITIONS REQUIRED FOR CURING ROOT
CROPS[3]

Crop	Temperature (°C)	Relative humidity (%)	Time (days)	Ref.
Potatoes	15—20	85-90	5—10	4,16
Sweet potatoes	30—32	85-90	4—7	17-19
Cassava	30—40	High	—	3
Aroids	Curing with smoke has been recorded			20

Table 4
PERCENTAGE OF WEIGHT LOSS DURING
STORAGE OF CURED AND UNCURED ROOT
CROPS[3]

Crop	Duration of storage (days)	Percentage weight loss		Ref.
		Cured	Uncured	
Yams	150	10.0	24.0	18
Sweet potatoes	113	17.0	42.0	19
Potatoes	210	5.0	5.4	21

thiabendazole, and benomyl. Formaldehyde and organomercurials have been used on roots not suitable for human and animal consumption. The use of liquid paraffin wax has been tested with cassava, extending its storage life to 3 or 4 weeks.

PROCESSED ROOT CROPS

Different Types of Processed Products

Roots and tubers are normally used as fresh vegetables for human consumption. The different storage methods are used to preserve an acceptable eating quality. Roots and tubers can also be processed in a variety of ways that will permit conservation of the product for long periods. Cassava and potatoes are processed to a large extent for the production of starch, flour or meal, and chips. Some of the principal processing methods are briefly described below, with special reference to cassava and potato products.

Starch — Cassava or potato starch may be obtained by small-scale, wet extraction processes or at the industrial level. For the production of cassava starch, the roots have to be washed, peeled, sliced, and put through a rasping or grating machine to produce a slurry or pulp. The pulp is then sieved to separate the fibrous material from the starch milk. This starch suspension is left in sedimentation tanks for 6 to 10 hr, after which the starch settles at the bottom and the liquid is drawn off the surface. The surface of the starch mass contains impurities that have to be removed. The starch mass is mixed with water and left to settle several times until the starch is of the quality required. The starch cake is then dried and pulverized. The processing of cassava for starch extraction leads to the availability of byproducts such as spent pulp or bran and the peels, which are normally used for animal feeding.

Tapioca — Is a product made from cassava starch cakes, broken up, and crumbled into coarse lumpy particles. These are used for cooking flakes, while for pearl and seed tapioca, the coarse material is granulated into globules or beads.[22] The tapioca is partially gelatinized in greased iron pans at 60 to 70°C. Once the product is properly prepared, there are few storage problems if protection is given against moisture and insect attacks. It was reported that both steamed and roasted tapioca grains store well in a variety of containers for 12 months and more and are insect resistant.[23]

Chips — They may be prepared mechanically or manually. The process requires washing and slicing or chipping into adequate sizes, followed by drying. The product may either be sun dried or dried artificially in any smokeless drier. If the product is properly dried, the chips will have a good keeping quality, but if stored too long, they are subject to attack by molds and insects.[24] Cassava chips made from bitter varieties store longer than sweet varieties.[25]

Pellets — In some countries, mainly Thailand, cassava pellets are produced for export to be used as animal feed. Meal is compressed into pellets with the addition of a binding agent, i.e., bentonite, molasses.

Meal or flour — The preparation of meal or flour is simply the process of grinding chips in hammer, stone, or cylindrical mills. Flour, like chips, generally stores well for long periods, adjusting to the humidity of the surrounding atmosphere, although eventually it is attacked by molds and insects.

Dehydrated products — Dehydration of fresh roots and tubers converts them into very rich starchy products with a high energetic value comparable to cereal grains. This process is used mainly with potatoes and sweet potatoes to give various products such as flakes, granules, and dices. Potato flakes are prepared by dehydrating cooked mashed potatoes and the use of additives (glycerol monopalmitate, skim milk solids, sodium sulfide, sodium bisulfite) to improve stability and texture. In the preparation of dehydrated diced potatoes, the enzymes have to be inactivated by blanching in steam or boiling water and then the product is dried. In the preparation of potato granules, cooked mashed potatoes are partially dried into granules to give a moist mix that can be finely pulverized.

Canned products — This process is mainly used with potatoes. The tubers are washed, peeled, and sliced if necessary. They are then put into cans of boiling water; or a 1.5 to 3.0% salt solution is added, and in some cases calcium salts (not more than 0.015%) are added, to improve texture. The cans are then heated to above 70°C, closed and cooked, normally to 20 to 55 min at 114-120°C, and then cooled immediately to 37°C.

Fermented products — Gari and farina are two fermented products used widely as human foodstuffs in West Africa and Brazil, respectively. The process varies to some extent from one locality to another; but in general the cassava is peeled, grated, and allowed to ferment spontaneously. The moist residual mass is dewatered by squeezing and then partially gelatinized and dried to form a granular product.[26,27] Storage problems in the case of gari seem to be related mainly to moisture content.[22] The moisture content permitting adequate storage is 12 to 13%.[28] Well-processed gari can be kept intact for several months.[29]

Poi — is a product prepared from taro on a commercial scale in Hawaii. The roots are milled and cooked in steam retorts. The resultant product is centrifuged and poured into plastic bags to be kept at room temperature. Under these conditions fermentation by *Lactobacillus* spp. is rapid and the pH drops to 3.8 to 4.0. Fresh and fermented poi can be canned satisfactorily. The fresh product is filled into cans at 76°C and retorted for 100 min at 100°C. The fermented product is heated to 93°C, filled hot into cans, and cooled immediately without further heat treatment.

Effect of Processing on Nutrient Values

Most roots and tubers are cooked, boiled, fried, or baked when used for human consumption. In most cases, these types of processing are used to change physical properties, i.e., texture, consistency, in order to improve palatability and/or acceptability, and in some others to improve digestibility and/or reduce toxic factors from the raw product.

In animal feeding, the main reason to cook the roots and tubers relates to the improvement in digestibility and the reduction of toxic factors. Sweet cassava and sweet potatoes do not need to be cooked when used in animal feeding whereas potatoes, bitter cassava, yams, and taro should be cooked or processed in some way in order to reduce their toxicity.

Cassava varieties are normally classified as sweet or bitter according to their cyanide content. Although there are no definite standards for this classification, cassava roots containing 0.02 to 0.03% hydrocyanic acid (HCN) or 200 to 300 ppm are considered bitter varieties. Most of the HCN or cyanide (CN) is found in the form of a cyanogenic glucoside known as linamarin. The concentration of linamarin, as evidenced by the CN liberated, is substantially higher in the peel of the roots than in the pulp.[30,31] Linamarin releases HCN on treatment with dilute acids; naturally, however, the release of HCN is due to the action of the enzyme linamarase, usually present in the tissues — notably in the peel — of the roots. The contact of the enzyme with the substrate linamarin normally occurs when the tissues are damaged mechanically, either by crushing or by destroying the cellular structure of the plant tissues.

In potatoes, the phenolic substances comprise lignins, coumarins, tannins, monohydric phenols, polyphenols, anthocyanins, and flavones. These substances are associated with the color of the raw product and with certain types of discoloration of processed potato products. In sound, healthy tubers there is no net oxidation of the phenolic substances to form discoloration of products whereas in injured tubers, the phenols are rapidly converted to colored melanins, which cause the change in color.

Considerable quantities of potatoes are utilized for the production of solanin, and some reports have indicated that its presence in the raw product causes a deleterious effect on consumption and digestibility.[32] The normal concentration of solanin in raw potatoes is between 0.01 to 0.1% on a dry matter basis, increasing to about 4.3% during germination. Potatoes containing more than 0.1% solanin are considered unacceptable for human consumption. It has been indicated that solanin inhibits the enzymes trypsin and chymotryspin.[33] Sprouts should be removed before the potatoes are consumed because of their high solanin content. Cooking reduces the effects of solanin and also converts the starch fraction of potatoes into a form that can be digested more easily.[34] Either steaming or boiling the potatoes in water is an effective mechanism for eliminating the alkaloid.

Yams are also used for human consumption, either boiled, baked, or fried. In West Africa, most yams are eaten as "fufu," a firm glutinous dough that has been boiled in water. Yams are sometimes dried to produce flour for industrial purposes, especially as a starting material for drugs. Yams contain oxalate crystals and also small quantities of alkaloids and antimetabolites. Especially wild varieties of yams contain the toxic alkaloid, dioscorine, plus tannins and saponins, which are responsible for the bitter taste of the raw product.[35] Some species (*Dioscorea dumetorum*) contain the toxic alkaloid dihydrodioscorine.[36] The toxic alkaloids in yams are readily water-soluble, which facilitates detoxification.

ANIMAL FEEDS FROM ROOT CROPS

Roots, due to their low protein content, have to be supplemented with high protein levels in order to obtain adequate performance in animal feeding. They can be used either fresh or processed through cooking, ensiling, or dehydration procedures.

Table 5

COMPARISON OF INTAKE AND PERFORMANCE OF FINISHING PIGS FED
EITHER SWEET OR BITTER FRESH CASSAVA[39]

	Experimental diets			
Parameter	Sweet cassava + prot. sup. *Ad libitum*	Sweet cassava + prot. sup.	Bitter cassava + prot. sup. *Ad libitum*	Bitter cassava + prot. sup.
Av daily gain (kg)	0.66	0.77	0.56	−0.08
Daily intake cassava (kg)	2.99	3.40	0.99	0.93
Daily intake prot. sup. (kg)	0.81	0.82	1.21	0.22
Total feed intake (kg)[a]	1.98	2.01	1.59	0.58
Protein (%)	14.1	13.3	23.5	13.3
Feed/gain	2.99	2.61	2.86	Neg.

[a]   Expressed to contain 10% moisture.

Cassava

Cassava has been extensively evaluated as a feedstuff for animals. It has been shown that when properly supplemented it constitutes an excellent energy source during the different stages of the life cycle in pigs.[37,38] Sweet and bitter varieties have been included in feeding systems developed at Centro Internacional de Agricultura Tropical (CIAT)[39,40] that make maximum use of this feedstuff. When fed free choice along with a protein supplement, daily consumption of fresh sweet cassava was 3.00 kg as compared to only 0.99 kg when bitter cassava (CMC-84) was fed (Table 5).[39] Pigs consuming sweet cassava (*Llanera*) also consumed an average of 0.81 kg of protein supplement. As the consumption of bitter cassava was low, these pigs compensated by consuming more protein supplement (1.21 kg). Low consumption of bitter cassava and the overconsumption of protein supplement resulted in a diet excessive in protein, besides being uneconomical.

Cassava meal was also prepared from these two varieties and fed in balanced diets to growing swine for 4 weeks. Pigs fed bitter cassava meal consumed less feed during the first week; this reduced consumption continued to a lesser degree throughout the 4-week trial. The lower level of consumption of this diet resulted in a reduction in average daily gain from 0.62 kg for sweet cassava to only 0.56 kg/day for those pigs fed bitter cassava (Table 6).[39] This reduction in gain and feed consumption was accompanied by an improvement in feed conversion efficiency. This suggests that the bitter cassava meal reduces the palatability of the diet; but once it is consumed, it has little, if any, detrimental effect on metabolic processes except those related to detoxification of residual HCN or HCN produced from beta-glucosides present in the cassava meal. During the processing of drying cassava, most of the HCN is released because linamarase and linamarin come into physical contact when the cassava roots are chopped for drying; thus meal prepared from bitter cassava roots has a relatively low HCN content (100 to 150 ppm on a dry matter basis). Although a composite diet including high levels (approximately 73%) of bitter cassava meal was consumed slightly less by growing pigs than a diet based on similar levels of sweet cassava meal, the difference in consumption was not as great as that observed with the fresh roots. Drying the roots greatly reduces the problem of limited consumption of fresh bitter roots by growing pigs.

Cassava silage offers practical advantages in animal feeding, being an inexpensive method of storing and preserving the product. Roots are preserved with the original moisture, and nutrient losses can be reduced to a minimum by proper fermentation,

Table 6
EFFECT OF SWEET AND
BITTER CASSAVA MEAL
AS THE MAJOR
CARBOHYDRATE IN
DIETS FOR GROWING
SWINE[39]

Parameter	Cassava meal	
	Sweet	Bitter[a]
Initial weight (kg)	38.9	39.3
Final weight (kg)	57.1	54.9
Av daily gain (kg)	0.62	0.56
Av daily feed (kg)	1.77	1.35
Feed/gain	2.86	2.43

[a]   Estimated to contain 150—200 mg
HCN/kg of fresh cassava.

Table 7
PERCENTAGE OF CONCENTRATION OF ORGANIC ACIDS
FROM CASSAVA SILAGE (CS) AND SWEET POTATO
SILAGE (SPS) AT WEEKLY INTERVALS[43]

Period	Acetic acid		Butyric acid		Lactic acid		pH	
	CS	SPS	CS	SPS	CS	SPS	CS	SPS
First week	0.52	0.52	0.45	0.03	0.09	0.05	5.10	4.95
Second week	0.48	0.32	0.36	0.21	0.09	0.06	4.40	5.75
Third week	0.61	0.28	0.49	0.22	0.10	0.05	4.30	4.60
Fourth week	0.69	0.28	0.26	0.02	0.09	0.02	4.95	5.10
Fifth week	0.46	0.21	0.31	0.06	0.05	0.02	4.70	5.20

preventing the entrance of air and water. Roots, due to their high starch content and rapidly available carbohydrates, ferment very quickly, facilitating the silage process. The concentration of soluble carbohydrates decreases rapidly during fermentation, favoring an increase in nonvolatile fatty acids, i.e., acetic, butyric, and lactic acids, with little reduction in energy and dry matter content. If the production of butyric acid is excessive, there is an increase in proteolysis; this is of minor importance, however, since the roots are low in protein and a pH below four prevents proteolysis.[41] In Table 7, the concentration of acetic, butyric, and lactic acid in cassava and sweet potato silage is compared.[42] Cassava silage had consistently higher contents of organic acids than sweet potato silage. Butyric acid was about three times greater and lactic acid about twice that in sweet potato silage. Minerals and vitamins are affected in a very small proportion and carotene has been shown to be effectively preserved by silage.[43] Experimental evaluation with cassava silage has demonstrated excellent performance results with pigs (Table 8).[40] When cassava silage is supplemented with adequate protein, the energy fraction from common cereals (corn, sorghum) could be completely replaced by cassava silage.

Potatoes

Potatoes are also used extensively for animal feeding in some countries as a fresh

Table 8

PERFORMANCE OF GROWING-FINISHING PIGS FED
CASSAVA SILAGE[40]

| | Daily feed consumption (kg) | | Daily gain | |
Criterion	Silage	Supplement	(kg)	Feed/gain
Growing pigs				
Cassava silage +				
30% protein supplement	1.9	0.97	0.65	2.98
40% protein supplement	2.3	0.70	0.71	2.60
Growing-finishing pigs				
Cassava silage + 40% protein supplement	3.8	1.01	0.77	3.25

Table 9

PERCENTAGE COMPOSITION OF POTATO MEAL,
FLAKES, SLICES, AND PULP[34]

Potato product	Dry matter	Crude protein	Crude fiber	Ether extract	Ash	Nitrogen-free extract
Meal	89.3	9.2	2.1	0.4	4.2	73.4
Flakes	90.0	8.1	2.7	0.3	3.2	75.7
Slices	89.7	10.2	2.0	0.4	4.2	72.9
Pulp	88.4	7.7	6.5	0.3	3.4	75.9

product or silage, cooked or dried, and in the form of a meal for pigs and poultry. Extensive experimental work with fresh potatoes and stored or ensiled potatoes has been conducted for animal feeding purposes.[44,45] Potatoes make excellent silage when adequate preservatives and roughages or grains were used as absorptive agents.[45] Starch waste has also been ensiled for several months and used for cattle feeding without observable signs of spoilage.[46] A large proportion of the potato crop is used for swine feeding. Precooked drum-dried fodder flakes have been popular, but sliced raw potatoes dried in rotary or airlift driers are also used widely. Ensiling steamed potatoes, the cheapest form of preservation, may result in loss of water-bound substances, which may reach 20 to 30%.[47] Raw potatoes in good condition may be consumed by pigs, but palatability and digestibility are inferior as compared with cooked or processed potatoes. Pigs fed a daily ration of 1.14 kg or a 19% protein concentrate and cooked potatoes to appetite performed at a rate and efficiency not different from pigs fed all-concentrate rations. Pigs, 14- to 15-weeks-old, consumed about 2.2 kg of cooked potatoes per day. Pigs fed raw instead of cooked potatoes gained about 30% less weight and required 40 days longer to reach market weight. In addition, the pigs fed raw potatoes required 0.66 kg more concentrate and 2.3 kg more potatoes to produce 1 kg of weight gain than those fed cooked potatoes.[32]

Potato pulp, a byproduct obtained in the manufacture of starch, can be fed to livestock either wet or dried. For every 100 kg of potato starch, about 20 kg of dried pulp (10% moisture) is produced. Both total pulp yield and composition will vary according to the method of starch extraction employed. Modern equipment is designed to increase the starch extraction rate, with a corresponding decrease in the amount of solids in the spent pulp. The composition and the digestibility for potato meal, flakes, slices, and pulp are included in Tables 9 and 10 which consider the values reported by several authors.[34,48,49]

Table 10
PERCENTAGE DIGESTIBILITY OF DRIED
POTATO BYPRODUCTS BY PIGS[48,49]

Potato product	Crude protein	Crude fiber	Organic matter	Nitrogen-free extract
Meal	39.9	64.8	89.2	96.4
Flakes	81.8	—	98.1	100.0
Slices	74.9	32.6	92.5	96.9
Pulp	27.0	91.0	91.2	91.2

Table 11
DIGESTIBILITY OF RAW OR COOKED SWEET
POTATOES[34]

Parameter	Raw potatoes	Cooked potatoes
Dry matter (%)	90.4	93.5
Energy (%)	89.3	93.0
Nitrogen (%)	27.6	52.8
Digestible energy (kcal/kg)	3373	3463

Table 12
SWEET POTATO MEAL AS A
SUBSTITUTE FOR CORN IN
FATTENING STEERS

	Average daily gain (kg)		
Corn	Sweet potatoes	Corn + sweet potatoes[a]	Ref.
1.18	1.08	1.29	50
0.88	0.79	0.77	51
0.86	0.87	0.88	51
1.09	0.95	1.10	52

[a] Cottonseed meal was used to adjust the protein content of the diet to that of the corn meal.

Sweet Potatoes

In animal feeding sweet potatoes can be used fresh, cooked, dried, or ensiled. Moldy tubers contain the metabolites ipomeamarone and isomeamaranol, which are toxic to the liver and other organs, also causing lung edema. Digestibility determinations have shown that the dry matter and energy digestibility of raw and cooked sweet potatoes are similar, but protein digestibility is almost two times higher for the cooked potatoes, as illustrated in Table 11.[34] Most feeding experiments have used sweet potato meal as a substitute for corn meal in fattening diets. The results of some experiments with steers, pigs, and chickens are summarized in Tables 12, 13, and 14.[50-54] In general, when the protein source is added to sweet potato meal to bring its protein content to that of the corn meal, the sweet potato meal has about 90% the value of corn meal for fattening steers. Pigs fed sweet potato meal instead of corn meal gained less weight, whereas the gradual replacement of sweet potato meal for corn resulted in better gains. In chickens, a maximum of about 25% sweet potato meal was used, without causing any significant reduction in growth rate or in feed conversion efficiency.

Table 13
DAILY WEIGHT GAINS OF
FINISHING PIGS FED SWEET
POTATO MEAL (SPM) AS A
SUBSTITUTE FOR CORN MEAL
(CM)[53]

| | Average daily gain (kg) | | |
CM	2 parts CM + 1 part SPM	1 part CM + 2 parts SPM	SPM
0.97	—	—	0.51
1.01	0.87	0.76	—
0.87	0.82	0.71	—
0.92	0.90	0.83	—
0.75	—	—	0.59

Table 14
SWEET POTATO MEAL AS A SUBSTITUTE FOR CORN
IN DIETS FOR CHICKENS[54]

| Ingredient, (kg) | | Weight gain after 10 weeks (kg) | Feed/gain | Efficiency index | Mortality (%) |
Corn	Sweet potato				
20	—	0.89	3.46	100	16
16	4.5	0.87	3.46	98	17
10	9.0	0.86	3.66	97	11
14	5.0	0.79	4.10	89	24
18	—	0.74	4.30	85	23

Yams and Taro

Other roots, including yams and taro, are used less as animal feedstuffs since so very little information is available as to their nutritional value. Both have corpuscles called raphid idioblasts in their structure, apparently consisting of a bundle of raphid crystals of calcium oxalate salts contained within a polysaccharide matrix. Ingestion of fresh plant material containing raphid idioblasts usually results in severe irritation of the digestive tract, especially the mouth. The oxalate crystals also irritate the intestinal mucosa, which is probably the main reason for the laxative effect observed in animals consuming the product raw. Three explanations have been given for this effect:

1.   Mechanical irritation by the calcium oxalate crystals[55,56]
2.   Forceful ejection by the idioblast of the crystals into the tissues[55,57]
3.   Chemical irritation by a curare-like substance associated with the crystal[58]

However, recent ultrastructural studies of raphid idioblasts have not related structural characters to acridity.[59-61] Other recent studies suggest that the physical characteristics of the crystals themselves are responsible for the irritation caused by uncooked tissues of *Colocasia* and *Xanthosoma*.[62] It has been suggested that the irritation was caused by a curare-like substance since raphides are ingredients of curare mixtures.[58] However, curare is primarily an alkaloid and several authors have failed to detect the presence of alkaloids in chemical analyses of these species.[62-64]

In recent studies conducted at CIAT, it was observed that pigs and rats consuming

Table 15
NUTRITIVE VALUE AND DRY MATTER
DIGESTIBILITY OF TWO SPECIES OF YAMS IN
DIETS FOR GROWING RATS[39]

Parameter[a]	*Dioscorea alata*		*Dioscorea esculenta*	
	Raw	Cooked	Raw	Cooked
Total weight gain (g)	39.44	39.84	6.06	49.64
Total feed intake (g)	272.70	234.36	163.56	257.40
Feed/gain	6.95	5.94	36.01	5.21
Protein efficiency ratio	1.42	1.68	0.38	1.92
Dry matter digestibility (%)	70.6	79.9	77.8	88.7

[a] Experimental period 21 days.

Table 16
NUTRITIVE VALUE OF RAW AND COOKED
YAM (*D. ALATA*) MEAL IN DIETS FOR
GROWING PIGS[40]

Parameter[a]	Control	Yam meal	
	Common corn	Raw sun dried	Cooked oven dried[b]
Daily gain (kg)	0.72	0.51	0.76
Daily feed (kg)	1.83	1.73	1.92
Feed/gain	2.53	3.40	2.53

[a] Experimental period 35 days.
[b] Boiled for 20 min, oven dried at 60°C.

raw yams developed gastrointestinal distension.[39,40] The gastrointestinal tract was engorged with undigested feed. This phenomenon was not observed in animals fed cooked yams. The results in Table 15 demonstrate that cooking reduces the effect of the deleterious factor present in raw yams and improved digestibility when two species (*Dioscorea alata* and *D. esculenta*) were fed to rats.[39] Raw yam meal was poorly utilized by growing pigs, whereas meal prepared from cooked roots was efficiently used when compared to a balanced corn diet (Table 16).[40] In another study the effect of cooking time on the nutritive value of yams (*D. alata*) was evaluated. Yams were boiled for 15, 30, and 45 min, respectively, and then oven dried at 60°C. The results in Table 17 indicate that cooking time had no effect on the growth rate of rats; a cooking period of 15 min appeared to be sufficient to eliminate the factor(s) responsible for the gastrointestinal distension observed when raw yams were fed.[40]

Some feeding trials with rats have been conducted at CIAT to evaluate the nutritive value of taro when fed either raw or cooked (Table 18).[40] Rats fed diets containing either sun-dried or oven-dried raw taro consumed less feed and grew at a slower rate than the control group. Gastrointestinal distension was not observed in rats fed raw taro-based diets.

SUMMARY

Root crops are produced throughout the world, notably in tropical regions. Roots

Table 17
EFFECT OF COOKING TIME ON THE NUTRITIVE
VALUE OF YAMS (*D. ALATA*) FOR RATS[40]

Parameter[a]	Raw yams	Cooking time (min)		
		15	30	45
Total weight gain (g)	118.6	122.3	124.8	124.9
Total feed consumed (g)	653.4	420.3	473.5	441.8
Feed/gain	5.51	3.44	3.79	3.54
Total weight of gastrointestinal tract (g)	23.9	10.6	13.9	11.2

[a]   Experimental period 28 days.

Table 18
EFFECT OF COOKING AND DRYING
METHODS ON THE NUTRITIVE VALUE OF
TARO[40]

Treatment	Total body gain (g)	Total feed intake (g)	Feed/gain
Control	131.5	439.7	3.64
Taro meal			
Cooked, oven dried[a]	118.0	427.4	3.62
Raw, sun dried	93.0	339.4	3.65
Raw, oven dried	93.8	357.4	3.81

[a]   Boiled for 30 min, oven dried at 60°C.

and tubers constitute important staples for people living in these areas and are basically grown for food. However, recent developments in crops such as cassava, potatoes, and sweet potatoes may lead to an increased production which would make feasible other alternative uses for roots and tubers, such as animal feedstuffs. This chapter reviews only some root and tuber crops of major importance in human and animal feeding, notably cassava, potatoes, sweet potatoes, yams, and taro.

Fresh roots and tubers contain high levels of moisture or water in their chemical composition and are, therefore, highly perishable after harvesting. Postharvest losses of perishable roots have been estimated at 25% of production. The principal causes of postharvest losses, as well as the main methods for storing fresh roots and tubers, are reviewed. The storage methods are essentially used to preserve the nutritive quality of roots and tubers for human consumption and to minimize postharvest losses.

Conservation of the product for long periods requires the application of different processing methods. The production of starch, chips, pellets, meal, or flour, and dehydration, canning, and fermentation processes are reviewed, with special reference to cassava and potato products. In addition to obtaining stable storable products, processing methods usually change the physical properties of roots and tubers, resulting in improved feeding value. In the case of bitter cassava roots and potatoes, a reduction of toxic factors in the raw products is achieved through processing.

The development of intensive animal production in tropical areas will depend on the production of agricultural commodities in these regions. Roots and tubers constitute valuable potential products to be used as animal feeds. Their nutritive value is based on the high-quality starch they contain. Extensive experimental evidence on the use of cassava, potatoes, and sweet potatoes suggests the possibility of partially or totally substituting the conventional energy-supplying feeds with roots and tubers, either fresh or processed. The use of root crops in animal feeding is reviewed.

ACKNOWLEDGMENT

The authors wish to express their sincere appreciation to Mrs. Trudy Martinez for her editing of this manuscript and to Mrs. Maruja Bejarano for her assistance in typing.

REFERENCES

1. **Coursey, D. G. and Haynes, P. H.**, Roots crops and their potential as food in the tropics, *World Crops,* 22, 261—265, 1970.
2. **Kay, D. E.**, Root crops, *Trop. Prod. Inst. Rep.,* Her Majesty's Stationery Office, London, 1973, 245.
3. **Booth, R. H.**, Postharvest deterioration of tropical root crops, *Trop. Sci.,* 16, 49—63, 1974.
4. **Burton, W. G.**, The potato, A survey of its history and factors influencing its yield, nutritive value, quality, and storage, in *European Association for Potato Research,* Wageningen, 1966, 382.
5. **Coursey, D. G. and Booth, R. H.**, The postharvest phytopathology of the perishable produce, *Rev. Plant. Pathol.,* 51, 751—765, 1972.
6. **McClelland, T. B.**, Report of horticulturist, *P.R. Agric. Exp. Stn. Bull.,* 11, 15, 1925.
7. **Porter, W. L. and Heinze, H.**, Changes in composition of potatoes in storage, *Potato Handb.,* 10, 5—10, 1965.
8. **Samotus, B. and Schwimmer, S.**, Predominance of fructose accumulation in cold-stored immature potato tubers, *J. Food Sci.,* 27, 1—4, 1962.
9. **Schwimmer, S. and Burr, H. K.**, Structure and chemical composition of the potato tuber, *Potato Handb.,* 12, 42—45, 1967.
10. **Brown, H. D.**, Nutritive value of potatoes, *Potato Handb.,* 5, 61—66, 1960.
11. **Booth, R. H., de Buckle, T. S., Cardenas, O. S., Gómez, G., and Hervas, E.**, Changes in quality of cassava roots during storage, *J. Food Technol.,* 11, 245—264, 1976.
12. **Talley, E. A., Fitzpatrick, T. J., Porter, W. L., and Murphy, H. J.**, Chemical composition of potatoes. I. Preliminary studies on the relationships between specific gravity and the nitrogenous constituents, *J. Food Sci.,* 26, 351—355, 1961.
13. **Czyhrinciw, N.**, Consideraciones sobre industrialización de ráices y tubérculos tropicales, *Rev. Fac. Agron., Univ. Cent. Venez.,* 5(2), 110—117, 1969.
14. **Martin, W. J.**, Effect of storage temperature on development of internal cork in sweet potato roots, *Plant Dis. Rep.,* 39, 619—621, 1955.
15. **Priestley, J. H. and Swingle, C. F.**, Vegetative propagation from the stand-point of plant anatomy, *U.S. Dep. Agric. Tech. Bull.,* 151, 1929.
16. **Booth, R. H. and Proctor, F. J.**, Considerations relevant to the storage of ware potatoes in the tropics, *Pest Artic. News Summ.,* 18, 409—432, 1972.
17. **Kushman, L. J. and Wright, F. S.**, Sweet potato storage, *U.S. Dep. Agric. Handb.,* No. 358, 1969.
18. **González, M. A. and Collazo de Rivera, A.**, Storage of fresh yam under controlled conditions, *J. Agric. Univ. P. R.,* 56, 46—56, 1972.
19. **Thompson, A. K.**, Storage and transport of fruit and vegetables in the West Indies, in *Proc. Seminar on Hort. Devpt. in the Caribbean,* University of West Indies, Trinidad, 1972, 170—176.
20. **Okamoto, S. and Izawa, G.**, Effect of mineral nutrition on metabolic changes induced in crop plant roots, *Soil Plant Food, Tokyo,* 9, 8—13, 1963.
21. **Smith, O.**, Studies of potato storage, *Cornell Univ. Agric. Exp. Stn. N.Y., Bull.* No. 553, 1933, 57.
22. **Ingram, J. S. and Humphries, J. R. O.**, Cassava storage, A Review, *Trop. Sci.,* 14(2), 131—148, 1972.
23. **Rajasekharan, N., Rao, N. G., Kapur, N. S., Batia, D. S., and Subrahmanyan, V.**, Keeping quality of tapioca and nutromacaroni, *Food Sci.,* 9(7), 240—243, 1960.

24. Chadha, Y. R., Sources of starch in Commonwealth territories. III. Cassava, *Trop. Sci.*, 3(3), 101—113, 1952.

25. Kerr, A. J., The storage of native food crops in Uganda, Cassava, *East Afr. Agric. J.*, 7(2), 75—76, 1941.

26. Schery, R. W., Manioc — a tropical staff of life, *Econ. Bot.*, 1(1), 20—25, 1947.

27. Akinrele, I. A., Cook, A. S., and Holgate, R. A., The manufacture of gari from cassava in Nigeria, in *Proc. Int. Congr. Sci. Technol.*, Vol. 4, 1966, 633—644.

28. Halliday, D., Qureshi, A. H., and Broadbent, J. A., Investigations on the storage of gari, *Ann. Rep. Nigerian Stored Prod. Res. Inst.*, Tech. Rep., No. 16, 131—141, 1967.

29. Jones, W. O., *Manioc in Africa*, Stanford University Press, Stanford, Calif., 1959, 108—113.

30. de Bruijn, G. H., The cyanogenic character of cassava (*Manihot esculenta*), in *Chronic Cassava Toxicity*, IDRC Monogr., 010e, Proc. Interdisciplinary Workshop, London, 1973, 43—48.

31. Wood, T. J., The cyanogenic glucoside content of cassava and cassava products, *J. Sci. Food Agric.*, 16, 300, 1965.

32. Braude, R. and Mitchell, K. G., Potatoes for fattening pigs. Comparison of cooked and raw potatoes, *Agriculture*, 57, 501—504, 1951.

33. Vogel, R., Trautschold, I., and Werle, E., *Natural Proteinase Inhibitors*, Academic Press, New York, 1968, 159.

34. Pond, W. G. and Maner, J. H., *Swine Production in Temperate and Tropical Environments*, W. H. Freeman, San Francisco, 1974, 646.

35. Henry, T. A., *The Plant Alkaloids*, 4th ed., Churchill Livingston, London, 1949.

36. Bevan, C. W. L. and Hirst, J., A convulsant alkaloid of *Dioscorea dumetorum*, *Chem. Ind.*, 4, 103, 1958.

37. Gómez, G., Camacho, C., and Maner, J. H., Utilización de yuca fresca y harina de yuca en alimentación porcina, in *Memoria Seminario Intern. de Ganaderia Tropical*, Acapulco, 1976, Imptesora Azteca, Mexico, 1976, 91—102.

38. Maner, J. H., Buitrago, J., and Jiménez, I., Utilization of cassava in swine feeding, *Proc. Int. Symp. on Trop. Root Crops*, Vol. 2, (Part 6), University of West Indies, St. Augustine, Trinidad, 1967, 62—71.

39. Centro Internacional de Agricultura Tropical, Swine Production Systems, Annual Report, Cali, Colombia, 1973, 120—144.

40. Centro Internacional de Agricultura Tropical, Swine Production Systems, Annual Report, Cali, Colombia, 1974, 153—195.

41. Oldfield, J. E., Effect of fermentation on the chemical and nutritional value of feeds, in *Effect of Processing on the Nutritional Value of Feeds*, Proc. Symp., Gainesville, Fla., National Academy of Sciences, Washington, D.C., 1973, 34—47.

42. Castillo, L. S., Aqlibut, F. B., Javier, T. A., Gerpacio, A. L., García, G. V., Puyaoan, R. B., and Ramin, B. B., Camote and cassava tuber silage as replacement for corn in swine growing-fattening rations, *Philipp. Agric.*, 47(9—10), 460—474, 1964.

43. Underwood, E. J. and Curnow, D. H., Vitamin A in the nutrition of sheep in western Australia, *Aust. Vet. J.*, 20, 248—253, 1944.

44. Allender, D. R., Potatoes for livestock feed, *U.S. Dep. Agric. Misc. Publ.*, 676, 1948.

45. Johnson, R. F., Rhinerhard, E. F., and Hickman, C. W., Sun-dried potatoes for fattening steers, *Idaho Agric. Exp. Stn. Bull.*, 201, 1—3, 1954.

46. Dickey, H. C., Brugman, H. H., Plummer, B. E., and Highlands, M. E., The use of byproducts from potato starch and potato processing, in *Proc. Int. Symp. Utilization Disposal of Potato Wastes*, Research and Productivity Council, New Brunswick, Canada, 1966, 106—121.

47. Talburt, W. F. and Smith, O., *Potato Processing*, AVI Publishing, Westport, Conn., 1967, 588.

48. Woodman, H. E. and Evans, R. E., Further investigations of the feeding value of artificially dried potatoes: the composition and nutritive value of potato cossettes, potato meal, potato flakes, potato slices, and potato dust, *J. Agric. Sci.*, 33, 1—14, 1943.

49. Kirsch, W. and Jantzon, H., Value of root vegetables as a swine feed including whole sugar-beet slices, molasses slices, dried molasses slices, steffen slices, and diffusion slices, *Biedermanns Zentralbl. Abt. B.*, 15, 206, 1943.

50. Southwell, B. L. and Black, W. H., Dehydrated sweet potatoes for fattening steers, *G. Coastal Plain Exp. Stn. Bull.*, 45, 1—24, 1948.

51. Singletary, C. C., McCraine, S. E., and Berwick, L., Dehydrated sweet potato meal for fattening steers, *La. Agric. Exp. Stn. Bull.*, 446, 1—7, 1950.

52. Darlow, A. E., Dried sweet potatoes as a replacement for corn in fattening beef cattle, *Okla. Agric. Exp. Stn. Bull.*, B-342, 1—15, 1950.

53. Edmond, J. B. and Ammerman, G. R., *Sweet Potatoes: Production, Processing, Marketing*, AVI Publishing, Westport, Conn., 1971.

54. Tillman, A. D. and Davis, H. J., Studies on the use of dehydrated sweet potato meal in chick rations, *La. Agric. Exp. Stn. Bull.,* 358, 1—12, 1943.

55. Haberlandt, G., *Physiological Plant Anatomy,* Macmillan, London, 1914, reprint ed., 1965, 777.

56. Black, O. F., Calcium oxalate in the dasheen, *Am. J. Bot.,* 5, 447—451, 1918.

57. Middendorf, E. A., Plants with blowguns, *Turtox News.,* 46, 162—164, 1968.

58. Barnes, B. A. and Fox, L. E., Poisoning with Diffenbachia, *J. Hist. Med. Allied Sci.,* 10, 173—181, 1955.

59. Arnott, H. J. and Pautard, F. G. E., Calcification in plants, in *Biological Calcification,* Schraer, H., Ed., Appleton-Century-Crofts, New York, 1970.

60. Ledbetter, M. C. and Porter, K. R., *Introduction to the Fine Structure of Plant Cells,* Springer-Verlag, Basel, 1970, 188.

61. Schotz, F., Diers, L., and Bathelt, H., Zur Feinstructure der Raphidenzellen. I. Die Entwicklung der Vakuden und der Raphiden, *Z. Pflanzenphysiol.,* 63, 91—113, 1970.

62. Sakai, W. S. and Hanson, M., Mature raphid idioblast structure in plants of the edible aroid genera *Colocasia, Alocacia,* and *Xanthosoma, Ann. Bot.,* 38, 739—748, 1974.

63. Enzumah, H., Miscellaneous tuberous crops in Hawaii, in *Proc. 2nd Int. Symp. Tropical Root and Tuber Crops,* Vol. 1 (Part 6), Plucknett, D., Ed., College of Tropical Agriculture, Honolulu, 1970, 166—171.

64. Standal, B. R., The nature of poi carbohydrates, in *Proc. 2nd Int. Symp. on Tropical Root and Tuber Crops,* Vol. 1 (Part 6), Plucknett, D., Ed., College of Tropical Agriculture, Honolulu, 1970, 146—148.

EFFECT OF PROCESSING ON NUTRIENT CONTENT OF FEEDS: SUGAR CROPS

Leo V. Curtin

INTRODUCTION

Although the development of the sugar industry — cane and beet — has been remarkable since 1850, the use of sugar byproducts for animal feeding was much slower in developing and on a more erratic basis.[1,2]

World sugar production is now approaching 85 million tons, which results in production of 53 million tons of bagasse (dry basis), 12 million tons of dried beet pulp, and 27 million tons of molasses.[3] The early history of the production of sugar products from sugar cane and sugar beets and general descriptions of the processes involved are given in the *Spencer-Meade Cane Sugar Handbook,*[4] *The Beet Sugar Story,*[5] and *The Molasses Story.*[6]

Except for molasses, detailed statistics are not available to show end-uses on a world basis. Very little bagasse is used for animal feed with almost 96% of the bagasse produced being burned as fuel to generate steam in the cane sugar factories.[3] Beet pulp is used exclusively as animal feed. Molasses has a more varied utilization, with the largest use being for animal feed, the U.S. alone consuming over 3 million tons of molasses yearly for this purpose.

MOLASSES

The original use of the term molasses was to describe the final effluent obtained in the preparation of sucrose by repeated crystallization from the juices of sugar cane or sugar beets. In recent years the term has been applied to sugar-containing products of other origins.

The use of molasses in livestock feeds goes back many years. One of the earliest reports in the U.S. showing the value of cane molasses in cattle feeds was published in 1980 by Texas workers.[7] Early nutrition writings at the end of the last century refer to feeding of molasses as a source of energy and to improve palatability. Molasses is extremely palatable and will induce cattle to eat lower quality feeds that might not otherwise be consumed.

In a comprehensive review of the literature, Scott[2] wrote, "The benefits resulting from the use of molasses in rations for farm animals have been discovered, lost in obscurity and rediscovered many times throughout the past one hundred and fifty years."

Types Of Molasses

The *Association of American Feed Control Officials* (AAFCO), *Official Publication*[8] and the *U.S.-Canadian Tables of Feed Composition*[9] describe the following types of molasses:

AAFCO, 1977	National Research Council, 1969
63.1 Beet molasses	Beet, sugar, molasses (ref. no. 4-00-668)
63.3 Citrus molasses	Citrus, syrup (ref. no. 4-01-241)
63.4 Cane molasses	Sugar cane, molasses (ref. no. 4-04-696)
63.5 Hemicellulose extract	Wood, molasses (ref. no. 4-05-502)
63.6 Starch molasses	—

Cane molasses is the byproduct of the manufacture of refining of sucrose from sugar cane. It must not contain less than 46% total sugars expressed as invert. If the moisture of the cane molasses exceeds 27%, the density determined by double dilution must not be less than 79.5° Brix.[8]

Beet molasses is defined by AAFCO[8] as a byproduct of the manufacture of sucrose from sugar beets. It must contain not less than 48% total sugars expressed as invert, and its density determined by double dilution must not be less than 79.5° Brix.

Citrus molasses is the partially dehydrated juices obtained from the manufacture of dried citrus pulp. It must contain not less than 45% total sugars expressed as invert and the density determined by double dilution must not be less than 71.0° Brix.[8]

Hemicellulose extract is a byproduct of the manufacture of pressed wood. It is the concentrated soluble material obtained from the treatment of wood at elevated temperature and pressure without use of acids, alkalis, or salts. It contains pentose and hexose sugars and has a total carbohydrate content of not less than 55%.[8]

Starch molasses is a byproduct of the manufacture of dextrose from starch derived from corn or grain sorghums in which the starch is hydrolyzed by use of enzymes and/ or acid. It must contain not less than 43% reducing sugars expressed as dextrose and not less than 50% total sugars expressed as dextrose. It shall contain not less than 73% total solids.[8] Starch molasses derived from corn oftentimes is referred to as Hydrol®, the trade name of one manufacturer.

Of the five types of molasses described, cane molasses and beet molasses are very much the dominant types from the standpoint of total production and use for livestock feeding. Supplies of citrus molasses, hemicellulose extract, and starch molasses are small, and all normally are used locally in the areas of production.

Recent figures from the U.S. Department of Agriculture (USDA), Foreign Agricultural Service, show that the total world production of cane and beet molasses for 1975 to 1976 was 26,302,000 t.[10] It has been estimated that approximately 57% of the world molasses production is cane molasses.[6] The U.S. is the largest consumer of molasses. The USDA Molasses Market News, Market Summary for 1975[11] reports that approximately 3,991,000 t of molasses were utilized in the U.S. in 1975; of this total, 78%, or 3,104,000 t, were used for animal feeding.

Nutritional Properties of Molasses

Literature on the feeding value of molasses is quite voluminous and is available in several reviews that have been written on the subject. No attempt will be made in this paper to cite all of the available references. The comprehensive review of molasses literature by Scott[2] has been mentioned previously.

The average composition of molasses products as adapted from information taken from several published sources,[9,12-16] National Molasses Company analyses, and private communications with others in the industry is summarized in Table 1. Considerable variation in composition exists among the different types of molasses. Since all molasses are byproducts, variations also exist in the composition of products within any type depending on origin, season, processing, and storage variables.

The term Brix is commonly used in the molasses trade to indicate specific gravity and is a close approximation of the total solids content. When Brix readings are used for pure sugar solutions, they indicate the percentage of sugar solids by weight, however in addition to sugar, molasses contains minerals, gums, nitrogenous materials, and other materials, so a Brix reading on molasses will vary somewhat from actual content of sugars or total solids. The total solids of the different types of molasses vary between 65% and 77%.

The total sugars or soluble carbohydrates are relatively high in all types of molasses and account for a major part of the feeding value of molasses. None of the molasses

Table 1
COMPOSITION OF MOLASSES PRODUCTS

	Cane	Beet	Citrus	Starch	Hemicellulose extract
Brix	79.5	79.5	71	78	65
Total solids (%)	75	77	65	75	65
Total protein (%)	3.0	6.0	7.0	0.5	0.5
Ash (%)	8.1	8.2	6.1	8.0	5.0
Total sugars (%)	48	48	45	50	55 (Total carbohydrate)
Calcium (%)	0.8	0.15	1.3	0.1	0.5
Phosphorus (%)	0.08	0.03	0.16	0.2	0.05
Magnesium (%)	0.35	0.23	0.14	—	0.07
Potassium (%)	2.4	4.7	0.09	0.02	0.04
Sulfur (%)	0.8	0.5	0.17	0.05	—
Sodium (%)	0.2	1.2	0.27	2.5	—
pH	5.5	7.8	5.0	5.0	5.5

Reproduced from *Effect of Processing on the Nutritional Value of Feeds,* 289—290, with the permission of the National Academy of Sciences, Washington, D.C.

Table 2
TRACE MINERALS IN MOLASSES

	Cane (ppm)		Beet (ppm)	
	Av	Range	Av	Range
Cobalt	2.4	1.5—4.2	2.5	2.1—2.9
Copper	14	6.6—38	8.5	4.2—15.7
Iron	297	145—640	164	114—205
Manganese	28	2.1—67	16	2.2—29
Zinc	13	7.5—37	65	4.1—264

Reproduced from *Effect of Processing on the Nutritional Value of Feeds,* 289—290, with the permission of the National Academy of Sciences, Washington, D.C.

types contain significant levels of protein, although beet molasses and citrus molasses are somewhat higher in protein than the other types.

All molasses products are low in phosphorus. Calcium is higher in cane and citrus molasses. It should be noted that cane and beet molasses are relatively high in magnesium, potassium, and sulfur and may account for some of the so-called unidentified factor activity for ruminants that has been attributed to these products.

The trace mineral content of molasses can be quite variable, but can be important in the feeding of livestock, particularly in areas where feeding of a good mineral mixture is not routine. In Table 2 are shown trace mineral data obtained by the National Molasses Company for cane molasses from 14 different areas of production and beet molasses from 5 different areas. Cane and beet molasses are particularly high in cobalt as compared to other natural feed ingredients. Note the wide range of analyses for each of the trace minerals in both cane and beet molasses.

The vitamin content of molasses is not of great significance because of the low and variable levels present; also, the water-soluble vitamins normally are not required in the rations of mature ruminants because of the synthesis of these vitamins by rumen fermentation. In Table 3 are shown vitamin data obtained by the National Molasses

Table 3
VITAMINS IN MOLASSES

	Cane (ppm)		Beet (ppm)	
	Av	Range	Av	Range
Pantothenic acid	8.5	0—16.3	14.3	7.3—18.4
Niacin	5.2	2.7—7.0	10.4	8.1—14.2
Riboflavin	< 0.05	—	< 0.05	—
Choline	696	0—2670	389	355—416
Biotin	0.36	.01—0.7	0.46	0.36—0.61

Reproduced from *Effect of Processing on the Nutritional Value of Feeds*, 289—290, with the permission of the National Academy of Sciences, Washington, D.C.

Company for cane molasses from 14 different areas of production and beet molasses from 5 different areas. The relatively high content of biotin in molasses was noted by Hegsted.[17] Jukes[18] also noted the high content of pantothenic acid in blackstrap molasses.

Processing Variables — Molasses

Since molasses products are byproducts and have a relatively low economic value as compared to the value of the primary product(s) produced, one would not expect the manufacturer to place much emphasis on changes in processing solely for the purpose of changing the characteristics of the molasses. The objective of the sugar industry is to obtain as high a sugar yield as possible, and thus produce molasses whose sugar content is as low as possible. Cane and beet molasses contain the greater portion of the nonsugars of the juices from which they are derived, together with that portion of the sucrose and reducing sugars that could not be removed by crystallization. It follows, therefore, that the composition must vary with the variety and maturity of cane, climatic and soil conditions, mailing conditions, storage, and other factors.

Variations in composition that would most affect the nutritional value are changes in Brix, total sugars, protein, and ash.

Effect of Brix

The broad range of Brix in molasses as it comes from the centrifugals after removal of crystallized sugar is 85 to 92° Brix. Water may be added to facilitate handling, but normally molasses will be stored and shipped as "high" Brix to avoid handling water. In the tropical or subtropical areas of production, the cane molasses is used as it comes from the centrifugals without addition of water. In the U.S. most cane and beet molasses is adjusted to 79.5° Brix by the addition of water prior to sale for feed use, to facilitate handling. Although the average Brix of molasses as produced will vary among different areas of production,[19] this seldom is a factor in the nutritional value of molasses since it normally is marketed on a Brix basis and standardized to a constant Brix before use.

Sugars

The sugar in beet juice and cane juice consists predominately of sucrose. During harvest and manufacture of sugar, some of the sucrose is hydrolyzed to produce glucose and fructose, which make up the major portion of the reducing sugars reported in analyses. The sugar in beet molasses is largely sucrose with very little reducing sugars, however the ratio of sucrose to invert sugar or reducing sugars in cane molasses is in the range of 1.5:1 to 2.5:1.[12] Molasses from beets that have been frozen or that have spoiled will have considerably higher levels of reducing sugars.[20] The rate of hydrolysis increases with rising temperature and with falling pH during manufacture of sugar.[20] The high and varying amounts of invert sugar found in different cane molasses, however, are due mainly to the reducing sugars in the raw sugar cane, and only to a small extent to the inversion of sucrose during sugar manufacture or molasses storage.[20] The microflora, which multiply rapidly under the tropical conditions, are involved to some extent in the formation of invert sugar. Since very little of the invert sugar is removed in sugar manufacture, these sugars are concentrated in molasses resulting in higher total sugars in the molasses.

It follows, therefore, that sugar mills that are operated most efficiently from the time cane is cut in the field until final crystallization of the sugar will remove more of the sugar from cane juice and thus produce molasses with lower total sugars.

Protein

The total nitrogen in cane molasses ranges from 0.4 to 1.5% and from this the "crude protein" is figured at 2.5 to 9.5% (N × 6.25).[4] Molasses produced from cane grown on organic or muck soils will have the higher content of nitrogen.[15,19,21] Richardson[19] found an average of 8.9% protein (N × 6.25) with a range of 6.4 to 15.4% in eight samples of Florida molasses. Molasses produced from cane grown on organic soil in Florida between 1960 and 1963 contained from 7.1 to 10.1% total protein.[21] A more typical range of protein content for cane molasses produced on mineral soils is 2 to 4%.

The nitrogen content of beet molasses ranges from 1.2 to 2.2% or 7.5 to 13.7% protein.[20] Data quoted by Olbrich[20] indicate that if rainfall is scanty during the growing season, the nitrogen content of beet molasses will be higher.

Ash

Many complete analyses of molasses ash have been published, but these serve only to show the composition of the particular sample, or particular average of samples, for the conditions prevailing. The amount and composition of the ash of cane and beet molasses are affected by the cane variety, the conditions under which it was grown (climate, soil), and by the methods employed in the sugar factory.[4,20] In general, cane molasses produces less ash than beet molasses. Cane molasses has a higher content of calcium and phosphorus, whereas beet molasses contains considerably more potassium.

Olbrich[20] noted that, depending on soil conditions, an increasing quantity of ash was found in beet molasses in the following order: clay soil, sandy soil, and marshy soil. If the beet finds little lime in the soil, it takes up magnesium instead. Beets grown on lime-rich soil may produce a magnesium content of zero in the molasses if pure and magnesium-free water is used for processing.

The low phosphorus content of molasses indicates that the greatest part of this mineral is precipitated out of the juice as insoluble calcium phosphate during the purification step of manufacture.

Meade[4] also notes that the amount of ash in molasses has increased during the past 30 years with heavier milling and because of certain varieties of cane, as well as with improved methods of molasses exhaustion.

Vitamins

The desugaring of sugar juice concentrates the heat and alkali-stable vitamins in the final molasses. Pantothenic acid seems to be quite sensitive to the various operations in sugar production. The broad variation ranges in the vitamin content of molasses, as shown in Table 3 and noted by others,[20] is quite understandable because many factors influence the amounts present, extending from the growth of the beets or cane, through the manufacturing process, until the end product is produced; even the latter can undergo changes during the time of storage.

Changes During Storage — Molasses

Because of the seasonal nature of production and the long distances between areas of production and consumption, cane and beet molasses may be stored for several months or even more than a year before final use.

It generally is agreed that all cane molasses will show a gradual decrease in total sugar content over months of storage. As noted by Meade,[4] Dr. C. A. Browne of the U.S. Bureau of Chemistry and Soils stored two samples of Cuban cane molasses for 21 years (1914 to 1935) and noted the following changes:

1. Sucrose decreased from average of 33.04 to 9.61%.
2. Reducing sugars increased from average of 22.10 to 28.85%.
3. Total sugars (as invert sugar) decreased from average of 56.88 to 38.96%. This is an average decrease of 0.85% total sugars per year.
4. Total solids decreased 3% or an average decrease of 0.14% total solids per year.

The decomposition was not biological because the samples in 1914, and on various subsequent occasions, showed no yeasts, molds, bacteria, or other organisms.

The continuous decomposition of molasses in storage is attributed by Browne[4] chiefly to the reaction of unstable organic substances (originally produced during sugar production) with further quantities of reducing sugars in the molasses, which results in the formation of dark-colored colloidal impurities of high carbon content. Reaction between the amino acids and reducing sugars of the cane juice also may play a minor part in the early stages of decomposition.

Fromen and Bowland[22] reported the following losses in total sugars of cane molasses in storage at a Puerto Rican sugar mill:

	1953	1954
High temperatures of storage	100—106°F	100—108°F
Total sugars, initial	59.6%	59.0%
Total sugars, end	56.0%	55.0%
Loss in total sugars	3.6%	4.0%
Average storage period	9.5 weeks	19.5 weeks

Olbrich[20] believes that the gradual changes in composition of molasses while in storage are due not only to chemical reactions, but also at least in part to the activity of microorganisms. Olbrich,[20] however, believes that the chemical reactions of most importance during the storage of molasses are those for which Honig[20] has set up standard values for the usual slow changes in the composition of cane molasses, and which are valid for 6 months storage at 30°C as follows:

	Av reduction (%)
Sucrose	0.5
Reducing sugars	0.7
Total sugars	1.2
Increase in nonfermentable reducing substances	0.6

The data reviewed suggest that the major loss of total sugars in molasses during storage occurs during the first few weeks of storage, after which the rate of destruction is decreased considerably. The faster rate of destruction during initial storage appears to be associated with the higher temperatures at which the molasses comes from the centrifugals and into storage.

To restrict the chemical changes that may occur in freshly centrifuged cane molasses produced under tropical conditions, the molasses should be cooled as much as possible before being placed in storage. The alkalinity and the low atmospheric temperatures during production improve the keeping qualities of beet molasses.

There have been several reports of molasses suddenly and violently decomposing into a carbonized solid state. Fromen and Bowland[22] reported on extensive investigations of this type reaction in Puerto Rico in 1953, as well as reviewing data for four similar occurrences at other locations, and concluded that decomposition of the violent, destructive type is associated with going directly into storage tanks from the centrifugals with molasses at a high temperature. The "safe" storage temperature apparently varies with molasses of different compositions. Fromen and Bowland[22] concluded that decomposition of the violent, destructive type will never happen to any molasses if the storage temperature is not allowed to exceed 105 to 110°F (40 to 43°C), and that recirculating molasses will prevent quiescent high-temperature danger zones caused by accidental local overheating.

Summary — Molasses

The term molasses is used in the animal feeding industry to include cane molasses, beet molasses, citrus molasses, starch molasses (Hydrol®), and hemicellulose extract (Masonex®). Cane and beet molasses are of greatest importance from the standpoint of total world production and animal feed usage. All products are high in total soluble carbohydrates, but they differ in types of sugars present and their content of other nutrients.

Since molasses products are byproducts and have relatively low value as compared to the value of the primary products produced, manufacturers place very little emphasis on changes in processing solely for the purpose of changing the characteristics of the molasses.

Factors that will affect the nutrients present in cane and beet molasses include variety, climatic conditions, type of soil, processing conditions during production of sugar, and storage. The effect of these variables on solids (Brix), sugars, protein, ash, and vitamins are reviewed.

The most significant effect of processing and storage variables on the nutritive value of cane and beet molasses is their effect on total sugars. The more modern and more efficiently operated mills will produce molasses with lower total sugars. Total sugars decrease during storage of molasses, and the rate of reduction of total sugars is increased with higher temperatures. Under normal operating and storage conditions, it appears that the major loss of sugars in storage occurs during the first few weeks of storage.

BAGASSE

Bagasse is the fibrous residue of the sugar cane stalk after crushing and extraction of the cane juice. Its composition varies according to the variety of cane, its maturity, the method of harvesting, and the efficiency of the milling plant. Bagasse can be separated into two fractions for feeding or industrial use — pith and fiber. Pith is the center, softer portion that is more absorbent, which contains most of the residual sugar juice, and is high in ash. The outside, fibrous portion contains a higher proportion of cellulose.

Analyses of whole bagasse, pith, ground fiber (chicken litter), and ammoniated bagasse reported by Florida workers[23] are as follows:

	Whole bagasse	Pith	Chicken litter (fiber)	Ammoniated bagasse
Dry matter	89.75	90.00	92.30	91.89
Crude protein	1.75	1.69	2.63	11.66
Ash	2.73	14.31	2.00	3.12
Crude fiber	35.88	28.82	41.35	39.30
Fat	0.85	1.23	0.65	1.01
Nitrogen-free extract (NFE)	48.45	43.95	45.67	36.79
TDN (total digestible nutrients)	20	20	20	20

Analysis of bagasse from eight cane-growing countries as reported by Paturau[1] shows that the solids in bagasse consist largely of cellulose, pentosans, and lignin. Much of the cellulose and pentosans will analyze as NFE in the proximate analysis procedure used for feeds as illustrated in the analysis reported by Kirk et al.[23]

The use of sugar cane bagasse as a feed for cattle has been investigated numerous times in many countries of the world.[24-27] On the whole, the results of several studies have not been satisfactory. In fact, the results of a number of these studies have shown that it takes more energy to digest bagasse than is obtained from the bagasse by the animals. Results of these studies have shown that when fed alone, bagasse or bagasse pith is not palatable and is not consumed in satisfactory amounts; molasses added to the bagasse or pith increases palatability up to levels of approximately 55% molasses. Bagasse pith when mixed with other ingredients, especially in concentrate feeds, was readily consumed.

In Florida experiments[24] it has been observed that a 4-year-old steer digested bagasse pith to the extent of 40 to 60% in different digestion trials, with an average of 53%. This compares with results with young animals where the digestibility averaged between 20 and 25%. Older animals apparently can digest bagasse with far more efficiency than younger animals.

Ammoniated bagasse was utilized satisfactorily as a source of nitrogen for protein synthesis by the cattle. A digestion trial with the ammoniated product in Florida indicated that it had a protein digestibility of approximately 63%.[24] The digestibility of the bagasse itself, as measured in terms of crude fiber digestion, was not greatly improved by the ammoniation. Urea added as a source of nonprotein nitrogen in place of ammonia appeared to be equally satisfactory as a source of nitrogen, improving the digestibility of crude fiber somewhat.[23,24]

Pelleted bagasse has been compared with cottonseed hulls as a source of roughage for fattening cattle at the Florida Agricultural Experiment Station.[28,29] When the bagasse was pelleted with 5 to 8% molasses, it mixed quite well with other feed ingredients. Although 15% of the bagasse pellets included in the complete ration appeared to be too high for optimum steer performance, 7.5 to 10% of the bagasse pellets did very well, and performance was comparable or slightly superior to cattle receiving the same level of cottonseed hulls in the ration. When either roughage was included at a level of 5% of the ration, it was difficult to keep the cattle on full feed because of lack of adequate roughage.

Research in Puerto Rico[30,31] has shown that bagasse can be used in dairy rations.

Although the use of bagasse for feeding to livestock is limited because of its low digestibility, palatability, and low bulk density, experimental work has shown that when properly supplemented and fed at low levels, it can be used effectively as a source of needed fiber and low cost energy for mature cattle. Its utilization is improved by addition of molasses and by pelleting to increase bulk density.

BEET PULP

In the manufacture of sugar from sugar beets, the beets are washed and cut into thin strips. Next the juice is thoroughly extracted with warm water, leaving the byproduct known as wet beet pulp. Wet beet pulp can be fed to cattle as is in areas surrounding a beet sugar mill, or more often is dried and sold as dried beet pulp for feeding to beef and dairy cattle. In some cases, wet beet pulp is combined with molasses before drying to form molasses beet pulp. These feeds are palatable, bulky, slightly laxative, and keep well in storage.

Dried beet pulp and dried molasses beet pulp are high in carbohydrates, but relatively low in protein and fat. Average analyses of the two products are shown in Table 4.[9,14]

Though it is high in fiber, the fiber in beet pulp is well digested. Because of the bulky nature and palatability, beet pulp is very popular in feeds for dairy cattle, and most of the supply is used for this purpose. It also is used extensively in the sugar beet districts as a substitute for part of the grain for fattening cattle or sheep.

Molasses beet pulp pellets are also produced. Pellets weigh about four times as much per cubic foot as ordinary dried molasses beet pulp and eliminate loss from blowing when fed in the open. As well, they facilitate shipment in bulk to other consuming areas.

In results of experiments compiled by Morrison,[14] these feeds have been about equal to barley or grain sorghum when replacing not more than one half the grain in rations for fattening cattle or lambs.

Table 4
COMPOSITION OF BEET PULP PRODUCTS

Analyses	Dried Beet Pulp (4-00-669)[a]		Dried Molasses Beet Pulp (4-00-672)[a]	
	U.S.-Canadian Tables[9]	Morrison[14]	U.S.-Canadian Tables[9]	Morrison[14]
Dry matter (%)	91.0	91.2	92.0	92.2
Ash (%)	3.6	3.5	5.7	5.8
Crude fiber (%)	19.0	19.6	16.0	15.2
Ether extract (%)	0.6	0.6	0.5	0.5
Nitrogen-free extract (%)	58.7	58.7	60.7	61.8
Crude protein (%)	9.1	8.8	9.1	8.9
Digestible protein (%)	4.1 (Cattle)	4.1	6.0 (Cattle and sheep)	5.9
Lignin (%)	8.0	—	—	—
Digestible energy (kcal/kg)	2889 (Cattle and sheep)	—	3002 (Cattle)	—
Metabolizable energy (kcal/kg)	2370 (Cattle and sheep)	—	2462 (Cattle)	—
Total digestible nutrients (%)	66 (Cattle and sheep)	68.7	68 (Cattle)	72.4
Calcium (%)	0.68	0.69	0.56	0.57
Phosphorus (%)	0.10	0.08	0.08	0.07
Iron (%)	0.03	0.03	—	—
Magnesium (%)	0.27	0.27	0.13	0.17
Zinc (%)	0.7	—	—	—
Potassium (%)	0.21	0.18	1.64	1.63
Niacin (mg/kg)	16.3	16.3	—	—
Pantothenic acid (mg/kg)	1.5	1.5	—	—
Riboflavin (mg/kg)	0.7	0.7	—	—
Choline (mg/kg)	829	—	—	—

[a]   NRC ref. no.

REFERENCES

1. Paturau, J. M., *By-Products of the Cane Sugar Industry,* American Elsevier, New York, 1969, 1—6, 25—26, 140—152.
2. Scott, M. L., Use of molasses in the feeding of farm animals, *Sugar Res. Found. Technol. Rep. Ser.,* 9, 1—153, 1953.
3. Paturau, J. M., Comparative economic value of by-products of the sugar industry, *F. O. Licht's Int. Molasses Rep.,* 13, 1—5, 1976.
4. Meade, G. P., *Spencer-Meade Cane Sugar Handbook,* John Wiley & Sons, New York, 1963, 267—284.
5. Anon., *The Beet Sugar Story,* U.S. Beet Sugar Association, Washington, D.C., 1959, 6—18, 44—54.
6. Anon., *The Molasses Story,* National Molasses Company, Willow Grove, Pa., 1970, 1—16.
7. Gully, F. A. and Carson, J. W., Feeding experiment, *Tex., Agric. Exp. Stn. Bull.,* 10, 31, 1890.
8. Anon., *Assoc. Am. Feed Control Off., Off. Publ.,* 105—106, 1977.
9. Anon., *U.S.-Canadian Tables of Feed Composition, Publ.* No. 1684, National Research Council, National Academy of Sciences, Washington, D.C., 1969, 25—26, 35, 81, 90.
10. Anon., World Sugar and Molasses Output Expands in 1975/76, 2-76, Foreign Agricultural Service, U.S. Department of Agriculture, Washington, D.C., 1976, 10-11.
11. Anon., *Molasses Market News, Market Summary 1975,* Agricultural Marketing Service, U.S. Department of Agriculture, Washington, D.C., 1976, 1-15.
12. Binkley, W. W. and Wolfrom, M. L., Composition of cane juice and final molasses, *Sugar Res. Found. Sci. Rep. Ser.,* 15, 1-23, 1953.
13. Hendrickson, R. and Kesterson, J. W., By-products of Florida citrus, composition, technology and utilization, *Fla. Agric. Exp. Stn. Bull.,* 698, 1-76, 1965.
14. Morrison, F. B., *Feeds and Feeding,* 22nd ed., Morrison, Ithaca, N.Y., 1959, 528—539, 1044, 1058—1059, 1099, 1107.
15. Allen, R. D., Feed ingredient analysis table: 1976 edition, in *Feedstuffs Reference Issue,* Miller, Minneapolis, 1976, 33-38.
16. Anon., Feed ingredient analyses, in *Feed Industry Red Book,* Communications Marketing, Edina, Minn., 1975, 118-122.
17. Hegsted, D. M., Mills, R. C., Briggs, G. M., Elvehjem, C. A., and Hart, E. B., Biotin in chick nutrition, *J. Nutr.,* 23, 175—179, 1942.
18. Jukes, T. H., Distribution of pantothenic acid in certain products of natural origin, *J. Nutr.,* 21, 193—200, 1941.
19. Richardson, L. R., Factors in developing grades and standards for blackstrap molasses for feed, *U.S. Dep. Agric. Mark. Res. Rep.,* 302, 1—42, 1959.
20. Olbrich, H., Molasses, in *Principles of Sugar Technology,* Vol. 3, Honig, P., Ed., Elsevier, New York, 1963, 511—697.
21. Chapman, H. L., Jr., Kidder, R. W., Koger, M., Crockett, J. R., and McPherson, M. K., Blackstrap molasses for beef cows, *Fla. Agric. Exp. Stn. Bull.,* 701, 1—31, 1965.
22. Fromen, G. and Bowland, E., *Rapid Deterioration and Destruction of Blackstrap Molasses in Storage,* Fajardo Sugar Co., Fajardo, Puerto Rico, 1955, 3—32.
23. Kirk, W. G., Chapman, H. L., Jr., Peacock, F. M., and Davis, G. K., Utilizing bagasse in cattle fattening rations, *Fla. Agric. Exp. Stn. Bull.,* 641A, 1—16, 1969.
24. Davis, G. K. and Kirk, W. G., Bagasse as a cattle feed, *Sugar J.,* 12—13, 40, 1958.
25. Pate, F. M. and Coleman, S. W., Use of Sugarcane and Sugarcane By-Products for Finishing Cattle on Pasture and in the Feedlot, in Proc. 10th Annu. Conf. Livestock Poul. Latin America, Institute of Food and Agricultural Sciences (IFAS), University of Florida, Gainesville, 1976, A-41—A-49.
26. Brown, P. B., Roughage — molasses rations for wintering beef cattle, *La. Agric. Exp. Stn. Bull.,* 617, 1—12, 1967.
27. Wayman, O. and Iwanaga, I., Further studies on the use of rations high in molasses for fattening beef cattle, *Hawaii Agric. Exp. Stn. 5th, Prog. Notes,* No. 110, 1—12, 1956.
28. Chapman, H. L., Jr., Sugarcane bagasse, a waste-disposal problem, *Proc. Soil Crop Sci. Soc. Fla.,* 31, 22—23, December 1971.
29. Chapman, H. L., Jr. and Palmer, A. Z., Bagasse pellets in beef cattle fattening rations, *Fla. Agric. Exp. Stn. Circ.,* S-216, 1—7, 1972.
30. Randel, P. F., Feeding lactating diary cows concentrates and sugarcane bagasse as compared with a conventional ration, *J. Agric. Univ. P.R.,* 50, 255—269, 1966.
31. Randel, P. F., Soldevila, M., and Salas, B., A complete ration composed of concentrates and sugarcane bagasse vs. a conventional ration of pangolagrass and supplemental concentrate for milk production, *J. Agric. Univ. P.R.,* 53, 167—176, 1969.

EFFECT OF PROCESSING ON NUTRIENT CONTENT OF FEEDS: AQUATIC PLANTS

James G. Linn

INTRODUCTION

Aquatic plants are rapidly becoming a major problem in the commercial and recreational use of inland waters. Increased eutrophication of many lakes and streams has resulted in abundant plant growth. Eradication of these plants is not only impossible but detrimental to the environment. Chemical control of plants increases the problem as dead plants sink to the bottom and decay causing more nutrients to be released. Biological controls are not well developed and not always feasible. Thus, harvesting of these plants may be the most effective and desirable method for control.

Use of these aquatic plants as a livestock feed could partially offset harvesting costs. Even when not removed for control purposes, exploitation of aquatic plants as a livestock forage in years of drought may be beneficial.

CHEMICAL COMPOSITION

Comprehensive nutrient composition data are available for only a few species of aquatic plants (Table 1). Boyd[1] reported aquatic plants dry matter contained as much or more crude protein, crude fat, and mineral matter than conventional forage crops. Linn et al.[2] reported the chemical analyses of 21 aquatic plant species and found many of the plants to be high in crude protein, low in crude fiber, and excellent sources of minerals. Hemicellulose, cellulose, and lignin analyses indicate they should be highly digestible by ruminants.

HARVESTING AND PROCESSING OF AQUATIC PLANTS FOR LIVESTOCK FEEDS

Most aquatic plants are commercially harvested by a self-propelled barge with an underwater sickle cutter bar in front. Behind the cutter is a conveyor that transports the cut vegetation to the deck of the barge. Two different types of harvesters are available depending on handling methods.[9] The first is an Aquamarine Harvester® that stores the cut vegetation on the deck. When the deck is full the load is transferred to a transport barge. The second is a Modified Grimwold-Thomas Harvester® that stores fresh cut vegetation in a bin at the stern and periodically transfers it to a trailering transport barge. However, before being put on the transport barge, all vegetation passes through a chopper to reduce the volume approximately 60%.

Once harvested, aquatic plants can be processed in several different ways. Most commonly aquatic plants are chopped or reduced in bulkiness to improve handling characteristics. Subsequently, the plants can be used fresh, dried, ensiled, or fractionated into a fibrous fraction and a liquid fraction.

Directly harvested fresh aquatic plants may be used as a forage for ruminant animals. However, high moisture content of the plants dilutes out nutrients and animals may not be able to consume large enough amounts to meet their nutrient requirements.

Drying large volumes of chopped aquatic plants with commercial dehydrators is economically inefficient. Removal of the high moisture content (approximately 90%) of aquatic plants requires considerable energy. Energy expenditures of 1500 to 2100 BTU/lb of water evaporated have been reported[10] for *Eichhornia crassipes* and *Hy-*

Table 1
CHEMICAL COMPOSITION OF 27 FRESH WATER AQUATIC PLANT SPECIES[a]

Plant identification	Dry matter	Crude protein	Ash	Ether extract	N-free extract	Crude fiber	Cellulose	Hemicellulose	Lignin	Ref.
Anacharis canadenis	—	14.7	4.5	1.1	68.4	11.4	22.2	0.8	3.8	2
Calla palustris	—	13.9	1.3	4.4	63.0	17.4	18.3	0.4	8.1	2
Carex lacustris	—	8.6	1.0	3.0	60.0	27.4	34.2	17.8	5.2	2
Carex stricta	—	9.9	0.9	0.9	58.9	29.4	34.7	16.8	8.1	2
Ceratophyllum demersum	14.3	17.0	2.2	1.5	64.1	15.2	27.3	6.7	8.1	2
	5.2	16.6	22.3	2.1	42.2	16.8	—	—	8.2	7
	—	21.7	20.6	6.0	—	—	27.9	—	—	3
Chara vulgaris	—	14.6	17.5	0.7	49.9	17.3	—	—	—	6
Eichhornia crassipes	6.0	7.9	5.6	0.1	77.6	7.6	12.6	1.4	2.3	2
Eleocharis smalli	—	15.9	17.0	3.5	—	—	28.0	—	—	4
Elodea canadensis	7.5	5.8	1.6	1.1	64.5	27.0	37.4	16.9	4.8	2
Hydrilla verticillata	8.0	26.8	21.9	3.5	32.4	15.4	—	—	—	8
Lemna minor	—	17.1	27.6	3.5	—	—	32.0	—	—	4
Myriophyllum exalbescens	—	17.9	1.6	2.2	66.5	11.8	14.2	17.0	4.6	2
	—	12.3	1.7	0.2	72.4	13.5	26.5	11.4	4.3	2
	—	17.0	10.9	1.3	57.1	13.7	—	—	—	6
Myriophyllum spicatum	13.3	25.8	13.8	2.5	43.7	14.1	22.3	3.9	7.4	8
Nuphar variegatum	—	15.7	1.0	2.5	57.8	23.1	15.7	3.1	5.0	2
Nymphaea tuberosa	—	19.9	0.8	2.4	60.6	16.3	27.1	0.6	9.9	2
Potamogeton amplifolius	—	14.4	2.4	1.5	66.0	15.7	—	—	—	7
	22.7	12.0	27.3	1.2	39.7	19.8	—	—	—	7
Potamogeton pectinatus	—	14.0	3.2	0.1	67.0	15.6	25.4	15.8	7.3	2
Potamogeton richardsonii	—	11.2	2.6	1.1	65.9	19.1	25.5	6.6	8.6	2
Sagittaria cuneata	—	21.8	1.9	1.4	57.6	17.3	26.4	3.9	3.1	3
Sagittaria latifolia	15.0	17.1	10.3	6.7	—	—	27.6	—	—	3
Sagittaria rigida	—	14.8	2.3	1.8	57.5	23.7	27.5	0.1	11.6	2
Sparganium eurycarpum	—	7.6	2.5	0.7	68.6	20.6	37.4	18.9	4.7	2
Sparganium fluctuans	—	13.2	1.0	1.7	69.7	14.4	16.7	24.1	3.4	2
Typha augustifolia	—	6.9	0.9	1.0	63.7	27.5	30.9	19.2	8.1	2
Vallisneria americana	9.7	15.2	3.1	1.0	53.5	27.3	25.5	12.24	3.3	2
	—	21.1	—	2.7	—	—	30.0	—	—	7
Vallisneria spirallis	5.2	15.2	15.6	4.3	49.1	15.8	—	—	—	8
Zizania equatica	—	9.9	2.4	1.1	59.2	27.4	41.7	8.3	12.1	2

[a] All analyses are on a dry matter basis (% of DM).

drilla species. Even at these rates of energy expenditures, the final moisture levels were too high for safe storage.

An alternative method would be field drying. Whole aquatic plants could be harvested and allowed to sun-cure similar to that used for conventional land forages. However, the rigid cell wall structure of aquatic plants doesn't readily release intercellular water and decomposition may occur before safe moisture levels for preservation are achieved.

Bagnall et al.[10] have done extensive work on the presss fractionation processing of water hyacinth and hydrilla. Fractionation of the plant material into a liquid phase and drier fibrous portion has a distinct advantage. The nutrients suspended in the liquid phase can be removed[11,12] with the resulting liquid being cleaner and returning less nutrients back to already polluted waters. The fibrous portion can be dried or ensiled for livestock feeding. However, this method of process does result in a lower quality press residue product because considerable amounts of nutrients are removed in the pressed juice. Nutrient losses of 15% of the dry matter, 15% of protein, and 50% of the ash content have been reported for *E. crassipes*.[10] Ensiling of aquatic plants could be a feasible alternative to two major problems in processing and feeding aquatic plants: (1) since ensiling is the process of conserving high moisture feedstuffs via acid production from fermentation, ensiling aquatic plants would reduce the amount of drying required for safe storage, and (2) ensiled forages are usually more acceptable and result in higher feed consumption by livestock, particularily ruminants, than dried feedstuffs. Thus, ensiling may be a method of overcoming acceptability problems observed when dried aquatic plants are fed.

Barnett[15] describes silage fermentation as a process which preserves the material by achieving a sufficient concentration of lactic acid in the mass to inhibit other forms of microbial activity. He delineates a desirable fermentation as consisting of four phases. In addition, he describes a further phase if a sufficient concentration of lactic acid is not attained in Phase 4. This fifth phase involves butyric acid-producing bacteria and results in deamination of amino acids and the fermentation of higher volatile fatty acids and amines. These phases are summarized in Figure 1.

Linn et al.[2] studied the silage making quality of an aquatic plant mixture (approximately 50% *Myriophyllum*, 30% *Ceratophyllum*, 10% *Potamogeton*, 5% *Vallisneria*, and 5% unknown). Chopped mixtures of aquatic plants were ensiled with the following treatment (% on a wet basis):

1. Control (no additions)
2. 0.44% formic acid
3. 2% acetic acid
4. 2% propionic acid
5. 2% of a 60% acetic-40% propionic acid solution
6. 2% of a 40% acetic-60% propionic acid solution
7. 5% corn grain
8. 50% chopped alfalfa
9. 75% chopped alfalfa
10. Sterilized aquatic plants
11. Sterilized aquatic plants plus 50% chopped alfalfa

As a comparison to a terrestrial forage, chopped alfalfa was ensiled with no additive and with the organic acids used to treat the aquatic plants. Chemical compositions of unensiled and acid-ensiled aquatic plant and alfalfa silages are presented in Table 2.

All acid-treated aquatic plant and alfalfa silages were higher in crude protein contents than their controls, indicating that acid additions decreased losses of crude pro-

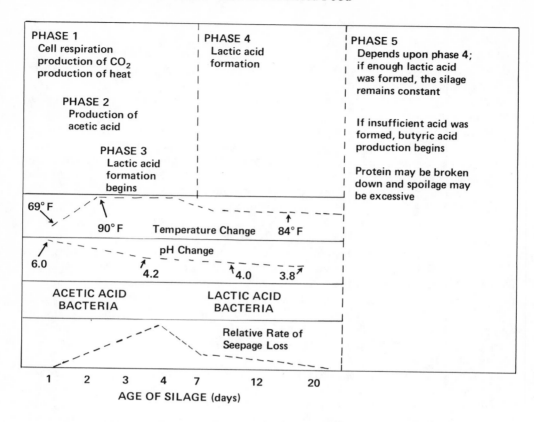

FIGURE 1.   A schematic representation of the five phases in silage fermentation.[15]

tein during fermentation. The rapid drop in pH due to the addition of acids may have resulted in the inhibition of protein breakdown by bacteria.

Ash contents of the unensiled and ensiled aquatic plant silages were exceptionally high, averaging 47.5%. Sand and soil associated with lake bottoms were the obvious contaminants.

The pH and organic acid contents of acid-ensiled aquatic plants are given in Table 2. The acid-ensiled and control aquatic plants continued into Phase 5 of fermentation, as classified by Barnett,[15] and produced undesirable silage: the pH was above 4.2, lactic acid content was only 0.2% and butyric acid was high. Formic acid was the most effective acid in reducing butyric acid production. Since all aquatic plant silages contained low amounts of lactic acid, butyric acid was assumed to be produced primarily from lactic acid. Fermentation of lactic acid by butyric acid-producing anaerobes occurs in the presence of hydrogen acceptors.[15] Some of the pigments found in terrestrial silages act as hydrogen acceptors. Aquatic plants may contain more of these hydrogen accepting pigments and therefore allow more butyric acid production to occur. As lactic acid diminishes, corresponding increases in pH occur, since lactic acid is a stronger acid than the other short chain fatty acids.

The organic acid contents and pH values of alfalfa silage (Table 2) were characteristic of a forage that has undergone proper fermentation. The control and acid-treated alfalfa silages had a pH below 4.2, were adequate in lactic acid content for preservation, and contained little or no butyric acid.

Chemical composition of the aquatic plants ensiled with 5% corn grain or 75% alfalfa and the effects of sterilization on aquatic plants are presented in Table 3. No large differences in composition were observed.

Table 2
CHEMICAL COMPOSITION OF UNENSILED AND ACID-ENSILED FORAGES[a]

Item	Unensiled	Control	Acid treatment[b]				
			0.44% Formic acid	2.0% Acetic acid	2.0% Propionic acid	1.2% Acetic acid plus 0.8% propionic acid	0.8% Acetic acid plus 1.2% propionic acid
			Aquatic Plants				
Dry matter, %	34.98	33.33	33.50	35.00	36.59	35.74	38.22
Crude protein, %	11.69	9.96	11.93	10.16	12.23	10.39	11.06
Ether extract, %	0.59	0.99	0.58	0.68	1.27	1.14	0.99
Crude fiber, %	11.63	9.59	11.58	9.79	10.93	10.86	12.10
Ash, %	50.88	47.88	52.36	48.43	42.40	46.40	44.46
Nitrogen-free extract, %	25.21	31.58	23.56	30.94	33.37	31.21	31.39
Cell wall constituents, %	61.99	60.68	63.09	56.99	56.21	64.31	61.76
Acid detergent fiber, %	57.70	46.88	52.42	51.15	47.42	51.84	53.79
Acid detergent lignin, %	9.21	9.84	10.42	8.60	8.63	8.27	12.42
Hemicellulose[c], %	4.29	13.80	10.67	5.84	8.79	12.47	7.97
Cellulose[d], %	48.49	37.04	42.00	42.55	38.79	43.57	41.37
pH	5.71	5.58	6.60	4.50	4.92	4.95	4.71
Organic acids, %							
Lactic	0.13	0.01	0.03	0.27	0.39	0.12	0.38
Acetic	0.40	5.76	0.64	8.28	1.48	2.63	2.96
Propionic	0	5.06	0.55	1.52	3.82	2.33	3.18
Butyric	0	1.66	0.24	0.72	0.61	0.66	0.50
Valeric	0	0.20	0.03	0.04	0.05	0.06	0.03
			Alfalfa Silage				
Dry matter, %	30.81	32.28	29.90	28.77	31.12	29.32	29.26
Crude protein, %	22.61	21.32	22.93	22.15	22.47	22.73	22.02
Ether extract, %	2.52	4.64	3.99	3.79	4.03	3.65	3.79
Crude fiber, %	22.48	28.71	26.61	29.28	25.42	28.43	26.52
Ash, %	9.23	9.06	10.25	10.08	9.35	9.34	9.11

Table 2 (continued)

CHEMICAL COMPOSITION OF UNENSILED AND ACID-ENSILED FORAGES[a]

Item	Unensiled	Control	Acid treatment[b]					
			0.44% Formic acid	2.0% Acetic acid	2.0% Propionic acid	1.2% Acetic acid plus 0.8% propionic acid	0.8% Acetic acid plus 1.2% propionic acid
Nitrogen-free extract, %	43.16	36.27	36.22	34.70	38.73	35.85	38.56
Cell wall constituents, %	45.03	46.68	47.18	50.68	45.88	46.70	48.42
Acid detergent fiber, %	31.83	36.19	35.14	37.20	33.26	34.68	35.24
Acid detergent lignin, %	6.79	7.77	7.57	8.34	7.08	7.00	7.48
Hemicellulose[c], %	13.20	10.49	12.04	13.48	12.62	12.02	13.18
Cellulose[d], %	25.04	28.42	27.57	28.86	26.18	27.68	27.76
pH	5.00	3.98	4.10	3.85	4.00	3.92	3.92
Organic acids, %							
Lactic	0.02	4.60	2.68	1.84	0.67	2.18	1.67
Acetic	0.70	5.16	1.03	5.17	0.97	3.28	3.58
Propionic	0.17	0.40	0.42	0	6.20	1.86	3.96
Butyric	0.09	0.23	0	0	0	0	0.40
Valeric	0.06	0.04	0	0.06	0.08	0	0.10

[a] All analyses are on a dry matter basis.
[b] Treatment percentages are on a wet basis.
[c] Cell wall constituents minus acid detergent fiber.
[d] Acid detergent fiber minus acid detergent lignin.

From Linn, J. G., Staba, E. J., Goodwich, R. D., Meiske, J. C., and Otterby, D. E., *J. Anim. Sci.*, 41, 605, 1975. With permission.

Table 3
CHEMICAL COMPOSITION OF UNENSILED AND ENSILED FORAGES (%)[a,b]

Item	Unensiled aquatic plants	95% Aquatic plants, 5% corn	25% Aquatic plants, 75% alfalfa	50% Aquatic plants, 50% alfalfa	Sterilized aquatic plants	50% Sterilized aquatic plants, 50% alfalfa
Dry matter	34.98	37.26	26.26	33.78	31.05	27.49
Crude protein	11.69	12.46	16.98	16.20	13.32	15.26
Ether extract	0.59	0.55	0.83	1.11	0.89	1.05
Crude fiber	11.63	9.12	28.42	19.86	13.17	17.91
Ash	50.88	41.14	31.80	33.57	52.58	35.10
Nitrogen-free extract	25.21	37.93	21.97	29.26	20.04	30.68
Cell wall constituents	61.99	59.66	56.95	56.85	57.86	55.16
Acid detergent fiber	57.70	44.09	49.23	46.80	56.48	48.05
Acid detergent lignin	9.21	10.07	11.78	10.89	12.35	9.55
Hemicellulose[c]	4.29	15.57	7.72	10.05	1.38	7.11
Cellulose[d]	48.49	34.02	37.45	35.91	44.13	38.50
pH	5.71	5.20	8.23	5.22	6.20	5.18
Organic acids						
Lactic	0.13	0.08	0.05	0.69	0.17	2.93
Acetic	0.40	1.56	1.15	1.33	2.91	7.36
Propionic	0	1.77	0.22	0.96	0.13	0.49
Butyric	0	1.26	0.64	0.35	0.10	0
Valeric	0	0.03	0	0.02	0	0

[a] All analyses are on a dry matter basis.
[b] All treatment percentages are on a wet basis.
[c] Cell wall constituents minus acid detergent fiber.
[d] Acid detergent fiber minus acid detergent lignin.

From Linn, J. G., Staba, E. J., Goodrich, R. D., Meiske, J. C., and Otterby, D. E., *J. Anim. Sci.*, 41, 606, 1975. With permission.

Aquatic plants ensiled with 5% corn or 50 to 75% alfalfa were similar in organic acid contents and pH values to those of the acid-ensiled aquatic plants (Table 3). The additions of corn or alfalfa as sources of soluble carbohydrates did not improve aquatic plant silage quality.

Additions of alfalfa to the sterilized aquatic plants produced a silage of more acceptable quality than did the addition of 50% alfalfa to unsterilized aquatic plants (Table 3). Sterilization of the aquatic plants either altered the chemical structure, enabling a more desirable fermentation to occur, or destroyed butyric acid-producing organisms, resulting in higher lactic acid concentrations.

FEEDING VALUE OF AQUATIC PLANTS

Several investigators have suggested aquatic plants may be useful as a feedstuff for livestock.[2,3,5,7,13,14] Aquatic plants have been studied as a feed source for fish,[27] swine,[16,28] and poultry.[29] Since ruminants can utilize fibrous feeds as energy sources and readily consume high moisture and fermented feedstuffs, the most promising area for feeding large amounts of aquatic plants is in ruminant diets.

Literature on feeding aquatic vegetation to ruminants is limited, but as early as 1918, Dutch workers[17,18] were proposing the idea of feeding fresh or ensiled aquatic plants to cattle and swine. However, the first reported feeding of aquatic plants didn't appear until 1936. Mrsic[19] reported that Yugoslavian farmers routinely used water plants as a source of forage during periods of drought. The plants were usually fed as dried fodder. Cattle readily consumed the fodder and showed no digestive disturbances. However, vegetation from marshy and warm waters was rejected by cattle due to a musty odor.

Dried Aquatic Plants

Crough[6] prepared a dehydrated aquatic plant meal from equal parts by weight of *Ceratophyllum demersum, Elodea densa,* and *Myriophyllum exalbescens.* The dehydrated meal contained more protein, phosphorus, and iron but less fiber and calcium than dehydrated alfalfa meal. The aquatic plant meal was unpalatable to cattle unless mixed with concentrate feeds at a level of only 8 to 10%. When offered as the principal forage for wintering beef cows and growing heifers, voluntary intake was less than half that expected when conventional forages are fed. Steers fed a ration that contained 8 to 10% aquatic meal gained 0.85 kg/day compared to 0.97 kg/day for steers fed a similar level of dehydrated alfalfa.

Hentges[20] fed two aquatic forages, *E. crassipes* (water hyacinth) and *H. verticillata* (Florida elodea) to yearling Angus and Hereford steers that initially weighed 295 kg and observed the following:

1. Cattle restricted to diets of dehydrated water hyacinth and Florida elodea plus 10% cane molasses consumed only 1.6 and 2.4 kg of air dry feed per head per day or less than 1% of their body weight.
2. Addition of 30% cane molasses increased intake to 3.0 kg for water hyacinth (1% of body weight) and to 4.3 kg (1.5% of body weight) for Florida elodea.
3. Pelleting of the forages increased intakes to 4.1 kg (1.5% of body weight) for water hyacinth and to 8.6 kg (2.9% of body weight) for Florida elodea.
4. Complete diets that contained 25 or 50% of the respective aquatic forages were fed *ad libitum* to yearling cattle. Highest intake of water hyacinth was observed (2.8% of body weight) when it comprised 25% of the ration. Diets that contained 50% Florida elodea exceeded 2.8% of body weight intake.
5. Yearling cattle fed aquatic forage diets for 9 months showed no apparent toxic effects or digestive disorders.

Hentges et al.[21] fed two aquatic plants, *H. verticillata* (Florida elodea) and *E. crassipes* (water hyacinth) and a land forage, coastal Bermuda grass, to steers in pelleted diets that contained 33% forage organic matter. Voluntary intake of steers was higher for the Bermuda grass than the Florida elodea, but consumption of the water hyacinth was similar to that for steers fed Bermuda grass. Apparent digestion coefficients for Bermuda grass, Florida elodea, and water hyacinth were, respectively: 72.4, 70.5, and 66.0% for organic matter; 65.2, 47.9, and 51.7% for crude protein; 37.3, 54.1, and 31.3% for cellulose. Estimated digestible energy in Bermuda grass, Florida elodea, and water hyacinth were 3.4, 3.2 and 3.0 Mcal/kg of organic matter, respectively. Stephens et al.[22] reported mean net retention of minerals for Bermuda grass, Florida elodea and water hyacinth were, respectively, 53.5, 15.5% and * for Ca; 42.3, 24.5, and 34.6% for P; 43.6, 53.5, and 50.9% for Na; 29.3, 23.4, and 7.4% for Mg; 48.7, 5.4, and 75.4% for Mn; 12.5, 11.6, and 7.0% for S; 38.5, 14.1, and 21.5% for Fe; 72.7, 32.8, and 65.2% for Cu; 68.9, 67.4, and 52.0% for Zn.

Two aquatic plants hydrilla (*H. verticillata* casp.) and water hyacinth (*E. crasspes* Mart.) and a land forage, coastal Bermuda grass, were fed at 33% of the organic matter in pelleted diets to yearling steers.[23] Mean daily dry matter intakes for the coastal Bermuda grass, hydrilla, and water-hyacinth diet were 24, 17, and 21 g/kg body weight. Intakes and retention of 10 dietary minerals as compared to NRC requirements are shown in Table 4. Although ash content of most aquatic plants is very high, making the material unacceptable to livestock, the data indicate that aquatic plants fed at 33% of the organic matter are capable of meeting ten or more mineral requirements of yearling steers.

A feeding study of fresh and processed water hyacinths was conducted at Iowa State University.[24] Three mature Charolais bulls fed fresh hyacinths plus 9.1 kg of a balanced maintenance ration consumed 25 kg of water hyacinth daily for 7 days.

Steers fed 20.6 kg (1.1 kg dry matter) of water hyacinth plus a basal ration for 28 days had a daily gain of 535.7 g. It was estimated that the basal ration contributed enough energy to support a gain of 245.1 g daily.

Heifers fed a basal ration plus 1.82 kg of dehydrated water hyacinth pellets gained 136.2 g/day compared to 440.4 g/day for heifers fed dehydrated alfalfa pellets. Pelleted water hyacinths appeared to be less palatable than alfalfa pellets as heifers would sort them out at feeding time.

Linn et al.[25] determined the digestibility of two aquatic plant species, *Myriophyllum exalbescens* and *Potomogeton pectinatus*. The plants were dried (55°C), ground, and then fed in a meal form. During the preliminary period, lambs fed rations containing only *M. exalbescens* or *P. pectinatus* consumed little dry matter. Addition of dehydrated alfalfa to these rations, 50% aquatic plants and 50% alfalfa, increased intake to a satisfactory level. High ash contents (25.1 to 27.6%) probably contributed to the unpalatability of the rations that contained only aquatic plants. Other workers[6,20,24] reported palatability problems, but did not speculate as to the cause.

The consumption of individual rations and dry matter, energy, and crude protein digestion coefficients are presented in Table 5. Dry matter digestion was significantly ($p < 0.01$) higher for sheep fed dehydrated alfalfa than those fed either aquatic plant forage. The higher ash content of the aquatic plants probably decreased dry matter digestion, as ash is a part of dry matter but is relatively indigestible.

The energy digestion coefficient for sheep fed *M. exalbescens* was significantly higher than that for either *P. pectinatus* ($p < 0.01$) or dehydrated alfalfa ($p < 0.05$). Energy digestion is an ash free value, and thus, the difference from dry matter digestion.

---

* Calcium value not reported for water hyacinth.

Table 4

DAILY INTAKES AND RETENTION OF MINERALS BY STEERS FED COASTAL
BERMUDA GRASS, HYDRILLA OR HYACINTH AQUATIC PLANTS

Mineral	Coastal Bermuda[a]			Hydrilla[a]			Hyacinth[a]		
	Intake	Retention	NRC Req.	Intake	Retention	NRC Req.	Intake	Retention	NRC Req.
Ca, g	34.8	19.0	14.0	204.0	31.3	14.0	46.7	9.8	14.0
P, g	15.3	6.5	14.0	11.6	2.9	14.0	17.6	6.1	14.0
Mg, g	12.3	3.6	5.0	18.7	3.9	5.6	15.6	1.1	5.1
Na, g	20.7	9.6	8.9	23.7	12.9	7.0	29.6	14.9	7.9
S, g	10.5	1.7	8.0	8.6	1.1	6.3	12.5	0.9	7.1
K, g	34.7	16.1	48.0	44.1	20.3	37.7	40.3	16.9	42.7
Fe, g	2.5	1.0	0.6	1.6	0.2	0.5	4.5	1.0	0.6
Cu, mg	41.3	28.3	48.2	18.0	6.0	37.7	41.3	26.5	42.7
Mn, mg	204.5	100.5	40.2	86.5	79.1	31.5	437.0	237.3	35.6
An, mg	29.3	20.2	80.4	137.7	21.3	62.9	23.0	17.0	71.2

[a]   Daily dry matter intakes (kg) for steers fed: coastal Bermuda, 5.60; hydrilla, 4.15; and hyacinth, 5.19.

From Stephens, E. L., Easley, J. F., Shirley, R. L., and Hentges, J. F. Jr., *Soil Crop Soc. Fla.*, 32, 30, 1973. With permission.

Table 5
DRY MATTER, ENERGY, AND CRUDE PROTEIN DIGESTION COEFFICIENTS OF DEHYDRATED ALFALFA AND TWO AQUATIC PLANTS

Item	Dehydrated alfalfa	*Potamogeton pectinatus*	*Myriophyllum exalbescens*
Number of lambs	4	4	4
Dry matter intake, g/day			
Dehydrated alfalfa	932.0	407.6	466.0
Aquatic plants		406.0	462.8
Total ration	932.0	813.6	928.8
Crude protein intake, g/day			
Dehydrated alfalfa	159.1	69.6	78.5
Aquatic plants		69.4	79.5
Total ration	159.1	139.0	158.0
Dry matter digestibility, %			
Ration	50.85	47.05	47.32
Forage[e]	50.85[a]	43.45[b]	43.77[b]
Energy digestibility, %			
Ration	49.23	48.43	51.12
Forage[e]	$49.23_c^{a,b}$	47.36_c	53.69_d
Crude protein digestibility, %			
Ration	58.30	51.02	52.16
Forage[e]	58.30[a]	44.11[b]	45.96[b]

[a,b] Means within a row with different superscript letters differ significantly ($p < 0.01$).

[c,d] Means within a row with different subscript letters differ significantly ($p < 0.05$).

[e] Digestibility of aquatic plants calculated by the equation:

$$\text{Digestion coefficient for aquatic plant} = \frac{\left(\begin{array}{c}\text{grams of nutrient}\\ \text{consumed from}\\ \text{alfalfa plus}\\ \text{aquatic plant}\end{array}\right)\left(\begin{array}{c}\text{digestion coef-}\\ \text{ficient for the}\\ \text{mixed ration}\end{array}\right) - \left(\begin{array}{c}\text{grams nutrient}\\ \text{supplied by}\\ \text{alfalfa}\end{array}\right)\left(\begin{array}{c}\text{digestion}\\ \text{coefficient}\\ \text{for alfalfa}\end{array}\right)}{\text{grams of nutrient supplied by aquatic plant}}$$

From Linn, J. G., Goodrich, R. D., Otterby, D. E., Meiske, J. C., and Staba, E. J., *J. Anim. Sci.*, 41, 612, 1975. With permission.

Crude protein digestion coefficients for *P. pectinatus* (44.11%) and *M. exalbescens* (45.96%) were significantly lower ($p < 0.01$) than that for dehydrated alfalfa (58.3%).

Ensiled Aquatic Plants

Baldwin[26] determined the voluntary feed intake and the digestibility of three water hyacinth press residue silages. Silage 1 contained 3 kg of dried citrus pulp plus 1 kg cane molasses; Silage 2 contained 3 kg dried citrus pulp; and Silage 3 contained 0.7 kg cane molasses per 100 kg of water hyacinth press residue (12% DM). Sheep fed Silage 2 had a higher mean daily dry matter intake per kilogram of body weight (8.7 g) than sheep fed either Silage 1 (7.8 g) or Silage 3 (7.5 g). Daily crude protein intake (gms per kilograms of body weight) was higher for Silage 1 (1.1) than for Silage 2 (0.92) or Silage 3 (0.80). Apparent digestion coefficients for Silages 1, 2, and 3, respectively, were 38.0, 35.7, and 29.3% for dry matter and 41.8, 25.6, and 29.9% for crude protein.

In a second experiment, Baldwin[26] compared the digestibility of two water hyacinth silages made with either 4 kg dried citrus pulp plus 0.5 kg cane molasses or 2 kg of dried citrus pulp plus 0.5 kg cane molasses per 100 kg of pressed plant residue (Trial 1). Water hyacinths for these treatments were collected from a sewage disposal lagoon. Also, a pangola grass and a fresh water lake water hyacinth silage were compared with 4 kg dried citrus pulp and 0.5 kg cane molasses per 100 kg fresh plant material (Trial 2). Chemical composition and apparent digestion coefficients of the four treatments are given in Table 6. No differences in composition or digestibility were observed between water hyacinth treatments harvested from the sewage lagoon (Trial 1). Pangola grass silage was higher in dry matter (22.4 vs. 12.1) and lower in crude protein (8.8 vs. 12.2), and ash (7.4 vs. 15.9) than the water hyacinth silage made from fresh water plant material. Daily dry matter intakes ($p < 0.05$), dry matter ($p < 0.01$), and crude protein ($p < 0.01$) digestibilities were higher for pangola grass silage than the water hyacinth silage (Table 6).

A mixture of aquatic plant species (approximately 50% *Myrrophyllum*, 30% *Ceratophyllum*, 10% *Potamogeton*, 5% *Vallisneria*, and 5% unknown) were used to determine the digestibility of large masses of aquatic plants commercially harvested from Minnesota lakes.[25] The following five ensiled rations were used in this trial (% dry matter basis):

1. 100% aquatic plants
2. 85.1% aquatic plants plus 14.9% ground corn grain
3. 52.2% aquatic plants plus 47.8% chopped alfalfa
4. 100% chopped alfalfa
5. 81.2% chopped alfalfa plus 18.8% ground corn grain

The aquatic plant mixture was field dried to reduce moisture content before ensiling. Composition of the ensiled rations are given in Table 7.

Apparent digestibilities and nitrogen retentions of aquatic plant and alfalfa silage rations are presented in Table 8. Dry matter and nitrogen intakes were lower for lambs fed the three rations which contained aquatic plants than for those fed the two alfalfa silage rations.

Apparent organic matter digestibility was lowest for the aquatic plant ration (54.0%) and highest for the alfalfa silage plus corn ration (68.2%). Additions of corn or alfalfa to the aquatic plants significantly ($p < 0.05$) increased organic matter digestibility.

Crude fiber digestibility was significantly ($p < 0.05$) higher for the aquatic plant ration than for the other four rations. Cell wall constituent digestibility was lowest for the aquatic plant ration. This ration also was highest in lignin (Table 7).

Apparent energy digestibilities did not differ significantly among the aquatic plant-containing rations, but were lower ($p < 0.05$) than those for the alfalfa silage or alfalfa silage plus corn rations.

Nitrogen digestibility was lower ($p < 0.05$) for aquatic plant rations than for the alfalfa silage or alfalfa silage plus corn rations. The addition of corn decreased, while the addition of alfalfa increased, the nitrogen digestibility of the aquatic plant-containing rations. Negative nitrogen retentions of lambs fed aquatic plant-containing rations reflected the low nitrogen and digestible energy intakes of these lambs. All lambs that were fed aquatic plant-containing rations lost weight, indicating that body protein breakdown occurred to meet energy needs.

Table 6

CHEMICAL COMPOSITION AND DIGESTION
COEFFICIENTS OF WATER HYACINTH AND
PANGOLA GRASS SILAGES[a]

| | | Trial 1 | | Trial 2 | |
		Water hyacinth[b]		Pangola grass[c]	Water hyacinth[c]
Item	Dried citrus pulp, kg	4	2	4	4
	Cane molasses, kg	0.5	0.5	0.5	0.5
Composition[d]					
	Dry matter, %	11.7	11.9	22.4	12.1
	Crude protein, %	18.5	18.5	8.8	12.2
	Ash, %	12.2	12.1	7.4	15.9
	pH	4.4	4.8	4.1	4.8
	Acetate, %	4.3	5.6	0.58	2.1
	Propionate, %	1.6	2.5	0.10	1.1
	Butyrate, %	5.6	2.7	1.2	2.0
	Lactate, %	0.48	0.67	1.53	0.36
Intakes, daily					
	Dry matter, g/kg BW	12.1	11.1	21.1[d]	11.0[e]
	Crude protein, g/day	72.3	64.6	64.5	46.8
Digestion coefficients					
	Dry matter, %	46.6	43.4	54.0[f]	35.0[g]
	Crude protein, %	51.0	52.3	76.1[d]	52.8[e]

[a]   All analyses are on a dry matter basis.
[b]   Amount per 100 kg of pressed plant residue.
[c]   Amount per 100 kg of fresh plant material.
[d,e]   Means with different superscript differ significantly ($p < 0.05$).
[f,g]   Means with different superscript differ significantly ($p < 0.01$).

From Baldwin, J. A., M.S.A. thesis, University of Florida, Gainesville,
1971.

SUMMARY

Although considerable variation in nutrient content occurs among aquatic plant species, average analyses indicate they contain adequate amounts to be considered as a
livestock feed. Because of the limited amount of research conducted, the complete
effects of processing on aquatic plant nutrient content is unknown. It can be assumed
that some of the same processing factors affecting conventional terrestrial forages will
affect aquatic plants. Specific factors such as ensiling forages of too high moisture
and the fractionation process result in large losses of nutrients in the effluent. The
opposite is also true, heating of feeds to high temperatures for drying purposes results
in nutrient complexes forming, (especially protein) reducing their availability.

Aquatic plants appear to have limited acceptability and lower nutrient availability
than conventional forages in ruminant rations. Ensiling of aquatic plants so that
proper fermentations occur improves intakes. Decreasing the pH by increased production of volatile fatty acids within the silage mass has generally been associated with
increased intakes. Another factor negating large intakes of aquatic plants is high ash
content. Fresh, directly harvested plants may contain 20% or more ash (dry matter
basis). Because of these factors, digestibility of aquatic plants and the performance of

Table 7

CHEMICAL COMPOSITION OF ENSILED RATIONS[a]

	Aquatic plants, 100%	Ration[b] Aquatic plants, 85.1%; corn grain, 14.9%	Aquatic plants, 52.2%; alfalfa, 47.8%	Corn grain, 18.8%; alfalfa, 81.2%	Alfalfa, 100%
Dry matter, %	44.44	44.57	34.90	33.50	29.21
Crude protein, %	10.48	7.45	12.06	19.10	21.08
Ether extract, %	1.41	0.87	1.70	4.57	4.79
Crude fiber, %	16.59	7.23	16.15	23.79	24.76
Ash, %	36.47	37.06	29.95	8.84	9.99
Nitrogen-free extract, %	35.05	47.39	40.14	43.70	39.38
Cell wall constituents, %	57.91	32.61	34.19	45.55	43.08
Acid detergent fiber, %	50.67	23.10	31.99	32.09	35.41
Acid detergent lignin, %	11.34	4.36	6.26	6.33	7.19
Hemicellulose, %[c]	7.24	9.51	2.20	13.46	7.67
Cellulose, %[d]	39.33	18.74	25.73	25.76	28.22

[a] All analyses are on a dry matter basis.

[b] All mixture percentages are on a dry matter basis.

[c] Cell wall constituents minus acid detergent fiber.

[d] Acid detergent fiber minus acid detergent lignin.

From Linn, J. G., Goodrich, R. D., Otterby, D. E., Meiske, J. C., and Staba, E. J., *J. Anim. Sci.*, 41, 613, 1975. With permission.

Table 8

DIGESTIBILITY AND NITROGEN RETENTIONS OF ENSILED RATIONS[a]

	Ration[b]				
	Aquatic plants, 100%	Aquatic plants, 85.1%; corn grain, 14.9%	Aquatic plants, 52.2%; alfalfa, 47.8%	Corn grain, 18.8%; alfalfa, 81.2%	Alfalfa, 100%
Number of lambs	4	4	4	4	4
Dry matter intake g/day	586.3	712.4	664.6	1099.8	875.1
Digestibility, %					
Dry matter	41.4	32.0	38.5	66.2	61.9
Organic matter	54.0[b]	66.8[c,e]	58.5[d]	68.2[a]	64.1[c]
Crude fiber	61.8[b]	50.2[c]	51.6[c]	50.2[c]	45.5[c]
Cell wall constituents	39.0[b]	53.0[c]	44.4[d]	56.0[c]	47.6[d]
Energy	51.4[b]	54.5[b]	49.6[b]	65.5[d]	60.9[c]
Nitrogen intake, g/day	9.8	8.5	12.8	33.6	29.5
Urinary nitrogen, % of intake	45.6	41.8	73.0	57.3	63.5
Nitrogen digestibility, %	33.0[b]	22.8[c]	40.9[d]	74.0[e]	74.3[e]
Nitrogen retention, g/day	−0.9[b]	−1.3[b]	−4.2[c]	5.6[b]	3.1[d]

[a]   All mixture percentages are on a dry matter basis.

[b,c,d,e]   Means within a row with different superscript letters differ significantly ($p < 0.05$).

From Linn, J. G., Goodrich, R. D., Otterby, D. E., Meiske, J. C., and Staba, E. J., *J. Anim. Sci.*, 41, 613, 1975. With permission.

animals fed these plants either dried or ensiled has been lower than animals fed conventional forages.

New or improved processing methods are required to improve nutrient availability. At the present, aquatic plants must be classified in the crop residue category. Thus, they will only be important as feedstuffs during times of feed shortages or when economics dictate their partial or total use in ruminant rations.

REFERENCES

1. Boyd, C. E., Some aspects of aquatic plant ecology, in *Reservoir Fishery Resources Symp.,* University of Georgia Press, Athens, 1967, 114.
2. Linn, J. G., Staba, E. J., Goodrich, R. D., Meiske, J. C., and Otterby, D. E., Nutritive value of dried or ensiled aquatic plants. I. Chemical composition, *J. Anim. Sci.,* 41, 601, 1975.
3. Boyd, C. E., Fresh-water plants: a potential source of protein, *Econ. Bot.,* 22, 359, 1968.
4. Boyd, C. E., The nutritive value of three species of water weeds, *Econ. Bot.,* 23, 123, 1969.
5. Boyd, C. E. and Blackburn, R. D., Seasonal changes in the proximate composition of some common aquatic weeds, *Hyacinth Control J.,* 8, 42, 1970.
6. Crouch, E. K., Dehydrated aquatic plant material for cattle feeding, *Feedstuffs,* 34, 43, 1964.
7. Gortner, R. A., Lake vegetation as a possible source of forage, *Science,* 80, 531, 1934.
8. Nelson, J. W. and Palmer, L. S., Nutritive value and chemical composition of certain fresh-water plants of Minnesota, *Univ. Minn. Agric. Exp. Stn., Tech. Bull.,* No. 136, 1939.
9. Koegel, R. G., Livermore, D. F., and Bruhn, H. D., Harvesting aquatic plants, *Agric. Eng.,* 56(3), 20, 1975.
10. Bagnall, L. O., Shirley, R. L., and Hentges, J. F., Jr., Processing Chemical Composition and Nutritive Value of Aquatic Weeds, Publ. No. 25, Water Resour. Res. Ctr., University of Florida, 1973.
11. Taylor, K. G. and Robbins, R. C., The amino acid composition of water hyacinth *(Eichhornia crassipes)* and its value as a protein supplement, *Hyacinth Control J.,* 7, 24, 1968.
12. Taylor, K. G., The Protein of Water Hyacinth *(Eichhornia crassipes)* and its Potential Contribution to Human Nutrition, M.S.A. thesis, University of Florida, Gainesville, 1969.
13. Bailey, T. A., Commercial possibilities of dehydrated aquatic plants, *Proc. Southern Weed Conf.,* 18, 543, 1965.
14. Boyd, C. E., Evaluation of some common aquatic weeds as possible feedstuffs, *Hyacinth Control J.,* 7, 26, 1968.
15. Barnett, A. J. G., *Silage Fermentation,* Academic Press, New York, 1954.
16. Ferle, F. R., Eine neue Futterpflanze, *Elodea canadensis* Rich., die Wasserpest, *Fuhlings. Landw. Ztg.,* 53, 549, 1904.
17. Anon., Aquatic plants which may be used as a food for cattle, Investigations in Holland, *Bull. Agric. Intelligence,* 9, 1079, 1918.
18. Anon., The Canadian water weed as a fodder plant, *J. Board Agric. (London),* 26, 321, 1919.
19. Mrsic, V., Lake vegetation as a possible source of forage, *Science,* 83, 391, 1936.
20. Hentges, J. F., Jr., Processed Aquatic Plants for Cattle Nutrition, paper presented at Aquatic Plant Conf., University of Florida, Gainesville, 1970.
21. Hentges, J. F., Jr., Salveson, R. E., Shirley, R. L., and Moore, J. E., Processed aquatic plants in cattle diets, *J. Anim. Sci.,* 34 (Abstr.), 360, 1972.
22. Stephens, E. L., Shirley, R. L., and Hentges, J. F., Digestion trials with steers fed aquatic plant diets, *J. Anim. Sci.,* 34 (Abstr.), 363, 1972.
23. Stephens, E. L., Easley, J. F., Shirley, R. L., and Hentges, J. F., Jr., Availability of nutrient mineral elements and potential toxicants in aquatic plant diets fed steers, *Soil Crop Soc. Fla.,* 32, 30, 1973.
24. Vetter, R. L., Preliminary Tests on the Feeding Value for Cattle Fresh and Processed Water Hyacinths, A. S. Leaflet R 169, Iowa State University, Ames, 1972.
25. Linn, J. G., Goodrich, R. D., Otterby, D. E., Meiske, J. C., and Staba, E. J., Nutritive value of dried or ensiled aquatic plants. II. Digestibility by sheep, *J. Anim. Sci.,* 41, 610, 1975.

26. **Baldwin, J. A.**, Utilization of Ensiled Water Hyacinths in Ruminant Diets, M.S.A. thesis, University of Florida, Gainesville, 1973.

27. **Liang, J. K. and Lovell, R. T.**, Nutritional value of water hyacinth in channel catfish feeds, *Hyacinth Control J.*, 9, 40, 1971.

28. **Combs, G. E., Jr. and Wallace, H. D.**, Use of dehydrated water hyacinth in swine diets, *Annu. Res. Rep.*, Al, 73-3, University of Florida, Gainesville, 1973.

29. **Muztar, A. J., Slinger, S. J., and Bukton, J. H.**, Nutritive value of aquatic plants for chicks, *Poult. Sci.*, 55, 1917, 1976.

EFFECT OF PROCESSING ON NUTRITIVE VALUE OF FEEDS: MEAT AND MEAT BYPRODUCTS

Geoffrey R. Skurray

INTRODUCTION

Meat and meat-byproducts have long been used in the manufacture of feeds for livestock and domestic animals. They are primarily a rich source of protein, but the contribution of B-vitamins, minerals, and energy are also important.[1] Meat and meat-byproducts are usually secondary dried products from abattoirs and depending on the raw material are called, bone meal, meat meal, blood meal, or feather meal, although canned pet food may contain large amounts of meat (skeletal muscle) from condemned animals.

Large variations in the composition and nutritional value of meat[2] and meat-byproducts[1,3-5] from different batches from the same abattoir and batches from different abattoirs have been widely reported. The variation in the nutritional value of the protein is not only due to the protein content of different meat products but also the concentration of indispensable amino acids (IAA) and their availability.[1,2,6,7]

Table 1 indicates that there is a wide variation in the composition and nutritive value of the different tissues of animals. In general these products such as meat meal and meat and bone meals that are prepared by dry rendering different animal tissues show the widest variations.

GROWTH-DEPRESSING FACTORS

Early studies on the nutritive value of meat and meat-byproducts were concerned with growth-depressing factors that were formed during processing or were present in the raw material. These factors are discussed below.

Toxins from Oxidized Lipid, Browning Products, and Nephrotoxic Amino Acids

There has been considerable discussion in the past two decades on the possible dangers associated with the presence of oxidized fats in animal diets.[1,17] However, under controlled experimental conditions it has been shown that it would be impossible to formulate a diet, using an oxidized meat meal sample, which would have a high enough peroxide value to cause growth depression.[21] Similarly Sathe and McClymont[22] studied the problem with both high- and low-quality meat meals and two antioxidants and concluded that oxidation in the course of processing or autoxidation during the storage were unlikely to be major determinants of the relative growth-promoting ability of meat-byproducts.

Browning products, formed by the reaction of amino acids with sugars have been shown to be toxic.[20,23,24] Lysinoalanine, a nephrotoxic amino acid formed upon alkaline treatment of protein, is widely distributed in processed high protein foods. However, only the free amino acid is toxic, so that its presence in protein is mitigated by the probability that it is not released by the normal digestive process.[29]

Atkinson and Carpenter,[25] however, showed that heat damaged tendon and meat meal stimulated growth in rats thus providing contradictory evidence of toxic factors in processed meat products.

Bacterial and Mycotoxin Contamination

Processing conditions have been shown to give rise to meat products contaminated

Table 1

VARIATION OF COMPOSITION AND NUTRITIVE VALUE OF MEAT AND MEAT-BYPRODUCTS

Product	Crude Protein (g%)	Ash (g%)	Available lysine (g/100 g crude protein)	Total indispensable amino acids (g/100 g crude protein)	Chemical Score	Net protein utilization[a]	Ref.
Beef muscle	45.2	7.5	9.2	43	86	75—89	1,2,8,9
Beef ribs	61	16	6.8	33	70	—	2,8
Beef neck	60	18	6.8	33	69	—	2,8
Pork feet	62	22	5.7	30	51	49	2,8,9
Beef liver	56—70	10—12	—	38	78	44	3,8,15
Sheep heads	43	38	5.0	41	70	—	2,7
Sheep stomach	61	7.7	4.7—7.1	44	80	—	7,14
Meat meal	40—70	4.8—32	3.2—6.6	9—43	36—68	30—70	1,2,4.6,7,10—13
Bone meal	27	58	3.0	32	10	30	10
Collagen	0	0	3.6	26	0	0	7,10
Blood meal	80	5	5.5—7.0	37	45	—	16

[a]   Crude protein in the test diet was 5.5—8.9%.

by *Salmonella* organisms through such external agents as flies, rats, dust, etc.[26,27] There is no critical evidence on whether differences in the types of *Salmonella* organisms are associated with variation in the nutritive value of meat-byproducts. It is expected that with modern factory practices of cleanliness, high processing temperatures, and the addition of low levels of antibiotic in feeds, the problem of bacterial contamination will not arise.[28]

Meat-byproducts that have been dried to approximately 10% moisture such as meat meals are stored at ambient temperature. These conditions permit fungal growth and aflatoxins have been isolated from meat meals.[63] Aflatoxins cause liver lesions and reduced growth rates or feed conversion efficiencies in poultry and pigs but there are no studies that have been published concerning the incidence or widespread effect of aflatoxins present in meat meals.[63]

Toxic Levels of Minerals and Vitamins

Aluminium and ferric ions are widely used to desludge abattoir wastes. The protein precipitate is then dried and added to meat meal. However, the concentration of these metal ions in finished feedstuffs would not be toxic to animals.[30]

Liver is a useful food that contains a high concentration of vitamins, IAA, and antianemic factors which make canned or dried liver an important supplement to animal feeds.[3,31] However, there are circumstances where liver can be harmful and result in deaths. The livers of certain animals are very rich in vitamin A and hypervitaminosis A has occurred in dogs and cats.[31] The large quantities of bones used in the manufacturing of meat meal and bone meal may have a depressing effect on the growth of animals. More than 1.2% of calcium in the diet depresses growth and appetite of rats, chicks, and pigs.[1,32,33] Parakeratosis may also develop in pigs on meat meal — wheat diets depending on the calcium, zinc, and phytate levels in the diet.[33,34]

The excessive dietary calcium levels present in diets when high levels of meat bones are used as a protein supplement to cereal diets have been shown to cause depression in growing chicks up to 6 weeks of age.[5,32,33] The depressing effect of calcium results from stimulation of microbiological degradation and utilization of amino acids and B-vitamins.[33]

Sathe and McClymont[32] have prevented the depression of growth by vitamin and antibiotic supplementation of diets containing 2.5% calcium, but not when the calcium exceeded this level. The problem of calcium toxicity is therefore only of importance with meat meals or meat-byproducts with an ash content exceeding 30%, and when meat product is the major protein rich supplement of the diet.

Contaminates and Additives

Nitrites, nitrates, and nicotinic acid have been used to preserve or enhance the red color of meat products fed to domestic animals. Nitrites together with alkylureas present in meat can lead to the formation of carcinogenic nitrosamides,[31] while sodium nicotinate has been reported to produce gastrointestinal symptoms in animals.[31]

Undercooked meat from animals that have eaten poisonous plant constituents has caused the deaths of dogs, sheep, goats, and humans.[31,34] Other contaminants that have been found in processed meat and meat-byproducts are radionuclides, pesticide, polycyclic hydrocarbons, antibiotics, and hormones.[31]

Recent studies on the effects of processing on the nutritional value of meat and meat products have shown that the IAA content and availability or digestibility of the IAA are the most important factors influencing nutritional values. Factors affecting the IAA content are the raw materials and processing conditions.

Table 2

EFFECT OF DIFFERENT RAW MATERIALS ON
THE GROWTH, FEED CONVERSION EFFICIENCY
(FCE), AND FEED INTAKE OF CHICKS[7]

Raw material	Average weight gain in 8 days (g)	FCE	Feed intake (g)
Sheep heads	54.0[a] a,[b] c	0.326 a	166
	45.6 b	0.258 b	178
	59.6 a,d	0.354 a	168
Calves head	49.6 b,c	0.299 b	166
Sheep trotters	58.2 a,d	0.329 a	177
	54.0 a,c	0.327 a	165
	63.3 d	0.358 a	177
Sheep heads and sheep guts (1:1)	56.5 a,c	0.316 a	179
Sheep guts	73.9 e	0.427 c	173
Sheep stomach	80.0 e	0.419 c	191
Beef neck trims	46.2 b	0.264 b	175
Control	81.0 c	0.501 d	162
Standard error of the mean	1.6	0.01	—

[a]   Average initial weight of chicks was 70.2 ± 2.7 g.
[b]   Values followed by the same letter are not significantly ($p < 0.05$) different.

RAW MATERIALS

Skurray and Herbert[7] studied the effects of different raw materials and processing conditions which are used in the preparation of meat and bone meals and meat meals (Tables 2 and 3). There was a significant correlation between weight gains and feed conversion efficiencies of chicks and the collagen content of the various materials (regression coefficient = −0.6, $p < 0.05$) even with different processing times and temperatures.

Collagen is derived from soft tissues of bone, skin, connective tissue, and tendon[10] and contains no tryptophan or cystine. Compared to sheep stomach tissue, it is deficient in lysine, histidine, threonine, isoleucine, valine, and tyrosine (Table 3).

Keratin protein from hair, hooves, feathers, and skin may be present to some extent in meat-byproducts for animal feeds. Keratin is deficient in tryptophan, methionine, lysine, and histidine, and unprocessed keratin is highly indigestible due to disulfide bonding between peptide chains.[38]

PARTIALLY DEFATTED CHOPPED BEEF

Partially defatted chopped beef and beef fatty tissue are products which, according to the USDA specifications, are produced in low temperature (50°C) rendering of beef tissues. These products have been shown to be nutritionally acceptable provided they are blended with meat, whey protein concentrate or vegetable protein food.[37] The protein efficiency ratio of these products is less than skeletal muscle (2.4 compared with 2.9) because of the high collagen content.[39]

Experiments have also shown that increasing the amount of collagenous tissues in the materials for meat meal based layer diets, reduces the egg production (see Table 4). Nevertheless, meat meals contribute calcium and phosphorus in sufficient amounts and in the correct ratio for egg production.

Table 3
TOTAL AMINO ACIDS (g/16 g OF N) OF MEAT MEALS PREPARED FROM DIFFERENT RAW MATERIALS UNDER DIFFERENT PROCESSING CONDITIONS[7]

(g/16 g N)

	Raw Material											
	Sheep heads	Sheep heads	Sheep heads	Calves heads	Sheep trotters	Sheep trotters	Sheep trotters	Sheep heads/ Sheep guts	Sheep guts	Sheep stomach	Beef neck trimmings	Ox-bone collagen
Indispensable amino acids(IAA)												
Lysine	6.0	6.1	5.9	6.4	4.6	4.4	4.7	6.1	7.2	8.8	6.0	3.6
Histidine	2.0	2.3	2.0	1.9	1.5	1.5	1.5	1.4	2.1	1.7	2.8	0.85
Tryptophan	0.45	0.50	0.52	0.10	0.11	0.10	0.09	0.09	0.05	0.80	0.80	0.0
Arginine	8.0	7.9	7.4	7.8	8.8	9.3	8.8	7.9	7.4	6.7	9.1	8.1
Threonine	3.7	3.0	3.2	3.6	3.3	3.7	3.9	4.6	4.0	3.8	3.6	2.4
Cystine	2.7	1.4	1.7	1.9	3.5	2.9	1.7	1.6	2.0	2.0	0.7	0.0
Valine	3.5	3.5	4.3	4.8	3.9	4.3	4.9	4.7	5.0	4.1	4.4	2.3
Methionine	0.80	0.41	0.40	0.31	0.34	0.48	0.61	1.2	0.52	0.81	1.2	0.70
Isoleucine	2.9	2.8	2.8	3.0	2.6	2.7	3.1	3.5	3.7	3.5	2.7	1.5
Leucine	5.9	6.2	5.9	5.8	5.3	6.2	6.4	6.7	6.9	6.6	6.2	3.4
Tyrosine	2.8	2.9	2.9	3.4	3.3	3.5	3.6	2.9	3.2	3.2	3.8	0.5
Phenylalanine	4.0	4.0	3.9	4.5	3.6	3.7	4.1	3.7	4.4	4.0	2.6	2.5
Aggregate	42.7	41.0	40.9	43.5	40.9	42.8	43.1	44.4	47.4	44.0	43.9	25.9

Table 3 (continued)
TOTAL AMINO ACIDS (g/16 g OF N) OF MEAT MEALS PREPARED FROM DIFFERENT RAW MATERIALS UNDER DIFFERENT PROCESSING CONDITIONS[7]

(g/16 g N)

	Raw Material											
	Sheep heads	Sheep heads	Sheep heads	Calves heads	Sheep trotters	Sheep trotters	Sheep trotters	Sheep heads/ Sheep guts	Sheep guts	Sheep stomach	Beef neck trimmings	Ox-bone collagen
Dispensable amino acids (DAA)												
Hydroxyproline	7.4	7.4	7.4	7.7	8.1	7.9	7.5	5.9	4.2	5.3	6.0	12.3
Aspartic acid	7.8	8.9	8.8	8.2	7.8	6.4	6.6	9.0	8.8	9.1	8.1	6.2
Serine	4.9	4.5	4.4	5.1	5.7	5.4	5.1	4.7	4.8	4.4	4.0	3.7
Glutamic acid	14.1	15.4	14.6	13.9	14.2	14.6	14.0	14.2	15.5	15.1	13.0	10.5
Proline	9.6	9.2	10.5	8.9	9.5	9.0	8.5	9.7	9.9	8.4	10.3	12.9
Glycine	13.8	14.0	14.7	14.1	14.7	12.5	14.0	10.6	8.7	9.4	11.3	22.2
Alanine	6.7	6.9	6.8	7.1	6.6	6.4	6.4	6.3	5.8	5.6	6.0	9.2
Aggregate	64.3	66.3	63.2	65.0	66.6	62.2	62.1	60.4	57.7	57.3	58.7	77.0
Ratio total IAA: total DAA	0.66	0.62	0.65	0.67	0.61	0.69	0.72	0.73	0.83	0.77	0.75	0.34
Estimated collagen content (g/100 g of protein)	33	33	33	37	42	40	35	14	7	9	10	100

Table 4
EFFECT OF RAW MATERIALS USED IN
MEAT MEAL MANUFACTURING ON EGG
PRODUCTION IN POULTRY[38]

Raw material	Egg production (eggs/day)	Egg production (eggs/kg feed)
Sheep heads and ribs	0.5 a[a]	2.7 c
Sheep rumen	0.8 b	4.0 d

[a] Values followed by the same letter were not significantly different ($p > 0.05$).

The effect of diluting skeletal protein of high nutritive value and high IAA content with collagen or keratin protein or low IAA content is thus of prime importance in the variation of the nutritional value of meat and meat-byproducts. Since the raw materials which have a high collagen content (heads, hooves, ribs) are also rich in bone minerals, particularly calcium and phosphorus, the wide variations in meat-byproduct composition may be due to the types of raw material.[1,7]

Hydroxypoline and 3-methylhistidine are amino acids which are present only in collagen[7,8] and skeletal muscle protein,[41] respectively, so that these amino acids would be an indication of the nutritive value of meat and meat-byproducts.

PROCESSING CONDITIONS

Aqueous Extraction

Collagen is soluble at low pH whereas the remaining meat proteins are insoluble in acid solutions.[42] Use of this property is made in the extraction of collagen for gelatin manufacturing using calcium hydroxide.

The nutritional value and IAA content of meat-byproducts produced from lungs, bones, and offal material has been increased by extracting the noncollagenous proteins from the raw materials with 0.1 N sodium hydroxide or phosphate buffers.[16,43]

The results are summarized in Table 5.

Dry Rendering

It has been well established that water-soluble vitamins and the availability of IAA are decreased markedly when meat proteins are cooked or dried.[25,44,45] It must be borne in mind that in order to obtain significant changes in nutrients, many of the reported studies involved processing at unrealistic temperatures for extended periods of time.[25,45] For example heating meat samples for 24 hr at 110°C results in a 20% loss of lysine and 34% loss in available lysine.[45] The normal time and temperature for meat meal production is 2 hr at 130°C.[46]

Mechanisms of Heat Damage

Minerals, water-soluble vitamins, and proteins may leach out of tissues during heating.[47] This would not be a problem in batch or continuous dry rendering or canning where all of the liquid is retained but during continuous wet rendering, a portion of the liquid that is leached out of the meat-byproduct tissues is separated from the meat meal.[40] Since collagen is the predominant hot-water soluble protein in animal tissues,[10] a significant amount of collagen is removed from the meat meal in this way. The nutritional value of meat meal protein produced by continuous wet rendering would thus be expected to be greater than that produced by batch or continuous dry rendering.[40]

Table 5
EFFECT OF ALKALINE EXTRACTION OF
MEAT-BYPRODUCTS ON NUTRITIONAL
VALUE

	Raw material	Extracted protein
Collagen (g/100 g protein)	13	5
Lysine (g/100 g protein)	5.0	6.2
Protein efficiency ratio (corrected to casein 2.5)	1.5	2.6

Table 6
CHANGES IN THE AVAILABLE IAA CONTENT AND
PERCENTAGE AVAILABILITY OF IAA (IN PARENTHESIS)
DURING BATCH DRY RENDERING OF MEAT-BYPRODUCTS[48]

Amino acid (g/16g N)	Time (min)						SEM[a]
	0	30	45	65	85	120	
Lysine	7.3 (87)	7.6 (94)	7.8 (96)	7.5 (92)	7.3 (91)	6.4 (85)	0.22
Histidine	4.1 (89)	3.5 (86)	3.9 (84)	4.0 (86)	3.9 (85)	3.4 (83)	0.23
Arginine	9.5 (64)	9.2 (91)	9.3 (92)	9.6 (95)	9.1 (90)	7.2 (81)	1.22
Threonine	3.4 (83)	3.7 (91)	3.8 (93)	3.9 (94)	3.8 (92)	2.8 (82)	0.15
Cystine	3.5 (83)	3.7 (88)	3.8 (89)	3.5 (86)	2.8 (84)	2.0 (77)	0.11
Valine	4.3 (79)	4.5 (81)	4.5 (83)	4.5 (85)	4.3 (85)	3.7 (92)	0.13
Isoleucine	2.2 (79)	2.2 (78)	2.0 (77)	2.1 (78)	2.2 (79)	2.0 (80)	0.07
Leucine	5.2 (90)	5.3 (91)	5.3 (92)	5.5 (94)	5.3 (92)	4.1 (85)	0.10
Tyrosine	2.1 (84)	2.1 (85)	2.2 (86)	2.2 (87)	2.1 (82)	1.8 (78)	0.08
Phenylalanine	3.0 (83)	3.0 (83)	3.0 (83)	3.1 (84)	3.0 (83)	2.3 (82)	0.09
Tryptophan	0.41 (77)	0.41 (77)	0.40 (76)	0.40 (75)	0.39 (73)	0.35 (71)	0.12
Aggregate	46.5	46.8	47.6	47.7	45.5	37.3	—

[a]   SEM — Standard error of the mean.

Protein Denaturation and Changes in IAA Availability

As processing temperatures approach 100°C at high water activities, meat proteins denature leading to an increase in the availability of IAA (Table 6) and nutritional value of the protein.[48] Presumably, the increase in the availability of IAA by heat at high water activities is due to either the cleavage of disulfide bonds of keratin,[5,48] or the denaturation of the protein allowing the proteolytic enzymes in the gut greater access to the peptide bonds.[45]

As processing time proceeds, water evaporates until an inversion of the two phases of fat and water occurs.[46] The decrease in the latent heat of evaporation allows the temperature of the meat-byproducts to increase rapidly.[46,48] Experiments have shown that this processing period of high temperatures (120°C) at 10 to 15% moisture leads

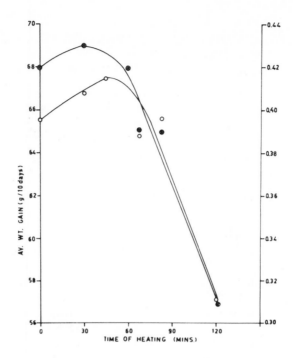

FIGURE 1. Effect of heating meat meals on the weight
gain (●) and feed conversion efficiency (FCE) and (O) of
chicks.[48]

to a marked decrease in the availability of IAA and the nutritional value of the protein
as judged by available IAA, growth rates, and feed conversion efficiency values of
chicks (Table 6 and Figure 1). The total IAA are also reduced significantly. It is important to note that prolonged heating of meat-byproducts for 3 hr at temperatures
of 145°C after the normal processing time of 2 hr does not reduce the nutritional value
of the protein.[49]

There is strong evidence that during normal commercial drying of meat-byproducts
the only significant reaction leading to the decrease in the digestibility of protein and
availability of IAA is the Maillard reaction (nonenzymic browning) between glucose
and the epsilon amino group of lysine. When sulfite ions are used to inhibit the Maillard reaction or yeast is used to remove the glucose in meat tissue by fermentation,
the available lysine and nutritional value of the protein to chick is the same in heated
(2 hr at 130°C) and unheated tissue[50] (see Table 7).

Similar results have been found with removal of glucose from dehydrated meat.[51]
The decrease in the total IAA during batch and continuous dry rendering does not
have any nutritional significance compared to the loss in their availability since sulfite
or yeast does not protect amino acids from destruction by heat but loss of the nutritional value of meat-byproducts is prevented by prior treatment with yeast or sulfite.[50]
A greater change in the total IAA is presumably required to decrease the nutritional
value of meat meals and Kondos and McClymont[52] have found that a difference of
5% in IAA content of meat meals could not be detected by chick growth studies.

The variation in the nutritional value of processed meat-byproducts which is associated with heat treatment may be related to the glucose content of the tissues which in
turn is a function of physical stress and the extent to which the animals have been
starved prior to slaughter.[50.]

The availability of IAA other than lysine is reduced during processing, suggesting

Table 7
EFFECTS OF THE MAILLARD REACTION ON THE NUTRITIONAL VALUE OF HEATED MEAT-BYPRODUCTS[50]

Meat-byproduct	Growth rates of chicks (g/10 days)	Available lysine (g/16g N)
Unheated	65.6	7.3
Heated	57.1	6.4
Sulfite-treated and heated	69.1	6.9
Yeast-treated and heated	64.2	6.8

Note: Heated at 130°C for 2 hr.

that the Maillard reaction may involve other IAA. Sulfite ions also protect arginine from becoming unavailable[53] and a mechanism for the reaction between methionine in a polypeptide chain and fructose has been found.[54] On the other hand, a decrease in the availability due to the Maillard reaction may lead to a decrease in the digestibility by proteolytic enzymes of other IAA in close proximity to the bound lysine.

The reduction of the nutritional value of meat-byproducts during processing may be inhibited by sulfite ions or yeast addition before heating. Since the Maillard reaction proceeds rapidly at temperatures above 100°C at low water activities,[48] it would be possible to prevent nutritional losses by drying meat-byproducts under vacuum so that the temperature of the products was less than 100°C when the water activity was low. However, the activation energy for the Maillard reaction is less than 200 KJ/mole, so that the reaction could still take place during extended storage at temperatures greater than 20°C.[55]

Losses of B-Vitamins

Besides the losses of B-vitamins during processing of meat and meat-byproducts by leaching, it has been shown that thiamin, p-amino-benzoic acid, and nicatinomide can participate in the Maillard reaction and become biologically unavailable.[44,56] However, with the common practice of adding these vitamins to feedstuffs, it is expected that the loss of these vitamins during processing is not important.

The nutritional consequences of the Maillard reaction involving IAA is that not only are the IAA and glucose addition compounds biologically unavailable to the animal for growth,[53,57] but the digestibility of the protein decreases leading to loss of IAA by bacterial deamination in the lower small intestine.[58] There is also evidence that these addition compounds may be absorbed and act as amino acid analogues, inhibiting IAA absorption across the mucosal barrier and reabsorption of IAA in the kidney.[59] However, these effects would seem to be of only marginal nutritional importance since of the total lysine ingested, the combined urinary losses of the addition compounds and free amino acids amount to 1.5%.[59]

Effect of Processing Temperatures on Composition

Meat-byproducts are a valuable source of saturated fat for the food industry and after most of the water has been evaporated during wet or dry rendering, a large proportion of the fat can be separated from the solid material. The effect of processing temperatures on meat meal composition is shown in Table 8. In general the higher the temperature of continuous[52] or batch[49] dry rendering, the greater the efficiency of fat extraction from meat-byproducts. From Table 8 it can be seen that the crude protein content of meat meals increases with decreasing fat content and increasing processing

Table 8
EFFECT OF PROCESSING TEMPERATURE ON
MEAT MEAL COMPOSITION[52]

Temperature (°C)	Crude protein (%)	Fat (%)	Calcium (%)
111	54.2	12.6	8.0
115	45.6	17.6	8.8
130	45.7	18.8	10.0
143	47.0	22.6	11.0

temperature. It is not always practical to determine the protein content of every batch of meat meal used in animal feeds and since the level of meat meal protein in feeds is positively correlated with the nutritional value of the feed,[50] processing temperatures are important in determining the nutritional value as well as the composition of meat meals.

PROCESSED MEAT-BYPRODUCTS IN FEEDSTUFFS FOR RUMINANTS

As previously discussed, heat processing decreases the digestibility of the protein of meat-byproducts, and hence, their nutritional value for monogastric animals. There is strong evidence that decreasing the digestibility of proteins fed to ruminants increases not only live weight gain but also milk and wool production.[61,62] The decrease in solubility of meat meal and formaldehyde-treated blood meal protein in rumen fluid leads to a decrease in degradation and loss of protein by the action of rumen microorganisms. This allows the protein to escape the rumen intact, therefore making it available for digestion in the abomasum and small intestine. Under low dietary protein conditions the response to feeding protected blood and meat meal proteins is an increase in feed intake without an increase in the efficiency of digestion or of the utilization of absorbed nutrients.[61]

SUMMARY AND CONCLUSIONS

Processed meat and meat-byproducts are primarily used in feedstuffs as a source of protein, although calcium and phosphorus are important nutrients supplied by meat meals in diets fed to laying birds. The wide variation in composition and nutritional value of these products is determined primarily by the composition and IAA content of the raw materials used in their manufacture. The dilution of skeletal muscle of high protein and IAA content and low ash content with connective and bone tissue of low IAA content and high ash content results in this variation. As a quality control measure it is suggested that the nutritional value of processed meat-byproducts be determined by their available lysine and hydroxyproline content.

The effect of heat on the nutritional value of meat and meat-byproducts is minimal under conditions used in the processing industry. Continuous dry and wet rendering lead to a reduction in available IAA by the Maillard reaction when temperatures of greater than 100°C are used at low moisture levels. This occurs in batch and continuous dry rendering when 80% of the moisture has been removed and a phase inversion from oil in water to water in oil occurs. This nutritional damage can be prevented by removing the glucose from the raw materials with sulfite ions, yeast fermentation, or glucose oxidase. The extraction of fat from meat-byproducts is facilitated at high rendering temperatures which can affect the final composition of the products. The effect of heat on meat or meat-byproducts protects the protein from degradation by rumen microorganisms resulting in improved growth rates, wool, and milk production. Alkaline extraction of meat-byproducts such as bones, rumen, and lung tissue results in an improvement in the nutritional value of the proteins since collagen is insoluble in alkaline solution. The feasibility of this method for commercial application has yet to be determined although this method is used in gelatin manufacturing. Toxic growth-depressing compounds formed during processing appear to be unimportant with modern processing techniques, but toxic compounds such as calcium, mycotoxins, metal ions from desludging processes, excessive vitamin A in liver tissue, and pesticide contaminants can lead to relatively low growth rates.

Strict quality control is therefore necessary to ascertain the nutritive value and safety in the use of processed meat and meat-byproducts while modifications to processing conditions and raw materials can retain the nutritional value of these foods and their uniformity of composition.

REFERENCES

1. Skurray, G. R., The nutritional evaluation of meat meals for poultry, *World's Poult. Sci. J.*, 30, 129—136, 1974.
2. Dvořák, Z. and Vognarová, I., Available lysine in meat and meat products, *J. Sci. Food. Agric.*, 16, 305—312, 1965.
3. Underwood, E. J., Conochie, J., Reed, F. M., and Smyth, R., The value of meatmeal and livermeal as sources of protein, *Aust. Vet. J.*, 26, 323—329, 1950.
4. Sathe, B. S., Cumming, R. B., and McClymont, G. L., Nutritional evaluation of meat meals for poultry. I. Variation in quality and its association with chemical composition and ash and lipid factors, *Aust. J. Agric. Res.*, 15, 200—213, 1964.
5. Summers, J. D., Slinger, S. J., and Ashton, G. C., Evaluation of meat meal as a protein supplement for the chick, *Can. J. Anim. Sci.*, 44, 228—234, 1964.
6. Sathe, B. S. and McClymont, G. L., Nutritional evaluation of meat meals for poultry. III. Association of chick growth with the bone, calcium, and protein contributed by meat meals to diets, and the effect of minerals and vitamin plus antibiotic supplementation, *Aust. J. Agric. Res.*, 16, 243—255, 1965.
7. Skurray, G. R. and Herbert, L. S., Batch dry rendering: influence of raw materials and processing conditions on meat meal quality, *J. Sci. Food Agric.*, 25, 1071—1079, 1974.
8. Dvořák, Z., The use of hydroxyproline analyses to predict the nutritional value of the protein in different animal tissues, *Br. J. Nutr.*, 27, 475—481, 1972.
9. Dvořák, Z. and Vagnarová, I., Nutritive value of the proteins of veal, beef, and pork determined on the basis of available essential amino acids or hydroxyproline analysis, *J. Sci. Food Agric.*, 20, 146—150, 1969.
10. Eastoe, J. E. and Long, J. E., The amino-acid composition of processed bones and meat, *J. Sci. Food Agric.*, 11, 87—92, 1960.
11. Stockland, W. L. and Meade, R. J., Meat and bone meals as sources of amino acids for the growing rat: use of a reference diet to predict amino acid adequacy by levels of plasma free amino acids, *J. Anim. Sci.*, 31, 1156—1167, 1970.
12. Grace, N. D. and Richards, E. L., The nutritive value of meat by-products, *J. Sci. Food Agric.*, 15, 711—716, 1964.
13. Bunyan, J. and Price, S. A., Studies on protein concentrates for animal feeding, *J. Sci. Food Agric.*, 11, 25—37, 1960.
14. Bremner, H. A., Batch dry rendering: the influence of controlled processing conditions on the quality of meat meal prepared from sheep stomachs, *J. Sci. Food Agric.*, 27, 307—314, 1976.
15. Seegers, W. H. and Mattill, H. A., The effect of heat and hot alcohol on liver proteins, *J. Biol. Chem.*, 110, 531—539, 1935.
16. Skurray, G. R., Recovery and Nutritional Value of Proteins Extracted from Ovine Offal, in *Proc. 2nd Biotech. Conf., Australia*, Melbourne, 1976, 32—35.
17. Colborn, L. R., Oxidized fats in animal feeds, *Vet. Rec.*, 65, 579—583, 1953.

18. McDonald, M. W. and Beilharz, R. C., The Toxic Factors in Low Quality Meat Meals, in *Proc. Poult. Sci. Conv., University of Sydney.*, 1959, 51—56.

19. Carpenter, K. J., Lea, C. H., and Parr, L. J., Chemical and nutritional changes in stored herring meal. IV. Nutritional significance of oxidation of the oil, *Br. J. Nutr.*, 17, 151—159, 1963.

20. Carpenter, K. J. and Booth, V. H., Damage to lysine in food processing: its measurement and its significance, *Nutr. Abstr. Rev.*, 43, 423—451, 1973.

21. Carpenter, K. J., L'Estrange, J. L., and Lea, C. H., Effects of moderate levels of oxidized fat in animal diets under controlled conditions, *Proc. Nutr. Soc.*, 25, 25—35, 1966.

22. Sathe, B. S. and McClymont, G. L., Nutritional evaluation of meat meals for poultry. V. Effect of addition of antioxidants during and after processing on growth-promoting value of high and low quality meat meals, *Aust. J. Agric. Res.*, 18, 183—191, 1967.

23. Petit, L. and Adrian, J., La réaction de Maillard: description et répercussions physiologiques, *Cah. Nutr. Diét.*, 11, 31—38, 1967.

24. Ferrando, R., Henry, N., and Vaiman, M., Toxicity of browning products, *World Rev. Nutr., Diet.*, 19, 71—122, 1974.

25. Atkinson, J. and Carpenter, K. J., Nutritive value of meat meals. I. Possible growth depressant factors, *J. Sci. Food Agric.*, 21, 360—365, 1970.

26. Gray, D. F., Harley, D. C., and Noble, J. L., The ecology and control of *Salmonella* contamination in bone meal, *Aust. Vet. J.*, 36, 246—250, 1960.

27. Leistner, L., Johantges, J., Deibel, R. H., and Niven, C. F., Jr., The Occurrence and Significance of *Salmonella* in Meat Meals and Animal By-product Feeds, Circ. No. 64, Report of the Meat Institute of Chicago, 1961.

28. Jackson, C. A. W., Lindsay, M. J., and Shiel, F., A study of the epizootiology and control of *Salmonella typhimurium* infection in a commercial poultry organization, *Aust. Vet. J.*, 47, 485—491, 1971.

29. Anon., Processed protein foods and lysinoalanine, *Nutr. Rev.*, 34, 120—122, 1976.

30. Underwood, E. J., *Trace Elements in Animal and Human Nutrition*, Academic Press, New York, 1975, 427—429.

31. Liener, I. E., *Toxic Constituents of Animal Foodstuffs*, Academic Press, New York, 1974, 2—12.

32. Sathe, B. S. and McClymont, G. L., Nutritional evaluation of meat meals for poultry. IV. Prevention of growth depression from high levels of bone or calcium by vitamin and antibiotic supplementation, *Aust. J. Agric. Res.*, 16, 491—503, 1965.

33. Batterham, E. S. and Holder, J. M., The effect of calcium level in wheat-animal protein diets, *Aust. J. Exp. Agric. Anim. Husb.*, 9, 43—36, 1969.

34. Oberleas, D., Muhrer, M. E., and O'Dell, B. L., Effects of phytic acid on zinc availability and parakeratosis of swine, *J. Anim. Sci.*, 21, 57—62, 1962.

35. Kondos, A. C. and McClymont, G. L., Effect of Protein, Vitamin, and Antibiotic Levels on the Depression of Chick Growth by High Calcium Diets and Mechanism of Action, in *Proc. Aust. Poult. Sci. Conv.*, Brisbane, 1967, 95—104.

36. Watt, J. M., and Breyer-Brandwyk, M. G., *The Medicinal and Poisonous Plants of Southern and Eastern Africa*, 2nd ed., Livingstone, Edinburgh, 1962, 3—5.

37. Cooke, G. W. J., Amino acid composition of collagen, *J. Agric. Sci.*, 48, 74—78, 1956.

38. Moran, E. T., Summers, J. D., and Slinger, S. J., Keratin as a source of protein for the growing chick, *Can. J. Anim. Sci.*, 45, 1257—1266, 1966.

39. Happich, M. L., Whitmore, R. A., Feairheller, S., Taylor, M. M., Swift, C. E., Naghski, J., Booth, A. N., and Alsmeyer, R. H., Composition and protein efficiency ratio of partially defatted chopped beef and of partially defatted tissue and combinations with selected proteins, *J. Food Sci.*, 40, 35—39, 1975.

40. Skurray, G. R. and Carroll, P. N., The effect of meat meal quality on egg production, *Worlds Poult. Sci. J.*, 34, 22—27, 1978.

41. Hibbert, I., and Lawrie, R. A., The identification of meat in food products, *J. Food Technol.*, 7, 333—335, 1972.

42. Haschemeyer, R. H., and Haschemeyer, A. E. V., *Proteins: A Guide to Study by Physical and Chemical Methods*, John Wiley & Sons, New York, 1973, 410—411.

43. Duerr, P. E. and Earle, M., The extraction of beef bones with water, dilute sodium hydroxide and dilute potassium chloride, *J. Sci. Food Agric.*, 25, 121—128, 1974.

44. Burger, I. H. and Walters, L. L., The effect of processing on the nutritive value of flesh foods, *Proc. Nutr. Soc.*, 32, 1—8, 1973.

45. Donoso, G., Lewis, O. A. M., Miller, D. S., and Payne, P. R., Effect of heat treatment on the nutritive value of proteins, *J. Sci. Food Agric.*, 13, 192—196, 1962.

46. Herbert, L. S. and Norgate, T. E., Heat and mass transfer in a batch dry rendering cooker, *J. Food Sci.*, 36, 294—298, 1971.

47. Lang, K., Influence of cooking on foodstuffs, *World Rev. Nutr. Diet.,* 12, 266—317, 1970.
48. Skurray, G. R. and Cumming, R. B., Physical and chemical changes during batch dry rendering of meat meals, *J. Sci. Food Agric.,* 25, 521—527, 1974.
49. Herbert, L. S., Dillon, J. F., MacDonald, M. W., and Skurray, G. R., Batch dry rendering: influence of processing conditions on meat meal quality, *J. Sci. Food Agric.,* 25, 1063—1070, 1974.
50. Skurray, G. R. and Cumming, R. B., Prevention of browning during batch dry rendering, *J. Sci. Food Agric.,* 25, 529—533, 1974.
51. Henrickson, R. L., Brady, D. E., Gehrke, C. W., and Brooks, R. F., Dehydrated pork studies. Removal of glucose by yeast fermentation, *Food Technol. (Chicago),* 9, 290—292, 1955.
52. Kondos, A. C. and McClymont, G. L., Continuous dry rendering: changes in available amino acids, *Aust. J. Agric. Res.,* 23, 721—730, 1972.
53. Mollah, Y., Sulphite Protection of Amino Acids from Loss in Availability, M. Rur. Sci. Thesis, University of New England, Armidale, Australia, N.S.W., 1969.
54. Horn, M. J., Lichtenstein, H., and Womack, M., Availability of amino acids, *J. Agric. Food Chem.,* 16, 741—745, 1968.
55. Labuza, T. P., Nutrient losses during drying and storage of dehydrated foods, CRC Crit. Rev. *Food Technol., (Chicago),* 3, 217—240, 1972.
56. Van der Poel, G. H., Participation of B-vitamins in nonenzymatic browning reactions, *Voeding,* 14, 452—455, 1956.
57. Sgarbieri, V. C., Amaya, J., Tanaka, M., and Chichester, C. O., Nutritional consequences of the Maillard reaction, *J. Nutr.,* 103, 657—663, 1973.
58. Skurray, G. R. and Cumming, R. B., Deamination of amino acids in the small intestine of chickens fed meat meal, *Poult. Sci.,* 54, 1689—1692, 1975.
59. Ford, J. E. and Shorrock, C., Metabolism of heat-damaged proteins in the rat, *Br. J. Nutr.,* 26, 311—322, 1971.
60. Sathe, B. S., Cumming, R. B., and McClymont, G. L., Nutritional evaluation of meat meals for poultry, *Aust. J. Agric. Res.,* 15, 698—718, 1964.
61. Leng, R. A., New Developments in the Nutrition of Sheep and Cattle, in *Proc. Aust. Poultry Stock Feed Conv.,* Melbourne, 1976, 101—114.
62. Hume, I., The proportion of dietary protein escaping degradation in the rumen of sheep fed on various protein concentrates, *Aust. J. Agric. Res.,* 25, 155—166, 1974.
63. Bryden, W. L., Aflatoxin and Animal Production, in *Proc. Aust. Poult. Sci. Conv. Hobart,* Australia, 47, 1974.

EFFECT OF PROCESSING ON THE NUTRITIVE VALUE OF FISH PRODUCTS USED FOR ANIMAL FEEDING

H. L. A. Tarr

INTRODUCTION

The general procedures employed in preservation of fish and related products destined for animal feeding do not differ radically from those used for preservation of similar products for human consumption. They involve application of heat, chilling or freezing, removal of water and, on occasion, certain chemical treatments. However, in general, the treatments used in preparation of fish for animal feeding are much more severe than those employed in preparation of human foods. Hence, there is a greater tendency for loss of nutritive value during handling, and by loss or destruction of vitamins and occurrence of carbonyl-amino reactions and lipid oxidation during processing and storage. This is particularly true of fish meals and related products where, especially in the past, excessive temperatures were used in their preparation. The importance of this is evident when it is realized that a very large portion of fish used for animal feeding is in the form of fish meal, and comparatively small amounts as canned or frozen products. Much of the literature dealing with the effect of processing procedures on the nutritive value of fish for human[1,2] and animal[3] feeding has been reviewed previously.

CONDITION OF FISH

This may have a significant bearing on physical losses during processing and the nutritive quality of the end products. It is influenced by factors such as fish maturity, degradation of fish by microorganisms, and enzymes.

Maturity

Changes in the composition of fish during maturation include marked increases in depot lipids (triglycerides), especially in pelagic species such as menhaden, herring, and sardines. As maturation proceeds the lipid content of the muscle decreases and there is a corresponding increase in gonad proteins and nucleic acids.[4] These differences could conceivably affect nutritive value of meals produced from the fish. However, comprehensive studies with herring indicated that this is not the case.[5] Composite samples of commercial herring meals produced between November and March were found to possess comparatively consistent composition and nutritive value as determined by their content of protein, fat, ash, water, minerals, vitamins of the "B complex", available lysine, and 13 amino acids, and by in vitro protein digestibility and supplementary protein value for chicks. Some of the results obtained in these experiments are given in Tables 1-4.

Bacterial Spoilage and Enzymatic Degradation

Bacterial spoilage of fish is caused largely by growth of psychrophylic bacteria of the genera *Pseudomonas* and *Achromobacter,* though other genera such as *Micrococcus, Flavobacterium,* and *Corynebacterium* are partly responsible. In fish reduction bacterial spoilage is not usually considered serious until it has reached a fairly advanced stage. When this occurs it becomes undesirable mainly because if contributes to the pollution hazard by production of malodors, and because of its tendency to cause softening of the fish and therefore poor recovery of the fish meal.

Table 1
AVERAGE PROXIMATE ANALYSIS OF MEALS PRODUCED FROM PACIFIC AND ATLANTIC HERRING

	Moisture (%)	Crude protein (%)	Ether extract (%)	Ash (%)	Ref.
Atlantic	6.7	73.6	8.89	10.4	34
Pacific	7.74	71.6	7.9	11.0	33
Mean literature values	6.93	73.6	—	—	36

Table 2
VITAMIN CONTENT OF PACIFIC AND ATLANTIC HERRING MEALS[a]

	Niacin (mg/100 g)	Riboflavin (mg/100 g)	Choline (mg/100 g)	Biotin (mg/100 g)	Pantothenic acid (mg/100 g)	Vitamin B$_{12}$ (mg/100 g)	Ref.
Pacific	8.38	1.16	580	0.06	2.10	0.053	33
Atlantic	10.3	0.62	500	0.041	1.29	0.027	34

[a] Practically all folic acid and much of the thiamin is destroyed by heat during the dyring procedure.

Table 3
MINERAL CONTENT OF ATLANTIC AND PACIFIC HERRING MEALS

	Ca (%)	P (%)	Na (%)	K (%)	Mg (%)	S (%)	Ref.
Atlantic	2.35	2.20	0.61	1.20	0.15	0.94	34
Pacific	2.09	2.12	0.528	1.14	0.129	0.372	33

	Fe (ppm)	Cu (ppm)	Mn (ppm)	Zn (ppm)	I (ppm)	Co (ppm)	Mo (ppm)	Ref.
Atlantic	117	4.4	—	112	—	—	—	34
Pacific	148	9.78	0.71	221	0.53	0.55	6.82	33

Spoilage that results from the action of proteolytic and lipolytic enzymes is very undesirable. These enzymes are largely liberated from the digestive tracts of fish post-mortem, particularly after heavy feeding. These enzymes can soften the visceral walls so that the viscera are exposed, and are, of course, very active when fish are stored at elevated temperatures. Lassen et al.[6,7] found that the yield of meal from sardines stored 120 hr at 20°C was 23% less than that prepared from fresh sardines, and that the volatile base and oil content was higher. The yield of oil was lower and it had poorer quality due to its high free fatty acid content and dark color. Also, the amount of press water obtained from the stored fish was greater, but it yielded condensed fish solubles of poor quality. The nutritive value of the meal produced from the spoiled fish was not appreciably decreased. The feeding trials were conducted some 25 years ago when supplementation with vitamins of the "B complex" did not meet today's requirements, and if further such experiments are carried out, use of more complete rations might yield different results.

Weight losses in fish held for reduction of the order of 15 to 20% have been reported frequently.[8-11] Such losses are, of course, largely due to elimination of water, but appreciable quantities of amino acids and peptides or proteins are included.[12-14]

Table 4

AMACINO ACID CONTENT OF PACIFIC
AND ATLANTIC HERRING MEALS
COMPARED WITH MEAN LITERATURE
VALUES (g/16 g of N)

	Atlantic[a]	Pacific[b]	Mean literature values[c]
Lysine	8.13	8.18	7.73
Histidine	3.17	2.66	2.41
Arginine	6.46	7.84	5.84
Aspartic	9.74	—	9.10
Threonine	4.51	4.00	4.26
Serine	3.97	—	3.82
Glutamic	12.8	—	12.77
Proline	4.42	—	4.15
Glycine	5.52	7.24	5.97
Alanine	6.53	—	6.25
Cysteine	0.935	—	—
Cystine	1.10	1.01	0.97
Valine	5.33	7.91	5.41
Methionine	2.74	2.71	2.86
Isoleucine	4.32	4.17	4.49
Leucine	7.88	7.20	7.50
Tyrosine	3.37	2.80	3.13
Phenylalanine	4.03	3.56	3.91
Tryptophan	1.39	0.905	1.15

[a]   Reference 34.
[b]   Reference 33.
[c]   Reference 36.

Preservation of Fish for Reduction

While whole fish and fish offals are often reduced to meal and oil without preservative treatments, conditions such as transportation and holding fish during gluts may make some form of preservation mandatory. Bacterial spoilage may be retarded by application of certain bacteriostatic agents, by chilling, or by both. It is usually impractical to freeze fish intended for reduction.

The Norwegian fishing industry has provided a classical example of the difficulty of holding large quantities of fish for reduction. Investigations of this problem, which were spread over some 20 years, have been summarized.[15,16] The preservatives used were sodium nitrite and formaldehyde. Sodium nitrite was first used experimentally as a fish preservative in 1939,[17,18] and after additional work[19] and submission of a brief[20] its use was approved for fresh fish preservation in Canada[21] in concentrations which could not exceed 0.02% of the weight of the fish. The Norwegian investigations showed that winter herring could be preserved by use of sodium nitrite alone, while actively feeding summer herring held at higher ambient temperatures required both preservatives, the formaldehyde assisting in maintaining flesh firmness.

In 1964 Ender and his collaborators identified a potent hepatoxic factor in certain herring meals as N-nitrosodimethylamine[22,23] which was formed by interaction of dimethylamine and nitrite. Since then numerous investigations have been carried out to determine the occurrence of N-nitrosocompounds in foods in view of their demonstrated carcinogenicity for certain animals. So far this has not been established for humans.[24] In this connection it is of interest that a very extensive investigation carried out by a number of Norwegian scientists failed to reveal any toxic effects resulting

Table 5
BACTERIAL COUNTS (MILLIONS PER GRAM AND
ORGANOLEPTIC RATING [QUALITY DECREASE FROM TEN
(NO UNDESIRABLE ODOR) TO ONE (PUTRID)] IN HERRING
HELD FOR REDUCTION

	Treatment				
Days stored	None	50 ppm CTC	1% Formaldehyde	1% NaNO₂	1% Formaldehyde + 1% NaNO₂
0	1.2 (9)	—	—	—	—
2	129 (3)	12 (6)	1.2 (7)	12 (4)	1.6 (8)
3	91 (2)	13 (7)	6.3 (5)	15 (4)	1.4 (7)
5	— (1)	6 (4)	1.0 (3)	6 (1)	9.0 (4)

Note: The dashes are used to indicate that at zero time storage the ratings were the same for treated and untreated samples.

from feeding herring meals prepared from nitrite-treated fish to livestock, including chickens, pigs, sheep, and cattle.[25] Also, as noted above, sodium nitrite was used quite extensively as a fresh fish preservative in Canada without any recorded adverse effects until it was superseded by chlortetracycline which was found to be a much more effective preservative.[26]

Because of their success in preservation of edible fish sodium nitrite,[27] chlortetracycline,[26] and refrigerated sea water[28] were studied to determine their effectiveness in preserving fish intended for reduction. Thus, the comparative effectiveness of 1 min immersion of whole nonfeeding winter-caught herring in various solutions of these preservatives was assessed.[27] The results (Table 5) showed that increase in bacteria was retarded by chlortetracycline, 1% formaldehyde, 1% formaldehyde plus 1% sodium nitrite, but not by 1% sodium nitrite alone. Development of spoilage as assessed by odor was, as with bacterial increase, retarded by all but the nitrite treatment alone. The lack of effectiveness of the nitrite treatment alone is not surprising since it has long been known that nitrite is not a particularly effective preservative unless the pH of the flesh is below 7.0, and preferably pH 6.5.

In further experiments[27] actively feeding summer herring were sprayed on a seine boat shortly after capture with 2 ppm of chlortetracycline, 0.05% formaldehyde, and a mixture of 0.05% formaldehyde plus 0.05% sodium nitrite. The fish were stored in insulated boxes on the vessel's deck. Simultaneously other herring were stored at 0.5°C in refrigerated sea water. The fish were weighed initially, and, after 5 days were rinsed with sea water and weighed again. The following weight losses were recorded: untreated, 19%; chlortetracycline, 15.9%; formaldehyde, 15.6%; formaldehyde plus sodium nitrite, 9.1%, and refrigerated sea water, 1.6%. The refrigerated sea water treatment was obviously most effective in preventing weight loss. Moreover, the effectiveness of this treatment in retarding both bacterial and enzymatic spoilage is well-documented, and herring have been held for up to 2 weeks in this medium without very serious spoilage. The cost of the refrigerated sea water treatment was calculated to be about 10 cents/ton, or about ¹/₈ that of the chemical treatments. However, the initial cost of the equipment is much greater.

METHODS OF UNLOADING AND REDUCTION

Conditions under which fish are unloaded from vessels, and under which they are subsequently reduced to yield meal and oil may occasion serious physical losses and, as far as reduction itself is concerned, marked lowering of nutritive value.

Unloading fish intended for reduction is often carried out with fewer precautions than are observed with fish destined for human food. However, improved methods of transporting fish at sea have tended to result in application of unloading methods which cause little or no physical damage to fish. A number of unloading "pumps" which will accept small or fairly large fish have been described.[29,30] Some of these have been used successfully in commerical operations and are ideally suited for unloading fish from tanks of refrigerated sea water with little or no physical damage.

The reduction process, especially if it is not carefully controlled, may result in loss of nutritive value. At one time batch, or "dry" rendering, was employed to a rather limited extent. The fish were cooked in steam-jacketed cookers and the water was removed, usually under vacuum, and the oil was pressed out. Meals prepared by this method seldom suffered from serious overheating. At the present time nearly all fish meals are made by wet reduction, in which the fish or fish wastes are cooked in continuous steam cookers and then pressed in a screw press. The resulting press cake is conveyed to a rotary drier and dried by means of hot air generated in oven-like devices: the flame drying procedure. The oil in the press liquid is separated by centrifugation and the aqueous fraction is dried to 25% or 50% solids content by multiple effect evaporators to yield condensed fish solubles. This soluble fraction is now usually sprayed back on the press cake and dried to yield "whole" meal rather than "press cake" meal. Meals from oily varieties of fish may contain 5 to 15% oil.[31,32] Overheating during drying is singly the chief cause of loss of nutritive value of fish meals.

EVALUATION OF FISH MEAL QUALITY

General

Many varieties of fish meal are available which vary according to the fish or fish offals used in their preparation and the processing conditions. Consequently it has proved impossible to establish a single satisfactory method of assessing fish meal quality, and it is customary to use several different tests for this purpose. However, where it is possible by virtue of the size of the operation and the variety and condition of the fish used to prepare meals of comparatively uniform composition the routine application of a number of assays should not be necessary. That this is possible has been demonstrated by studies of meals produced from Pacific[33] and Atlantic[34] herring, which demonstrated that these possess very similar chemical composition and nutritive value (Tables 1-4). Additional data showed that the following average values were obtained, respectively, for Atlantic and Pacific herring meals: pepsin digestibility (% of crude protein), 94.5, 92.4; available lysine (g/100 g protein), 7.8, 7.0; supplementary protein value (8% level), 111, 122. In view of these facts considerable effort has been directed toward devising analytical methods which are comparatively simple to perform and will detect decreases in nutritive value such as those occasioned by overprocessing.

Proteins and Amino Acids

Meals derived from whole teleost fish usually contain about 70% protein, most of which is derived from muscle proteins which are of high nutritive value, and little from stroma proteins which are of comparatively poor nutritive value. Meals prepared from fish wastes have a high ash content and a comparatively low protein content. The muscle of elasmobranch fish such as dogfish contains a significant amount of urea and a high level of stroma proteins and the meal prepared from such fish has been shown to be of poor nutritional quality.[35] Fish meals are used almost entirely as protein supplements capable of supplying essential amino acids, particularly lysine and methionine, and incidentally as sources of lipids, mineral salts, vitamins, and "uncharacterized growth factors''. Hence the protein fraction must be highly digestible and the essential amino acids nutritionally available. The total amino acid content of a fish meal may be determined with considerable accuracy by chemical hydrolysis followed by determination of the free amino acids thus liberated by ion exchange or gas chromatography, or somewhat less readily and accurately by microbiological assays. However, this technique does not measure the availability of the amino acids to an animal in instances where the fish meal has received serious overheating or has been stored for prolonged periods. The measurement of the nutritional availability of amino acids in fish meals has proved a difficult problem and no single satisfactory assay method has been worked out. The whole problem has been reviewed by Miller[36] and only certain procedures will be discussed.

Early work on the determination of the availability of amino acids by methods not involving tedious biological assays with animals depended on the use of comparative assays of amino acids liberated from proteins by chemical or enzyme hydrolysis. Low results in the enzyme procedure indicated poor availability of certain amino acids.[37,38] This method was tedious and not easily standardized because of the difficulty of obtaining consistent results in the enzyme hydrolysis and microbiological assay procedures. Since amino acid assays are in general too time-consuming for routine analysis simpler procedures of determining protein availability were sought.

Probably the earliest test was that of a determination of the in vitro digestion of the protein of fish meal by pepsin.[39,40] The method appears to determine the "solubilization" of the feed protein, and has been used very frequently to determine the protein nutritive value in conjunction with other tests, particularly those involving determination of the biological value of the protein by methods involving animal feeding. A reasonably good correlation between in vitro pepsin digestibility tests and determinations of biological value has been reported for fish meals by several investigators.[34,41,42]

For some 15 years the "available lysine" method of Carpenter[43,44] has been studied and used for fish meals and other proteins. It has proved very useful but, as Miller[36] has pointed out, it is only reliable as an indicator of loss of availability of lysine in fish meals or other protein feeds whose nutritive value has been damaged significantly by heating, but not with those whose nutritive value has suffered only marginal impairment. Also, not only have there been variations in results obtained between different laboratories, but "within-laboratory" variations have also been high. Presently available lysine determinations can only be relied on to separate fish meals of low nutritional quality from high quality meals. Attempts are being made to improve the sensitivity of the method.[45,46] Unfortunately, attempts to use microorganisms to determine the nutritive value of protein feeds have not proved promising.[36,47]

Lipids

The determination of lipids in fish meals is complex because not only does their total concentration vary, but there are several different classes of lipids of different solubility in organic solvents and these may or may not be oxidized or polymerized. The problems involved have been discussed by a number of investigators.[39,41,48-51] For many years extraction with diethyl ether was used extensively, and this method proved reasonably useful with freshly prepared fish meals which had not been subjected to severe heating during drying. However, this solvent always leaves some unextracted lipid, especially with overheated meals or those subjected to prolonged storage. Consequently considerable attention has been directed toward devising solvent systems that extract all or nearly all the lipids. A system using acetone and acetone-HCl has been used frequently, but in recent years a system employing chloroform and methanol has been employed with increasing frequency. It is rapid and simple. It tends to wet dry samples, disrupt lipid-protein bonds, and the chloroform not only dissolves free lipids but also those that are liberated. The method has the additional advantage that it extracts many of the highly oxidized lipids from overheated or long-stored meals.

Vitamins and Minerals

A knowledge of the vitamin content of fish meals is of some value in formulating animal rations. However, the monetary value of vitamins of the "B complex" in fish meals such as herring meal is low except with vitamin B_{12}.[52] Folic acid and thiamin are destroyed or almost destroyed during processing. Table 2 lists the "B vitamin" content of herring meals.

Biological Methods

In 1919 the protein efficiency ratio (PER) method for determining nutritive value of proteins by animal feeding experiments was introduced.[53] Since that time a number of biological assays employing different animals have ben described including: biological value (BV), true digestibility (TD), net protein utilization (NPU), net protein ratio (NPR), supplementary protein value (SPV), gross protein value (GPV), and metabolizable energy determinations (ME). The basal rations, protein or lipid levels, and animals employed have varied. It has been reported, using 23 widely different proteins feeds in rat studies, that PER values were highly correlated with NPR, NPU (nitrogen balance), and NPU (carcass) methods.[54] A survey of a large number of fish meals using chicks showed that GPV values based on weight gain and feed efficiency, were significantly correlated, as were respective estimates of SPV. However, correlations between GPV and SPV determinations were not statistically significant on either basis.[33] The complexity of the situation regarding application and interpretation of biological assays is apparent.

EFFECT OF REDUCTION AND STORAGE ON NUTRITIONAL QUALITY

The Protein Fraction

Heating fish during the initial cooking procedure where the water concentration is high and the temperature is not excessive has no adverse effect on quality. On the other hand the heating that occurs during drying when the moisture content is quite low, or that which occurs during oxidation of meals held in large piles after drying or during prolonged storage at ambient temperatures may seriously decrease the nutritive value.

Though it was long suspected that overheating of fish meals during drying considerably lowered their nutritional quality, it was not until 1949 that such treatment was shown to lower the availability of essential amino acids to chicks, whereas no such decrease occurred in meals dried by the wet rendering method or in those dried without excessive heat in dry rendering.[37] Subsequently the effect of heat on herring meals produced under controlled conditions was the subject of extensive investigations.[38,55,56] In these studies press cake prepared from fresh herring under commercial conditions was dried in rapidly moving air at 37 to 43°C in order to produce "ideal" low temperature dried herring meal. Portions of meals produced in this manner were heated in a revolving stainless steel drum for 1, 2, or 3 hr at 149°C, and in some experiments most of the lipids were removed by low temperature extraction with hexane. The meals were stored at −25°C until used for chemical or nutritional studies. Commercial "fair average quality" dry rendered herring meals were used for comparison. The meals employed are listed in Table 6.

In one experiment[55] each of the rations, the composition of which is given in Table 7, was fed to duplicate groups of 20-day-old New Hampshire chicks. The chicks were

Table 6
LIST OF HERRING MEALS USED IN CHEMICAL AND BIOLOGICAL ASSAYS (TABLES 7 TO 13)

Meal Number	Month Produced	Description[a]	Meal Number	Month produced	Description
1	November	C	6	November	LTH 60 min
2	February	C	7	November	LTH 180 min
3	February	C	8	November	LTUH
4	November	LTUH	9	February	LTUH
5	November	LTH 30 min	10	February	LTH 180 min

[a]   C: commercial flame dried; LT: low temperature dried; H: heated at 159°C; UH: unheated.

Table 7
GROWTH OF CHICKS ON 20% PROTEIN RATIONS CONTAINING HERRING MEALS (TABLE 6) OR SOYBEAN MEAL AS PRINCIPAL PROTEIN SOURCES

Ingredients	Rations (lb)				
Ground yellow corn	61.22	76.89	77.53	78.00	76.88
Soybean meal	32.59	—	—	—	—
Herring meal 4 (LTUH)	—	18.92	—	—	—
Herring meal 5 (LTH 30 min)	—	—	18.28	—	—
Herring meal 6 (LTH 60 min)	—	—	—	17.81	—
Herring meal 2 (C)	—	—	—	—	19.13
Bone meal	2.0	—	—	—	—
Premix[a]	—	4.19	4.19	4.19	4.19
Average weight chicks after 4 weeks	360	366	389	386	304

[a]   Premix: limestone 1.0, dried distillers' solubles 2.0, iodized salt 0.5, feeding oil (3000 A, 400 D) 0.25, choline chloride (25%) 0.44 lb; manganese sulfate 10 g, nicotinic acid 0.45 g, calcium pantothenate 0.25 g, riboflavin 0.1 g.

From Tarr, H. L. A., Biely, J., and March, B. E., *Poult. Sci.,* 33, 242-250, 1954. With permission.

weighed each week for 4 weeks and the weights are recorded in the table. Much poorer growth resulted with the commercial meal than with the low temperature meal, and the nutritive value of the latter was somewhat improved by heating 30 or 60 min. Similar results with somewhat lower overall growth rates were obtained when the rations were adjusted to contain 18 instead of 20% protein. When the rations fed in the foregoing experiment were supplemented with a mixture of vitamins of the "B complex" it was found that the commercial meal supported chick growth as effectively as the low temperature dried meals (Table 8).

The effect of more severe heating on several of the low temperature meals was examined, and the results are recorded in Table 9. With the exception of low temperature

Table 8
EFFECT OF A MIXTURE OF VITAMINS OF THE "B COMPLEX" ON THE NUTRITIVE VALUE OF DIFFERENT HERRING MEALS

	Average weight of chicks at 4 weeks (g)	
Protein supplement	Basal ration	Basal ration plus vitamins[a]
Soybean meal	364	372
Herring meal 4 (LT)	345	381
Herring meal 4 (LTH 30 min)	396	378
Herring meal 4 (LTH 60 min)	396	383
Herring meal 2 (C)	297	380

[a]   Thiamin hydrochloride 0.5, pyridoxine hydrochloride 0.091, inositol 5.0, p-aminobenzoic acid 4.54, folic acid 0.227, menadione 0.0227, and alpha tocopherol 0.186, g/100 lb.

From Tarr, H. L. A., Biely, J., and March, B. E., *Poult. Sci.,* 33, 242-250, 1954. With permission.

Table 9
EFFECT OF SEVERE HEATING ON THE NUTRITIVE VALUE OF VARIOUS HERRING MEALS (TABLE 6)[a]

Protein supplement	Lb	Average weight chicks (4 weeks)
Soybean meal	32.59	300
Herring meal 2 (C)	19.13	232
Herring meal 1 (C)	18.80	241
Herring meal 3 (C)	18.38	240
Herring meal 8 (LTUH)	18.90	296
Herring meal 8 (LTH 60 min)	17.87	297
Herring meal 8 (LTH 180 min)	17.56	248
Herring meal 9 (LTUH)	18.90	238
Herring meal 9 (LTH 60 min)	17.87	288
Herring meal 9 (LTH 180 min)	17.56	86

[a]   The rations contained 3.19 lb/100 lb of a premix similar to that used in Experiment 1 (Table 7) and ground yellow corn as in that Experiment. The rations had 20% protein.

From Tarr, H. L. A., Biely, J., and March, B. E., *Poult. Sci.,* 33, 242-250, 1954. With permission.

meal No. 12 (February produced) the low temperature meals (unheated or heated 1 hr) were of higher nutritive value than the commercial meals. Heating 3 hr reduced the nutritive value of the low temperature meals, particularly that of No. 12. A further experiment was carried out under similar conditions except that the chicks were weighed after 2 weeks and that in one test L + lysine was added to the ration. The results (Table 10) showed that the 3 hr heating lowered the nutritive value of the low temperature meals, particularly that of No. 14, a February produced meal. Lysine addition affected only a minor improvement in nutritive value. Using a ration similar to that employed in the foregoing experiment and containing a commercial meal (No.

Table 10
EFFECT OF HEATING 3 HR AT 149°C,
AND ADDITION OF L-LYSINE ON
THE NUTRITIVE VALUE OF
HERRING MEALS

Herring meal supplement	Weight of chicks after 2 weeks (g)
No. 4 LTUH	131
No. 7 LTH 3 hr	119
No. 9 LTUH	120
No. 10 LTH 3 hr	55
No. 10 LTH 3 hr + L-lys-ine (1 g/lb)	63

From Tarr, H. L. A., Biely, J., and March, B. E.,
Poult. Sci., 33, 242-250, 1954. With permission.

Table 11
EFFECT OF VITAMINS OF THE "B COMPLEX" IN
IMPROVING THE NUTRITIVE VALUE OF
COMMERCIAL HERRING MEAL (NO. 2)

	Average weight of chicks after 4 weeks (g)
Basal ration	270
Basal ration + vitamin mix	316
Basal ration + folic acid	291
Basal ration + thiamin hydrochloride	274
Basal ration + pyridoxine hydrochloride	269
Basal ration + menadione	264
Basal ration + inositol	259
Basal ration + alpha tocopherol	246
Basal ration + *p* − aminobenzoic acid	274

From Tarr, H. L. A., Biely, J., and March, B. E., *Poult. Sci.,* 33, 242-
250, 1954. With Permission.

3) an experiment was carried out to determine whether one or more of the "B complex" vitamins was the limiting factor in determining the nutritive value of commercial herring meal. The results (Table 11) showed that, of the vitamins tested, only folic elicited a good growth response, though this was not as high as that effected by a mixture of vitamins. Additional experiments showed that good natural sources of vitamins of the "B complex" when added to the basal ration used in the preceding experiment occasioned a growth response in chicks which was almost as great as that from folic acid alone. While folic acid deficiency was undoubtedly responsible for much of the decrease in nutritive value of the commercial herring meals some experiments showed that several other B vitamins caused a slight improvement in growth obtained with the basal ration. The finding that heating for 30 or 60 min actually improved the nutritive value of the low temperature meals remains unexplained.

Experiments were undertaken to determine the reason for the decrase in nutritive value of the meals heated for 3 hr in the foregoing experiments.[38] Samples of herring press cake, low temperature dried meals (heated and unheated), and commercial meals were hydrolyzed chemically and enzymatically and the free amino acids thus liberated were determined by microbiological assays. The difference between the results obtained by the two methods was taken as a measure of the biological activity of the meals. The results of this experiment (Table 12) showed that the amino acid composition of the various samples as determined after chemical hydrolysis was very similar. This was also true of the enzymically hydrolyzed meals except in the case of those heated 3 hr at 149°C. Such heating decreased the availability of all the essential amino acids in the meals, the effect being much more pronounced with the meal produced in February than with that produced in November. These results parallel those obtained in the foregoing feeding studies with chicks. The February produced meals were derived from herring which contained comparatively mature gonads, and analyses showed that they contained considerably higher levels of ribose and deoxyribose than did the November-produced meals. The possibility that Maillard reactions occurring between liberated pentose and amino acids might be implicated in the decrease in nutritive value of the February-produced meals was considered but no evidence in support of this suggestion was obtained.

The Lipid Fraction

It is only during the past two decades that the role of the lipid fraction of fish meals in determining their nutritive value and stability during processing and storage has been examined in depth. Investigations have covered the areas of the composition of the lipid fraction, its nutritive value and how the fraction reacts during processing and storage.

The first investigation of the role of the lipid fraction in determining the nutritional quality of herring meals appears to have been made about 1954.[56] The herring meals used were prepared as in previous investigations,[55] and some of them were extracted with hexane in order to remove most of the lipid. In later studies chloroform-methanol was used since it is a much more effective extractant.[57,58] Low temperature herring meals treated in different ways were fed to chicks at an 18% level using a basic ration of ground corn. The results of this experiment (Table 13) showed that the removal of lipids by hexane extraction had no effect on the nutritive value of heated or unheated meals except with the unextracted meal heated for 2 hr and all the meals after heating for 3 hr. The addition of fresh herring oil added to replace the extracted lipid was without effect. When 2% fresh herring oil or herring oil heated for 30 or 105 hr at 110°C was added to a ration containing unextracted or extracted herring meal chick growth was repressed, but with addition of a supplement of vitamins of the "B complex" this inhibition was no longer apparent. As long as the ration used was supplemented with vitamins it was found that both unextracted and extracted meals could be stored for a year at −25°, 21°, or 37°C without appreciable effect on nutritive value.

Most of the research during the next decade involved various studies the purpose of which was to determine the cause of loss of nutritive value of fish meals heated at comparatively high temperatures and how this could be overcome. Several studies involved the use of antioxidants, many of which had already been investigated during studies of control of rancidity in frozen fish, or in model systems. Knowledge gained in these studies formed the groundwork for numerous investigations on fish meals.

Early work on oxidation of lipids in frozen fish showed that it could be largely overcome by exclusion of oxygen, as by storage in nitrogen[59] and that the application of certain antioxidants also offered promise.[59-61] Subsequently it was shown that typical Maillard type browning reactions occurred in fish muscle heated 1 hr at 120°C[62] and that D-ribose, formed rapidly post-mortem by nucleosidase action was largely responsible.[63-65] At about this time it was demonstrated that shaking an emulsion of cod liver oil, or linoleic acid, with protein and a hematin catalyst for 1 or 2 days at 37°C

Table 12

EFFECT OF HEAT ON THE AVAILABILITY OF ESSENTIAL AMINO ACIDS IN HERRING MEALS PRODUCED IN NOVEMBER AND FEBRUARY (g/16 g N)

	Arginine		Histidine		Isoleucine		Leucine		Lysine		Methionine		Threonine		Valine		Tyrosine		Tryptophan		Phenylalanine	
	C^a	E^b	C^a	E^b	C^a	E^b	C^a	E^b	C^a	E^b	C^a	E^b	C^a	E^b	C^a	E^b	C^a	E^b	C^a	E^b	C^a	E^b
November produced																						
Whole herring	6.7	6.5	—	—	—	—	—	—	11.8	9.1	3.5	3.8	—	—	—	—	—	—	—	—	—	—
Herring press cake	7.0	6.1	—	—	—	—	—	—	12.9	12.0	3.6	4.2	—	—	—	—	—	—	—	—	—	—
No. 4 LTUH	6.3	4.3	2.8	1.8	4.9	4.1	7.6	8.0	11.8	7.7	3.1	2.6	4.1	2.4	5.3	3.5	2.8	3.1	1.5	2.0	3.7	3.3
No. 5 LTH 30 min	6.4	4.9	2.6	1.9	4.3	3.5	7.4	6.6	10.8	7.9	3.0	2.5	3.9	2.3	5.4	3.8	2.6	3.1	0.65	1.9	3.8	3.4
No. 6 LTH 60 min	6.2	4.8	2.5	1.7	5.6	3.5	7.7	5.8	11.2	8.1	3.4	2.8	3.8	2.1	5.2	3.6	2.6	3.0	0.75	2.1	3.6	3.2
No. 7 LTH 180 min	6.5	2.3	2.2	0.2	4.4	0.9	7.0	0.8	10.5	1.1	2.6	0.9	3.4	0.26	5.5	0.4	2.6	0.4	0.6	0.18	3.9	0.35
No. 1 C	6.1	5.5	2.2	0.8	4.6	3.9	8.9	7.8	10.4	9.5	2.2	2.8	3.9	2.5	5.6	4.2	2.6	2.8	0.7	1.7	3.6	3.3
February produced																						
No. 9 LTUH	6.6	4.2	2.2	0.9	4.9	4.5	7.0	6.0	12.6	10.0	2.8	1.8	3.5	4.2	5.0	4.1	2.4	3.4	0.6	1.7	4.3	3.3
No. 13 LTH 60 min	7.1	7.0	2.2	1.1	4.9	3.8	7.7	5.2	11.9	9.2	2.6	1.9	3.9	3.6	5.3	3.3	3.0	2.5	0.63	1.5	3.8	2.7
No. 10 LTH 180 min	7.1	2.2	2.1	0.2	4.6	0.8	6.9	0.7	9.1	0.5	2.1	0.5	3.6	0.13	5.1	0.25	3.0	0.3	0.6	0.3	3.6	0.35
No. 4 C	6.6	5.0	2.1	1.0	4.6	4.4	5.9	5.6	12.2	12.0	2.4	2.3	3.3	4.8	4.8	3.7	3.3	3.6	0.62	1.6	3.3	3.6

a   C: chemical hydrolysis.
b   E: enzyme hydrolysis.

From Bissett, H. M. and Tarr, H. L. A., *Poult. Sci.*, 250-254, 1954. With permission.

Table 13
AVERAGE WEIGHTS OF CHICKS FED UNEXTRACTED AND LIPID-EXTRACTED LOW TEMPERATURE HERRING MEALS AS PROTEIN SUPPLEMENTS

No.	Treatment of meal	Average weight of chicks (g)[a]
1	Unextracted, UH	88
2	Unextracted, H 60 min	88
3	Unextracted, H 120 min	71
4	Unextracted, H 180 min	56
5	Extracted, UH	90
6	Extracted, H 60 min	91
7	Extracted, H 120 min	90
8	Extracted, H 180 min	54
9	No. 5, fresh herring oil added[b]	90
10	No. 6, fresh herring oil added	88
11	No. 7, fresh herring oil added	86
12	No. 8, fresh herring oil added	58

[a] Two groups of 16 birds fed for 10 days before weighing.
[b] Added in concentration similar to that of the lipid that had been extracted.

From Biely, J., March, B. E., and Tarr, H. L. A., *Poult. Sci.,* 34, 1274—1279, 1955. With permission.

resulted initially in formation of yellow-brown products by interaction of oxidized oil and protein, and eventually in development of insoluble dark brown copolymers rich in oxygen and nitrogen.[66] Oxidation of oils in fish meals and fish flesh was shown to be catalyzed by hematin compounds and suppressed by certain antioxidants.[67] It was suggested that browning of fish meals and other fish products is due to a combination of reactions including carbonyl-amino reactions between proteins and carbonyl groups of oxidized oils and oxypolymerization.[68] On the more practical side it was demonstrated that heating of piled fish meal could be controlled effectively by application of an oil soluble antioxidant.[69,70]

Storage of herring meal for 1 year at 20° to 25°C caused some loss in available lysine, but not in its nutritional value for chicks. Both available lysine and nutritional value were, however, seriously decreased by heating at 85 or 100°C and by overheating in bulk commercial storage.[71] That the marked loss in nutritive value with attendant browning that occurs in overheated fish meals is largely due to Maillard reactions is acknowledged.[72] However, the suggestion that the carbonyl component of such reactions might arise from D-ribose formed in degradation of ribosenucleic acid[38] is felt to be incorrect.[72] It is practically certain that reactions between oil oxidation products and amino groups of proteins or amion acids at comparatively high temperatures are largely responsible. The ribose-protein reaction is of importance only with the moist flesh of certain fish heated at temperatures similar to those used in canning procedures, or with freeze-dried fish.[73] The reason for this difference is that most fish meals are manufactured from oily fish and the comparatively high oil content combined with the low water content favors Maillard reaction between protein and products of oxidized oil at high drying temperatures or during prolonged storage. With fish flesh of high water content the small concentrations of ribose are very reactive at 120°C or in freeze-dried flesh.

Table 14
AVERAGE WEIGHTS OF CHICKS AFTER 5
WEEKS ON A RATION CONTAINING 10% OF
SEVERAL LIPID FRACTIONS WITH AND
WITHOUT VITAMINS OF THE "B COMPLEX"

Lipid fraction added	Control ration (g)	Ration with vitamins (g)
None	356	408
Herring oil	168	370
Extract of fresh herring meal	147	421
Extract of herring meal stored 7 months	161	396
Extract of BHT treated meal stored 7 months	117	396

From March, B. E., Biely, J., Clagget, F. G., and Tarr, H. L. A.,
Poult. Sci., 41, 873—880, 1962. With permission.

THE NUTRITIVE VALUE OF THE LIPID FRACTION

The lipid fraction of meals prepared from fatty fish such as anchovies, sardines, herring, menhaden, and mackerel represents about 5 to 10% of the product, and up to about 10 years ago the nutritive value of this fraction had not been studied. The following section deals with the effects of processing, storage, and antioxidant treatment on the nutritive value of this fraction in the intact meals and after solvent extraction.

The first detailed investigation of the nutritive value of a lipid fraction involved extraction of herring meals with a 2:1 chloroform-methanol mixture and removal of the solvent. This procedure proved very effective in extracting lipid, but the fractions extracted from overheated and highly oxidized meals were viscous brown masses and consequently were difficult to incorporate in the rations used.[74] Extracts were prepared from the following:

1. Freshly manufactured fair average quality commercial herring meal.
2. As above but after the meal had been stored at ordinary warehouse temperature 7 months.
3. As above but treated immediately after manufacture with 0.11% butylatedhydroxytoluene (BHT) antioxidant.

Each of the lipid fractions was fed to chicks in 10% concentration to replace cellulose in a ration containing corn, wheat, soybean meal, and herring meal with and without a mixture of vitamins of the "B complex". The results of the experiment are recorded in Table 14. It will be seen that all the lipid fractions suppressed growth, especially those from the antioxidant-treated meals. That this effect is due to destruction of vitamins of the "B complex" by the various lipid fractions is evident since, within limits of significance, addition of vitamin mixture to the rations prevented growth depression.

In a further experiment chloroform-methanol extracts of a commercial herring meal, treated as noted in Table 15, were fed at 10% level as in the foregoing experiment. Using either growth rate or efficiency of feed utilization as criterion, none of the lipid extracts of the meals showed toxicity when fed at the rate of 10% of the diet. Except

Table 15
AVERAGE WEIGHTS, FEED EFFICIENCY, AND LIPID UTILIZATION BY CHICKS FED 10% OF VARIOUS LIPID SUPPLEMENTS

	Average weight (g)	Feed efficiency[a]	Utilization of lipid (%)	Calculated utilization of total meal lipid[b] (%)	Utilization relative to herring oil[c] (%)
Freshly prepared meals					
None	274	2.93	—	—	—
Herring oil	321	2.30	94	—	—
Hydrogenated vegetable oil	328	2.41	77.1	—	—
Extract of normal meal	332	2.36	79.6	76	80
Extract of BHT treated meal	319	2.24	85.5	86	91
Extract of chilled meal	303	2.58	73.8	74	79
Meals stored 2 months					
None	362	2.70	97.2	—	—
Herring oil	359	2.30	88	—	—
Hydrogenated vegetable oil	363	2.27	—	—	—
Extract of normal meal stored at 25.5°C	343	2.49	77.1	62	64
Extract of BHT treated meal stored at 25.5°C	346	2.44	89.6	88	91
Extract of meal stored at −20°C	338	2.59	78.1	74	76
Meals stored 11 months					
None	348	2.84	87	—	—
Herring oil	371	2.44	87.9	—	—
Hydrogenated vegetable oil	378	2.34	75.8	—	—
Extract of normal meal stored at 25.5°C	368	2.62	71.4	59	67
Extract of BHT treated meal stored at 25.5°C	344	2.57	76.1	72	82
Extract of meal stored at −20°C	344	2.80	63.3	59	67

[a] Feed consumed/ gain in weight.
[b] Assumed zero utilization of unextractable fat.
[c] Utilization of total meal fat divided by utilization of herring oil times 100%.

From March, B. E., Biely, J., Clagget, F. G., and Tarr, H. L. A., *Poult. Sci.,* 41, 873—880, 1962. With permission.

in one instance addition of the lipid extracts to the ration improved feed efficiency, but they were not as well utilized as herring oil. Addition of BHT slightly improved the feed efficiency of the stored normal meals. Lipid extracts from the meals stored at −20°C were less well-utilized than those extracted from meals stored at 25.5°. In Table 15 the percentage utilization of the herring meal lipid fractions was calculated on the basis of the 14.1% total lipid content of the original herring meal and assuming a zero value for any lipid remaining after chloroform-methanol extraction. It will be seen that the lipid fraction of the freshly prepared meal was 80% as well utilized as herring oil, and that, after 11 months storage, this had dropped to 67%. Comparative values

Table 16

AVERAGE WEIGHTS OF CHICKS, FEED CONVERSION, AND CALORIC
VALUE OF DIETS IN TESTS TO COMPARE THE METABOLIZABLE ENERGY
VALUES OF HERRING MEAL LIPIDS

	Extracts from freshly manufactured meals			Extracts from meals stored 5 months			Extracts from meals stored 9 months		
Lipid fed	Average weight g	Feed gain	ME cal/lb diet	Average weight g	Feed gain	ME cal/lb diet	Average weight g	Feed gain	ME cal/lb diet
None	287	2.31	1118	386	2.31	1115	314	2.38	1124
Herring oil	319	1.93	1502	386	2.11	1471	332	2.06	1457
Extract of herring meal without BHT	281	2.07	1345	367	2.28	1350	328	2.15	1269
Extract of herring meal with 0.07% BHT	302	2.07	1367	381	2.17	1375	329	2.14	1364
Extract of herring meal with 0.15% BHT	300	2.06	1351	352	2.16	1417	331	2.11	1318

From March, B. E., Biely, J., Claggett, F. G., and Tarr, H. L. A., *Poult. Sci.*, 44, 679—685, 1965. With permission.

for the BHT stabilized meal were 91 and 82%, respectively. This protective action of BHT was substantiated in further experiments.

In view of the above findings regarding the nutritive value of the lipid fraction of herring meal and of its probable protection by BHT, the investigation was extended to determine the metabolizable energy (ME) and supplementary protein value of such meals. In these experiments the antioxidants BHT and Ethoxyquin (EQ) (1,2-dihydro-6-ethoxy-2,2,4-trimethylquinoline) were used.[75] The rations employed were designed to: (1) determine the ME of the lipid fractions, (2) determine the ME of the intact herring meals, and (3) determine the protein nutritive quality of the meals.[75]

In the first experiment commercial whole herring meal, with and without 0.07 and 0.15% of BHT was stored at 25°C. Lipid fractions were extracted from the freshly prepared meal and after the various samples had been stored 5 or 9 months. The lipid fractions were fed at a 10% level. The ME of the intact meals was determined by feeding them at a 25% level immediately after preparation and after storage for 13 months. The results of this experiment are given in Table 16. Feed efficiency was slightly improved by the addition of herring oil or the lipid fractions to the fresh or stored meals. The growth rates of chicks fed fresh and stored meals indicated that the lipid fractions did not depress growth when fed at a 10% level even after the meals were stored 9 months, thus confirming previous work.[58] The ME of the lipid fraction of the antioxidant treated stored meal was higher than that of the extract of the untreated meal.

In the second experiment four lots of commercial herring meals were studied. They were whole meals and press cake meals, and were untreated or treated with 0.05% EQ. The meals were: A and B series prepared from comparatively immature December herring, and were whole and press cake meals, respectively; C and D series, similar to A and B but made from comparatively mature February herring; A_1, B_1, C_1, and D_1 were untreated and A_2, B_2, C_2, and D_2 were treated with EQ; A_3, B_3, C_3, and D_3 had EQ added 3 ro 4 days after manufacture. The ME values of these herring were determined, using in addition the 13-month stored meals from the preceding experiment (Table 16). The results (Table 17) showed that the EQ treated meals had higher ME values than did the untreated meals after 11 or 13 months storage. The meals were all

Table 17
ME VALUES OF UNTREATED AND ANTIOXIDANT TREATED HERRING MEALS AFTER STORAGE (CAL/LB DRY WT)

Experiment 1

Meals stored 13 months

Untreated meal	1445
0.07% BHT	1535
0.15% BHT	1530

Experiment 2

	Series A stored 13 months	Series B stored 13 months	Series C stored 11 months	Series D stored 11 months
Untreated meal	1555	1410	1430	1400
0.05% EQ added immediately	1675	1785	1710	1620
0.05% EQ added 3—4 days after manufacture	1740	1735	1650	1580

From March, B. E., Biely, J., Claggett, F. G., and Tarr, H. L. A., *Poult. Sci.,* 44, 679—685, 1965. With permission.

tested for available lysine at intervals of 2 months up to 12 months and the results indicated that the antioxidant treatment had little or no effect.

The meals from both experiments were tested at 8% level for their value as protein supplements in a ration containing 85% ground wheat and the results are recorded in Table 18. The growth response of the chicks to the antioxidant treated herring meals which had been stored several months was better than that of the untreated meals. It was not, however, statistically significant until the meals had been stored 9 to 11 months.

It has been shown that the protein quality and energy value of Peruvian anchovy meal is markedly improved by EQ treatment.[76,77] It has also been demonstrated that the protein quality of mackerel meal as determined by NPU and PER assays is improved 15 and 16% respectively by such treatment. The ME was increased 5.3% and chick growth rates 1.2%.[78]

It has been recognized for some time that antioxidant treated fish meals prepared from oily fish may cause off flavors in broiler flesh.[79] Mackerel meals were prepared with EQ, vitamin E, or a combination of these, and stored for up to 9 months. It was found high dietary levels of fish meal reduced the organoleptic quality of broilers, and that EQ, although it reduced the carcass lipid content, caused an increase in the polyenoic fatty acid content and a marked decrease in quality. Under limited conditions vitamin E improved carcass quality.[80] Similar studies were carried out with stabilized and unstabilized herring meals.[81]

Recent investigations have included chemical and nutritional studies of herring meals during storage at different temperatures with and without antioxidant treatment.[82] The meals were prepared from fresh herring in a pilot plant with and without 0.25% EQ and were stored for 10 months at −20° and 21°C. The EQ treatment retarded oxidative changes resulting in loss of solubility, decrease in iodine value, and formation of peroxides and malonaldehyde in the lipid fraction. Biological assays for ME were carried

Table 18
COMPARATIVE WEIGHTS OF CHICKS FED UNTREATED AND ANTIOXIDANT TREATED HERRING MEALS STORED FOR DIFFERENT LENGTHS OF TIME (AVG WT (G) OF CHICKS AFTER 4 WEEKS)

Experiment 1

	Meals stored for 13 months
Untreated meal	100
0.07% BHT	102
0.15% BHT	107
Antioxidant effect	Not significant

Experiment 2

Meal No.	Meals stored for 1 month	Meals stored for 5 months	Meals stored for 11 months
A_1	100	100	100
A_2	101	105	105
A_3	98	108	101
B_1	100	100	100
B_2	101	106	108
B_3	102	106	108
Antioxidant effect	Not significant	Not significant	Significant $p < 0.01$

Meal No.	Meals stored for 0 months	Meals stored for 9 months
C_1	100	100
C_2	102	107
C_3	103	106
D_1	100	100
D_2	103	111
D_3	100	109
Antioxidant effect	Not significant	Significant $p < 0.01$

From March, B. E., Biely, J., Claggett, F. G., and Tarr, H. L. A., *Poult. Sci.*, 44, 679—685, 1965. With permission.

out as in previous work.[75] The results (Table 19) showed that either EQ treatment or storage at −20°C protected the availability of the energy supplying nutrients of the meal. The body weight gains for the chicks fed the different meals as protein supplement and the efficiency of conversion of the diets are recorded in Table 20. After 5 months storage the growth response to the untreated meal stored at 21°C was lower at all levels than that of the EQ treated meals stored at 21 or −20°C. The differences in growth rate were reflected in efficiency of feed conversion. After 10 months storage a similar pattern of response to the different meals was evident.

Table 19
ME VALUES OF HERRING MEALS STORED FOR 4 AND 9 MONTHS

Treatment	Storage temperature °C	ME (cal/g dry wt)	
		4 Months	9 Months
None	21	3395[a]	3280[a]
0.25% EQ	21	3910[b]	3935[b]
None	−20	3850[b]	3905[b]
0.25% EQ	−20	3890[b]	3950[b]

[a,b] Values in each column with the same superscript are not statistically different, $p < 0.01$.

From El-Lakany, S. and March, B. E., *J. Sci. Food Agric.*, 25, 899, 1974. With permission.

Table 20
GROWTH RESPONSE OF CHICKS TO STORED HERRING MEALS FED TO SUPPLY 4, 7, or 11% SUPPLEMENTARY PROTEIN

Treatment of meal	Storage temp °C	Meals supplying 4% protein		Meals supplying 7% protein		Meals supplying 11% protein	
		Gain (g)	Gain/Feed	Gain (g)	Gain/Feed	Gain (g)	Gain/Feed
Storage for 5 months							
None	21	63[a]	0.34[a,b]	81[a]	0.44[a]	100[b]	0.52[b]
0.25% EQ	21	71[a]	0.36[a]	90[a]	0.45[a]	106[a]	0.56[a,b]
None	−20	68[a]	0.34[b]	87[a]	0.45[a]	106[a]	0.54[a,b]
0.25% EQ	−20	67[a]	0.34[b]	89[a]	0.43[a]	107[a]	0.54[a,b]
Storage for 10 months							
None	21	65[b]	0.36[c]	85[a]	0.46[a]	107[a]	0.53[a]
0.25% EQ	21	73[a]	0.37[b,c]	98[a]	0.47[a]	116[a]	0.54[a]
None	−20	74[a]	0.38[a]	95[a]	0.46[a]	116[a]	0.55[a]
0.25% EQ	−20	72[a]	0.37[a,b]	96[a]	0.48[b]	111[a]	0.56[a]

[a,b,c] Values in each column with the same superscript are not significantly different, $p < 0.05$.

From El-Lakany, S. and March, B. E., *J. Sci. Food Agric.*, 25, 899, 1974. With permission.

REFERENCES

1. Tarr, H. L. A., *Nutritional Evaluation of Food Processing: Effect of Processing on Fish Products,* Harris, R. S. and von Loesecke, H., Eds., John Wiley & Sons, New York, 1960, 283—304.
2. Tarr, H. L. A., *Fish as Food: Changes in Nutritive Value Through Handling and Processing Procedures,* Vol. 2, Borgstrom, G., Ed., Academic Press, New York, 1962, 235—266.
3. Tarr, H. L. A. and Biely, J., *Effect of Processing on the Nutritional Value of Feeds: Effect of Processing on the Nutritional Value of Fish Meal and Related Products,* National Academy of Sciences, Washington, D.C., 1973, 252—281.
4. Tarr, H. L. A., *Marine Biology, Proc. 20th Annual Colloquium,* Oregon State College, Corvallis, 1970, 36—50.
5. March, B. E., Biely, J., and Tarr, H. L. A., *J. Fish. Res. Board Can.,* 20, 229—238, 1963.
6. Lassen, S., Bacon, E. K., and Dunn, H., *J. Poult. Sci.,* 28, 134—140, 1949.
7. Lassen, S., Bacon, E. K., and Dunn, H., *J. Ind. Eng. Chem.,* 43, 2082—2087, 1951.
8. Papenfuss, H. J., *Fischereiforschung,* 5, 38—41, 1962.
9. Osterhaut, L. E., *Can. Fish.,* 50(10), 18, 1963.
10. Wiechers, S. G. and Laubscher, H., 16th Ann. Rep., Fish, Ind. Res. Inst., Capetown, S. Africa, 97—99, 1970.
11. Wiechers, S. G. and Pienaar, A. G., 14th Ann. Rep., Fish. Ind. Res. Inst., Capetown, S. Africa, 14—26, 1960.
12. Dyer, W. J. and Dyer, F. E., *Fish. Res. Board Can. Prog. Rep. Atl. Coast Stn.,* 40, 3, 1947.
13. Cutting, C. L., *Fishing News,* 1975, 10—13, 1951.
14. Barker, R. and Idler, D. R., *Fish. Res. Board Can. Prog. Rep. Atl. Coast Stn.,* 104, 16—18, 1955.
15. Carpenter, G. A. and Olley, J., Preservation of Fish Wastes for Reduction, Aberdeen, Scotland, Tech. Paper No. 2, Torry Res. Stn., 1960.
16. Sand, G., A Review of the Use of Chemical Preservatives in Norweigan Fish Meal and Oil Industry, *Int. Assoc. Fish Meal Manuf.,* News Summary, No. 19, 1966, 50.
17. Tarr, H. L. A. and Sunderland, P. A., *Fish. Res. Board Can. Prog. Rep. Pacific Coast Stn.,* 104, 14—17, 1939.
18. Tarr, H. L. A., *Nature (London),* 147, 417—418, 1941.
19. Tarr, H. L. A. and Carter, N. M., *J. Fish. Res. Board Can.* 6, 63—73, 1942.
20. Carter, N. M. and Tarr, H. L. A., Unpublished submission to the Fish. Res. Board Can., 1941.
21. Pugsley, L. I., *Can. Food Ind.,* 28(11), 20—22, 1957.
22. Ender, F., Havre, G., Helgebostad, A., Koppang, N., Madsen, R., and Ceh, L., *Naturwissenschaften,* 51(24), 637, 1964.
23. Ender, F., N-Nitrosodimethylamine, the Active Principle, in Cases of Herring Meal Poisoning, Proc. 4th Int. Cong. World Assoc., Buiatrics, Zurich, 1966 (Abstr. in Fish Meal, A Comprehensive Bibliography), Department of Commerce, Washington, D.C., 1970.
24. Ravesi, E. M., *Mar. Fish. Rev.,* 38(4), 24—30, 1976.
25. Heen, E., Bakken, H., Stormorken, H., Dybing, O., Flatla, V. L., Ulvelsi, O., Naerland, G., Hvidsten, H., Husby, M., Njaa, R., Utne, F., Braekkan, O. R., Minsas, J., Sand, G. and Brierem, K., Sodium nitrite as a preservative for herring. Feeding experiments with herring meal from preserved herring and investigations on the effect of sodium nitrite on farm animals, *Fiskeridir. (Norway), Skr. Ser. Teknol. Unders.* 3(4), 96, 1954.
26. Tarr, H. L. A., *Fish as Food: Chemical Control of Microbiological Deterioration,* Vol. 1, Borgstrom, G., Ed., Academic Press, New York, 1961, 639—680.
27. Claggett, F. G., *Fish Res. Board Can.,* Circ. No. 37, 1968.
28. Roach, S. W., Harrison, J. S. M., and Tarr, H. L. A., Storage and transport of fish in refrigerated sea water, *Fish Res. Board Can.,* Bull. No. 126, 1961.
29. Roach, S. W., *Trade News,* 18(6), 4—7, 1965.
30. Payne, R. L., Some methods for moving whole fish, address to Society of Naval Architects and Marine Engineers Pacific Northwest Section, (Mimeo.), 1967.
31. Butler, C., Fish Reduction Processes, Fishery Leaflet No. 126, Fish and Wildlife Service, Department of Interior, 1949.
32. Snyder, D. G., *The Encyclopedia of Marine Resources,* Firth, F. E., Ed., Van Nostrand Reinhold., New York, 1969, 261—264.
33. March, B. E., Biely, J., and Tarr, H. L. A., *J. Fish. Res. Board Can.,* 20, 229—238, 1969.
34. Power, H. E., Savagaon, K. A., March, B. E., and Biely, J., *Fish. Res. Board Can.,* Tech. Rep. No. 114, 1969; *Feedstuffs,* 41(47), 48, 1969.
35. March, B. E., Biely, J., Claggett, F. G., and Tarr, H. L. A., *Poult. Sci.,* 50, 1072—1076, 1971.
36. Miller, E. L., Available amino acid content of fish meals, *FAO Fisheries Rep.,* No. 92, 1970.
37. Clandinin, D. R., *Poult. Sci.,* 28, 128—133, 1949.

38. Bissett, H. M. and Tarr, H. L. A., *Poult. Sci.*, 33, 250—254, 1954.
39. Lovern, J. A., *Fish. News Int.*, 3, 206, 209—210, 1964.
40. Lovern, J. A., Pirie, R., and Olley, J., *Fish. News Int.*, 3, 310, 312, 314, 1964.
41. Olley, J. and Payne, P. R., *Fish. News Int.*, 6(1)34—35, 1967.
42. March, B. E., Biely, J., Bligh, E. G., and Lantz, A. W., *J. Fish. Res. Board, Can.*, 24, 1291—1298, 1967.
43. Carpenter, K. J. and Ellinger, G. M., *Poult. Sci.*, 34, 1451—1452, 1955.
44. Carpenter, K. J., *Biochem. J.*, 77, 604—610, 1960.
45. Roach, A. G., Sanderson, P., and Williams, D. R., *J. Sci. Food Agric.*, 18, 274—278, 1967.
46. Matheson, N. A., *J. Sci. Food Agric.*, 19, 492—497, 1967.
47. Boyne, A. W., Price, S. A., Rosen, G. D., and Stott, J. A., *Br. J. Nutr.*, 21, 181—187, 1967.
48. Olley, J., *Fish. News Int.*, 5(5), 36, 1966.
49. Olley, J., Pirie, R., and Stephen, E., *Fish News Int.*, 5(7), 42—44, 1966.
50. Parks, P. F. and Hummel, M. E., *J. Assoc. Off. Agric. Chem.*, 48, 781—785, 1965.
51. Lee, C. F., Ambrose, M. E., and Smith, P., *J. Assoc. Off. Agric. Chem.*, 49, 946—949, 1966.
52. Tarr, H. L. A., *Fish. Res. Board Can.*, Circ. No. 24, 1960.
53. Osborne, T. B., Mendel, L. B., and Perry, E. L., *J. Biol. Chem.*, 37, 223—241, 1919.
54. Henry, K. M., *Br. J. Nutr.*, 19, 125—133, 1965.
55. Tarr, H. L. A., Biely, J., and March, B. E., *Poult. Sci.*, 33, 242—250, 1954.
56. Biely, J., March, B. E., and Tarr, H. L. A., *Poult. Sci.*, 34, 1274—1279, 1955.
57. March, B. E., Biely, J., and Tarr, H. L. A., *Fish. Res. Board Can., Prog. Rep. Pac. Coast. Stn.*, 108, 24—26, 1957.
58. March, B. E., Biely, J., Claggett, F. G., and Tarr, H. L. A., *Poult. Sci.*, 41, 873—880, 1962.
59. Tarr, H. L. A., *J. Fish. Res. Board Canada*, 7, 237—247, 1948.
60. Tarr, H. L. A., *Nature (London)*, 154, 824—826, 1944.
61. Tarr, H. L. A., *J. Fish. Res. Board Can.*, 7, 137—154, 1947.
62. Tarr, H. L. A., *J. Fish. Res. Board Can.*, 8, 74—81, 1950.
63. Tarr, H. L. A., *Nature (London)*, 171, 344—345, 1953.
64. Tarr, H. L. A., *Food Technol. (Chicago)*, 8, 15—19, 1954.
65. Tarr, H. L. A., *Biochem. J.*, 59, 386—391, 1955.
66. Tappel, A. L., *Arch. Biochem. Biophys.*, 54, 266—280, 1955.
67. Brown, W. D., Venolia, A. W., Tappel, A. L., Olcott, H. S., and Stansby, M. E., *Commer. Fish. Rev.*, 19(5a), 27—31, 1957.
68. Venolia, A. W., Tappel, A. L., and Stansby, M. E., *Commer. Fish. Rev.*, 19(5a), 32—34, 1957.
69. Mead, T. L., *Feedstuffs*, 28(20), 15—22, 1956.
70. Mead, T. L. and McIntyre, R. T., *Proc. Gulf Caribbean Fish. Inst.*, 10, 86—91, 1957.
71. Lea, C. H., Parr, L. J, and Carpenter, K. J., *Br. J. Nutr.*, 14, 91—113, 1960.
72. Carpenter, K. J., Morgan, C. B., Lea, C. H., and Parr, L. J., *Br. J. Nutr.*, 16, 451—465, 1962.
73. Tarr, H. L. A. and Gadd, R. E. A., *J. Fish. Res. Board Can.*, 22, 755—760, 1965.
74. March, B. E., Biely, J., Claggett, F. G., and Tarr, H. L. A., *Poult. Sci.*, 41, 873—880, 1962.
75. March, B. E., Biely, J., Claggett, F. G., and Tarr, H. L. A., *Poult. Sci.*, 44, 679—685, 1965.
76. Bürke, R. P. and Maddy, K. H., *Kraftfutter*, 49, 66—70, 1966.
77. De Groote, G., *Feedstuffs*, 40(51), 26—27, 54, 56, 61, 1968.
78. Opstvedt, J., Nygård, E., and Olsen, S., *Acta Agric. Scand.*, 20, 185—192, 1970.
79. Aure, L., *Arsberet. Vedkomm. Nor. Fisk.*, 3, 17—24, 1957.
80. Opstvedt, J., Olsen, S., and Urdahl, N., *Acta Agric. Scand.*, 20, 174—184, 1970.
81. Opstvedt, J., Nygård, E., and Olsen, S., *Acta Agric. Scand.*, 21, 125—133, 1971.
82. Safaa, E.-L. and March, B. E., *J. Sci. Food Agric.*, 25, 899—906, 1974.

EFFECT OF PROCESSING ON NUTRIENT CONTENT OF FEEDS: MILK BYPRODUCTS

J. H. B. Roy

INTRODUCTION

The effect of processing on the nutritive value of milk and milk products as human food has been reviewed in another section. This short review on the effect of processing of milk byproducts for animal feed should be considered as a supplement to that section.*

Milk byproducts are mainly used in the nutrition of the young farm animal, in particular the calf. Milk substitutes for use directly after the colostrum-feeding period consist of skim milk into which fat of animal and/or vegetable origin has either been homogenized before drying, or the fat is mixed into dried skim milk powder together with emulsifying agents by a dry-blending process. In addition, a proportion, usually not more than 15%, of dried whey may be included in such milk substitutes.

SKIM MILK POWDER

The processing treatment of the skim milk can have a profound effect on the health and performance of the neonatal calf. The detrimental effect is associated with the time-temperature relationship used in the preheating process before spray-drying or in roller-drying. Severe heat treatment results in denaturation of the whey proteins, α-lactalbumin, β-lactoglobulin, and the small amounts of immunoglobulins present. It also causes a reduction in the ionizable calcium and the release of SH-groups.

The effect of inclusion of a severely preheated spray-dried skim milk powder in milk substitutes for use by the neonatal calf depends on the immune status of the calf and the level of infection in the environment. It has been shown that, if a succession of newborn calves are introduced into a calf house, the growth rate of each calf is less than its predecessor, and after a time there will be an increased incidence of diarrhea during the first 3 weeks of life and eventually deaths will occur,[1] associated with the dominance of enterotoxemic strains of *Escherichia coli*.[2] Under clean environmental conditions in a calf house, which has remained empty for 2 to 3 months, the only effect of feeding a diet containing a severely preheated skim milk powder will be to reduce digestibility of nutrients and weight gain of the calves. Under higher levels of infection, there will be an increased incidence of diarrhea if the calves receive a diet containing a severely preheated skim milk powder, and once deaths have begun to occur, the mortality rate will be higher.[3]

The detrimental effects are associated with poor coagulation of milk in the abomasum,[4] and reduced acid, rennin, and pepsin secretion in the abomasum.[5] As a result of reduced proteolysis in the abomasum, there is an increased passage of undigested protein into the duodenum[4,6] and reduced pancreatic protease secretion.[6] However, the efficiency of retention of the digested protein, and thus its biological value is unaffected.[7]

The detrimental effects of severely heat-treated skim milk powders for calves have been confirmed in Australia[8] an in Canada,[9] but in New Zealand with calves reared out-of-doors the effect was small.[10] Reduced performance with severely heat-treated skim milk powder has also been found in lambs in the U.K.,[11] and in piglets the detri-

---

* Readers should refer to chapter by B. A. Rolls in Volume I.

Table 1
EFFECT OF VARIOUS HEAT TREATMENTS ON THE RATIO OF NONCASEIN N TO TOTAL N IN MILK, AND THE EFFECT OF THESE MILKS ON THE HEALTH AND PERFORMANCE OF THE CALF

Treatment of milk	Ratio of noncasein N to total N and effect on calf
—	No demonstrable effect on calf
Raw	0.25
Holder pasteurized (63°C for 30 min)	0.23
Spray-dried skim (preheating temperature 77°C for 15 sec)	0.22
—	Detrimental to the calf especially during the first 3 weeks of life
Spray-dried skim (preheating temperature 74°C for 30 min +)	0.15
Roller-dried skim (110°C)	0.13
UHT sterilized (135°C for 1-3 sec)	0.11

From Roy, J. H. B., *Proc. Nutr. Soc.,* 28, 160, 1969. With permission.

mental effect appeared to be restricted to the first 7 days of life.[12] All species appear to adapt to lack of coagulation of milk powder as they grow older.

In the U.S., there is an ADM1 grading system[13] of milk powder: high-heat powder (<1.5 mg whey protein nitrogen (WPN) per gram powder); medium-heat powder (>1.5 mg <6.0 mg WPN per gram); and low-heat powder (>6.0 WPN per gram) but no such system exists in Europe.[14] These values for WPN, produced from the difference between the total nitrogen and the values obtained by saturated NaCl precipitation of the casein together with the denatured whey proteins, do not include proteose peptone N and are thus 17.3% less[15,16] than the values obtained by the acid precipitation of casein in the method of Rowland[17] used in the U.K.

The dividing line between suitable and detrimental skim milk powders for inclusion in milk substitute diets for the newborn calf is between 160 and 180 mg noncasein N (NCN) per gram total N.[18] A value of 170 mg NCN per gram total N is equivalent to 10.1 mg NCN per gram dry matter. Subtraction of 3.3 mg nonprotein N per gram dry matter[17] gives a value of 6.8 mg WPN per gram dry matter, equivalent to an ADM1 value of 5.6 mg WPN per gram dry matter (having made an allowance for the proteose-peptone N not included in the ADM1 grading) or 5.4 mg WPN per gram powder (96% dry matter). Thus, an ADM1 low-heat powder would fulfill the necessary specification for a powder, suitable for the neonatal calf.

Effect of various heat treatments on the ratio of noncasein N to total N in milk and the effect of these milks on the health and performance of the calf are given in Table 1. The results of a study of the amount of denaturation that was found in skim milk powders, commercially available in the U.K. between 1965 and 1967 are given in Table 2, which gives some idea of the range of quality of products available for incorporation into milk substitute diets for calves.

WHEY POWDER

Since whey powder is normally included as only a small proportion of the diet of the neonatal calf there is no evidence as to whether mildly preheated whey powder is a superior product for the calf. However, partially delactosed whey and whey from cottage and cream cheese are acid and cause a low pH when used in milk substitute

Table 2
RATIOS OF NONCASEIN N TO
TOTAL N IN COMMERCIAL SKIM
MILK POWDERS PRODUCED AT U.K.
CREAMERIES (1965-1967)

Creamery	Type of dryer	Ratio of noncasein N to total N
A	Spray	0.22
B	Spray	0.22
C	Spray	0.16
D	Spray	0.15
E	Spray	0.12
F	Spray	0.11
G	Spray	0.11
H	Roller	0.16
I	Roller	0.13

diets. There are some indications that neutralization of such whey powders result in greater milk intake and higher weight gains.[19]

REFERENCES

1. Roy, J. H. B., Palmer, J., Shillam, K. W. G., Ingram, P. L., and Wood, P. C., The nutritive value of colostrum for the calf. X. The relationship between the period of time that a calfhouse has been occupied and the incidence of scouring and mortality in young calves, *Br. J. Nutr.*, 9, 11—20, 1955.
2. Wood, P. C., The epidemiology of white scours among calves kept under experimental conditions, *J. Pathol. Bacteriol.*, 70, 179—193, 1955.
3. Roy, J. H. B., Nutrition and management factors affecting perinatal mortality in calves, in *Perinatal Ill-Health in Calves*, Rutter, J. M., Ed., Commission of the European Communities, Coordination of Agricultural Research, Brussels, 1975, 125—140.
4. Tagari, H. and Roy, J. H. B., The effect of heat treatment on the nutritive value of milk for the young calf. VIII. The effect of the preheating treatment of spray-dried skim milk on the pH and the contents of total, protein and nonprotein nitrogen of the pyloric outflow, *Br. J. Nutr.*, 23, 763—782, 1969.
5. Williams, V. J., Roy, J. H. B., and Gillies, C. M., Milk-substitute diet composition and abomasal secretion in the calf, *Br. J. Nutr.*, 36, 317—335, 1976.
6. Ternouth, J. H., Roy, J. H. B., and Siddons, R. C., Concurrent studies of the flow of digesta in the duodenum of exocrine pancreatic secretion of calves. II. The effects of addition of fat to skim milk and of 'severe' preheating treatment of spray-dried skim milk powder, *Br. J. Nutr.*, 31, 13—26, 1974.
7. Shillam, K. W. G. and Roy, J. H. B., The effect of heat treatment on the nutritive value of milk for the young calf. V. A comparison of spray-dried skim milks prepared with different preheating treatments and roller-dried skim milk, and the effect of chlortetracycline supplementation of the spray-dried skim milk, *Br. J. Nutr.*, 17, 171—181, 1963.
8. Johnson, R. J. and Leibholz, J., The flow of nutrients from the abomasum in calves fed on heat-treated milks, *Aust. J. Agric. Res.* 27, 903—915, 1976.
9. Lister, E. E. and Emmons, D. B., Quality of protein in milk replacers for young calves. II. Effects of heat treatment of skim milk powder and fat levels on calf growth, feed intake and nitrogen balance, *Can. J. Anim. Sci.*, 56, 327—333, 1976.
10. Donnelly, P. E., Dean, R. J., and Kevey, C., The effects of heat treatment during processing in calf milk replacer quality, *Proc. N. Z. Soc. Anim. Prod.*, 36, 87—92, 1976.
11. Penning, I. M., Penning, P. D., and Treacher, T. T., The effect of quality and quantity of milk protein on feed digestibility, growth rate and nitrogen retention in lambs given milk substitutes, *Proc. Br. Soc. Anim. Prod.*, 3(Abstr.), 99, 1974.

12. Braude, R., Newport, M. J., and Porter, J. W. G., Artificial rearing of pigs. III. The effect of heat treatment on the nutritive value of spray-dried whole-milk powder for the baby pig, *Br. J. Nutr.,* 25, 113—125, 1971.
13. Anon., Standards for grades of dry milks including methods of analysis, Bull. 916, American Dry Milk Institute Inc., Chicago, 1971.
14. Knipschildt, M. E., Recent developments in milk drying techniques, *J. Soc. Dairy Technol.,* 22, 201—213, 1969.
15. Harland, H. A. and Ashworth, U. S., The preparation and effect of heat treatment on the whey proteins of milk, *J. Dairy Sci.,* 28, 879—886, 1945.
16. O'Sullivan, A. C., Whey protein denaturation in heat processing of milk and dairy products, *J. Soc. Dairy Technol.,* 24, 45—53, 1971.
17. Rowland, S. J., The determination of the nitrogen distribution in milk, *J. Dairy Res.,* 9, 42—46, 1938.
18. Roy, J. H. B., Diarrhea of nutritional origin, *Proc. Nutr. Soc.,* 28, 160—170, 1969.
19. Gorrill, A. D. L. and Nicholson, J. W. G., Effects of neutralizing acid whey powder in milk replacers containing milk and soybean proteins on performance and abomasal and intestinal digestion in calves, *Can. J. Anim. Sci.,* 52, 465—476, 1972.

Specific Nutrients and Nonnutrients

EFFECT OF PROCESSING ON NUTRITIVE VALUE OF FEEDS: LIPIDS*

Henry L. Fuller

INTRODUCTION

Differences in the nutritional value of fats arise from (1) the inherent or original value of fats from different sources, (2) characteristics arising from processing, as such, and (3) qualities imparted by mishandling either in storing, transporting, mixing, or handling by the final user. To understand how the nutritional value of fats might differ requires first a description of the various fats offered to the feed trade and secondly, an appreciation of the nutritional values and quality factors of fats.

The Association of American Feed Control Officials[1]** has defined the various classifications of fats available to, and acceptable by, the feed trade and has identified each with the permanent number assigned to it by the Subcommittee on Feed Composition, Committee on Animal Nutrition, National Research Council (NRC). Eight categories of feed grade fats are thus designated as follows:

33.1	Animal fat	NRC 4-00-409
33.2	Vegetable fat or oil	NRC 4-05-077
33.3	Hydrolyzed fat or oil, feed grade	NRC 4-00-376 and 4-05-076
33.4	—Ester—	(No NRC number)
33.5	Fat product, feed grade	NRC 4-08-071
33.6	Corn endosperm oil	NRC 4-02-852
33.7	Vegetable oil refinery lipid, feed grade	NRC 4-05-078
33.8	Corn syrup refinery insolubles, feed grade	NRC 4-02-893

Most of the fat sold to the feed trade would technically fall into two or possibly three categories: animal fat, hydrolyzed animal and vegetable fat, and fat product. Virtually all are blends rather than pure fats.

"Feed grade animal fat" is a term applied to various mixtures of fats produced by the rendering industry. At one time the term "tallow" was applied loosely to all animal fats but is no longer applicable, since the amount of pure tallow going into the feed trade is small. Raw material inputs vary from one plant or geographical region to another and consist principally of "shop fat and bone" (trimmings from meat counter in local shops), packing house offal, poultry offal, fallen animals, and restaurant grease (spent cooking oil). In a survey of 40 renderers throughout the U.S. conducted by the Fats and Proteins Research Foundation,[2] restaurant grease contributed about 40% of the total input and fallen animals only 10% (Table 1).

Although the input materials for production came from varied sources the consistency of source usage by the individual rendering plants was very high (approximately 98%). Only 1 renderer in 40 indicated that he varied his input materials.

Thus, any attempt to describe a "typical" or "average" feed grade animal fat would not accurately reflect the individual renderer's input ingredients as shown by the large standard deviation and range figures. These values indicate that some renderers are using only one raw material or a combination of two ingredients (e.g., 100% restaurant

---

* Article submitted 1976.

** The reader is referred to current issues of the *Official Publication* of the AAFCO for timely revisions in these definitions. A set of "tentative" definitions for fats published in 1981 is expected to become "Official" in the near future.

Table 1

UTILIZATION OF RAW MATERIALS BY RENDERERS IN
PRODUCING FEED GRADE ANIMAL FAT (% OF TOTAL)

	Shop fat and bone	Packing house offal	Fallen animals	Poultry offal	Restaurant grease	Other sources
Mean	20.92	18.60	9.82	6.33	40.41	4.43
Standard deviation	25.62	22.64	21.61	16.06	43.60	14.19
Median	7.50	5.00	0.0	0.0	14.50	0.0
Range	78.00	85.00	98.50	80.00	100.00	67.00

From Boehme, W. F., *J. Am. Oil Chem. Soc.*, 51, 526A, 1975. With permission.

grease or 90% restaurant grease and 10% poultry offal). Further confirmation of the exclusive use of some raw materials over others is shown by the "0.0" medians in Table 1 for fallen animals and poultry offal indicating that only 50% of all the renderers used either of these two materials in their production. The results of the study showed that there is a high consistency in the usage of raw material sources for producing feed grade fat by individual renderers; however, the raw material inputs vary greatly with the geographic location of the rendering plant.

The rendering process has been described in detail by the writer[3] as follows:

Most processing of animal fats is now accomplished by dry rendering in contrast to the old wet-rendering process. Dry rendering is done in steam-jacketed cookers. The fat is released as cell walls rupture because of escaping moisture. The operation is completed when the moisture in the material has been reduced to 5 to 7%. Ideally, this requires 1.5 to 4 hr with the temperature in the cooker reaching 115.5 to 121.1°C. During the cooking operation, the material is agitated constantly to prevent burning against the steam-jacketed shell of the cooker. Often a part of the fat is recycled so that the introduction of hot fat to the fresh raw material accelerates the cooking process. This will depend on the style of cooker used. There is some increase in the use of continuous cookers as opposed to the batch type, and these usually call for some recycling of fat. Such a process may darken the fat somewhat as a result of successive heating and cooling, but it has been our experience that this color does not reflect nutritional quality.

The cooked meat, or tankage, still containing considerable fat is dropped out of the cooker onto percolating pans that have screen bottoms allowing the free fat to drain off. While still hot the residue is transferred to a pressing operation of hydraulic or expeller type, more frequently the latter. The hydraulic press consists of a barrel or cylinder made of heavy iron bars spaced so the fat can drain off. The tankage is loaded into the barrel in layers separated by metal plates that help to equalize the pressure throughout the charge and permit the fat to flow from the center outward. The pressure is built up gradually and allowed to stand at full pressure until as much of the fat as possible is drained off. This operation generates very little heat, and the fat coming out should be as good as that which went in.

The expeller or screw press simply forces the tankage through narrow openings that squeeze out the fat. In this process the constant grinding action increases the amount of fines that pass out with the fat. The amount of fines will be further increased if the tankage is overcooked, because it then becomes dry and does not form a cake. A good cake acts as a filter both in the hydraulic and screw-press operations. Improper operation of the expeller may also result in overheating or uneven heating of the fat. Although it may darken the fat somewhat, this amount of heat should not lower its nutritional value.

Settling tanks are usually employed to remove fines, moisture, or other impurities. These are equipped with heating coils and valves for drawing off the settled fat at a point above any possible settlings. Troublesome fines that refuse to settle may have to be removed by washing or coagulation. The presence of fines is not particularly detrimental nutritionally but may be troublesome in nozzles or valves in storage and handling equipment.*

* Reproduced from *Effect of Processing on the Nutritional Value of Feeds,* page 135, with the permission of the National Academy of Sciences, Washington, D.C.

Virtually all of the vegetable fats or oils that find their way into the feed trade are byproducts of the refining process, the primary products of which are cooking oils and toilet soap. In the refining process, crude fats or oils are saponified with NaOH or KOH. Residues from this process consisting of some neutral fats, free fatty acids, and all of the unsaponifiable material are reacidulated and sold as "acidulated _____ soapstock" (with the appropriate source indicated), or blended with animal fats and sold as "hydrolyzed animal and vegetable fat." By definition "hydrolyzed fat, or oil" consists predominately of free fatty acids, must contain 85% total fatty acids (as opposed to 90% for either animal or vegetable fats), and may contain up to 6% unsaponifiable material (as opposed to 2.5% for animal fat and 2.0% for vegetable fat).

In some instances, the oil refinery byproducts are methylated and sold as "methyl esters of _____" (with appropriate origin indicated).

To understand how processing or handling can affect the nutritional value of fats requires an understanding of those values and how they might vary inherently in fats from different sources. The nutritional value of fat lies in the purpose for which it is used, since successive increments may be used to meet different needs of the formula. Fats serve a threefold purpose in feeds:

1. They improve physical characteristic and palatability of feed
2. They provide essential fatty acids (EFA)
3. They provide a concentrated source of energy

The first increment of supplemental fat is programmed into the formula with a value placed on all three of these purposes and would be pulled into the formula at almost any price. In practice this would be accomplished by simply specifying a minimum level of 1 to 1½% of fat in the formula. After this level has been reached the programmer cannot continue to place a value on those physical factors and further increments of fat must be justified on the basis of their contribution of essential fatty acids and energy (2 and 3, above). Fat will continue to be brought in by the computer on the basis of these combined values until the EFA requirements specified in the formula have been met. Depending upon the other ingredients in the diet and the age and class of animal to be fed, this need may vary from nil to 1 or 1½%. Like any essential nutrient, once this requirement is met, additional quantities of EFA will serve no useful purpose as such and additional increments of fat will be valued solely for their energy contribution.

In contrast to the inherent nutritional values of fat, i.e., EFA and energy, there are a number of quality factors which relate more closely to processing or handling of fats than to origin. These are usually identifiable by their negative values and include stability, corrosiveness, and the presence of undesirable substances. Although not inherent in the fat these negative quality factors are influenced in some measure by the nature of the fat itself.

PHYSICAL FACTORS

The nonnutritional or physical factors listed as purpose (1) above, include improvement of palatability, prevention of particle separation, reduction of dust losses, lubrication of feed mixing and pelleting equipment, and improvement of handling qualities. Most of these functions can be performed equally well by any type of feed grade fat provided it is stable and nonreactive (or noncorrosive). Rancid fat is both unpalatable and reactive so it is unacceptable in this regard. Fats that are too high in moisture or free fatty acids may be corrosive to metals in the feed mixing and handling facilities.

Table 2
FATTY ACID COMPOSITION OF SOME COMMON FATS AND OILS[a]

Type of fat	14:0	14:1	16:0	16:1	18:0	18:1	18:2	18:3	Ref.
Vegetable fats									
Corn oil	—	—	10.5	—	1.8	26.5	58.3	2.8	4
Soybean oil	0.5	—	14.3	0.5	1.6	27.4	45.6	10.2	4
Cottonseed oil[b]	0.8	—	22.8	1.0	2.5	19.4	53.2	0.2	5
Linseed	—	—	6.0	0.1	2.1	20.3	14.6	56.7	4
Animal fats									
Beef tallow	2.6	1.5	21.9	9.4	20.1	37.9	6.3	0.1	5
Lard	0.9	—	24.4	6.5	10.6	38.4	19.3	—	4
Animal grease[c]	1.1	—	25.1	1.0	16.2	46.9	8.7	0.4	7
Poultry fat	1.4	0.2	21.4	6.8	5.9	39.5	23.5	1.0	6
Marine oils									
Menhaden[d]	11.9	0.4	23.2	16.4	5.6	15.3	2.7	1.9	4
Hydrolyzed fats									
ACSS[b]	0.4	—	18.2	0.7	3.4	20.5	51.3	2.5	5
HAV[c]	1.2	—	25.1	—	17.0	43.9	11.7	0.2	7

[a]   Number of C atoms:number of double bonds.
[b]   Cottonseed oil and acidulated cottonseed soapstock (ACSS) were found to contain 0.2 and 2.9%, respectively, of cyclopropene fatty acids (malvalic and sterculic).
[c]   Composition of animal grease and hydrolyzed animal and vegetable fat (HAV) will depend upon the particular blend employed.
[d]   Menhaden oil also contained about 22% of 20 and 22 C polyunsaturated fatty acids.

Reproduced from *Effect of Processing on the Nutritional Value of Feeds,* page 131, 1973, with the permission of the National Academy of Sciences, Washington, D.C.

ESSENTIAL FATTY ACIDS

For all practical purposes a discussion of EFA can be limited to linoleic acid (18:2) for all livestock and poultry. Arachidonic acid (20:4) is formed from 18:2 in the animal body and therefore could spare the latter in the diet. It is often listed as one of the EFA; however, it is not present in sufficient quantities in feed ingredients to be considered in practical feed formulation. Also, there are certain long chain, highly unsaturated fatty acids in fish oils that may spare linoleic acid in the diet, but again, these would not usually be present in sufficient quantities to warrant practical consideration.

Added fat should provide enough linoleic acid to supplement that which occurs naturally in the basal portion of the diet and bring it up to the required level. The level of supplementation at which this occurs will depend upon the level specified in the formula, the linoleic acid content of the other ingredients in the diet and of the fat being added.

The vegetable fats are considerably higher in linoleic acid (18:2) than are the animal fats (Table 2). Whether this has any practical significance will depend on the basal diet formula. In high corn diets the 18:2 content would probably be sufficient for all classes of livestock and poultry (60% corn × 3.9% fat × 58.3%; 18:2 = 1.36%; 18:2 in the diet). Where the major grain component is sorghum, wheat, or barley — these grains have relatively lower levels of total fat than does corn — a supplemental source of 18:2 *may* be necessary.

ENERGY

The greatest demand for fat is for its contribution of energy. It is a concentrated source having 2¼ times as much usable energy per unit of weight as carbohydrates or proteins, thereby leaving more room in the feed bag for flexibility of the feed formula, or in keeping with modern concepts of feed formulation, permitting greater density of energy and all of the nutrients. Feed grade fats are evaluated on the basis of metabolizable energy (ME). ME is defined as the gross energy (heat of combustion) minus the unabsorbed portion when fed to chicks under prescribed conditions. It is determined by the following general formula:

$$\text{ME diet} = \text{GE diet} - (\text{GE excreta} + 8.22 \text{ N})$$

where ME diet = metabolizable energy per gram diet dry matter; GE diet = gross energy per gram diet dry matter; GE excreta = gross energy in excreta per gram diet dry matter (determined by using an inert marker such as chromic oxide in the diet); and N = nitrogen retention per gram diet dry matter (8.22 is the heat of combustion or uric acid per gram of nitrogen. This term corrects classical ME to nitrogen equilibrium).

To determine the ME of an individual ingredient, the test ingredient is incorporated into the basal diet at a prescribed level. This diet then becomes the best diet and is fed to a second group of animals. The ME is then determined on both the basal diet and the test diet. The difference in ME of the two diets is presumed to represent the ME of the test ingredient when extrapolated to 100%. This method implies that the absorbability of the basal portion of the test diet has not changed and therefore, the difference in gross energy of the excreta of the basal and that of the test diet represents the unabsorbed portion of the test substance. The measurement of ME of an ingredient, therefore, is not a direct measure of the ME of the test substance itself but rather a measure of the effect of the test substance on the ME of the entire diet, extrapolated to 100%.

All triglycerides have essentially the same gross energy value, about 9.4 kcal/g; therefore, differences in ME arise entirely from differences in absorbability, by definition. Factors influencing absorbability have been studied extensively by workers at Cornell University.[8] They include:

1. Chain length of fatty acids
2. Degree of unsaturation
3. Degree of esterification (FFA level)
4. Position of fatty acid on the glyceride moiety
5. Ratio of unsaturated to saturated fatty acids

All of these factors interact with one another so that differences attributable to any single factor are difficult to measure under practical conditions. The work of Sibbald et al.[9] illustrates this interaction. They demonstrated a synergistic effect between two sources of fat (Table 3). Tested individually soybean oil and tallow had ME values of 8.46 and 6.94 kcal/g, respectively. When fed together in a 1:1 ratio, the mixture was found to provide 8.41 kcal ME/g. If the soybean oil remained unchanged, the tallow would have to contain 8.36 kcal/g under these conditions.

Renner and Hill[10] showed that palmitic (16:0) and stearic (18:0) acids were utilized poorly when fed as free fatty acids in the absence of other fats; however, when they were fed as intact triglycerides, utilization was greatly improved. Young and Garrett[8] demonstrated the beneficial effect of oleic and linoleic acids on the absorbability of 16:0 and 18:0. The combination was less dramatic than that of oleic acid alone and much less than that of monolein. Absorbability of these long chain fatty acids was greatly improved when the ratio of unsaturated to saturated fatty acids was greater than 1.4 to 1 and when the ratio of 16:0 to 18:0 was 1.5 to 1 or greater.

Table 3
METABOLIZABLE ENERGY VALUES OF TALLOW, UNDEGUMMED SOYBEAN OIL, AND A 50:50 MIXTURE OF THE TWO FATS

(kcal/g)

| Protein level | 24.4% | | 34.0% | | |
Fat level	10%	20%	10%	20%	Average
Kind of fat					
Tallow	6.02	7.24	6.79	7.69	6.94
Soybean oil	8.65	8.22	8.94	8.02	8.46
Mixed fat	8.11	8.31	8.82	8.41	8.41

From Sibbald, I. R., Slinger, S. J., and Ashton, G. C., *Poult. Sci.*, 40, 303, 1961. With permission.

Table 4
REPORTED VALUES FOR ME OF VARIOUS FATS

(kcal/kg)

Refined soy	Pr. or F.[a] tallow	FGAF[b]	Lard	HAV[c]	Ref.
8536	7493—7579	7453	7055	—	13
—	7018	6461	—	7106	17
8290	7620	7550—8480	—	—	9
—	7600	7660	—	—	18
9020	6600—7590	7656	8800—9200	8140	7, 8
—	7920	8440	—	6987	19
9262	6292	7260	8756	—	20
Average 8777	7287	7610	8478	—	

[a]   "Prime" or "fancy" tallow.
[b]   Reported as feed-grade tallow or grease.
[c]   Hydrolyzed animal and vegetable fat.

In a more recent report Young[11] concluded:

These experiments reveal several factors which are important in evaluating or preparing a good quality feed grade fat.

The total fatty acids should be made up of not more than 50% free fatty acids; the remaining fatty acids should be in a triglyceride form. Higher free fatty acid levels in the fat interfere with efficient pelleting, but more important the triglycerides are needed as a source of 2-monoglycerides in the intestinal lumen to facilitate the maximum absorption of the saturated fatty acids.

The ratio of palmitic acid to stearic acid should equal or be greater than 1.5 to 1. Not only is stearic acid poorly absorbed, but it interferes with the absorption of palmitic acid.

The ratio of unsaturated to saturated fatty acid should be equal to or greater than 1.4 to 1. The unsaturated acids along with the 2-monoglycerides are required for the formation of micelles which solubilize the saturated fatty acids in the intestinal lumen.

These reports reflect the interaction of factors affecting absorption, and therefore the metabolizable energy of fats. They also reflect more nearly, practical feeding conditions; whereas, tabular values are usually based upon a set of conditions which tend to isolate many of the individual factors affecting absorption.

In spite of the criticism of ME as a valid measure of the energetic contribution of fats, it is still the most widely used evaluation by the feed industry. Table 4 lists some of the ME values for various fats that have been reported in the scientific literature where two or more fats have been compared. The averages are not statistically defensible because of missing values, but should serve as a guide.

Many reports have compared caloric efficiency and efficiency of body weight gains of diets containing a wide variety of fats.[12-15] Little or no differences among fats have been found in these important criteria except for hydrogenated coconut oil which was relatively poorer than other fats.[12] DeGroote et al.[13] obtained a significantly higher tissue energy gain on the prime tallow diet relative to the lard and soybean oil diet. The findings of Volker and Amich-Galli[16] suggest that losses of ME occurring during metabolism are lower for animal fats than for vegetable fats, thus bringing their net or productive energy closer together.

Both EFA and energy levels are related more to the origin of fats than to their processing; however, if blending is included in the collective term "processing" then a great deal can be accomplished in offering to the feed industry, fats with satisfactory EFA levels and maximum metabolizable energy.

Stability

Oxidative rancidity of fats results in peroxide formation which can be measured under prescribed conditions to determine "peroxide value" (PV). Freshly produced fats of whatever origin normally have low PVs. The tendency to oxidize, or instability, is determined by the active oxygen method (AOM) which measures the time required, in hours, to produce a peroxide value of 20 when a sample of fat is heated with a stream of air bubbling through it under prescribed conditions. A PV of 20 was selected for the end point because the fat will begin to have a rancid odor at higher values. This value applies to animal fats only, and is not applicable to vegetable fats where a higher peroxide value must be used as the end point. Vegetable fats do not develop a rancid odor until the peroxide value reaches 75 or more so that odor is not a good indication of high peroxide values in vegetable fats. For this reason, the peroxide value should be checked chemically instead of relying upon odor if any vegetable fats are used in feeds, because rancid fat, whatever the source, is unacceptable.

Reactivity is related to stability. Fats which are unstable to oxidation and too high in free fatty acids or moisture are corrosive and may react with metal surfaces in feed milling and handling equipment. The metals, especially copper (or brass which contains copper) act as catalysts to further accelerate the breakdown.

Stability can be assured with adequate levels of a suitable antioxidant added to the fat while it is still fresh from the extraction or rendering process. This is routine practice among reputable fat processors and should be a part of the specifications for feed grade fat. Fresh animal fats even when unstabilized will usually run 20 AOM hours or more, whereas stabilized fats will usually test 40 to 50 hr.

Free Fatty Acids

Free fatty acids (FFA) result from hydrolysis of neutral fats (triglycerides) or phosphoglycerides. FFA in themselves are not harmful to animals and do not lower the quality of fats that are to be used in feed except for a slight reduction in absorbability of long chain saturated fatty acids (palmitic and stearic), which may have little significance under practical feeding conditions. Siedler et al.[21] tested 3 or 6% stabilized feed grade animal fats or 3% stabilized free fatty acids in high energy broiler rations. The rates of gain of all groups tested showed no significant differences in two experiments. Feed and caloric efficiency showed that utilization of the free fatty acids was at least equal to all other materials tested. They concluded that fats properly stabilized are utilized equally well regardless of the free fatty acid content of the fats.

The Association of American Feed Control Officials definitions of acceptable feed ingredients include one for hydrolyzed animal and/or vegetable fats.[1] These products consist primarily of fatty acids and are produced by the alkaline hydrolysis and re-acidulation of fats and oils as in the manufacture of soaps. There is no inherent differ-ence in the fatty acids produced in this manner and those occurring naturally. They are just as reactive, and in the case of fatty acids of vegetable origin perhaps more reactive because the degree of unsaturation is usually greater. If not properly stabilized they will likewise undergo oxidation rapidly.

The important thing to consider is not the *level* of FFA but rather how they occurred and what happens to them after formation. "FFA" and "rancidity" are *not* synony-mous. Nor is absorption of oxygen necessarily accompanied by rancidity. Drying oils, such as linseed oil, for instance, absorb oxygen and become hard. Feed fats, depending upon the degree of unsaturation, usually absorb oxygen more slowly but frequently become rancid in the process. It is the oxidative decomposition following hydrolysis that is objectionable.

Time, temperature, and moisture all favor hydrolysis of neutral fats into FFA and glycerol. The process is also accelerated by the more reactive metals such as copper. These factors also favor oxidative decomposition of the FFA into peroxides and hy-droperoxides. Since the FFA and the oxidative process tend to accelerate in the man-ner of a chain reaction, being "autocatalyzed" by both the FFA and the peroxides themselves. This chain reaction can be checked by refrigeration, elimination of mois-ture, exclusion of oxygen, and by the use of a suitable antioxidant.

Cleanliness and Purity

Undesirable substances are for the most part included in the trade term "MIU" (moisture, insoluble impurities, and unsaponifiable matter).

Moisture is a diluent and reduces the energy value of the fat by the proportion of its presence to the total. Of greater concern is its contribution to the instability of the fat and to its reactivity with metals. Hathaway[22] reported that the corrosion rate on mild steel is doubled when the moisture level of fat is increased from 0.5 to 3.0%. In the case of solvent-extracted fat the value for "moisture" will include any residual solvent such as hexane. This would constitute an explosion hazard and should be spot-ted by the processor before the shipment leaves his plant.

Insoluble impurities consist mostly of "fines" which are undesirable primarily be-cause of the trouble they may cause in nozzles or valves in fat handling equipment.

Unsaponifiables consist of materials which are soluble in fat solvents but will not react with NaOH or KOH to produce soap. In this category are hydrocarbons, waxes, and tars which are removed in the refining process in the preparation of edible oils and soaps. They have little food value. Included also are sterols, cholesterol esters, phospholipids, fat-soluble vitamins, and pigments. Unsaponifiable matter that occurs normally in fat is not categorically harmful;[22] however, it is in this fraction that toxic substances may appear. Chlorinated hydrocarbons (PCBs, etc.) and the "chick edema factor" were found to accumulate in the unsaponifiable fraction.

PCBs and related toxic substances may be present in animal tissues prior to render-ing or extracting. Since these are fat-soluble materials they will be concentrated in the fat fraction. With the recent tolerance levels in animal tissues imposed by the Food and Drug Administration (FDA), the possibility of such occurrences are much less likely. PCBs are also forbidden to be used in machinery that is used in processing foods or feeds so that contamination with this class of compound is highly unlikely.

REFERENCES

1. Association of American Feed Control Officials, Inc., *Official Publication,* 1974, 184.
2. Boehme, W. F., Composition of feed grade animal fat, *J. Am. Oil Chem. Soc.,* 51, 526A, 1975; personal communication.
3. Fuller, H. L. Effect of processing animal and vegetable fats on their stability and nutritional value, in *Effect of Processing on the Nutritional Value of Feeds,* National Academy of Sciences, Washington, D.C., 1973, 131—141.
4. Edwards, H. M., Jr., Fatty acid composition of feeding-stuffs, *Ga. Agric. Exp. Stn. Tech. Bull.,* 36, 1—34, 1964.
5. Edwards, H. M., Jr., Ashour, A. A., and Nugara, D., Effect of Nutrition on the Body Composition of Broilers, Proc. Ga. Nutr. Conf., Atlanta, February 1971, 42—57.
6. Edwards, H. M., Jr., Fatty acid composition of poultry offal fat, *Poult. Sci.,* 40, 1770—1771, 1961.
7. Young, R. J., Fats and Fatty Acids in Animal Nutrition, Proc. Md. Nutr. Conf., Washington, D.C., 1965, 61—71.
8. Young, R. J. and Garrett, R. L., Effect of oleic and linoleic acids on the absorption of saturated fatty acids in the chick, *J. Nutr.,* 81, 321—329, 1963.
9. Sibbald, I. R., Slinger, S. J., and Ashton, G. C., Factors affecting the metabolizable energy content of poultry feeds. II. Variability in the M.E. values attributed to samples of tallow and undegummed soybean oil, *Poult. Sci.,* 40, 303—308, 1961.
10. Renner, R. and Hill, F. W., Factors affecting the absorbability of saturated fatty acids in the chick, *J. Nutr.,* 74, 254—258, 1961.
11. Young, R. J., Evaluation of the Nutritional Value of Feed Grade Fat, Proc. Ark. Nutr. Conf., Fayetteville, 1968, 53—71.
12. Carew, L. B., Jr., Hopkins, D. T., and Nesheim, M. C., Influence of amount and type of fat in metabolic efficiency of energy utilization by the chick, *J. Nutr.,* 83, 300—306, 1964.
13. DeGroote, G., Reyntens, N., and Amich-Galli, J., Fat studies. II. The metabolic efficiency of energy utilization of glucose, soybean oil, and different animal fats by growing chicks, *Poult. Sci.,* 50, 808—819, 1971.
14. Veen, W. A. G., Grimbergen, A. H. M., and Stappers, H. P., The true digestibility and caloric value of various fats used in feeds for broilers, *Arch. Geflugelk.,* 38, 213—220, 1974.
15. Fuller, H. L. and Rendon, M., Energetic efficiency of different dietary fats for growth of young chicks, *Poult. Sci.,* 56, 549—557, 1977.
16. Volker, L. and Amich-Galli, J., The Comparative Values for Metabolizable Energy and Net Energy for Production Resulting From Addition of Tallow and Other Fats to Broiler Rations, National Renderers Assoc., European Office, Rome, Italy, 1967.
17. Matterson, L. D., Potter, L. M., and Stutz, M. W., The metabolizable energy of feed ingredients for chickens. *Conn. Agric. Exp. Stn. Storrs Res. Rep.,* 7, 1—11, 1965.
18. Lewis, D. and Payne, C. G., Fats and amino acids in broiler rations. VI. Synergetic relationships in fatty acid utilization, *Br. Poult. Sci.,* 7, 209—218, 1966.
19. Cullen, M. P., Rasmussen, O. G., and Wilder, O. H. M., Metabolizable energy value and utilization of different types and grades of fat by the chick, *Poult. Sci.,* 41, 360—367, 1962.
20. Renner, R. and Hill, F. W., The utilization of corn oil, lard, and tallow by chickens of various ages, *Poult. Sci.,* 39, 849—854, 1960.
21. Siedler, A. J., Scheid, H. E., and Schweigert, B. S., Effects of different grades of animal fats on the performance of chicks, *Poult. Sci.,* 34, 411—414, 1955.
22. Hathaway, H. D., The selection of commercial fat sources for use in livestock and poultry feed, in *Proc. 2nd Nutr. Conf. Feed Manuf.,* Butterworths, London, 1968, 22—42.

EFFECT OF PROCESSING ON NUTRITIVE VALUE OF FEEDS: PROTEIN

P. J. de Wet

INTRODUCTION

Food proteins have always played an important role in human nutrition to which heat in one way or another is applied before consumption. In an industrialized world, more and more of the proteins for human consumption are processed in factories. Because of the variety of proteins consumed by the human, any damage due to processing cannot be detected easily.

With a vast increase in the agricultural production of oil seeds primarily for their oils, large quantities of plant protein sources become available for use in animal nutrition. The plant proteins were used to produce palatable and highly nutritive animal proteins like poultry, pork, eggs, milk, and meat. Poultry and pigs are kept under artificial circumstances under which they become totally dependent on balanced rations provided by the human. It became apparent quite early that the various proteins differed in their nutritional value; also, the same kind of protein — from different sources — may not have the same beneficial effects. It was soon proved that heat was the most important factor in processing of proteins, especially the plant proteins.

Because of the vast field of processing in protein nutrition no attempt will be made here to cover that of human nutrition. Research carried out with rats, mice, and chickens as simple stomach animals will give an indication òf what can be expected in human nutrition. Only the literature concerning mainly soybean protein which is the most important plant protein and, to a lesser extent, fish meal as animal protein source is covered here. In the text, reference is made to textbooks which will give the reader more information concerning the well-known plant proteins, and particulars of the processes used in the manufacturing of proteins. Emphasis is laid on the beneficial effect of heat and the influence on the amino acid content, nutritional effects, and the mechanisms involved effecting the nutritional properties of a protein.

INFLUENCE OF HEAT ON THE NUTRITIVE VALUE OF PROTEINS

As early as 1917 Mendel and Osborne[1] gave evidence of the effects of heat on the nutritive value of proteins when they observed that soybeans would not support the growth of rats unless they had been cooked for 3 hr in a steam bath. In another experiment[2] they showed that steaming of cottonseed for a sufficient length of time rendered it harmless to animals, thereby destroying the toxic effect. However, when the heating time was prolonged, the nutritive value of cottonseed was reduced. Since these results of Osborne and Mendel, numerous publications have appeared over a period of 60 years. Two opposing effects of heat treatment became apparent, i.e., improvement of the nutritional value and loss under various conditions. These experiments also indicated the importance of moisture in the heat treatment of plant proteins to prevent or minimize damage.

Plant Proteins
Improvement of Nutritional Value

A variety of plant proteins are used today in animal nutrition. The importance of each differs from country to country depending on the suitability of the environment of a country for the cultivation of one or a variety of plant proteins. From the literature it is apparent that soybean protein received considerable attention on this aspect. No attempt will be made to cover each type but attention will be drawn to those aspects where differences may exist with regard to the reaction to heat or nutritional effects. For information on each type of plant protein, the reader is referred to an excellent account.[3]

Since the work of Osborne and Mendel draws the attention to the beneficial effects of heating,[1,2] numerous reports on the subject appeared in the literature confirming the superiority of heat-processed soybean meal. Shrewsbury et al.[4] called attention to the poor results frequently reported when farm animals are fed rations containing soybeans. Their results showed that pigs respond in the same manner as rats and that the cooking of soybeans made them acceptable as protein supplements. Similar results were shown for rats,[5-9] chickens,[7,10-12] and swine.[13,14]

In general these findings showed that the degree of improvement in the nutritive value is dependent on the temperature, duration of heating, and moisture content. Hayward et al.[6] found that commercial soybean oil meal processed at 105°C for 2 min or hydraulic pressed meal cooked at 82°C for 90 min contained proteins similar in nutritive value to raw soybeans. Commercially produced soybean meal which had been prepared at medium or high temperatures, 112 to 130°C and 140 to 150°C for 2½ min or hydraulic pressed meals cooked at 105°C and 121°C for 90 min contained protein with twice the nutritive value of raw soybeans or of low temperature meals. They found that digestibility increased by 3% and the biological value by 12%. According to them it was possible that the heat caused some essential protein fraction, which was unavailable in the raw soybean, to become available for absorption and metabolic use. Johnson et al.[15] also drew attention to a fraction of the raw protein which is not available for utilization in the body and which becomes available for metabolic use after heating.

Several studies were undertaken to establish the optimum conditions of heat treatment with autoclaving.[16-19] From these data it is evident that heat treatment together with autoclaving over short periods (15 to 30 min) improves the nutritive value of soybean meal as illustrated in Figure 1 as presented by Klose et al.[19]

Temperatures of 110°C for 30 min,[16] 128°C; 2½ min,[17] 1.345 kg/cm^2 pressure; 15 min,[18] 100°C, 110°C, and 120°C; 30 min[12] indicate the optimum conditions for improvement of protein value.

Utilization by Ruminants

In ruminants the feed which is ingested, is first subjected to fermentive digestion by microorganisms and protozoa. During fermentation the proteins are to a large extent degraded to fatty acids and ammonia. The amount degraded depends on various factors, the most important of which is protein solubility. The reader is referred to various articles in textbooks on the subject, for example the one by McDonald.[20]

In lambs Miller and Morrison[21] found that the digestibility of soybean protein increased from 62.9% for raw soybeans, 69% for solvent extracted soybean meal to 70.6% for heat treated solvent process soybean meal. Nitrogen retention averaged 18.2, 23.7, and 26.3%, respectively, for the three treatments. Contrary to data obtained with nonruminants, these data showed that additional heat treatment of solvent process soybean meal results in no appreciable improvement in the protein but rather in higher digestibility. Chalmers et al.[22] demonstrated in 1954 the degradational effect of the microbes in the rumen by infusing casein into either the rumen or duodenum. Heat treating the casein at 105°C can to a certain extent overcome this degradation as shown by an increased N-balance of +0.20 g/day per sheep for commercial casein to +1.42 g/day for treated casein. Less ammonia was found in the rumen. They later

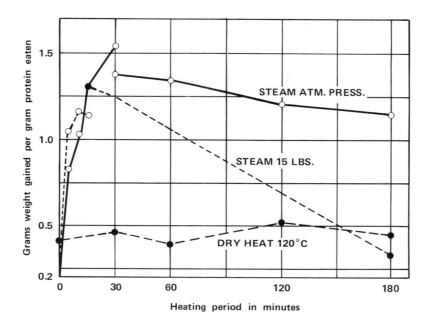

FIGURE 1.   The effect and extent of heat treatment on nutritional value of soybean, ● = Experiment 1; ○ = Experiment 2. Test period, 42 days; 12 rats per group; 80 g average initial weight. (Reprinted from Klose, A. A., Hill, B., and Ferold, H. L., *Food Technol.*, 2, 201, 1948, Copyright© by Institute of Food Technologists.)

found similar results with toasting of groundnut meal, i.e., increased nitrogen retention and lower ammonia levels although digestibility of the protein was affected.[23] No influence on digestibility was experienced when soybean meal was either heated or gossypol added. However, the combination of the two significantly reduced protein digestibility. Urinary nitrogen was increased when gossypol was added but nitrogen retention was reduced significantly only when the protein-gossypol mixture was autoclaved.[24] Sherrod and Tillman[25] found with lambs that fecal nitrogen increased, urinary nitrogen decreased, and nitrogen retention increased with increased times of autoclaving (45 min and 90 min); NH₃ levels were also reduced. Autoclaving for 45 min improved gains and feed efficiencies. Feeding the same proteins to rats, reduced gains with increased times of autoclaving. The optimum nutritive level for nonruminants should, therefore, be different from that for ruminants. In a further experiment with sheep in which cottonseed meal was autoclaved at 120°C for 20, 60, 120, 180, and 240 min, the greatest N-retention was found at 60 min autoclaving. Decreased retention took place after longer periods of heating as illustrated in Figure 2.[26,27]

Steaming soybean meal for 15 min at 120°C decreased the solubility of the protein from 61.2 to 13.1%. Ammonia liberation in the rumen of the sheep showed a striking decrease.[28] The nitrogen digestion and balance results are given in Table 1.

Heating soybean meal at 149°C for 4 hr improved the nutritive value to a point where the gains of young lambs on 12% protein level in the ration were comparable to those obtained at 17% unheated soybean meal level.[29] Solubility of the soybean protein was reduced from 72 to 35%. When full-fat, steam-heated soybeans were fed to sheep, digestibility of the protein was not affected by length of heat treatment. A significant linear increase in average daily gain was found as heating time of the soybeans increased from 0 to 30 min. During the same experiment with rats, a significant improvement in rate of gain, feed conversion, and protein conversion was found when

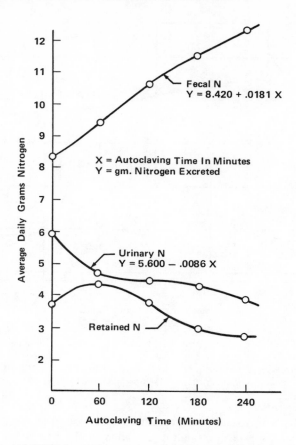

FIGURE 2.   Effects of heating cottonseed meal upon the nitrogen excretion and retention patterns of sheep. (From Sherrod, L. B. and Tillman, A. D., *J. Anim. Sci.*, 23, 510, 1964. With permission.)

heated for 15 min. Extending the time beyond this reduced the nutritional value.[30] Heat treatment of soybean meal reduced the rate of degradation of the protein by the microbes in the rumen of lambs as indicated by a low level of plasma urea and a significant increase in total nitrogen reaching the abomasum.[31] Infusion of heated soybean meal into the abomasum showed that heat had no effect on the digestibility of the protein and, therefore, had no effect on the utilization of protein posterior to the rumen.[31] In a comparison of a commercial groundnut meal, heat treated groundnut meal (110°C for 30 min, then at 100°C for 24 hr) and fish meal fed to early weaned calves, nitrogen retention and live weight gain differed significantly between diets, being highest on the fish meal diet and least on the commercial groundnut diet. Differences between diets in nitrogen utilization were due almost entirely to differential urinary nitrogen losses.[32]

Digestibility studies were undertaken by Schoeman et al.[33] on five different dietary sources of protein viz.: groundnut oil cake (Gro), sunflower oil cake (Sun), lupin seed (Lup), white fish meal (Fish), and cottonseed (Cot). The protein sources were dry heated at 149°C for 4 hr and steam heated at 1.0546 kg/cm² 121°C for 1 hr. In vivo rumen digestibilities in fistulated sheep were carried out over 12 and 24 hr followed by in vitro pepsin hydrolysis. The results are summarized in Table 2.

In general all the treatments resulted in significant (p ⩽ 0.01) decreases in protein

Table 1
MEAN VALUES FOR FOUR RAMS FOR NITROGEN DIGESTION AND BALANCE WITH DIETS CONTAINING THE VARIOUS SOYBEAN MEALS

	Meal		
	F[a]	UT[b]	T[c]
Apparent digestibility of total N in diet (%)	46.1	48.5	50.4
Apparent digestibility of N contained in hay and meal (%)	55.3	58.2	61.6
Daily N balance g/sheep	0.62	1.65	2.32
Daily N retention g/$W^{0.74}$	0.031	0.082	0.12
N-retention as % of total N intake	3.3	9.0	12.4
N-retention as % of N contained in hay and meal	4.0	10.4	15.1
N-excretion in urine as % of dig. N contained in hay and meal	93.0	83.7	74.9

[a] F = no heat.
[b] UT = commercial, untoasted soybean meal.
[c] T = steaming, 15 min, 120°C.

From Tagari, H., Ascarelli, I., and Bondi, I., *Br. J. Nutr.,* 16, 237, 1962. With permission.

Table 2
PERCENTAGE OF UNDIGESTED PROTEIN AFTER 12 AND 24 HR IN THE RUMEN AND PERCENTAGE PROTEIN HYDROLYZED BY PEPSIN AFTER 6 HR[33]

	Undigested protein in the rumen (%)						Protein hydrolyzed by pepsin after 6 hr in vitro (%)		
	Control		Steam		Dry				
Source	12 hr	24 hr	12 hr	24 hr	12 hr	24 hr	Control	Steam	Dry
Gro[a]	20.7	1.16	55.7	29.9	66.7	44.8	99.4	93.4	81.0
Sun[b]	36.5	10.1	54.3	25.4	83.8	73.0	98.6	83.9	69.2
Lup[c]	23.4	12.4	55.0	47.5	48.3	37.6	93.6	93.0	92.7
Fish[d]	45.9	41.1	50.0	47.2	65.2	64.8	85.7	70.9	51.2
Cot[e]	27.4	20.0	43.2	33.5	67.4	55.0	91.9	47.6	57.7

Note: Table compiled from other tables.

[a] Gro = groundnut oil cake.
[b] Sun = sunflower oil cake.
[c] Lup = lupin seed.
[d] Fish = fish meal.
[e] Cot = cottonseed oil cake.

digestion in the rumen. Differences between steam and dry heating were also significant except for lupin seed. These results showed that heat treatment can be an effective method for preventing severe degradation in the rumen and thereby increasing the efficiency of protein utilization by the ruminant.

Attention is drawn to the different effects of wet and dry heat on the utilization of

Table 3
IMPROVEMENT OF OVERHEATED
SOYBEAN MEAL BY AMINO ACID
SUPPLEMENTATION

Supplement to basal diet[a]	7-day gain[b] %
None	29.7
L-lysine (0.65%)	46.7
DL-methionine (0.34%)	23.9
Lysine + methionine	82.6
Heated soybean flakes (steamed 4 min at 1.345 kg/cm²)	85.8

[a] Basal diet — ground corn 42.5%; wheat bran 10%; wheat middlings 10%; alfalfa meal 5%; overheated soybean meal 28%; bone meal 2%; iodized salt 1.5%; fish oil (1500 A, 200 D); 0.5% $MnSO_4$; vitamins.

[b] Groups of eight white leghorn cockerels, selected and started on test rations after 2 weeks on practical ration.

From Clandinin, D. R., Cravens, W. W., Elvehjem, C. A., and Halpin, J. G., *Poult. Sci.*, 26, 150, 1947. With permission.

proteins by the ruminant on the one hand as indicated above and by the simple stomach animal, on the other. In the latter case, steam treatment is far superior to dry heat in growth studies with rats.[19] This phenomenon can be explained by the following: (1) prevention to a greater extent of degradation in the rumen by dry heat; (2) the lower percentage hydrolyzed by pepsin of protein which is dry heat treated; and (3) the sensitivity of the simple stomach animal to deficiencies in the diet, essential amino acids which can more easily be damaged under dry heat conditions.

Addition of Amino Acids and Digestibility of Protein

The low feeding value of raw plant proteins was also thought to be due to deficiencies in certain essential amino acids. It was thus shown that addition of cystine to unheated soybean meal improves the nutritive value to the same extent as proper heating.[5,34] Addition of methionine to raw soybean rations had the same beneficial effect,[7,35-38] although Clandinin et al.[18] showed that addition of methionine alone to overheated soybean meal had no effect. They showed, however, that supplements of lysine or both lysine and methionine increased growth rate in chickens considerably, the latter being superior to lysine alone, reaching the same growth rates as properly heated soybean meal. These results are summarized in Table 3.

Evans and McGinnis[12] also measured the growth rate of chickens given a ration containing 29% raw or heated soybean meal, supplemented in one of the treatments with 0.2% methionine. Growth rates are given in Table 4.

From these data one can conclude that time of heating as well as temperature up to 120°C will improve the nutritive value of raw soybean meal. Severe heating (130°C and 60 min) reduces it considerably. Addition of methionine improves responses significantly regardless of the nature of the meal, e.g., 58 to 149 g gain for raw soybean and 154 to 260 g for 30 min at 110°C.

Fischer and Shapiro[39] tried to counteract the growth retardation of raw soybean meal with extra protein and energy. They showed that 27% raw soybean meal in 3-week-old chicks is the equivalent of 27% heated soybean protein without the extra oil.

Table 4
INFLUENCE OF HEAT AND SUPPLEMENTATION WITH METHIONINE ON THE NUTRITIVE VALUE OF SOYBEAN PROTEIN

Treatment of raw soybean meal				
Time (min)	Temperature (°C)	Supplement (%)	Average 4-week gain (g)	Gain per g protein
0	—	0	58	1.14
30	110	0	154	1.66
30	120	0.2% choline	162	1.66
30	130	0	106	1.43
60	130	0.2% choline	69	1.09
0	—	0.2	137	1.90
30	110	0.2	260	2.19
30	130	0.2	223	1.95

From Evans, R. J. and McGinnis, J., *J. Nutr.*, 31, 449, 1946. With permission.

At equivalent levels of energy and protein intake, however, the raw soybean diets were inferior to heated soybean diets.

Methionine content of soybean meal heated to produce maximum nutritive value, was virtually unchanged compared to unheated soybean protein.[9,12,35,40,41] The difference between the digestibility of unheated and heated soybean meal in the rat was too small to account for the market differences in the biological value.[6,8,9] From nitrogen,[15,48] sulfur,[42] and methionine[9] absorption in the rat it was revealed that more than twice as much nitrogen was absorbed from heated protein as from raw protein.[42] Retention of nitrogen and sulfur from raw soybean protein was significantly less than from heated protein.[15] Johnson et al.[15] indicated further that the slight availability of the fraction of the protein containing the methionine was responsible for the poor growth responses obtained with raw soybean meal. They also came to the conclusion that the lower nutritive value of unheated soybean meal was not the result of incomplete digestion of the protein, but rather that the methionine was absorbed in a form which could not be effectively utilized for growth.

In contrast to the small differences in digestibility of unheated and heated protein in rats, the digestibility in chicks and absorption of methionine was significantly less in the unheated meal.[12,43-49] True digestibility of heated soybean protein (88.3%) was higher than the raw soybean protein (64.3%) while the biological values did not differ much (45.6 against 49.2).[47] These values showed a close resemblance with the digestibilities found by Nesheim and Garlick[48] with laying hens (85 against 54%). Nitzan[47] came to the conclusion that the low nutritional value of raw soybean meal could be attributed to the low digestibility which would lead to a reduction in nitrogen retention and growth rate and this lower digestibility would enhance the effect of methionine.[50]

Trypsin Inhibitor

As pointed out previously, the growth-promoting value of soybeans is greatly enhanced by heating. This increase was attributed to the general raising of the level of sulfur-bearing amino acids,[7,35] the undigestibility of a protein fraction of raw soybean meal, and the prevention of the conversion of a part of feed protein and endogenous protein to an unaccessible or a protein of lower digestibility due to the influence of raw soybean meal.[44]

The possibility of the nutritional significance of enzymatic inhibitors which modify the action of enzymes acting in the digestive tract has been mentioned.[51-53] The growth-retarding factor extractable from raw soybeans[53] exhibited a decreased gain in chickens receiving a basal ration of 23% autoclaved soybean meal.[54] Efficiency of gain (gain per gram feed) was 0.145 g compared to 0.287 g for the control group (without protein inhibitors). When autoclaved protein inhibitor was fed to a group of chickens receiving animal proteins in the basal ration, it was found that the unheated protein inhibitor extracted from soybeans, had the same growth retarding effect while the autoclaved inhibitor had no effect. Efficiencies of gain were: inhibitor 0.277 g; autoclaved inhibitor 0.366 g.[54]

These results proved two important points:

1.   That the protein inhibitor present in raw soybeans retards the proteolytic action in the small intestines.
2.   The activity of the protein inhibitor is destroyed by heat, explaining the beneficial effect of heat on the feeding value of soybean meal.

These observations by Ham and Sandstedt[53,54] were confirmed by a number of investigations by Kunitz and other workers[55-61] who succeeded in isolating and crystallizing a trypsin inhibitor. Growth inhibition found with chicks[54,62] was also shown by rats.[62-64] As little as one fourth of the dietary protein provided as raw soybean meal caused the growth inhibition to take place but it was further shown that crude trypsin[65] as well as crystalline trypsin[66] was able to reverse the growth effect of raw soybean meal. Several workers provided evidence that the inhibition of raw soybean meal is due to a general interference with protein digestion. Subsequent release of amino acids in heated protein, especially essential amino acids like methionine, play an important role in growth.[40,41,67,68]

Chernick et al.[69] were the first to show that raw soybean meal does not only affect the proteolytic activity in the intestines of chicks, but causes an increase in the size and proteolytic activity of the pancreas. This result was confirmed in later work.[70] Alumot and Nitzan[45,46] found that the proteolytic activity in the intestine of chickens is totally inhibited by raw soybean meal. At 4 weeks proteolysis reached a normal level. The hypertrophy of the pancreas which was also observed by them was ascribed to increased enzymatic activity to overcome the inhibitory effect of the soybean meal.[45]

The hypertrophy of the pancreas found in chickens up to the age of 6 weeks[45,46] was absent in the case of laying hens which received a ration of raw soybean meal over a period of 8 weeks.[46] Also, no statistical difference was found in nitrogen retention, feed conversion, or egg production. These authors came to the conclusion that raw soybean meal posesses a protein fraction which cannot be digested by the young chicken but a part is absorbed which causes the hypertrophy of the pancreas and reduces growth rate. In their later work, Lepkovsky et al.[71] came to the same conclusion based on a large percentage of nitrogen in the lower gut of chickens and the lower nitrogen absorption from raw soybean meal. Other workers found the same trend but they disagreed with the mechanism of tryptic inhibitor in chickens[46,47,71,72] and rats.[73,74]

Feeding two crystalline trypsin inhibitors (α-chymotrypsin inhibitor and Kunitz soybean inhibitor), together with heated soybean meal, Getler et al.[72] showed that growth inhibition is small in comparison with either a raw soybean ration or a raw and heated soybean ration (2:1), the latter two showing no difference. They also found that ethanol extracted soybean meal reduces growth rate, and postulated that raw soybean meal possesses more than one growth inhibiting factor. They proposed the following mechanism through which raw soybean meal may cause reduced growth rates:

1.   Trypsin inhibitors stimulate the hypertrophy of the pancreas and synthesis of proteolytic enzymes. They also cause a lower proteolytic activity in the intestines,

especially after intake of raw soybean meal and before pancreas secretion is stimulated.[45,47,71]

2. Raw soybean includes a protein fraction which can only be digested after heat treatment. This phenomenon can be explained by the lower digestibility of raw soybean meal[47,71] and the greater extent of growth inhibition by raw soybean meal supplemented with crystalline trypsin inhibitors.[72]

The presence of a protein fraction of low digestibility formed in the intestines, was also shown by Bilorai and Bondi[75,76] and it is this fraction which stimulates the pancreas and leads to lower growth rates.

The influence of raw soybean meal on pancreatic activity was based on indirect conclusions but several factors may have a detrimental effect on the utilization of raw soybean meal.[71] For this reason the main pancreatic duct of chicks and rats was connected to an external polyethylene tubule by Lepkovsky.[71] According to him the low nutritional value of raw soybeans is due to the detrimental effect of the trypsin inhibitors on tryptic proteolysis and not a decrease of proteolytic activity. This is accompanied by the formation of a tryptic inhibitory complex which cannot be digested by the proteolytic enzymes and is therefore excreted.

In earlier work[64,77-82] a toxic component was also isolated which was responsible, according to Liener,[82] for about one half of the growth inhibition caused by raw soybean meal. This component was called soybean hemagglutin by Liener which should be the same as the one referred to by Laufer et al.[77] as soyin.

Further evidence of an undigestible protein complex was presented by Beilorai et al.[83] based on the amount of free essential amino acids in the intestine of chickens. Using yttrium-91 (^{91}Y), Beilorai et al.[84] showed that 90% of the protein of heated soybean meal was digested in the intestines of chickens; 70% in the duodenum and 20% in the rest of the digestive tract. Chickens which received raw soybean meal also digested 70% in the duodenum but nothing in the lower parts although the growth inhibitory effect was greater than the 20% difference in digestibility. They explained this difference in terms of the energy necessary for the increased secretion of endogenous nitrogen. Work with ^{14}C casein[85] confirmed these results[84] and showed that at any stage three times more undigestible protein was present in the lower part of the digestive tract in comparison with chickens receiving heated soybean meal.

Kakade et al.[86] showed that trypsin inhibitors, four of which were already isolated and characterized,[87] are responsible for 38% of the difference in protein efficiency ration between raw and heated soybean meal; 41% of pancreas hypertrophy; and 43% of the difference in the in vitro digestibility of raw and heated meal. They alleged further that trypsin inhibitors are responsible for 40% of the growth inhibitory effect, pancreas hypertrophy and a lower in vitro digestibility. The remaining 60% is caused by the formation of an undigestible protein complex.

Further evidence was provided by Lyman et al.[88] on the mechanism of trypsin inhibitors. Since the production of trypsin and chymotrypsin is regulated by a negative flowback system from the pancreas, they showed that the secretion of pancreatic juices is stimulated by trypsin inhibitors which are able to inactivate trypsin efficiently and chymotrypsin cannot, therefore, be activated. By means of the negative flowback system, the pancreas is stimulated to produce trypsin and chymotrypsin, thus completing the cycle.

To overcome the detrimental effect of the trypsin inhibitors present in raw soybean meal and other plant proteins, evidence from the literature indicates that heat is necessary to destroy the heat labile trypsin inhibitors.

Adverse Effects of Overheating

Although heat treatment has undoubtedly a beneficial effect on soybean utilization,

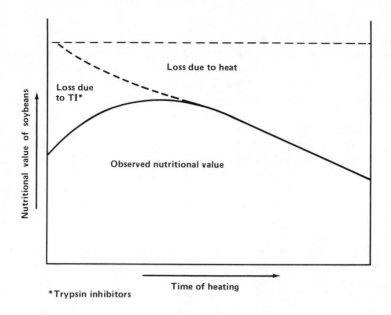

FIGURE 3.   The influence of time of heating on the nutritional value of soybeans. (Modified from Borchers, R., *Fed. Proc.*, 24, 1494—1497, 1965. With permission.)

excessive heating during commercial processing may have a harmful effect on the nutritive value.[6,12,16,18] This effect apparently involves changes in the lysine and methionine content of the protein since both of these amino acids are required to restore the nutritive value of soybean protein.[18,19,89,90] Destruction of a number of amino acids occurs with severe heat treatment, particularly lysine, arginine, tryptophan, and cystine.[12,40,88,89,91,92]

Digestibility of overheated soybean meal is also depressed[12,43,93] and is revealed by a slower release of amino acids from the protein.[40,94] Maximum liberation of lysine and arginine from soybean meal occurred at 15 to 30 min at 1.345 kg/cm² steam pressure and after 60 min at 0.358 kg/cm² pressure. Longer heating times resulted in a decreased liberation.[94] Soybean meal heated at 110°C for 30 min showed a greater release of amino acids than raw soybean meal but heated for 4 hr at 121°C reduced amino acid availability.[95] The lower nutritive value of overheated soybean meal is accentuated by the destruction of a third of the cystine, which normally saves methionine to a certain extent[91] and the destruction of nearly half of the lysine.[92] The influence of heat on the nutritional value of soybean meal is summarized in Figure 3.[96]

The above review focused attention mainly on soybean protein but other plant proteins are affected to the same general extent. Information on the different plant proteins can be found in the various chapters of textbooks mentioned earlier.[3]

Animal Proteins

Animal proteins are of special interest in nutrition because of their high biological values and their effectiveness in supplementing proteins of lower nutritional value. From the literature it is evident that despite their importance, animal proteins have received less attention than plant proteins. In this section only a brief account will be given of the effect of heat on animal proteins used in human nutrition. Attention will be drawn mainly to the influence of heat in animal nutrition because in most cases animals were used to measure changes in the quality of proteins.

Early investigators like Jarussowa[97] and Scheunert and Bischoff[98] concluded from their experiments that boiled or autoclaved meat diets gave growth and reproduction equal to that obtained with fresh meats. These results were later confirmed by Scheunert and Venus[99] using lower levels of protein in the diet. A number of investigations since that time supported the view that moderate moist heat treatment, such as might be encountered in home cooking (even roasting) and commercial processes, does not significantly lower the nutritional value of meat proteins.[100-104] When autoclaving time was increased from 7 min to 1 hr, less gain per gram of protein consumed was shown by rats.[105] Morgan and Kern[105] concluded that length of exposure and increased temperatures will damage animal proteins but Rice and Beuk[106] suggested that a certain critical temperature of probably 100°C must be exceeded before biologically detectable heat damage occurs. Thus, Seegers and Mattill[107,108] found that the proteins of beef liver had only a slightly lower biological value when heated for 2 weeks at 100°C, whereas heating at 120°C for 72 hr reduced the biological value to less than half that of the unheated product.

Fish meals are also affected in a similar way to meat when severe heat treatments are used, such as flame-drying vs. steam-drying or vacuum drying.[109] Biological values of 70 (menhaden fish meal), and 78 (haddock), for flame-dried and 76 and 85, respectively, for vacuum-dried fish meals were found. Similar results were found by Clandinin,[110] with chick feed vacuum-dried and stack-dried at 85°C and 104°C. Growth increases over a 3½ week period were 227, 287, and 144 g, respectively. Heating herring meal at 100°C for 60 min did not affect its nutritional value but increasing the time to 180 min did.

Doubt was expressed if severe heating has any practical significance because of a relatively high protein intake in human nutrition.[106] However, at the critical levels at which fish meal is used in animal nutrition (pigs and poultry) due to the price factor, difference in quality between meals may become important. Thus, Miller[111] demonstrated the differences which occurred between various fish meals, presented in Table 5.

When fish meal was dried under controlled laboratory conditions (acetone-dried, extracted protein and dried at maximum 50°C) a considerable increase in NPU values was found (± 80). Increase in drying temperatures from 50 to 120°C leads to a lowering in the NPU value from 83.0 to 63.0. Increase of water content from 0 to 9% and 17% H_2O reduced the NPU values from 65.0 (24 hr at 0%) to 43.0 (9%) and 40.0 (17%), respectively. This reduction according to Miller[111] was caused by the Maillard reaction.

A difference of 10% in the gross protein value between commercial fish meal and fish meal dried at 55°C was also found.[112] When a fillet with 2 to 3% moisture was heated under nitrogen for 31 hr at 100°C the feed value was not significantly affected. When the moisture content was increased to 11% and heated at 105°C for 36 hr, the gross protein value (GPV) was reduced by approximately 28%. Supplementing this meal with 0.021% L-lysine, increased the GPV significantly. According to Carpenter et al.[112] the reduction in GPV cannot be explained solely in terms of the Maillard reaction.

Another problem with fish meals is the oxidation of fats and oils that takes place during storage. This leads to a reduction in available lysine of 9 to 12% after 12 months. Addition of antioxidants (e.g., butylated hydroxy-toluene) not only reduced the oxidation during storage (20 to 25°C) considerably, but also increased the available lysine.[113] When herring meal was heat-damaged during storage, a reduction of 60% of the available lysine took place as well as a significant reduction in GPV.[114] A slight effect of oxidation on the available lysine was found (4% reduction) when herring meal was stored at fairly low temperatures (10°C), but when heated subsequently (30 hr at 100°C) under nitrogen, available nitrogen decreased a further 12% although fresh

Table 5
THE PERCENTAGE OF NET
PROTEIN UTILIZATION
(NPU) OF DIFFERENT FISH
MEALS

Origin	Type	NPU
United Kingdom	W.F.	51.0
	W.F.	57.3
	W.F.	58.2
	W.F.	61.4
South Africa	Sardine meal	33.0
	W.F.	68.3

Note: WF = white fish meal.

From Miller, D. S., *J. Sci. Food Agric.*, 7, 337, 1956. With permission.

herring meal under similar conditions did not show any significant loss. Increase of the temperature to 130°C, caused significant losses in available lysine.[114] Lea et al.[114] came, therefore, to the conclusion that two reactions with lysine take place, viz. an oxidative fat dependent one under more drastic heat conditions and longer times, and the well-known Maillard reaction. The severity of the latter type is increased by the presence of moisture[111,112] lowered in the case of the first.[113,115]

The sulfur amino acids are also damaged by heat treatment and can be used as an indication of the nutritional quality of the protein. Availability of methionine and methionine and cystine in tuna fish meal heated for 3 hr at 202°C decreased by 41 and 33%, respectively, although these severe conditions are not found under commercial circumstances.[116] Similar results were found by several investigators.[117-121] The lower nutritional value of commercial fish meals was ascribed to the lower digestibility of lysine and methionine[117] but Miller et al.[118] were of the opinion that part of the methionine in heat-damaged fish meal becomes biologically inaccessible due to a decrease in the biological value.[122] This unused methionine was excreted in the urine while inaccessible lysine due to heat treatment was mainly in the feces. Bjarnason and Carpenter[123] found an increase of 52% in fecal lysine with only 3% increase in urine lysine when heated bovine plasma (27 hr at 145°C) was fed to rats. Ford and Shorrock[120] found that peptide-bound amino acids excreted in the urine increased from 18.6 to 48.8 μ mol per rat per day when fed cod meal heated for 20 hr at 135°C. There was also a change in the amino acid content of these peptides. The most obvious effect was an increase in lysine from 2.98 to 20.30 μ mol per rat per day. Free amino acids also increased from 53.7 μ mol to 114.4 μ mol. They pointed out, however, these quantities actually formed a very small part of total nitrogen absorbed; 0.2% of the total of 0.6 g nitrogen absorbed per rat per day was excreted in the urine. A summary of the percentage losses of essential and semiessential amino acids due to heat treatment under various conditions is given in Table 6.

Three comments can be made on the results in Table 6:

1.   Cystine is damaged at lower temperatures than lysine
2.   Spontaneous heating during storage and flame drying cause greater losses of lysine than fairly severe heating
3.   Presence of sugars cause considerable destruction of lysine at lower temperatures[124]

Table 6

THE PERCENTAGE LOSS OF ESSENTIAL AND SEMIESSENTIAL AMINO ACIDS IN HEAT TREATED PROTEINS

Material	Treatment	Lysine	Cystine	Methionine	Arginine	Histidine	Threonine	Valine	Isoleucine	Leucine	Phenylalanine	Tyrosine	Tryptophan	Ref.
Cod protein	27 hr 116°C	6	61	0	0	0	0	0	0	0	0	0	0	124
Fish meal	36 hr 105°C	11	64	0	0	12	0	0	0	0	0	6	—	124
Herring meal	12 hr 121°C	12	—	0	2	7	8	3	0	0	0	0	—	126
	Flame drying (105°C) vs. vacuum drying	25	—	0	9	5	11	4	1	2	8	—	0	110
	Spontaneous heating during storage	30	37	13	1	20	14	12	11	11	9	15	27	127
Bovine plasma albumin	27 hr 115°C	4	50	0	0	0	3	0	0	0	0	9	—	123
	27 hr 145°C	15	92	0	11	17	32	0	12	0	0	0	—	123
Pig protein	24 hr 110°C	20	44	16	2	8	13	11	3	20	10	11	5	122
Cod meal + glucose, 10%	27 hr 85°C	31	33	9	18	0	0	0	0	0	0	0	0	124

Chemical and Physical Changes

Only a brief account will be given here of the chemical and physical changes induced by heat in proteins. The reader is referred to textbooks on the subject.[3,128]

One must distinguish between *inactivation* or *destruction* of an amino acid. Thus, Evans and Butts[92] did not observe any modifications in the amino acid content determined after acid hydrolysis, but a reduction in the enzymatic release of lysine, histidine, and cystine. Beuk et al.[129] found that only cystine is partially destroyed, whereas most amino acids are not due only to the effect of heat but also to the presence of carbohydrates (glucose) as indicated in the work of Patton.[130] Refluxing soybean globulin in 5% glucose destroyed lysine, arginine, and tryptophan[130] the same amino acids which are destroyed by overheating soybean meal.[40] Evans and associates[91,92,131-134] showed that none of the amino acids was significantly destroyed by autoclaving (1.345 kg/cm², 4 hr) soybean protein alone; but when mixed with sucrose prior to autoclaving, over 40% of lysine and arginine was destroyed. They ascribed this to the free amino groups because 45% of each free amino acid added to the protein, except cystine, was destroyed. The amount of all amino acids released by enzymatic digestion (inactivated), was reduced, some to a greater extent than others. For example, 30% lysine was inactivated without sucrose and 84% with sucrose; arginine 8 and 55%; cystine 14 and 86%; and methionine 6 and 44%.

From these results Evans pointed out that the destruction of amino acids involves a reaction of protein-bound amino acids and sucrose to give a complex which is either resistant to acid hydrolysis, or, which is more likely, cannot be utilized by the animal. Liener[3] suggested that this sucrose was inverted to an acyclic aldose isomer in equilibrium with cyclic forms of glucose and fructose. Functional side chain nitrogen characterize the amino acids which are destroyed in this manner, namely lysine, arginine, histidine, and tryptophan. The simultaneous inactivation of several amino acids suggests that new enzyme-resistant cross-links are formed between the side-chains of the same protein molecule, so that part of the latter becomes biologically unavailable. A series of possible cross-links are shown by Mauron[136] although it is uncertain whether cross-links are formed upon heating. Bjarnason and Carpenter[135] demonstrated a cross-link between the ε-amino group of lysine and the γ-amide group of glutamin upon heating with the liberation of ammonia. However, it was shown by Mauron[136] and later by Waibel and Carpenter[137] that ε-(γ-glutamyl)-L-lysine was just as effective as lysine itself as a lysine source in the rat. The possibility still remains that the ε-γ-cross-links in a protein render the α-α-peptide links of the main chain inaccessible to proteolytic enzymes which lead to biological inactivation of the amino acids in that part of the protein. Support for this possibility is found in the results of Ford and Shorrock[120] who showed that lysine, aspartin, and glutamin form 70% of the peptides excreted in the urine of rats fed heated cod fillets (135°C, 20 hr).

As shown earlier, cystine is the only amino acid which is always partly destroyed when protein is heated in the presence of moisture and is accompanied by H_2S liberation.[135] Another possibility is that the sulfur atom of cystine is liberated by β-elimination with the formation of dehydroalanine in the peptide chain. The dehydroalanine can easily be hydrolyzed to a pyruvic acid derivative and an amide group.

The biological availability of methionine may be affected by heat processing although little information with regard to the nature of the modification exists. Methionine sulfoxide is a possible form which may be biologically unavailable as Miller and Samuel[138] found that its potency as a methionine source is only of the order of 25%. Ellinger and Palmer[139] found that free methionine sulfoxide is used by the rat but protein-bound sulfoxide is used to a lesser degree. No other oxidative reaction is known leading with certainty to the inactivation of amino acids by heating proteins.[128]

THE MAILLARD REACTION

It has been known for a long time that proteins and carbohydrates react under a variety of conditions leading to the well-known browning of foodstuffs which occurs during heating or storage. Maillard in 1912 disclosed a reaction between amino acids and reducing sugars which was accompanied by a darkening of the reaction mixture. This reaction is popularly referred to as the Maillard Reaction. It was well-established that oxidation does not form part of the process, hydrogen and oxygen are lost to form water and CO_2 is split off. In spite of all the data available today the Maillard reaction is not yet fully understood.

The first reaction scheme put forward was that by Hodge[140] and given below.

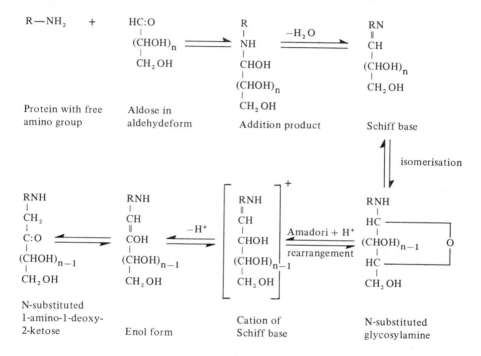

The 1-amino-1-deoxy-2-ketose is a necessary step in the browning reaction but it was observed that during the Amadori rearrangement a stage is gone through that is extremely prone to browning.[141] Mauron[128] is of the opinion that the enol form of the aminodeoxyketose may well be the required intermediate. Taking account of some of the research since 1953 Mauron[128] proposed an extension of Hodges' scheme through further breakdown of the aminodeoxyketose by dehydration and scission of the enol form and then by Strecker degradation to → aldehyde + amino compound to → aldols + aldimines + ketimines and to → melanoidins.[140] The reaction between certain proteins or model polypeptides and glucose can occur without visible browning under dry conditions. In this case the in vitro digestibility was considerably reduced compared to a system in which the reactants were dissolved in water.[142,143]

It must be remembered that the inactivation of amino acids is not only dependent on the presence of sugars, but also on the formation of linkages between the free carboxylic groups of glutamic or aspartic acids and the free amino groups of lysine and arginine or the imidazole group of histidine.

The nutritional consequence of the Maillard reaction is firstly the binding of lysine as an essential amino acid. It was shown by Mauron[136] that deoxyfructose is not degraded in the gut but absorbed by the intestine, passed into the bloodstream, and finally excreted without having been utilized. Lysine is, therefore, biologically unavailable although acid hydrolysis recuperates 50% of the lysine from deoxyfructose. Secondly, blocking of ε-amino group of lysine inhibits trypsin action. Thirdly, the Maillard reaction may produce biologically active substances which can enhance or reduce growth and influence palatability.[128]

CONCLUSIONS

Heat is likely to cause changes in the nutritive value of proteins. These changes or effects can be either beneficial or detrimental. Under moderate conditions of heating, the nutritive value of most plant proteins is increased. This is particularly so with soybean meal, which is widely used in animal and human nutrition. This is ascribed to the destruction of heat-labile antinutritional factors such as trypsin inhibitor, hemagglutinin, saponin, etc., in soybean oil meal, and gossypol in cottonseed meal. In ruminant nutrition these factors do not play a role and the beneficial effect of heating is ascribed to a reduction in the solubility of the protein, thereby causing a greater resistance to degradation followed by deamination of the protein through the action of microorganisms in the rumen.

Overheating in the processing of protein sources causes profound changes in the protein molecule itself. These lead to a retardation of enzymatic digestion, especially trypsin, of the protein as a whole but in particular to certain parts in the polypeptide chain. The effect is a consequent reduction in the amount of essential amino acids released for absorption into the bloodstream. Lysine is the essential amino acid involved in the formation of peptide linkages by two possible ways: the interaction of side-chain amino groups of lysine or arginine, and the interaction of these amino groups with reducing sugars. Under severe heat treatment this last reaction can proceed to a point where lysine can no longer be completely recovered, even with acid hydrolysis.

The reduced digestibility of protein fractions induced by heat leads to their excretion, thus depriving the animal of essential amino acids, for example, methionine becomes deficient in an otherwise sufficient protein source with lysine reaching critical levels of availability.

One should, therefore, be aware of the conditions under which a protein source must be processed so that optimum use can be made of it in nutrition. We cannot allow the nutritive quality of our protein sources to be impaired as this may lead to loss in protein quantity in a protein hungry world.

REFERENCES

1. Osborne, T. B. and Mendel, L. B., The use of soybeans as food, *J. Biol. Chem.*, 52, 369, 1917.
2. Osborne, T. B. and Mendel, L. B., The use of cottonseed as food, *J. Biol. Chem.*, 29, 289, 1917.
3. Liener, I. E., Effect of heat on plant proteins, in *Processed Plant Protein Foodstuffs*, Altschal, A. M., Ed., Academic Press, New York, 1958, Chap. 5.
4. Shrewsbury, C. L., Vestal, C. M., and Hauge, S. M., The effect of yeast and casein supplements to corn and soybean rations when fed to rats and swine, *J. Agr. Res.*, 44, 267, 1932.
5. Mitchell, H. H. and Smuts, D. B., The amino acid deficiencies of beef, wheat, corn, oats and soybeans for growth in the white rat, *J. Biol. Chem.*, 95, 263, 1932.
6. Hayward, J. W., Steenbock, H., and Bohstedt, G., The effect of heat as used in the extraction of soybean oil upon the nutritive value of the protein of soybean oil meal, *J. Nutr.*, 11, 219, 1936.

7. Hayward, J. W. and Hafner, F. H., The supplementary effect of cystine and methionine upon the protein of raw and cooked soybeans as determined with chicks and rats, *Poult. Sci.,* 20, 139, 1941.

8. Mitchell, H. H., Hamilton, T. S., Beadles, J. R., and Simpson, F., The importance of commercial processing for the protein value of food products, *J. Nutr.,* 29, 13, 1945.

9. Melnick, D., Oser, B. L., and Weiss, S., The rate of enzyme digestion of protein as a factor in nutrition, *Science,* 103, 326, 1946.

10. Hayward, J. W., Halpin, J. G., Holmes, C. E., Bohstedt, G., and Hart, E. B., Soybean oil meal prepared at different temperatures as a feed for poultry, *Poult. Sci.,* 16, 3, 1937.

11. Mattingly, J. P. and Bird, H. R., Effect of heating, under various conditions, and of sprouting on the nutritive value of soybean oil meals and soybeans, *Poult. Sci.,* 24, 344, 1945.

12. Evans, R. J. and McGinnis, J., The influence of autoclaving soybean oil meal on the availability of cystine and methionine for the chick, *J. Nutr.,* 31, 449, 1946.

13. Vestal, C. M. and Shrewsbury, C. L., The nutritive value of soybeans with preliminary observations on the quality of pork produced, *Am. Soc. Anim. Prod. Rec. Proc. Ann. Meet.,* 127, 1932.

14. Becker, D. E., Adams, C. R., Terrill, S. W., and Meade, R. J., The influence of heat treatment and solvent upon the nutritive value of soybean oil meal for swine, *J. Anim. Sci.,* 12, 107, 1953.

15. Johnson, L. M., Parsons, H. T., and Steenbock, H., The effect of heat and solvents on the nutritive value of soybean protein, *J. Nutr.,* 18, 423, 1939.

16. Parsons, H. T., Effect of different cooking methods on soybean proteins, *J. Home Econ.,* 35, 211, 1943.

17. Bird, H. R. and Burkhardt, G. J., Factors affecting the nutritive value of soybean oil meals and soybeans for chickens, *Md., Agr. Exp. Stn., Ann. Rep.,* Bull. No. A27, 35, 1943.

18. Clandinin, D. R., Cravens, W. W., Elvehjem, C. A., and Halpin, J. G., Deficiencies in over-heated soybean oil meal, *Poult. Sci.,* 26, 150, 1947.

19. Klose, A. A., Hill, B., and Ferold, H. L., Food value of soybean protein as related to processing, *Food Technol. (Chicago),* 2, 201, 1948.

20. McDonald, I. W., Physiology of digestion, absorption, and metabolism in the ruminant, in *Nutrition of Farm Animals of Agricultural Importance,* Vol. 17, Cuthbertson, Sir, D., Ed., Pergamon Press, London, 1969, 87.

21. Miller, J. J. and Morrison, F. B., Effect of heat treatment and oil extraction on the utilization and digestibility of soybean protein in lambs, *J. Agri. Res.,* 68, 35, 1944.

22. Chalmers, M. I., Cuthbertson, D. P., and Synge, R. L. M., Ruminal ammonia formation in relation to the protein requirements of sheep. I. Duodenal administration and heat processing as factors influencing fate of casein supplements, *J. Agri. Sci.,* 44, 254, 1954.

23. Chalmers, M. I., Jayasinghe, T. B., and Marshall, B. M., The effect of heat treatment in the processing of groundnut meal on the value of the protein for ruminants with some additional experiments on copra, *J. Agri. Sci.,* 63, 283, 1964.

24. Tillman, A. D. and Kruse, K., Effect of gossypol and heat on the digestibility and utilization of soybean protein by sheep, *J. Anim. Sci.,* 21, 290, 1962.

25. Sherrod, L. B. and Tillman, A. D., Effects of varying the processing temperature upon the nutritive values for sheep of solvent-extracted soybean and cottonseed meals, *J. Anim. Sci.,* 21, 901, 1962.

26. Sherrod, L. B. and Tillman, A. D., Further studies on the effects of different processing temperatures on the utilization of solvent-extracted cottonseed protein by sheep, *J. Anim. Sci.,* 23, 510, 1964.

27. Danke, R. J., Sherrod, L. B., Nelson, E. C., and Tillman, A. D., Effects of autoclaving and steaming of cottonseed meal for different lengths of time on nitrogen solubility and retention in sheep, *J. Anim. Sci.,* 25, 181, 1966.

28. Tagari, H., Ascarelli, I., and Bondi, I., The influence of heating on the nutritive value of soybean protein by sheep, *Br. J. Nutr.,* 16, 237, 1962.

29. Glimp, H. A., Karr, M. R., Little, C. O., Woolfolk, P. G., Mitchell, G. E., Jr., and Hudson, L. W., Effect of reducing soybean protein solubility by dry heat on the protein utilization of young lambs, *J. Anim. Sci.,* 26, 858, 1967.

30. Dysli, R. R., Ammerman, C. B., Loggins, P. E., Moore, J. E., and Arrington, L. R., Effects of steam-heating upon the nutritive value of full-fat soybeans for sheep and rats, *J. Anim. Sci.,* 26, 618, 1967.

31. Hudson, L. W., Glimp, H. A., Little, C. O., and Woolfolk, P. G., Ruminal and postruminal nitrogen utilization by lambs fed heated soybean meal, *J. Anim. Sci.,* 30, 609, 1970.

32. Whitelaw, F. G., Preston, T. R., and Dawson, G. S., The nutrition of the early weaned calf. II. Comparison of a commercial groundnut meal, heat-treated groundnut meal and fishmeal as the major protein source in the diet, *Anim. Prod.,* 3, 127, 1961.

33. Schoeman, E. A., de Wet, P. J., and Burger, W. J., The evaluation of the digestibility of treated proteins, *Agroanimalia,* 4, 35, 1972.

34. Shrewsbury, C. L. and Bratzler, J. W., Cystine deficiency of soybean protein at various levels, in a purified ration and as a supplement to corn, *J. Agric. Res.,* 47, 889, 1933.

35. Almquist, H. J., Mecchi, E., Kratzer, F. H., and Grau, C. R., Soybean protein as a source of amino acids for chicks, *J. Nutr.*, 24, 385, 1942.
36. Fisher, H. and Johnson, D., Jr., The effectiveness of essential amino acid supplementation in over-coming the growth depression of unheated soybean meal, *Arch. Biochem.*, 77, 124, 1958.
37. Borchers, R., Counteraction of the growth depression of raw soybean oil meal by amino acid supplements in weanling rats, *J. Nutr.*, 75, 330, 1961.
38. Jimenez, A. A., Perry, T. W., Pickett, R. A., and Beeson, W. M., Raw and heated soybeans for growing-finishing swine and their effect on fat firmness, *J. Anim. Sci.*, 22, 471, 1963.
39. Fisher, H. and Shapiro, R., Counteracting the growth retardation of raw soybean meal with extra protein and calories, *J. Nutr.*, 80, 425, 1963.
40. Riesen, W. H., Clandinin, D. R., Elvehjem, C. A., and Cravens, W. W., Liberation of amino acids from raw, properly heated, and overheated soybean oil meal, *J. Biol. Chem.*, 167, 143, 1947.
41. Ingram, G. R., Riesen, W. H., Cravens, W. W., and Elvehjem, C. A., Evaluating soybean oil meal protein for chick growth by enzymatic release of amino acids, *J. Poult. Sci.*, 28, 898, 1949.
42. Carroll, R. W., Hensley, G. M., and Graham, W. R., The site of nitrogen absorption in rats fed raw and heated soybean meals, *Science*, 115, 36, 1952.
43. Evans, R. J., McGinnis, J., and St. John, J. L., The influence of autoclaving soybean oil meal on the digestibility of the proteins, *J. Nutr.*, 33, 661, 1947.
44. Bouthilet, R. J., Hunter, W. L., Luhman, C. A., Ambrose, D., and Lepkovsky, S., The metabolism of raw soybeans in birds with colostomies, *Poult. Sci.*, 29, 837, 1950.
45. Alumot, E. and Nitzan, Z., The influence of soybean antitrypsin on the intestinal proteolysis of the chick, *J. Nutr.*, 73, 61, 1961.
46. Saxena, H. C., Jensen, L. S., and McGinnis, J., Influence of age on utilization of raw soybean meal by chickens, *J. Nutr.*, 80, 391, 1963.
47. Nitzan, Z., The effect of heating soybean meal on the apparent digestibility and metabolism of protein, methionine, and lysine by cockerels, *Poult. Sci.*, 1036, 1965.
48. Nesheim, M. C. and Garlich, J. D., Digestibility of unheated soybean meal for laying hens, *J. Nutr.*, 88, 187, 1966.
49. Combs, G. E., Conness, R. G., Berry, T. H., and Wallace, H. D., Effect of raw and heated soybeans on grain, nutrient digestibility, plasma amino acids and other blood constituents of growing swine, *J. Anim. Sci.*, 26, 1067, 1967.
50. Getler, A., Birk, Y., and Bondi, A., A comparative study of the nutritional and physiological significance of pure soybean trypsin inhibitors and of Ethanol-extracted soybean meals in chicks and rats, *J. Nutr.*, 91, 388, 1967.
51. Kneen, E. and Sandstedt, R. M., An amylase inhibitor from cereals, *J. Am. Chem. Soc.*, 65, 1247, 1943.
52. Bowman, D. E., The ether soluble fraction of navy beans and the digestion of starch, *Science*, 98, 308, 1943.
53. Ham, W. E. and Sandstedt, R. M., A proteolytic inhibiting substance in the extract from unheated soybean meal, *J. Biol. Chem.*, 154, 505, 1944.
54. Ham, E., Sandstedt, R. M., and Mussehl, F. E., The proteolytic inhibiting substance in the extract from unheated soybean meal and its effect upon the growth in chicks, *J. Biol. Chem.*, 161, 635, 1945.
55. Kunitz, M., Crystallization of a trypsin inhibitor from soybeans, *Science*, 101, 668, 1945.
56. Kunitz, M., Crystalline soybean trypsin inhibitor, *J. Gen. Phys.*, 29, 149, 1946.
57. Kunitz, M., Crystalline soybean trypsin inhibitor. II. General properties, *J. Gen. Physiol.*, 30, 291, 1947.
58. Bowman, D. E., Differentiation of soybean anti tryptic factors, *Proc. Soc. Exp. Biol. Med.*, 63, 547, 1946.
59. Borchers, R., Ackerson, C. W., and Sandstedt, R. M., Trypsin inhibitor. III. Determination and heat destruction of the trypsin inhibitor of soybean, *Arch. Biochem.*, 12, 367, 1947.
60. Bowman, D. E., Further differentiation of bean trypsin inhibiting factors, *Arch. Biochem.*, 16, 109, 1948.
61. Bowman, D. E., Fractions derived from soybeans and navy beans which retard tryptic digestion of casein, *Proc. Soc. Exp. Biol. Med.*, 57, 139, 1944.
62. Borchers, R., Ackerson, C. W., Mussehl, F. E., and Moehl, A., Growth inhibiting properties of soybean trypsin inhibitor, *Arch. Biochem. Biophys.*, 19, 317, 1948.
63. Klose, A. A., Hill, B., and Fevold, H. L., Presence of a growth inhibiting substance in raw soybean, *Proc. Soc. Exp. Biol. Med.*, 62, 10, 1946.
64. Liener, I. E., Deuel, H. J., Jr., and Fevold, H. L., The effect of supplemental methionine on the nutritive value of diets containing concentrates of the soybean trypsin inhibitor, *J. Nutr.*, 39, 325, 1949.

65. Almquist, H. J. and Merrit, J. B., Effect of soybean anti trypsin on growth of the chick, *Arch. Biochem. Biophys.*, 35, 352, 1952.

66. Almquist, H. J. and Merrit, J. B., Effect of crystalline trypsin on the raw soybean growth inhibitor, *Proc. Soc. Exp. Biol. Med.*, 83, 269, 1953.

67. Hou, H. C., Riesen, W. H., and Elvehjem, C. A., Influence of heating on the liberation of certain amino acids from whole soybeans, *Proc. Soc. Exp. Biol. Med.*, 70, 416, 1949.

68. Liener, I. E. and Fevold, H. L., The effect of the soybean trypsin inhibitor on the enzymatic release of amino acids from autoclaved soybean meal, *Arch. Biochem.*, 21, 395, 1949.

69. Chernick, S. S., Lepkovsky, S., and Chaikoff, I. L., A dietary factor regulating the enzyme content of the pancreas. Changes induced in the size and proteolytic activity of the chick pancreas by the ingestion of raw soybean meal, *Am. J. Phys.*, 155, 33, 1948.

70. Lyman, R. L. and Lepkovsky, S., The effect of raw soybean meal and trypsin inhibitor diets on pancreatic enzyme secretion in the rat, *J. Nutr.*, 62, 269, 1957.

71. Lepkovsky, S., Furuta, F., Koike, T., Hasegaura, N., Dimick, M. K., Krause, K., and Barnes, F. J., The effect of raw soybeans upon the function of the pancreas of intact chickens and of chickens with ileostomies, *Br. J. Nutr.*, 19, 41, 1965.

72. Getler, A., Birk, Y., and Bondi, I., A comparative study of the nutritional and physiological significance of pure soybean trypsin inhibitors and of Ethanol-extracted soybean meals in chicks and rats, *J. Nutr.*, 91, 388, 1967.

73. Haines, P. C. and Lyman, R. L., Relationship of pancreatic enzymes secretion of growth inhibition in rats fed soybean trypsin inhibitor, *J. Nutr.*, 74, 445, 1961.

74. De Meulenaere, H. J. H., Studies on the digestion of soybeans, *J. Nutr.*, 82, 197, 1964.

75. Beilorai, R. and Bondi, A., Relationship between "antitryptic factors" and some plant feeds and products of proteolysis precipitable by trichloro acetic acid, *J. Sci. Food Agric.*, 14, 124, 1963.

76. Beilorai, R., Comparative digestibility of groundnut and soybean meal in vitro and in chicks, *J. Sci. Food Agric.*, 20, 345, 1969.

77. Laufer, S., Tauber, H., and Davis, C. F., The amylolytic and proteolytic activity of soybean seed, *Cereal Chem.*, 21, 267, 1944.

78. Liener, I. E., The intraperitoneal toxicity of concentrates of the soybean trypsin inhibitor, *J. Biol. Chem.*, 193, 183, 1951.

79. Liener, I. E. and Pallansch, M. J., Purification of a toxic substance from defatted soybean flour, *J. Biol. Chem.*, 197, 29, 1952.

80. Pallansch, M. J. and Liener, I. E., Soyin, a toxic protein from the soybean, *Arch. Biochem. Biophys.*, 45, 366, 1953.

81. Almquist, H. J. and Merrit, J. B., Effect of soybean anti trypsin on experimental amino acid deficiency in the chick, *Arch. Biochem. Biophys.*, 31, 450, 1951.

82. Liener, I. E., Soyin, a toxic protein from soybean. I. Inhibition of rat growth, *J. Nutr.*, 49, 527, 1953.

83. Beilorai, R., Harduf, Z., and Alumot, E., The free amino acid pattern of the intestinal contents of chicks fed raw and heated soybean meal, *J. Nutr.*, 102, 1377, 1972.

84. Beilorai, R., Tamir, M., Alumot, E., Bar, A., and Hurwitz, S., Digestion and absorption of protein along the intestinal tract of chicks fed raw and heated soybean meal, *J. Nutr.*, 103, 1291, 1973.

85. Tamir, M., Beilorai, R., and Alumot, E., Labelled casein as an indicator of inhibited protein digestion in chicks fed raw soybean meal, *J. Nutr.*, 104, 648, 1974.

86. Kakade, M. L., Hoffa, D. E., and Liener, I. E., Contribution of trypsin inhibitors to the deleterious effects of unheated soybeans fed to rats, *J. Nutr.*, 103, 1772, 1973.

87. Schingoethe, D. J., Tidemann, L. J., and Uckert, J. R., Studies in mice on the isolation and characterization of growth inhibitors from soybean, *J. Nutr.*, 104, 1304, 1974.

88. Lyman, R. L., Olds, B. A., and Green, G. M., Chymotrypsinogen in the intestine of rats fed soybean trypsin inhibitor and its inability to suppress pancreatic enzyme secretions, *J. Nutr.*, 104, 105, 1974.

89. Fritz, J. C., Kramke, E. H., and Reed, C. A., Effect of heat treatment on the biological value of soybeans, *Poult. Sci.*, 26, 657, 1947.

90. McGinnis, J. and Evans, R. J., Amino acid deficiencies of raw and overheated soybean oil meal for chicks, *J. Nutr.*, 34, 725, 1947.

91. Evans, R. J., Groschke, A. C., and Butts, H. A., Studies on the heat inactivation of cystine in soybean oil meal, *Arch. Biochem.*, 30, 414, 1951.

92. Evans, R. J. and Butts, H. A., Studies on the heat inactivation of lysine in soybean oil meal, *J. Biol. Chem.*, 175, 15, 1948.

93. Evans, R. J. and McGinnis, J., Cystine and methionine metabolism by chicks receiving raw or autoclaved soybean oil meal, *J. Nutr.*, 35, 477, 1948.

94. Clandinin, D. R. and Robblee, A. R., The effect of processing on the enzymatic liberation of lysine and arginine from soybean oil meal, *J. Nutr.*, 46, 525, 1952.

95. Smith, R. F. and Scott, H. M., Use of amino acid concentrations in blood plasma in evaluating the amino acid adequacy of intact proteins for chick growth. II. Free amino acid patterns of blood plasma of chicks fed sesame and raw, heated, and overheated soybean meals, *J. Nutr.*, 86, 45, 1965.

96. Borchers, R., Raw soybean growth inhibitors, *Fed. Proc.*, 24, 1494, 1965.

97. Jarussowa, N., Der Einfluss des Kochens auf den Nährwert der Nahrung, *Biochem. Z.*, 207, 395, 1929.

98. Scheunert, A. and Bischoff, H., Uber den Nährwert Reiner Fleischkost, Hergestellt aus Rohem, Gekochten und Autoklavierten Muskel Fleisch bei Ratten, *Biochem. Z.*, 219, 186, 1930.

99. Scheunert, A. and Venus, C., Uber den Nährwert des Muskel Fleisches für Wachstum und Fortpflanzung, *Biochem. Z.*, 252, 231, 1932.

100. Seegers, W. H., Schultz, H. W., and Mattill, H. A., The biological value of heattreated proteins, *J. Nutr. Abstracts*, 11, 5, 1936.

101. Swanson, P. P. and Nelson, P. M., Biological value of autoclaved pork muscles, *Iowa, Agric. Exp. Stn. Res. Bull.*, Part I, 167, 1938.

102. Mayfield, H. L. and Hedrick, M. T., The effect of canning, roasting, and corning on the biological value of the proteins of western beef, finished on either grass or grain, *J. Nutr.*, 37, 487, 1949.

103. Mitchell, H. H., The nutritional effects of heat on food proteins with particular reference to commercial processing and home cooking, *J. Nutr.*, 39, 413, 1949.

104. Wilder, O. H. M. and Krybill, H. R., Effect of cooking and curing on the lysine content of pork luncheon meal, *J. Nutr.*, 33, 235, 1947.

105. Morgan, A. F. and Kern, G. F., The effect of heat upon the biological value of meat protein, *J. Nutr.*, 7, 367, 1934.

106. Rice, E. E. and Beuk, J. F., Effects of heat upon the nutritive value of protein, *Adv. Food Res.*, 4, 233, 1953.

107. Seegers, W. H. and Mattill, H. A., The effect of heat and hot alcohol on liver proteins, *J. Biol. Chem.*, 110, 531, 1935.

108. Seegers, W. H. and Mattill, H. A., The nutritive value of animal tissues in growth, reproduction, and lactation. III. The nutritive value of beef heart, kidney, round, and liver after heating and after alcohol extraction, *J. Nutr.*, 10, 275, 1935.

109. Maynard, L. A. and Tunison, A. V., Influence of drying temperature upon digestibility and biological value of fish proteins, *Ind. Eng. Chem.*, 24, 1168, 1932.

110. Clandinin, D. R., The effects of methods of processing on the nutritive value of herring meals, *Poult. Sci.*, 28, 128, 1949.

111. Miller, D. S., The nutritive value of fish proteins, *J. Sci. Food Agric.*, 7, 337, 1956.

112. Carpenter, K. J., Ellinger, G. M., Munro, M. I., and Rolfe, E. G., Fish products as protein supplements to cereals, *Br. J. Nutr.*, 11, 162, 1957.

113. Lea, C. H., Parr, L. J., and Carpenter, K. J., Chemical and nutritional changes in stored herring meal, *Br. J. Nutr.*, 12, 297, 1958.

114. Lea, C. H., Parr, L. J., and Carpenter, K. J., Chemical and nutritional changes in stored herring meal. II. *Br. J. Nutr.*, 14, 91, 1960.

115. Carpenter, K. J., Morgan, C. B., Lea, C. H., and Parr, L. J., Chemical and nutritional changes in stored herring meal. III. Effect of heating as controlled moisture contents of the binding of amino acids in freeze-dried herring press cake and in related model systems, *Br. J. Nutr.*, 16, 451, 1962.

116. Ousterhout, L. E., Grau, C. R., and Lundholm, B. D., Biological availability of amino acids in fish meals and other protein sources, *J. Nutr.*, 69, 65, 1959.

117. Miller, E. L. and Carpenter, K. J., Availability of sulphur amino acids in protein foods. I. Totale sulphur amino acid content in relation to sulphur and nitrogen balance studies with the rat, *J. Sci. Food Agric.*, 15, 810, 1964.

118. Miller, E. L., Carpenter, K. J., and Milner, C. K., Availability of sulphur amino acids in protein foods. III. Chemical and nutritional changes in heated cod muscle, *Br. J. Nutr.*, 19, 547, 1965.

119. Ford, J. E. and Salter, D. N., Analysis of enzymically digested food proteins by sephadex-gel filtration, *Br. J. Nutr.*, 20, 843, 1966.

120. Ford, J. E. and Shorrock, C., Metabolism of heat-damaged proteins in the rat. Influence of heat damage on the excretion of amino acids and peptides in urine, *Br. J. Nutr.*, 26, 311, 1971.

121. Nesheim, M. C. and Carpenter, K. J., The digestion of heat damaged protein, *Br. J. Nutr.*, 21, 399, 1967.

122. Donoso, G., Lewis, O. A. M., Miller, D. S., and Payne, P. R., Effect of heat treatment on the nutritive value of proteins. Chemical and balance studies, *J. Sci. Food Agric.*, 13, 192, 1962.

123. Bjarnason, J. and Carpenter, K. J., Mechanisms of heat damage in proteins. I. Models with acelated lysine units, *Br. J. Nutr.*, 23, 859, 1969.

124. **Miller, E. L., Hartley, A. W., and Thomas, D. C.**, Availability of sulphur amino acids in protein foods. IV. Effect of heat treatment upon the total amino acid content of cod muscle, *Br. J. Nutr.,* 19, 565, 1965.

125. **Ellinger, G. M. and Boyne, F. B.**, Amino acid composition of some fish products and casein, *Br. J. Nutr.,* 19, 587, 1965.

126. **Smith, R. E. and Scott, H. M.**, Measurement of the amino acid content of fish meal proteins by chick growth assay. I. Estimation of amino acid availability in fish meal proteins before and after heat treatment, *Poult. Sci.,* 44, 401, 1965.

127. **Boge, G.**, Amino acid composition of herring and herring meal. Destruction of amino acids during processing, *J. Sci. Food Agric.,* 11, 362, 1960.

128. **Mauron, J.**, Influence of industrial and household handling on food protein quality, in *Protein and Amino Acid Functions,* Vol. 2, Bigwood, E. J., Ed., Pergamon Press, Oxford, 1972, Chap. 9.

129. **Beuk, J. F., Chornock, F. W., and Rice, E. E.**, The effect of severe heat treatment upon the amino acids of fresh and cured pork, *J. Biol. Chem.,* 175, 291, 1948.

130. **Patton, A. R., Hill, E. G., and Foreman, E. M.**, The effect of browning on the essential amino acid content of soy globulin, *Science,* 108, 659, 1948.

131. **Evans, R. J. and Butts, H. A.**, Inactivation of amino acids by autoclaving, *Science,* 109, 569, 1949.

132. **Evans, R. J. and Butts, H. A.**, Studies on the heat inactivation of methionine in soybean oil meal, *J. Biol. Chem.,* 178, 543, 1949.

133. **Evans, R. J. and Butts, H. A.**, Heat inactivation of the basic amino acids and tryptophan, *Food Res.,* 16, 415, 1951.

134. **Evans, R. J., Butts, H. A., and Bandemer, S. L.**, Heat inactivation of threonine glycine and the acidic amino acids, *Arch. Biochem. Biophys.,* 32, 300, 1951.

135. **Bjarnason, J. and Carpenter, K. J.**, Mechanisms of heat damage in proteins. II. Chemical changes in pure proteins, *Br. J. Nutr.,* 24, 313, 1970.

136. **Mauron, J.**, *Internat. Z. Vitaminforsch.,* 40, 209, 1970.

137. **Waibel, P. E. and Carpenter, K. J.**, Mechanisms of heat damage in proteins. III. Studies with ε-(γ-l)-glutamyl-L-lysine, *Br. J. Nutr.,* 27, 509, 1972.

138. **Miller, D. S. and Samuel, P.**, Methionine sparing compounds, *Proc. Nutr. Soc.,* 27, 21A, 1968.

139. **Ellinger, G. M. and Palmer, R.**, The biological availability of methionine sulphoxide, *Proc. Nutr. Soc.,* 28, 42A, 1969.

140. **Hodge, J. E.**, Dehydrated foods. Chemistry of browning reaction in model systems, *J. Agri. Food Chem.,* 1, 928, 1953.

141. **Heyns, K. and Paulsen, H.**, Veränderungen der Nahrung durch industrielle und haushaltmässige Verarbeitung, Wissenschaftliche Veröffentlichungen der Deutschen Gesellschaft für Ernährung, Vol. 5, D. Steinkopf, Darmstadt, Germany, 15, 1960.

142. **Schroeder, L. J., Iacobellis, M., and Smith, A. H.**, In vitro digestibility studies on model peptides heated with glucose, *J. Nutr.,* 55, 97, 1955.

143. **Schroeder, L. J., Iacobellis, M., and Smith, A. H.**, The influence of water and pH on the reaction between amino compounds and carbohydrates, *J. Biol. Chem.,* 212, 973, 1955.

EFFECT OF PROCESSING ON THE AVAILABILITY AND NUTRITIONAL VALUE OF VITAMINS IN ANIMAL FEEDS

Milton L. Scott

INTRODUCTION

Many factors, including those imposed by nature, may alter the vitamin content of cereal grains, oilseed meals, mill byproducts, forages, animal protein materials, and other feedstuffs which are used as basic ingredients in the production of formula feeds. Some processes to which feedstuffs are subjected destroy vitamins; others enhance the stability of some vitamins; still others improve the availability of the vitamins in certain feedstuffs.

To meet the vitamin requirements of simple-stomached animals, particularly poultry and swine, vitamin values are determined for each of the feed ingredients used in the basic diet. The contributions that the various ingredients make toward meeting the animal's total vitamin requirements are calculated; the basic formula then may be supplemented, to ensure its adequacy, either with special vitamin-rich ingredients or with a vitamin premix.

It is well recognized that the vitamin levels shown in tables of vitamin composition of feedstuffs are average values, and that the actual vitamin content of each feedstuff varies over a fairly wide range. It is important to the nutritionist, therefore, to understand all the factors that may contribute to variations in the vitamin content of each of the feedstuffs common in diets for poultry, swine, and other livestock. This information, together with a knowledge of all the factors that may alter the vitamin requirements of the animals, helps the nutritionist to determine the levels of vitamin fortification needed in each particular feed.

AGRONOMIC EFFECTS

Factors affecting the vitamin content of growing plants include soil fertility, plant nutrients added in the form of fertilizers, and growing conditions of the crops (amount and intensity of sunshine, weather temperature, rainfall or irrigation schedules, day length during the growing season, and total length of the growing season). Workers at the U.S. Plant, Soil, and Nutrition Laboratory at Cornell University have shown that the fertilization of boron-deficient alfalfa with boron results in a higher carotene content in the alfalfa. In forage crops, the factors that favor the production of lush green plants also favor the production of many vitamins, particularly β-carotene, vitamin E, and vitamin K. With tomatoes, the vitamin C content depends primarily upon the intensity of sunlight striking the fruits in the immediate preharvest period.

HARVESTING CONDITIONS

The vitamin content of feedstuffs often is affected by harvesting conditions. When the growing season ends before full maturity of corn is achieved, the vitamin content is greatly reduced and it retains a very high moisture content for a prolonged period before harvest. If high-moisture corn is subjected to alternate periods of freezing and thawing, the corn kernels ferment and the vitamin content, particularly of vitamin E and cryptoxanthin, is reduced.

Certain legumes such as alfalfa and soybeans contain an enzyme, lipoxidase, that unless quickly inactivated readily destroys much of their carotene and xanthophyll content. Mangelson et al.[36] showed that alfalfa mechanically dehydrated at 350°F within 1 hr of its cutting produces meal containing 2.5 times the carotene of the meal obtained from sun-ripened plants.

Blaylock et al.[10] showed that lipoxidase is completely destroyed by commercial alfalfa dehydration. No riboflavin, pantothenic acid, niacin, or folic acid is lost during mechanical dehydration. Field-cured alfalfa is lower in riboflavin, and when exposed to rain loses large amounts of pantothenic acid and niacin.

POSSIBLE EFFECTS OF FUNGI ON VITAMINS

Unfavorable harvest conditions in the northeastern U.S. in 1965 and 1972 brought about a large infestation of "field" fungi in the corn crops. Infected corn caused markedly reduced growth rates in pigs and increased the amount of feed required per pound of gain for broilers, turkeys, and ducks.[52] Combs[13] has separated the problems associated with molds and mycotoxins into three interrelated groupings:

1. Effects of toxic metabolites of the molds
2. Effects due to modification of nutrient composition of feed ingredients
3. Effects due to modification of nutrient utilization by animals

Aflatoxins produced in certain feedstuffs by growth of *Aspergillus flavus* also seriously interfered with growth, particularly in turkeys and ducks. Most of these were the effects of toxins produced by the fungi.

Southern leaf blight also damages the corn kernels and results in a decrease in nutrient value of the corn. Corn showing a high incidence of this *Helminthosporium* disease was found to contain less than 40% the amount of vitamin E present in normal corn.[58] Lower than normal levels of vitamin E in some samples of milo and soybean meal also have been attributed to fungal destruction of α-tocopherol. Although hemorrhagic disorders often are found accompanying feeding of these fungi-infested feedstuffs, there has been no direct evidence that the fungi themselves are directly responsible for vitamin K loss in the feedstuffs. Some isolates of *Fusarium moniliforme* produced a heat-labile thiaminase activity which caused polyneuritis in chicks fed the fungal-treated diets.[20] Thiamin appeared to be utilized by *Aspergillus flavus*[18] and by *A. oryzae*[5] during growth and production of aflatoxin B$_1$. Sherby et al.[50] found no effect on growth, feed to gain ratio or mortality in feeding corn infested separately with three fungal isolates, a *Fusarium* and two *Aspergilli*. Scott[47] found at least 46 strains of 5 genera of molds isolated from cereals and legumes which produced growth depression in young ducklings.

LOSSES OF VITAMINS DURING STORAGE OF PLANT FEEDSTUFFS

Wilder and Bethke[57] showed that the destruction of carotene in dehydrated alfalfa meal was only 10% in 6 months at storage temperatures of −23 to −26°C; the loss was 60 to 72% over the same period of storage at room temperature. There was a 50% loss in six months at 1 to 6°C. Carotene values were completely preserved by storage of the alfalfa in vacuo or under nitrogen. The stability of carotene in stored alfalfa was improved by the addition of antioxidants to alfalfa meal.[24,39] Since Quackenbush et al.[45] showed that tocopherols influenced the stability of carotene, it is likely that antioxidants also protect tocopherols as well as the carotenoids.

EFFECTS OF PROCESSING ON VITAMIN CONTENT OF ANIMAL PRODUCTS

Early studies on vitamin K proved that exposing fish meal and rice bran to the action of microorganisms increased the content of vitamin K. Almquist et al.[1] showed that amounts of vitamin K increased appreciably as a consequence of the putrefaction of ether-extracted, vitamin K-free meal, and they identified some of the microorganisms responsible for the synthesis. Ansbacher et al.[3] found that the vitamin B_{12} content of fish meals and solubles depended greatly upon the degree of putrefaction in these materials. Thus, trends in recent years toward preparation of better quality fish meals and fish solubles under conditions where putrefaction is prevented have actually resulted in lower levels of vitamin K and vitamin B_{12} than are present when a considerable degree of putrefaction occurs.

To the extent that fish meals or any other products are preserved by use of nitrites or sulfates, however, there may be considerable interference with vitamin nutrition. Whenever nitrates or particularly nitrites become a contaminant of ruminant feeds, vitamin A nutrition is seriously hampered.

Sulfur dioxide forms sulfite in solution that in turn destroys thiamin by cleaving the molecule into the inactive thiazole and pyrimidine fractions. It also enhances the peroxidative effects of polyunsaturated fatty acids upon vitamin E.[14,40]

Many raw fish contain a potent thiaminase that also cleaves thiamin into the inactive thiazole and pyrimidine moieties. This thiaminase is destroyed upon cooking or upon use of the heat needed to dehydrate fish meal.[22,23]

Prolonged heat-treatment destroys the fat-soluble vitamins and thiamin, pantothenic acid, folic acid, and biotin. This was found to be especially true if the material contained high levels of polyunsaturated fat.[6,19,27,49] Waisman et al.[56] found that chicks maintained on a heated diet required additional levels of folic acid, pantothenic acid, and biotin.

Further losses of all the water-soluble vitamins occurs during cooking if the cooking water is drained off rather than being dried back with the feed material. Thus the B-vitamin content of fish stick-water is considerably higher than that of fish meal, and the final fish meal dried without the fish solubles fraction is very low in water-soluble vitamin content.[44] Booher et al.[11] compiled a list of the vitamin values of foods as affected by processing and other variants.

Biely et al.[8] showed that the addition of herring oil to a diet marginal in vitamins depressed growth but that the growth rate was normal if the vitamin content of the diet was adequate. In later work, March et al.[38] found that treatment of the herring fish meal with butylated hydroxytoluene (BHT) did not decrease its deleterious effect on peroxidation of certain vitamins. Studies reported by March and Biely[37] showed that ethoxyquin greatly preserved the metabolizable energy and digestible protein values of herring and other fish meals. This antioxidant also has been shown to preserve the vitamin content of the fish meals and to prevent destruction of vitamins by polyunsaturated fish oils in feeds.

Most samples of meat and bone meal contain little or no thiamin. Apparently the processing conditions used in commercial manufacture of meat meals result in almost complete destruction or loss of this vitamin.

FACTORS AFFECTING STABILITY OF VITAMINS

Peroxidizing Polyunsaturated Oils

Unsaturated oils (e.g., cod liver, corn, soybean, sunflower-seed, and linseed oils) increase the requirements for vitamins A, E, D, and K as well as biotin. This is especially true if these oils are allowed to undergo oxidative rancidity in the diet or are in the process of peroxidation when consumed by the animal. If the oils become completely rancid before ingestion, the vitamins present in the oil and in the feed that contains the rancidifying oil are destroyed. If they are undergoing oxidative rancidity when consumed, they apparently can cause destruction of body stores of vitamin E, thereby producing encephalomalacia in growing chicks, very poor hatchability in breeding hens, and steatorrhea in mink, swine, and other animals. These effects have been prevented by adding vitamin E and an effective antioxidant such as ethoxyquin.[26,43,48]

Pelleting

When feeds are pelleted, destruction of vitamin A, vitamin E, and vitamin K may occur, especially if the diet contains an amount of antioxidant insufficient to prevent the accelerated oxidation of these vitamins under conditions of moisture and high temperature.[9] Charles and Huston[12] also have found that pelleting of the feed caused considerable destruction of vitamin K. In each test, however, menadione dimethyl-pyrimidino-bisulfite exhibited significantly higher stability than other forms of vitamin K to the stress of pelleting. Menadione sodium bisulfite (water-soluble) appeared to be the least stable. Sources of menadione sodium bisulfite complexes were intermediate in these stability tests.

Pelleting of feed may have a beneficial effect on the availability of some vitamins. Much of the nicotinic acid and biotin in feedstuffs may be present in bound form. The nicotinic acid in wheat middlings was found by Heuser and Scott[28] to be almost completely unavailable. Kodicek[33] has done considerable work on the availability of niacin in feedstuffs. It is well-recognized that the usual dehydration process employed for drying eggs does not destroy the avidin of egg white sufficiently to prevent its reaction with biotin. When eggs are used as a feed ingredient, therefore, they must be heat-treated sufficiently to destroy the avidin, in order to prevent biotin deficiency.[54]

Lih et al.[35] showed that bound pantothenic acid was apparently as available as free pantothenic acid in the diet of rats. Cropper and Scott[15,16] found that bound folic acid (folic acid conjugates) in dried brewer's yeast was as effective as free folic acid for growth and prevention of folic acid deficiency in chicks and poults.

Kratzer and Williams[34] found that linseed oil meal in the diet of chicks caused a marked increase in the requirement for vitamin B_6. Water treatment of linseed meal appeared to destroy its antipyridoxine activity.

The human niacin deficiency disease, pellagra, does not exist in Mexico, because the Mexicans treat their cornmeal with limewater before making it into tortillas. This treatment releases the nicotinic acid so that it is nutritionally available. The possibility that steam and pelleting processes also release some nicotinic acid and some biotin from bound forms in animal feeds has not been adequately investigated.

Pelleting can have a beneficial effect on vitamin E and vitamin B_{12} nutrition if the diet contains raw soybeans or other raw beans. Hintz and Hogue[29] showed that raw kidney beans (*Phaseolus vulgaris*) markedly increased the amount of vitamin E needed to prevent nutritional muscular dystrophy in the chick. Raw kidney beans also increased the incidence and severity of the stiff lamb disease. The improvement shown when the beans were cooked indicated that the vitamin E and selenium in the basic ingredients were then more readily utilized.

Edelstein and Guggenheim[17] showed that unheated soybean flour decreased vitamin B_{12} absorption. They found a decrease of injected ^{57}Co-labeled vitamin B_{12} in kidney, liver, and spleen when the diet of rats contained unheated soybean flour. Frolich[21] observed rapid depletion of vitamin B_{12} reserves in chickens fed an underheated, low-quality soybean oil; thus they had a greater need of vitamin B_{12} in the diet.

Since nutritional interrelationships exist between vitamin B_{12} and folic acid, pantothenic acid, choline, and other vitamins, unheated soybean protein may affect the nutritional requirements for a number of different vitamins.

Effects of Minerals and Chemicals on Vitamin Stability

Many studies have shown that vitamins A, D, and E are relatively unstable in mixed feed, especially in the presence of trace minerals (e.g., manganese and iron), and even more so as the heat and length of time in storage increase.[4,7,25,46] Iron salts are particularly destructive of vitamin E. Ferric chloride in ether solution has been used to destroy vitamin E for experiments on diets low in vitamin E and on vitamin E nutrition.

Bleaching agents such as nitrogen trichloride and chlorine dioxide, at concentrations needed to bleach flour, will destroy much of the tocopherols in the flour. Baking will destroy another 47% of the tocopherols remaining in the treated flour.[41]

STABLE FORMS OF VITAMINS A, D, AND E

Pure vitamin A alcohol (retinol) and pure α-tocopherol are exceedingly unstable. Retinyl acetate and retinyl palmitate are much more stable than the pure vitamin. In like manner, α-tocopheryl acetate is very stable in comparison with pure α-tocopherol. Thus, the ester forms of vitamin A and vitamin E ordinarily are used in commercial feeds. It has been impossible to stabilize vitamin D_3 (cholecalciferol) by formation of chemical esters or other derivatives of this vitamin.

Because these vitamins have great nutritional importance, many commercial producers of vitamins have developed ways to increase the stability of vitamins A, D, and E. Small beads produced by the mechanical envelopment of minute droplets of the vitamin in stable fat or gelatin protect most of the vitamin from contact with oxygen in the feed before digestion in the intestinal tract. The induction period that precedes active oxidation of the vitamin also can be greatly prolonged through the addition of effective antioxidants. The antioxidant most commonly used is 6-ethoxy-1,2-dihydro-2,2,4-trimethylquinoline (ethoxyquin). Where the two processes are combined, the "stabilized" bead containing an effective antioxidant protects these vitamins for storage periods of 4 to 8 weeks without much loss of vitamin potency.

The instability of vitamin D in peroxidizing diets often is overlooked. Studies in the author's laboratory have shown that even very high dietary levels of vitamin D are completely destroyed in diets containing high levels of peroxidizing polyunsaturated fatty acids; the chicks consequently suffer severe rickets by 3 weeks of age. When the diet is supplemented with 125 mg ethoxyquin per pound of feed, a normal level of vitamin D prevents rickets.

EFFECTS OF RADIATION

Riboflavin is stable to most factors involved in processing. It is very readily destroyed, however, by irradiation with either visible or ultraviolet light. Pyridoxine and ascorbic acid also are destroyed by light.[30,31] Therefore, premixes or feeds containing these vitamins must be protected from light and radiation.[51] X-irradiation, which is sometimes used to sterilize foods, has a very destructive effect on ascorbic acid, thiamin, and pyridoxine.

EFFECTS OF COLLOIDS IN THE FEED OR IN THE INTESTINAL TRACT

Many colloidal materials are known to be excellent adsorbing agents for most vitamins. Studies by Hunt et al.[32] indicated that bone, bentonite, and soft clay rock phosphate interfered with the availability of riboflavin. A deficiency was produced by these colloidal materials only when they were added to diets containing barely adequate amounts of the vitamin. Apparently the deleterious effect of the adsorbents was overcome by using higher levels of vitamins in the feed.

EFFECTS OF PROCESSING

Many methods developed for commercial processing result in decreases in the vitamin content of the grains and oilseed meals used to formulate feeds. The outer bran layer of cereal grain is higher in most vitamins than the inner floury portion.[2] Thomas et al.[53] have shown that the vitamin E content of fresh endosperm flour is relatively high, containing about 80% of the vitamin E content of the whole wheat kernel. When coarse grinders are used in flour milling, less vitamin E is lost than when finer flour is ground. More enzymes that destroy vitamin E may be released. When wheat germinates, the vitamin E falls rapidly, and can soon drop as low as 30% of the original value. Since bran is relatively indigestible and therefore low in energy for simple-stomached animals, this portion of the byproducts of grain milling usually is added to the feeds for ruminants.

In 1938, Norris[42] reported that additional vitamin E was not needed in poultry rations because the expeller soybean meal and the high levels of alfalfa meal in those rations contained plenty of this vitamin. A marked reduction of the vitamin E and vitamin K content of feeds for poultry and other livestock has appeared now, however, with changes in processing such as the solvent extraction of soybean meals.

SUMMARY

Many factors affect vitamin content and availability in feedstuffs. Some are agronomic; others involve variations that occur during harvesting, storage, and processing of feedstuffs. The feed formulation and methods of manufacture and handling of premixes and finished feeds also affect the vitamins. The nutritionist must consider each and all of these variables in his formulation of feed.

Dietary levels of unstabilized polyunsaturated oils and iron salts may have detrimental effects. Mold in feeds and exposure of feeds to light or other sources of radiation are detrimental.

Addition of vitamin E and antioxidants helps to prevent oxidative destruction of several of the vitamins. Use of vitamins A and E esters and incorporation of these into "stabilized" beadlets have markedly improved vitamin A and E nutrition of poultry and livestock.

The gradual reduction of levels of many vitamins in feeds by methods of preparation, storage, etc. has been balanced by the remarkable achievements in the chemical and microbiological synthesis of vitamins on a commercial scale which makes it possible to supplement feeds at low cost with pure forms of the needed vitamins.

REFERENCES

1. Almquist, H. J., Pentler, C. F., and Mecchi, E., Synthesis of the antihemorrhagic vitamin by bacteria, *Proc. Soc. Exp. Biol. Med.*, 38, 336—341, 1938.
2. Andrews, J. S., Boyd, H. M., and Terry, D. E., The riboflavin content of cereal grains and bread and its distribution in products of wheat milling, *Cereal Chem.*, 19, 55, 1942.
3. Ansbacher, S., Hill, H. H., Jr., Tiernan, J. W., Downing, J. F., and Caldwell, J. H., Jr., Microbially synthesized APF (animal protein factor), *Fed. Proc.*, 8, 180, 1949.
4. Baird, F. D., Ringrose, A. T., and MacMillan, M. J., The stability of vitamins A and D in mixed feed ingredients. I. Vitamin D, *Poult. Sci.*, 18, 35, 1939.
4a. Baird, F. D., Ringrose, A. T., and MacMillan, M. J., The stability of vitamins A and D in mixed feed ingredients. II. Vitamin A, *Poult. Sci.*, 18, 441, 1939.
5. Basappa, S. C., Jayaraman, A., Sreenivasamurthy, A., and Parpia, H. A. A. B., Effect of B-group vitamins and ethyl alcohol on aflatoxin production by *Aspergillus oryzae, Indian J. Exp. Biol.*, 5, 262—264, 1967.
6. Bauernfeind, J. C., Norris, L. C., and Heuser, G. F., The pantothenic acid requirements of chicks, *Poult. Sci.*, 21, 142, 1942.
7. Bethke, R. M., Record, P. R., and Wilder, O. H. M., The stability of carotene and vitamin A in a mixed ration, *Poult. Sci.*, 18, 179—187, 1939.
8. Biely, J., March, B. E., and Tarr, H. L. A., The nutritive value of herring meals, *Poult. Sci.*, 45, 1274—1279, 1955.
9. Bierer, B. W. and Vickers, C. L., The effect of pelletizing on vitamins A and E, *J. Am. Vet. Med. Assoc.*, 133, 228, 1958.
10. Blaylock, L. G., Richardson, L. R., and Pearson, P. B., The riboflavin, pantothenic acid, niacin and folic acid content of fresh, dehydrated and field-cured alfalfa, *Poult. Sci.*, 29, 692—695, 1950.
11. Booher, L. E., Hartzler, E. R., and Hewston, E. M., A compilation of the vitamin values of foods in relation to processing and other variants, *U.S. Dept. Agric. Circ.*, No. 638, 1942.
12. Charles, O. W. and Huston, T. M., Stability studies of vitamin K materials subjected to pelleting and storage, XIV World's Poultry Congr. Sci. Comm. II, Madrid, Spain, 1970, 693—697.
13. Combs, G. F., Jr., Problems with molds and mycotoxins in poultry feeds, *Proc. 1975 Cornell Nutr. Conf.*, Ithaca, New York, 1975, 109—115.
14. Cremer, H. D. and Hoetzel, D., Sulfite antagonism of thiamine *in vivo*; an antithiamine effect, *Bibl. Nutr. Dieta*, 8, 1826, 1966.
15. Cropper, W. J. and Scott, M. L., Nature of the blood folates in young chicks and poults fed pteroylmonoglutamic acid or pteroylheptaglutamates, *Proc. Soc. Exp. Biol. Med.*, 122, 817—820, 1966.
16. Cropper, W. J. and Scott, M. L., Studies on folic acid nutrition in chicks and poults, *Br. Poult. Sci.*, 8, 65—74, 1967.
17. Edelstein, S. and Guggenheim, K., Causes of the increased requirement for vitamin B_{12} in rats subsisting on an unheated soybean flour diet, *J. Nutr.*, 100, 1377—1382, 1970.
18. Eldridge, D. W., Nutritional Factors Influencing the Synthesis of Aflatoxins B_1 by *Aspergillus flavus*, M.Sci. thesis, Auburn University, Ala., 1964, 47.
19. Elvehjem, C. A., Kline, O. L., Keenan, J. A., and Hart, E. B., A study of the heat stability of the vitamin B factors required by the chick, *J. Biol. Chem.*, 99, 309, 1932.
20. Fritz, J. C., Pla, G. W., Mislivec, P. B., and Dantzman, J., Toxicogenicity of moldy feed for young chicks, *Poult. Sci.*, 50, (Abstr.) 1577, 1971.
21. Frolich, A., Relation between the quality of soybean oil meal and the requirements of vitamin B_{12} for chicks, *Nature (London)*, 173, 132, 1954.
22. Fujita, A. and Tashiro, T., Studies on thiaminase. III. Nature of "cothiaminase", *J. Biol. Chem.*, 196, 305, 1952.
23. Goldbeck, C. G., Some studies on the content of thiamine and antithiamine factor in fishery products, *Commer. Fish. Rev.*, 9, 13—21, 1947.
24. Gordon, R. S. and Machlin, L. J., A method of evaluation of antioxidants based on vitamin A protection, *Poult. Sci.*, 38, 1463, 1959.
25. Halverson, A. W. and Hendrick, C. M., Some factors affecting the stability of carotene in mixed feeds, *S. D. Agric. Exp. Stn. Tech. Bull.*, 14, 1955.
26. Harms, R. H., Douglas, C. R., and Waldroup, P. W., Ethoxyquin and vitamin E studies in poultry, *Q. J. Fla. Acad. Sci.*, 27, 131, 1964.
27. Heller, C. A., McCay, C. M., and Lyon, C. B., Losses of vitamins in large-scale cookery, *J. Nutr.*, 26, 377, 1943.
28. Heuser, G. F. and Scott, M. L., Studies in duck nutrition, *Poult. Sci.*, 32, 137—143, 1953.
29. Hintz, H. F. and Hogue, D. E., Kidney beans (*Phaseolus vulgaris*) and the effectiveness of vitamin E for prevention of nutritional muscular dystrophy in the chick, *J. Nutr.*, 84, 283, 1964.

30. Hochberg, H., Melnick, D., and Oser, B. L., On the stability of pyridoxine, *J. Biol. Chem.*, 155, 129, 1944.

31. Holmes, A. D. and Jones, C. P., Effect of sunshine upon the ascorbic acid and riboflavin content of milk, *J. Nutr.,* 19, 201, 1945.

32. Hunt, C. H., Bentley, O. G., Hershberger, T. V., and Moxon, A. L., Effect of certain adsorbents and mineral mixtures on the availability of riboflavin and other B-vitamins in the rations, *Ohio Agric. Exp. Stn. Res. Bull.,* 748, 1954.

33. Kodicek, E., The availability of bound nicotinic acid to the rat, *Br. J. Nutr.,* 14, 13, 25, 35, 1960.

34. Kratzer, F. H. and Williams, D. E., The improvement of linseed oil meal for chick feeding by the additions of synthetic vitamins, *Poult. Sci.,* 27, 236—238, 1948.

35. Lih, H., King, T. E., Higgins, H., Baumann, C. A., and Strong, F. M., Growth promoting activity of bound pantothenic acid in the rat, *J. Nutr.,* 44, 361, 1951.

36. Mangelson, F. L., Draper, C. I., Greenwood, D. A., and Crandall, B. H., The development of chicks fed different levels of sun-cured and dehydrated alfalfa and the vitamin A and carotene storage in their livers, *Poult. Sci.,* 28, 603—609, 1949.

37. March, B. E. and Biely, J., Nutritional evaluation of fishmeals for poultry feeding, Proc. 1967 Cornell Nutr. Conf., Ithaca, New York, 1967, 123—129.

38. March, B. E., Biely, J., Claggett, F. E., and Tarr, H. L. A., Nutritional and chemical changes in the lipid fraction of herring meals with and without antioxidant treatment, *Poult. Sci.,* 41, 837—880, 1962.

39. Matterson, L. D., Potter, L. M., Pudelkiewicz, W. J., Carlson, D., and Singsen, E. P., Studies on the utilization of carotenoid pigments from alfalfa in the presence of antioxidants, *Poult. Sci.,* 35, 1156, 1956.

40. Miller, R. F., Small, G., and Norris, L. C., Studies on the effect of sodium bisulfite on the stability of vitamin E, *J. Nutr.,* 55, 81—95, 1955.

41. Moore, T., Sharman, I. M., and Ward, R. J., The destruction of vitamin E in flour by chlorine dioxide, *J. Sci. Food Agric.,* 8, 97, 1957.

42. Norris, L. C., Is additional vitamin E needed in poultry rations?, *Flour Feed,* 39, 8, 9, 22, 23, 1938.

43. Opstvedt, J., Olsen, S., and Urdahl, N., Influence of residual lipids on the nutritive value of fish meal. I. The effects of adding 1,2-dihydro-6-ethoxy-2,2,4-trimethylquinoline to fish meal on protein quality, energy value and vitamin E status in chickens, *Acta Agric. Scand.,* 20, 174—184, 1970.

44. Pratt, J. M. and Biely, J., A note on the value of stickwater meal as a riboflavin supplement in poultry rations, *Poult. Sci.,* 24, 377, 1945.

45. Quackenbush, F. W., Cox, R. P., and Steenbock, H., Tocopherol and the stability of carotene, *J. Biol. Chem.,* 145, 169—177, 1942.

46. Reid, B. L., Daugherty, H. K., and Couch, J. R., The stability of vitamin A in mixed feeds and premixes, *Poult. Sci.,* 34, 603, 1955.

47. Scott, de B., Toxigenic fungi isolated from cereal and legume products, *Mycopathol. Mycol. Appl.,* 25, 213—222, 1965.

48. Scott, M. L., Nesheim, M. C., and Young, R. J., *Nutrition of the Chicken,* M. L. Scott and Associates, Ithaca, N.Y., 1969.

49. Scott, M. L., Norris, L. C., Maynard, L. A., and Spector, H., The nutritive value of army rations as determined with growing chicks. Methods for evaluation of nutritional adequacy and status—A symposium, Advis. Bd. Quartermaster Res. and Development, Commissariat on Foods, U.S. Department of the Army, University of Chicago, 1954, 164—178.

50. Sherby, T. F., Templeton, G. E., and Stephenson, E. L., Effect of heat-treated mold infested corn on broiler chick performance, *Poult. Sci.,* 50, 1629—1630, 1971.

51. Stamberg, O. E. and Petersen, C. F., Photolysis of riboflavin in poultry feeds, *Poult. Sci.,* 25, 394—395, 1946.

52. Teague, H. S., Field fungi and the nutritional value of corn, Proc. 1966 Cornell Nutr. Conf., Ithaca, New York, 1966, 97—102.

53. Thomas, B., Feldheim, W., and Rothe, M., Vitamin E and corn, *Ernaehrungsforschung,* 11, 603, 1957.

54. Tully, W. C. and Franke, K. W., A nutritional disease demonstrating a feed deficiency in dried eggs, *Poult. Sci.,* 13, 343—347, 1934.

55. U.S. Plant, Soil and Nutrition Laboratory Staff, The effect of soils and fertilizers on the nutritional quality of plants, U.S. Department of Agriculture *Res. Serv. Agric. Inf. Bull.,* No. 299, 1965, 1—24.

56. Waisman, H. A., Mills, R. C., and Elvehjem, C. A., Factors required by chicks maintained on a heated diet, *J. Nutr.,* 24, 187—198, 1942.

57. Wilder, O. H. M. and Bethke, R. M., The loss of carotene in machine dried alfalfa meal under variable conditions of storage, *Poult. Sci.,* 20, 304—312, 1941.

58. Kurnick, A. A., personal communication, Hoffman-LaRoche, Inc., Nutley, N.J., 1972.

EFFECT OF PROCESSING ON THE BIOAVAILABILITY OF MINERALS IN FEED

James C. Fritz

INTRODUCTION

Processing of feed includes many diverse operations.[1] Feed processing may alter the total quantity of an essential nutrient, or in a few cases, it may change the biological availability of a nutrient.[2] Purposes of processing include:[3]

1. Isolation of a specific part
2. Improvement in acceptability (palatability)
3. Alteration of particle size
4. Improvement in digestibility
5. Extention of shelf life or preservation
6. Alteration in nutrient makeup
7. Detoxification

Many conditions influence the bioavailability of a required nutrient.[4] Among these are

1. Digestibility of the food that supplies the nutrient
2. Chemical form or combination of the element
3. Particle size
4. Interaction with other nutrients
5. Chelation
6. The effect of processing
7. The body's need for the nutrient

Some of these factors can be controlled to assure optimum utilization of the food and feed supply.

If a given foodstuff cannot be digested, there will, of course, be little or no utilization of any nutrients that the foodstuff contains. In this connection it is necessary to consider species differences. For example, a high fiber hay may contain much iron, but that iron will not be available to single stomach species that consume the hay although it would be a good source of iron for ruminant animals.

Another example is seen in the choice of protective coatings used to insulate nutrients. These must be selected so that they will be removed in the digestive tract and thereby make the nutrients available to the animal or bird that consumes the coated material.[5]

Particle size is another factor in digestibility and solubility, and thus in bioavailability. The particle size is especially important when relatively insoluble materials are used in food or feed fortification for technological reasons.[6,7] The influence of the chemical form or combination of the element is probably dependent upon the solubility in the digestive tract. Obviously, if the compound cannot be made soluble under the conditions in the gastrointestinal tract, the compound will be excreted in the feces and the nutrients will not become available to the animal.

Interaction with other nutrients has an important bearing on the extent to which an essential nutrient will be utilized. An excess of one mineral element in the digestive tract may interfere with the utilization of another essential nutrient. Two elements may compete for binding sites on carrier proteins. Another possibility is that the mineral present in excess may combine with another mineral to form a compound which is excreted, thus causing a deficiency of the second mineral. An example can be noted in the production of rickets in the rat in the vitamin D assay.[8] The excess calcium is excreted as an insoluble calcium phosphate. This, in turn, produces a low phosphorus rickets, which can be cured by either vitamin D or additional phosphorus. It is obviously impossible to state a specific requirement for calcium, phosphorus, or vitamin D unless the quantities of all three in the diet are considered. Excess calcium is recognized as increasing the dietary need for many nutrients.[9-11]

Chelation is poorly understood. It may, however, have a marked effect on biological availability. The mineral element is bound to an organic ligand which may make the mineral either more or less available to the animal that consumes the chelate. Scott et al.[12] recognized three types of chelates in biological systems:

1. Chelates that serve to transport metal ions, such as the amino acids.
2. Chelates essential in metabolism, such as heme, the chelate portion of hemoglobin.
3. Chelates which interfere with nutrition by making an essential element less available biologically, such as phytic acid.

O'Dell[13] has shown that the zinc requirements depend upon the phytate content of the diet, and that in a low zinc diet based on isolated soybean protein 100 ppm of ethylenediaminetetraacetic acid (EDTA) was approximately equivalent to 8 ppm of added zinc. Unfortunately, a chelating agent that makes one mineral element more available may make other essential elements less available. EDTA has been shown to make iron and manganese less available to the chick.[14] Some observations on this point with chicks fed a diet low in several essential minerals, with and without EDTA, are shown in Table 1.

One effect of grinding to reduce particle size is to make the relatively insoluble sources of nutrients more available. This is illustrated by the observations on bioavailability of powdered metallic iron shown in Table 2.[6]

Another possible effect of grinding may be the addition of mineral elements to the material being ground. Ammerman et al.[15] have shown that grinding citrus pulp in a Wiley mill will significantly increase the content of iron, zinc, copper, manganese, and sodium but will have little or no effect on the content of calcium, phosphorus, magnesium, and potassium. Data are shown in Table 3.

Processing may change the form of a mineral element. Sometimes this increases the bioavailability of the relatively insoluble sources of the element. It has been shown that relatively unavailable iron compounds added to liquid foods may be changed into forms with higher bioavailability by processing or by prolonged storage.[16-18]

Species differences, and the body's need for an essential nutrient, have a marked bearing on the quantity of that nutrient which will be utilized,[11] but have relatively little bearing on the effect of processing. We, of course, recognize that the laying hen has a large need for dietary calcium for egg shell formation.[5,11] In a similar vein we can note the high requirement for manganese by the young of avian species.[19]

CALCIUM AND PHOSPHORUS

Calcium is generally well-utilized. Little difference in utilization is noted if the level of vitamin D is adequate and if the calcium-to-phosphorus ratio is reasonable for the species and feeding purpose.[20] Chick studies by Dilworth et al.[21] indicated the following relative values for availability of several calcium sources:

Table 1
EFFECT OF ADDING EDTA TO A DIET LOW IN
CALCIUM, IRON, AND MANGANESE[a]

Chemical added to basal diet (ppm)		Chick wt[b] (g at 3 weeks)	Toe ash (%)	Hemoglobin[b] (g/100 ml)	Perosis score[b]
None	—	302 ± 40	10.08	6.30 ± 0.47	0.27 ± 0.33
EDTA	100	262 ± 41	9.92	5.51 ± 0.40	0.50 ± 0.56
	200	244 ± 34	10.20	5.28 ± 0.56	0.68 ± 0.53
	400	288 ± 35	10.29	5.12 ± 0.39	0.64 ± 0.37
	800	211 ± 15	9.23	3.48 ± 0.55	1.32 ± 0.68
Na₂EDTA	100	280 ± 52	10.26	5.88 ± 0.39	0.55 ± 0.27
	200	250 ± 53	10.13	5.96 ± 0.66	0.95 ± 0.41
	400	284 ± 27	10.73	5.88 ± 0.39	0.68 ± 0.24
	800	275 ± 33	9.90	4.68 ± 0.73	0.90 ± 0.49
	1600	221 ± 6	9.92	3.55 ± 0.29	1.20 ± 0.84
CaNa₂EDTA	100	256 ± 56	10.22	6.14 ± 0.37	0.73 ± 0.74
	200	256 ± 61	9.78	5.66 ± 0.93	0.59 ± 0.49
	400	267 ± 49	10.30	5.95 ± 0.51	0.82 ± 0.71
	800	268 ± 35	10.36	4.72 ± 0.36	1.11 ± 0.65
	1600	229 ± 30	9.99	3.58 ± .30	1.00 ± 0.35

Least significant differences:[42]

P = 0.05	41		0.49	0.56
P = 0.01	59		0.70	0.79

[a] The basal diet contained 0.32% Ca, 13.5 mg Fe/kg, and 28.7 mg Mn/kg. Reduction of available calcium should have reduced bone ash, reduction in available iron should have reduced hemoglobin, and reduction in available manganese should have increased perosis score (0 = normal; 4 = severe deformity with slipped tendons).

[b] Average ± standard deviation.

From Fritz, J. C., Pla, G. W., and Boehne, J. W., *Poult. Sci.,* 50, 1444, 1971. With permission.

Calcium carbonate, USP	100
Low fluorine rock phosphate	90
Defluorinated rock phosphate	92—95
Soft rock phosphate	68

It has long been suspected that older individuals absorb less calcium,[22,23] but recent reviews have not yielded definitive data on the subject.[24]

Phosphorus is utilized less efficiently, and in a much more variable manner than calcium.[20] Some data from Food and Drug Administration (FDA) tests with different sources of phosphorus are shown in Table 4.[25] Similar utilization data had been previously shown by Gillis et al.[26] Utilization by other domestic species generally falls in line with these chick data.[20] Summers et al.[27] found that steam pelleting of plant sources increased the availability of phosphorus for growing chicks by 19 to 29%. For pigs, pelleting was less effective than the addition of inorganic phosphorus.[28]

MAGNESIUM

Concern has been expressed about possible magnesium deficiency in human nutrition[29] and in the nutrition of domestic animals.[30] Loosli[20] pointed out that soluble magnesium salts are readily absorbed, but that the percentage of the available magnesium that was absorbed decreased with age:

Table 2
INFLUENCE OF PARTICLE SIZE ON
BIOAVAILABILITY OF REDUCED IRON

Production method and test laboratory	Particle size (μ)	Relative biological value
Electrolytic (AOAC	7—10	63.5
collaborative study)	27—40	38
Electrolytic (FDA)	0—10	76
	10—20	75
	20—40	48
	>40	45
Hydrogen reduction	10—20	54
(FDA)	325[a]	34
	100[a]	18
Carbon monoxide reduction	7—10	36
(University of Guelph)	14—19	21
	27—40	13
Carbonyl iron (FDA)	<4	69
	3—5	69
	4—8	64

[a]   Mesh size.

From Fritz, J. C., Pla, G. W., and Rollinson, C. L., *Bakers Dig.*, 49(2), 46, 1975. With permission.

Table 3
MINERAL CONTAMINATION OF FEED BY GRINDING

Element	Unit	Unground	Ground
Calcium	%	1.47 ± 0.123	1.54 ± 0.073
Potassium	%	1.12 ± 0.052	1.08 ± 0.060
Sodium	%	0.099 ± 0.0096	0.110 ± 0.0160
Magnesium	%	0.124 ± 0.0053	0.123 ± 0.0048
Phosphorus	%	0.110 ± 0.0074	0.105 ± 0.0046
Iron	ppm	94.0 ± 13.55	134.8 ± 11.67
Zinc	ppm	11.4 ± 1.23	17.7 ± 1.67
Copper	ppm	6.0 ± 0.70	9.7 ± 1.53
Manganese	ppm	5.8 ± 0.44	6.7 ± 0.46

From Ammerman, C. B., Martin, F. G., and Arrington, L. R., *J. Dairy Sci.*, 53, 1514, 1970. With permission of The American Dairy Science Association.

Young milk-fed calves, rats, and guinea pigs	70 + %
Older calves	30—50%
Mature ruminants	10—40%

It is reported that several factors increase magnesium availability, including adding glucose, molasses, or starch, cooking the concentrate mixture, and grinding hay, while protein supplements tend to decrease it.[20]

Recently Guenter and Sell[31] proposed a test for measuring the bioavailability of magnesium. The test involves feeding chicks, and using magnesium sulfate as the reference standard. Magnesium, supplied as magnesium sulfate, was utilized to the extent of 57.4%, and this was assigned a relative availability index of 100.

Table 4
RELATIVE BIOLOGICAL VALUE OF PHOSPHORUS
SUPPLEMENTS AS DETERMINED BY CHICK ASSAY,
USING CALCIFICATION AND GROWTH AS THE
CRITERIA OF RESPONSE[25]

Phosphorus source	Calcification	Growth
Diammonium phosphate	123	126
Monocalcium phosphate	112	103
Monoammonium phosphate	103	104
Monosodium phosphate[a]	100	100
Disodium phosphate	97	101
Potassium phosphate	96	102
Dicalcium phosphate (hydrated, feed grade)	94	97
Dicalcium phosphate (anhydrous)	90	91
Curacao phosphate	85	78
Tricalcium phosphate	84	84
Bone meal (steamed)	82	91
Soft rock phosphate	62	35
Calcium phytate	44	38

[a]   Reference standard.

Loosli concludes that feed processing is not known to change the utilization of magnesium.[20] He had previously noted that dry lot feeding, low temperature, added potassium chloride, or the presence of ammonia may depress utilization of dietary magnesium. He raises the question about whether the processing of feeds containing nonprotein nitrogen might affect magnesium availability.

SODIUM, POTASSIUM, AND CHLORIDE

Sodium is the principal cation in extracellular fluid and potassium is the principal cation in intracellular fluid. Together they have a major influence on osmotic pressure and fluid volume in the body.[29] The limited information on the bioavailability of these elements was reviewed by Peeler.[30]

Most sodium and potassium salts are soluble and readily available, but are under homeostatic control. Nott and Combs[32] found the sodium in defluorinated phosphate to be only 83% as available to the chick as sodium supplied in the form of sodium chloride. Dilworth et al.[33] found that particle size of sodium chloride is important in its utilization by the chick. Salt retained on U.S.B.S. sieves sizes 18 through 25 gave better results than did either larger or smaller particle sizes.

Chloride is the most important anion in the regulation of fluid and electrolyte balance.[29] Nesheim et al.[34] stressed the importance of balance between sodium and potassium.

There are no data on the effect of processing on the bioavailability of these elements. However, the observations by Dilworth and co-workers[33] would certainly indicate that any processing that would change particle size might be expected to alter the bioavailability.

TRACE ELEMENTS

In addition to the nutrients required in substantial quantities a wide variety of other elements, known as trace elements, are present in the animal body. These are obtained chiefly from one source — the food or feed that is consumed. We recognize that in a

Table 5

TRACE MINERAL CONTENT OF WHEAT AND WHEAT PRODUCTS (ppm)

Material	Cobalt	Copper	Iodine	Iron	Manganese	Selenium	Zinc
Wheat, hard	0.05	5	—	40	38	0.50	24
Wheat, soft western	0.03	4	0.04	43	35	0.28	22
Wheat bran	0.10	12	0.08	150	108	0.64	88
Wheat feed flour	0.01	3	—	10	5	—	9
Wheat flour, patent	—	2	—	8	4	0.47	6
Wheat germ meal	0.04	12	—	94	160	—	125
Wheat middlings, flour	0.11	13	0.11	95	63	—	77
Wheat middlings, standard	0.10	13	0.11	100	101	0.80	109
Wheat millrun	0.10	12	0.09	100	103	—	100
Wheat red dog	0.04	3	—	55	35	—	16

Reproduced from *Effect of Processing on the Nutritional Value of Feeds,* page 114, with the permission of the National Academy of Sciences, Washington, D.C.

few instances substantial quantities can be taken in via the water that is consumed or the air that is breathed. During the past few years, a number of additional elements have been found to be essential[35] and the number may be expected to increase in the future. Part of this is due to better methods of analysis, and to the discovery of essential functions for elements that were known to exist in the animal body but for which no nutritional role had been recognized.

At the present time, the following are considered essential nutrients:

Arsenic	— Essential for optimum growth of chicks and rats.
Chromium	— A part of the glucose tolerance factor.
Cobalt	— A part of the vitamin B_{12} molecule.
Copper	— Essential for hemoglobin formation.
Fluorine	— Required for normal bone and tooth structure.
Iodine	— A part of thyroid enzymes and necessary for prevention of goiter.
Iron	— A component of the hemoglobin molecule.
Manganese	— Necessary for normal bone formation and prevention of perosis.
Molybdenum	— A part of xanthine oxidase.
Nickel	— Required for optimum growth of chicks and rats.
Selenium	— Needed for absorption of fats and fat-soluble factors.
Tin	— Required for optimum growth and feathering of chicks.
Vanadium	— Present in many enzyme systems.
Zinc	— A part of dehydrogenases and needed to prevent parakeratosis.

Some authorities would add others to this list, and it is likely that we shall all do so in the years ahead. However, in regard to processing, we know virtually nothing about its effect on the bioavailability of even the most well-recognized of these elements. In many instances we may assume that what little we do know about the effect of processing on one essential element will apply to others. The factors that we have earlier noted as possibly affecting bioavailability apply to the trace elements.

Much of what we know about the influence of processing on minerals can be illustrated by observations on the iron nutrition of man and animals. The addition of minerals from contamination has already been noted in the case of grinding and mixing equipment.[15] Some other examples can be cited. The nutrition surveys in Ethiopia showed a high iron content in teff — a small seed that constitutes the chief cereal product in the Ethiopian diet. A later study by Hofvander[36] showed that at least ¾ of the iron in commercial teff was due to contamination with high-iron soil during primitive threshing operations. This iron was mostly in the oxide form and had low bioavailability.

Table 6
EFFECT OF INCORPORATION OF IRON INTO FOODS AND OF PROCESSING ON BIOAVAILABILITY OF IRON FROM KNOWN SOURCES

Iron source	Form added to test diet	Relative biological value (vs. FeSO$_4$ = 100)
Ferric ammonium citrate	Iron compound alone	107
	Dissolved in fluid whole milk	89
Ferric orthophosphate	Iron compound alone	14
	Dry farina No. 1	28
	Dry farina No. 2, first test	24
	Cooked farina No. 2, first test	25
	Dry farina No. 2, second test	26
	Cooked farina No. 2, second test	17
	Dry farina No. 3	32
	Cooked farina No. 3	35
	Ready-to-eat breakfast cereal	7
	Vitamin and mineral tablet	8
Ferric pyrophosphate	Iron compound alone	45
	Meal replacement No. 1	27
	Meal replacement No. 2	66
Ferripolyphosphate	Iron compound alone	80
	Dissolved in fluid whole milk	70
Ferripolyphosphate-whey protein complex	Iron compound alone	94
	Dissolved in fluid whole milk	94
Ferrous fumarate	Iron compound alone	95
	CSM[a]	91
	Biscuits baked from CSM	98
Ferrous sulfate	Iron compound alone	100
	Baked biscuits	89
	Dissolved in evaporated milk	110
	Dissolved in skim milk	95
Reduced iron	Powdered iron alone	37
	Enriched flour	32
	Ready-to-eat breakfast cereal	43

[a]   Corn-soy-milk mixture distributed by Agency for International Development (AID).

From Fritz, J. C. and Pla, G. W., *J. Assoc. Off. Anal. Chem.*, 55, 1128, 1972. With permission.

A somewhat different situation exists in southern Africa where the Bantu consume large quantities of a native beer that is brewed in iron containers.[37] The iron is dissolved from these vessels by the acid beer, and is in a highly available form. Galvanized equipment similarly contributes zinc to acid solutions.

When wheat is milled a large portion of the trace minerals present is removed along with the bran and middlings. These parts, plus the germ, are usually used in animal feeds and make the contribution of trace minerals much better. Some data are summarized in Table 5.

Processing may affect bioavailability as well as the total quantity of iron present. The work of Hodson[16] and of Theuer and associates[17, 18] showed that iron added as ferric orthophosphate to liquid foods was slowly changed into more available ferrous forms and it then became more biologically available.

In human foods, other ingredients present and the processing involved during normal food preparation have relatively little effect on the bioavailability of the iron. Pertinent observations are summarized in Table 6.

Table 7
IRON CONTENT OF SOYBEAN PRODUCTS

Material	Iron content (ppm)	Relative biological value
Raw soybean, yellow variety	113	54
Soybean meal, 44% protein	193	60
Dehulled soybean meal, 49% protein	113	58
Soy flour, 51.5% protein	88	47
Soy protein concentrate, 67.8% protein	155	69
Sodium soy isolate (proteinate)	192	55
Isolated soy protein, isoelectric	135	60
Raw soybeans, black Wilson variety	79	81
Soybeans, autoclaved black Wilson variety	83	62

Reproduced from *Effect of Processing on the Nutritional Value of Feeds,* page 116, with the permission of the National Academy of Sciences, Washington, D.C.

Table 8
MINERAL CONTENT OF CHICK STARTERS IN MASH AND CRUMBLED FORMS

Feed	Form	Iron (ppm)	Relative biological value[a]	Copper (ppm)	Manganese (ppm)	Zinc (ppm)	Calcium (%)	Manganese (%)
A	Mash	134	58	6.3	109	41	1.23	0.177
	Crumbles	133	58	6.6	125	46	1.21	0.178
B	Mash	114	64	138	97	98	1.12	0.151
	Crumbles	119	51	135	73	112	0.82	0.152

[a]   Relative biological value of iron (vs. $FeSO_4$ = 100).

Reproduced from *Effect of Processing on the Nutritional Value of Feeds,* page 116, with the permission of the National Academy of Sciences, Washington, D.C.

Since soybeans and soybean products are better than average sources of trace elements[38] and since several reports showed good bioavailability of iron from soybeans and soybean products,[17,39,40] it was desirable to determine whether or not routine processing altered the bioavailability of iron. Some pertinent data on the iron content of soybean products are shown in Table 7. While some variations are shown among samples, these data show little effect on bioavailability from the processing involved.

Pelleted feeds have many advantages. Practically all commercial broiler starters are now crumbled pellets. In an attempt to study the effect of pelleting and crumbling feeds, two manufacturers furnished feed in both mash and crumbled forms, made from the same lots of ingredients. Several mineral elements were determined by atomic absorption methods, and the bioavailability of the iron was determined by the chick hemoglobin repletion test.[41] It may be concluded from these data, summarized in Table 8, that pelleting and crumbling do not affect the trace mineral content of the feed.

CONCLUSIONS

While the possibility exists that some processing may alter the bioavailability of the minerals, most normal processing of feed has relatively little effect on the nutritional value of the nutrients. Exceptions must be noted where particle size is reduced or where the minerals are changed from a relatively insoluble form to one with greater solubility. A much greater possibility exists that the total content of mineral elements may be increased by contamination from soil or from grinding and mixing equipment.

REFERENCES

1. **Harris, L. E., Asplund, J. M., and Crampton, E. W.,** An international feed nomenclature and methods for summarizing and using feed data to calculate diets, *Utah Agric. Exp. Stn. Bull.,* No. 479, 1968.
2. **Fritz, J. C.,** Effect of processing on the availability and nutritional value of trace mineral elements, in *Effect of Processing on the Nutritional Value of Feeds,* National Academy of Sciences, Washington, D.C., 1973, 109—118.
3. **Harris, L. E. and Crampton, E. W.,** NRC names for feed processes and their use in evaluating the nutrient content of feeds, in *Effect of Processing on the Nutritional Value of Feeds,* National Academy of Sciences, Washington, D.C., 1973, 1—22.
4. **Fritz, J. C.,** Bioavailability of Mineral Nutrients, *Chem. Technol.,* 6, 643—648, 1976.
5. **Titus, H. W. and Fritz, J. C.,** *The Scientific Feeding of Chickens,* 5th ed., The Interstate, Danville, Ill., 1971, 205—206.
6. **Fritz, J. C., Pla, G. W., and Rollinson, C. L.,** Iron for enrichment, *Bakers Dig.,* 49(2), 46—49, 1975.
7. **Harrison, B. H., Pla, G. W., Clark, G. A., and Fritz, J. C.,** Selection of iron sources for cereal enrichment, *Cereal Chem.,* 53, 78—84, 1976.
8. **Anon.,** *Official Methods of Analysis,* Association of Official Analytical Chemists, Washington, D.C., 1975, 851—856.
9. **Tucker, H. F. and Salmon, W. D.,** Parakeratosis or zinc deficiency disease in the pig, *Proc. Soc. Exp. Biol. Med.,* 88, 613—616, 1955.
10. **Roberson, R. H. and Schaible, P. J.,** The effect of elevated calcium and phosphorus levels on the zinc requirement of the chick, *Poult. Sci.,* 39, 837—840, 1960.
11. **Underwood, E. J.,** *Trace Elements in Human and Animal Nutrition,* 3rd ed., Academic Press, New York, 1971.
12. **Scott, M. L., Nesheim, M. C., and Young, R. J.,** *Nutrition of the Chicken,* M. L. Scott, Ithaca, N.Y., 1969, 262—264.
13. **O'Dell, B. L.,** Effect of dietary components upon zinc availability, *Am. J. Clin. Nutr.,* 22, 1315—1322, 1969.
14. **Fritz, J. C., Pla, G. W., and Boehne, J. W.,** Influence of chelating agents on utilization of calcium, iron, and manganese by the chick, *Poult. Sci.,* 50, 1444—1450, 1971.
15. **Ammerman, C. B., Martin, F. G., and Arrington, L. R.,** Mineral contamination of feed samples by grinding, *J. Dairy Sci.,* 53, 1514—1515, 1970.
16. **Hodson, A. Z.,** Conversion of ferric to ferrous iron in weight control dietaries, *J. Agric. Food Chem.,* 18, 946—947, 1970.
17. **Theuer, R. C., Kemmerer, K. S., Martin, W. H., Zomas, B. L., and Sarett, H. P.,** Effect of processing on availability of iron salts in liquid infant formula products. Experimental soy isolate formulas, *J. Agric. Food Chem.,* 19, 555—558, 1971.
18. **Theuer, R. C., Martin, W. H., Wallender, J. F., and Sarett, H. P.,** Effect of processing on availability of iron salts in liquid infant formula products. Experimental milk based formulas, *J. Agric. Food Chem.,* 21, 482—485, 1973.
19. **Committee on Poultry Nutrition,** *Nutrient Requirements of Poultry,* NAS-NRC Publ. 1345, National Academy of Sciences, Washington, D.C., 1966.
20. **Loosli, J. K.,** Effect of processing on the availability and nutritional value of calcium, phosphorus, and magnesium supplements, in *Effect of Processing on the Nutritional Value of Feeds,* National Academy of Sciences, Washington, D.C., 1973, 91—108.
21. **Dilworth, B. C., Day, E. J., and Hill, J. E.,** Availability of calcium in feed grade phosphates to the chick, *Poult. Sci.,* 43, 1132—1134, 1964.
22. **Anon.,** Adaptation to low calcium diets, *Nutr. Rev.,* 11, 274—276, 1953.

23. Morrison, F. B., *Feeds and Feeding,* 22nd ed., Morrison Publishing, Ithaca, N.Y., 1956, 105—114.

24. Irwin, M. I. and Keinholz, E. W., A conspectus of research on calcium requirements of man, *J. Nutr.,* 103, 1019—1095, 1973.

25. Fritz, J. C., Availability of mineral nutrients, in *Proc. Md. Nutr. Conf.,* Washington D.C., 1969, 1—15.

26. Gillis, M. B., Norris, L. C., and Heuser, G. F., Studies on the biological value of inorganic phosphates, *J. Nutr.,* 52, 115—125, 1954.

27. Summers, J. D., Slinger, S. J., and Cisneros, G., Some factors affecting the biological availability of phosphorus in wheat byproducts, *Cereal Chem.,* 44, 318—323, 1967.

28. Bayley, H. S. and Thompson, R. G., Phosphorus requirements of growing pigs and effect of steam pelleting on phosphorus availability, *J. Anim. Sci.,* 28, 484—491, 1969.

29. *Recommended Dietary Allowances,* 8th ed., NAS-NRC, National Academy of Sciences, Washington, D.C., 1974.

30. Peeler, H. T., Biological availability of nutrients in feeds: Availability of major mineral ions, *J. Anim. Sci.,* 35, 695—712, 1972.

31. Guenter, W. and Sell, J. L., A method for determining "true" availability of magnesium from foodstuffs using chicks, *J. Nutr.,* 104, 1446—1457, 1974.

32. Nott, H. and Combs, G. F., Availability of sodium in defluorinated rock phosphates, *Poult. Sci.,* 48, 482—485, 1969.

33. Dilworth, B. C., Schultz, C. D., and Day, E. J., Salt utilization studies with poultry. I. Effect of salt sources, particle size, and insolubles on broiler performance. II. Optimum particle size of salt for young chicks, *Poult. Sci.,* 49, 183—192, 1970.

34. Nesheim, M. C., Leach, R. M., Ziegler, T. R., and Serafin, J. A., Interrelationship between dietary levels of sodium, chlorine, and potassium, *J. Nutr.,* 84, 361—366, 1964.

35. Miller, W. J., Newer candidates for essential trace elements, *Fed. Proc.,* 33, 1747, 1974.

36. Hofvander, Y., Hematological investigations in Ethiopia with special reference to a high iron intake, *Acta Med. Scand., Suppl.,* 494, 1968.

37. Walker, A. R. P., Some aspects of nutrition research in South Africa, *Nutr. Rev.,* 14, 321—324, 1956.

38. National Research Council, *United States-Canadian Tables of Feed Composition,* 2nd rev. ed., National Academy of Sciences, Washington, D.C., 1970.

39. Layrisse, M., Iron absorption from food, *Pan Am. Health Organ. Sci. Publ.,* No. 184, 1969.

40. Fritz, J. C., Pla, G. W., Roberts, T., Boehne, J. W., and Hove, E. L., Biological availability in animals of iron from common dietary sources, *J. Agric. Food Chem.,* 18, 647—651, 1970.

41. Pla, G. W. and Fritz, J. C., Availability of iron, *J. Assoc. Off. Anal. Chem.,* 53, 791—800, 1970.

42. Snedecor, G. W., *Statistical Methods,* 5th ed., The Iowa State College Press, Ames, 1956.

43. Fritz, J. C. and Pla, G. W., Application of the animal hemoglobin repletion test to measurement of iron availability in foods, *J. Assoc. Off. Anal. Chem.,* 55, 1128—1132, 1972.

EFFECT OF PROCESSING ON NUTRIENT CONTENT AND VALUE OF FEEDS: NATURAL TOXINS AND INHIBITORS

M. L. Kakade

The presence of naturally occurring toxic substances in various plant and animal products has long been recognized by man. It appears though, as the result of trial and error testing over the centuries, he has not only learned to select those foods which are nontoxic, but he has also devised ways and means of eliminating these toxic substances from them. For example, age-old techniques such as cooking, soaking, germination, fermentation, and other common means of preparation of foods have proven to be effective in either destroying or removing many of these toxicants.

Although there is a number of excellent monographs and review articles on the subject matter,[1-7] efforts are made here to summarize and tabulate the data concerning the effects of processing on these naturally occurring toxicants. The emphasis has been directed toward the toxicants occurring in plants because it is believed plant products will play a major role in supplying the protein as well as caloric needs of an ever-increasing world population.

Every attempt has been made to define the conditions or the principles of detoxification of the processes. However, the reader is urged to refer to the original articles for further details relative to experimental design and conditions. It is hoped that the information contained in this chapter will serve as a guideline to expand our food and feed supply through new or improved methods of processing as man's continued survival may depend upon further exploration of new sources of plant and animal products.

REFERENCES

1. Gortzea, I. and Sutzescu, P., *Natural Antinutritive Substances in Foodstuffs and Forages*, S. Karger, Basel, 1968.
2. Liener, I. E., *Toxic Constituents of Plant Foodstuffs*, Academic Press, New York, 1969.
3. Liener, I. E., *Toxic Constituents of Animal Foodstuffs*, Academic Press, New York, 1974.
4. Toxicants Occurring Naturally in Foods, *National Academy of Sciences*, Washington, D.C., 1973.
5. Liener, I. E., Toxic factors in edible legumes and their elimination, *Am. J. Clin. Nutr.*, 11, 281—298, 1962.
6. Mickelsen, O. and Yang, M. G., Naturally occurring toxicants in foods, *Fed. Proc.*, 25, 102—123, 1966.
7. Kakade, M. L. and Liener, I. E., The increased availability of nutrients from plant foodstuffs through processing, in *Man, Food, and Nutrition*, Rechcigl, M., Jr., Ed., CRC Press, Boca Raton, Fla., 1973, 231—241.

Table 1

ELIMINATION OR DESTRUCTION OF NATURALLY OCCURRING TOXICANT IN FOOD AND
FEEDSTUFFS BY PROCESSING

Toxicant	Common occurrence	Nutritional or biological effects	Processing treatment	Ref.
Protease inhibitors	Soybean, navy bean, lima bean, kidney bean, chick pea, cow pea, lentil, peanut, beet, potato, rice, wheat, corn, eggs	Decreased protein digestibility pancreatic hypertrophy; increased requirement for sulfur amino acids	Heat especially moist 100—120°C for 15—30 min	1—3
Amylase inhibitor	Wheat, rye, barley, navy beans	Interference with digestion of carbohydrate	Cooking	3
Hemagglutinins	Seeds belonging to *Leguminosae* (i.e., soybeans, kidney beans, navy beans, peanuts, etc.)	Interference with absorption of nutrients from intestines	Heat especially moist 100—120°C for 15—30 min	4
Gossypol	Cottonseed	Increase requirement for protein and iron	Moist heat at 120°C for 15 min	2, 5
			Solvent extraction, e.g., alcohol-acetone	2, 5
			Phytase treatment — a culture filtrate of *Aspergillus ficcum*	6
			Fungal treatment — a strain of *Diplodia*	7
			Addition of ferrous or different amine to form unabsorbable complexes	2, 5, 8
Osteolathyrogens	*Lathyrus odoratus* (sweet pea)	Defect in collagen formation	Repeated washings with boiling water	9
Neurolathyrogens	*L. sativus* (chickling vetch), *Vicia sativa* (common vetch)	Neurotoxic effects	Repeated washings with boiling water	9
Cyacasin	Cycads	Interference with synthesis of RNA	Repeated washings with boiling water	10
Oxalic acid	Rhubarb, spinach, tea, cocoa	Increased requirement for calcium and vitamin D	Repeated washings with boiling water	2

Goitrogens	Interference with utilization of dietary iodine	Cabbage, turnip, rutabaga, mustard seed, rape seed, crambe	Autolysis under moist condition and subsequent removal of autolyzed products by steam distillation as in the case of mustard seeds, or acetone extraction, e.g., crambe, or by simple water extraction as in the case of rape seed	11—19
Phytic acid	Interference with absorption of calcium, zinc, magnesium; increased requirement for vitamin D	Seeds of mono- and dicotylendons (i.e., wheat, corn, rice, soybean, navy bean, peanut, etc.)	Autolysis — use of endogenous enzyme under optimum conditions	20, 21
			Use of exogenous source of phytase, e.g., yeast fermentation or a culture filtrate of *Aspergillus ficcum*	6, 22, 23
			Prolonged boiling in 1% HCl solution	2
Cynogens	Inhibition of cytochrome oxidase	Cassava, lima beans, sorghum, kernels of such fruits as apricot, peach, cherry, plums, almonds	Autolysis under moist conditions and subsequent volatilization of cyanide by cooking process	24—28
Saponins	Nonspecific interference with protein digestion, limit food intake	Soybean, navy bean, jack bean, alfalfa	Moist heat	2
Oligosaccharides	Flatulence	Most common legumes	Hydrolysis in acid solution	29—32
			Hydrolysis of oligosaccharides by endogenous or exogenous galactosidase	
			Immobilization of protein by heat or isoelectric pH, precipitation, or hot alcohol and subsequent removal of oligosaccharides by water extraction as is done commercially in the production of soy protein concentrate and isolate	29

Table 1 (continued)
ELIMINATION OR DESTRUCTION OF NATURALLY OCCURRING TOXICANT IN FOOD AND FEEDSTUFFS BY PROCESSING

Toxicant	Common occurrence	Nutritional or biological effects	Processing treatment	Ref.
Antipyridoxine factor	Linseed meal	Increased requirement of pyridoxine	Autoclaving	2
Avidin	Eggs	Deficiency of biotin	Cooking at 80°C for 5 min	2
Antithiamin factor	Raw fish, bracken fern	Increased requirement of thiamin	Cooking or steaming	2
Ascorbic acid oxidase	Cabbage, cucumbers, pumpkin, lettuce, spinach, tomatoes, beets, potatoes, carrots, peaches	Destruction of vitamin C	Cooking or blanching	2
Antivitamin E	Kidney beans, alfalfa, peas	Antagonistic toward vitamin E	Alcohol extraction	2
Lupine alkaloids	Lupine seeds	—	Boiling and soaking in water	33
Ricin	Castor bean meal	—	Steaming at 125°C for 15 min	34
Unknown	Tung meal	—	Extraction with organic solvents prior to autoclaving	35
Unknown	Kidney bean, black bean, jack bean, sword bean	—	Soaking prior to autoclaving	36
Tannins	Sorghum, barley, carob, rape seed, lespedeza forage, oak bark and leaves	Complex with proteins	Extraction with hot water or methanol	37—39
Allergens	Various plant and animal products	Nausea, vomiting, diarrhea	Autoclaving at 120°C for 30 min	40

REFERENCES

1. Liener, I. E. and Kakade, M. L., Protease inhibitors, in *Toxic Constituents of Plant Foodstuffs,* Liener, I. E., Ed., Academic Press, New York, 1969, 8—68.
2. Gantzea, I. and Sutzescu, P., Natural antinutritive substances, in *Foodstuffs and Forages,* S. Karger, Basel, 1968.
3. Kakade, M. L. and Liener, I. E., The increased availability of nutrients from plant foodstuffs through processing, in *Man, Food, and Nutrition,* Recheigl, M., Jr., Ed., CRC Press, Boca Raton, Fla., 1973, 231—241.
4. Jaffe, W. G., Hemagglutinins, in *Toxic Constituents of Plant Foodstuffs,* Liener, I. E., Ed., Academic Press, New York, 1969, 69—101.
5. Bernardi, L. C. and Goldblatt, L. A., Gossypol, in *Toxic Constituents of Plant Foodstuffs,* Liener, I. E., Ed., Academic Press, New York, 1969, 212—266.
6. Rojas, S. W. and Scott, M. L., Factors affecting the nutritive value of cottonseed meal as a protein source in chick diets, *Poult. Sci.,* 48, 819—835, 1969.
7. Baugher, W. L. and Campbell, T. C., Gossypol detoxification by fungi, *Science,* 164, 1526—1527, 1969.
8. Mayorga, H., Gonzalez, J., Menchu, F., and Rolz, C., Preparation of a low free gossypol cottonseed flour by dry and continuous processing, *J. Food Sci.,* 40, 1270—1273, 1975.
9. Sarma, P. S. and Padmanaben, G., Lathyrogens, in *Toxic Constituents of Plant Foodstuffs,* Liener, I. E., Ed., Academic Press, New York, 1969, 267—291.
10. Yang, M. G. and Mickelsen, O., Cycads, in *Toxic Constituents of Plant Foodstuffs,* Liener, I. E., Ed., Academic Press, New York, 1969, 159—167.
11. Van Etten, C. H., Goitrogens, in *Toxic Constituents of Plant Foodstuffs,* Liener, I. E., Ed., Academic Press, New York, 1969, 103—142.
12. Mustakas, G. C., Griffin, E. L., Jr., Gastrock, E. A., D'Aquin, E. L., Keating, E. J., and Patton, E. L., Enzymatic process for mustard seed to produce oil, meal, and allyl isothiocyanate, *Biotechnol. and Bioeng.,* 5, 27—39, 1963.
13. Mustakas, G. L., Kirk, L. D., and Griffin, E. L., Jr., Mustard seed processing. Bland protein meal, bland oil, and allyl isothiocyanate as a byproduct, *J. Am. Oil Chem. Soc.,* 39, 372—377, 1962.
14. Bell, J. M., Youngs, C. G., and Downey, R. K., A nutritional composition of various rapeseed and mustard seed solvent-extracted meals of different glucosinolate composition, *Can. J. Anim. Sci.,* 51, 259—269, 1971.
15. Korsrud, G. O. and Bell, J. M., Effects of various heat and moisture treatments on myrosinase activity and nutritive value of solvent extracted crambe seed meal, *Can. J. Anim. Sci.,* 47, 101—107, 1967.
16. Van Etten, C. H., Daxenbichlers, M. E., and Wolff, I. A., Natural glucosinolates (thioglucosides) in goods and feeds, *J. Agric. Food Chem.,* 17, 483—491, 1969.
17. Ballester, D., Rodrigo, R., Nakouzi, J., Chichester, C. O., Yanez, E., and Monckeberg, F., Rapeseed meal. III. A simple method for detoxification, *J. Sci. Food Agric.,* 21, 143—144, 1970.
18. Tookey, M. L., Van Etten, C. H., Peters, J. E., and Wolff, I. A., Evaluation of enzyme-modified, solvent extracted crambe seed meal by chemical analyses and rat feeding, *Cereal Chem.,* 42, 507—514, 1965.
19. Van Etten, C. H., Gegne, W. E., Robbins, J. J., Booth, A. N., Dexenbichlex, M. E., and Wolff, I. A., Biological evaluation of Crambe seed meals and derived products by rat feedings, *Cereal Chem.,* 46, 145—155, 1969.
20. Becker, K., Olson, A. C., Frederick, D. P., Kon, S., Gumbmenn, M. R., and Wagner, J. R., Conditions for the autolysis of alpha-galactosides and phytic acid in California small white beans, *J. Food Sci.,* 39, 766—769, 1974.
21. Okubo, K., Waldrop, A. B., Iacobucci, G. A., and Myers, D. V., Preparation of low-phytate soybean protein isolate and concentrate by ultrafiltration, *Cereal Chem.,* 52, 263—271, 1975.
22. Reinhold, J. G., Phytate destruction by yeast fermentation in whole wheat meals, *J. Am. Diet. Assoc.,* 66, 38—41, 1975.
23. Nelson, T. S., Shieh, J. N., Wodzinsky, R. J., and Ware, J. H., The availability of phytate phosphorus in soybean meal before and after treatment with a mold phytase, *Poult. Sci.,* 47, 1842—1848, 1968.
24. Schab, R. and Yanna, S., An improved method for debittering apricot kernels, *J. Food Sci. Technol.,* 10, 51—59, 1973.
25. Montgomery, R. D., Cyanogens, in *Toxic Constituents of Plant Foodstuffs,* Liener, I. E., Ed., Academic Press, New York, 1969, 143—157.

26. Akinrele, I. A., Fermentation of cassawa, *J. Sci. Food Agric.*, 15, 589—594, 1964.

27. Rahmen, S. A., De, S. S., and Subrahmanyan, V., Effect of different treatments on the removal of hydrocyanic acid from the burme bean (*Phaseolus lunatus* linn), *Curr. Sci.*, 11, 351—352, 1947.

28. Wood, T. D., The cyanogenic glucoside content of cassava and cassava products, *J. Sci. Food Agric.*, 16, 300—305, 1965.

29. Rackis, J. J., Oligosaccharides of legumes alpha-galactosidase and the flatus problem, *Am. Chem. Soc. Symp. Ser.*, 15, 207—222, 1975.

30. Sugimoto, H. and Van Buren, J. P., Removal of oligosaccharides from soy milk by an enzyme from *Aspergillus saitoi*, *J. Food Sci.*, 35, 655—660, 1970.

31. Thananunkul, D., Tanaka, M., Chichester, C. O., and Lee, T.- C., Degradation of raffinose and stachyose in soybean milk by α galactosidase and *Martierella vinacca*, Entrapment of α galactosidase within polyacrytamide gel, *J. Food Sci.*, 41, 173—175, 1976.

32. Calloway, D. H., Hickey, C. A., and Murphy, E. L., Reduction of intestinal gas-forming properties of legumes by traditional and experimental food processing methods, *J. Food Sci.*, 36, 251—255, 1971.

33. Tannous, R. Z., Shadarevian, S., and Cowan, J. W., Rat studies on quality of protein and growth inhibiting action of alkaloids of lupine (*Lupinus termis*), *J. Nutr.*, 94, 161—165, 1968.

34. Bolley, D. S. and Holmes, R. L., Inedible oilseed meals, in *Processed Plant Protein Foodstuffs*, Altschul, A. M., Ed., Academic Press, New York, 1958, 829—857.

35. Liener, I. E., Miscellaneous toxic factors, in *Toxic Constituents of Plant Foodstuffs*, Liener, I. E., Ed., Academic Press, New York, 1969, 410—448.

36. Liener, I. E., Toxic factors in edible legumes and their elimination, *Am. J. Clin. Nutr.*, 11, 281—298, 1962.

37. Tamir, M. and Alumot, E., Inhibition of digestive enzymes by condensed tannis from green and ripe carobs, *J. Sci. Food Agric.*, 20, 199—202, 1969.

38. Tamir, M. and Alumot, E., Carob tannis — growth depression and levels of insoluble nitrogen in the digestive tract of rats, *J. Nutr.*, 100, 573—580, 1970.

39. Vapar, Z. and Clandinin, D. R., Effect of tannins in rapeseed meal on its nutritional value for chicks, *Poult. Sci.*, 51, 222—228, 1972.

40. Perlman, F., Allergens, in *Toxic Constituents of Plant Foodstuffs*, Liener, I. E., Ed., Academic Press, New York, 1969, 319—348.

Table 2
EFFECT OF HEAT ON TRYPSIN INHIBITOR (TI) AND HEMAGGLUTINATING (HU) ACTIVITY OF SOME PLANT FOODSTUFFS

Product	Heat treatment	Loss of TI (%)	Loss of HU (%)	Ref.
Navy beans	Steam at 120°C for 5 min	82	100	1
(*Phaseolus vulgaris*)	Steam at 120°C for 15 min	89	100	
	Steam at 120°C for 30 min	89	100	
	Steam at 120°C for 60 min	0	100	
Faba beans	Steam at 121°C for 20 min	89	99	2
(*Vicia faba*)	Steam pelleting at 70°C	17	None	
	Extruded at 130°C	78	89	
	Extruded at 152°C	89	93	
	Microwave at 101°C for 20 min	83	52	
	Microwave at 107°C for 30 min	89	52	
Kidney beans	Steam at 121°C for 5 min	74	97	3
(*Phaseolus vulgaris*)	Steam at 121°C for 20 min	91	99	
Guar bean	Steam at 121°C for 5 min	—	100	
(*Cyamopsis tetragonoloba*)	Steam at 121°C for 20 min	—	100	
Lentil	Steam at 121°C for 5 min	—	94	
(*Lens esculenta*)	Steam at 121°C for 20 min	—	100	
Pea	Steam at 121°C for 5 min	73	100	
(*Pisum sativum*)	Steam at 121°C for 20 min	90	100	
Cow pea	Steam at 121°C for 5 min	70	100	
(*Vigna sinensis*)	Steam at 121°C for 20 min	89	100	
Chick pea	Steam at 121°C for 5 min	77	—	
(*Cicer arientinum*)	Steam at 121°C for 20 min	93	—	
Soybeans	Dry heat at 100°C for 30 min	—	4	4
(*Glycine max*)	Moist heat at 121°C for 30 min	—	100	
Natal beans	Dry heat at 100°C for 30 min	—	18	
(*Phaseolus vulgaris*)	Dry heat at 100°C for 18 hr	—	62	
	Moist heat at 121°C for 30 min	—	100	
	Dry heat at 100°C for 30 min	—	39	
Umzumbi	Dry heat at 100°C for 30 min	—	100	
(*Phaseolus vulgaris*)	Dry heat at 100°C for 18 hr	—	100	
	Moist heat at 121°C for 30 min	—	100	
Ground nut	Dry heat at 125°C for 30 min	27	—	5
(*Arachis hypogaea*)	Dry heat at 125°C for 5 hr	89	—	
	Dry heat at 140°C for 2 hr	92	—	
	Dry heat at 150°C for 1 hr	96	—	
	Moist heat at 108°C for 15 min	100	—	
	Moist heat at 108°C for 45 min	100	—	
Cow pea	Steam at 121°C for 15 min	10	—	6
(*Vigna sinensis*)	Cook at 90—95°C for 45 min	52	—	
	Toast at 210°C for 30 min	45	—	
	Toast at 240°C for 30 min	23	—	
	Extrusion cooking	19	—	

REFERENCES

1. **Kakade, M. L. and Evans, R. J.,** Nutritive value of navy beans (*Phaseolus vulgaris*), *Brit. J. Nutr.,* 19, 269—276, 1965.
2. **Marquardt, L. D., Campbell, L. D., and Ward, T.,** Studies with chicks on the growth depressing factor(s) in fava beans (*Vicia faba* var. *minor*), *J. Nutr.,* 106, 275—284, 1976.
3. **Tannons, R. I. and Ullah, M.,** Effects of autoclaving on nutritional factors in legume seeds, *Trop. Agric. Trinidad,* 46, 123—129, 1969.

4. DeMuelenaere, H. J. H., Effect of heat treatment on the hemagglutinating activity of legumes, *Nature*, 201, 1029—1030, 1964.
5. Woodham, A. A. and Dawson, R., The nutritive value of groundnut protein. I. Some effects of heat upon nutritive value, protein composition, and enzyme inhibitory activity, *Br. J. Nutr.*, 22, 589—600, 1968.
6. Elias, L. G., Hernandez, M., and Bressani, R., The nutritive value of precooked legume flour processed by different methods, *Nutr. Rep. Int.*, 14, 385—403, 1976.

Table 3

EFFECT OF GERMINATION ON THE TRYPSIN INHIBITOR AND HEMAGGLUTINATING ACTIVITIES OF SOME PEAS AND BEANS[a]

Trypsin Inhibitor Activity

Germination in Days	1	2	3	4	5	6	7	8	9	10	Ref.
Navy bean (*Phaseolus vulgaris*)	9	8	(3)	(5)	—	—	—	—	—	—	1
Soy bean (*Glycine max*)	0	2	6	8	—	—	—	—	—	—	2
Kidney bean (*P. vulgaris*)	—	—	—	(45)	—	—	—	(70)	—	—	3
Moth bean (*P. aconitifolius*)	0	—	40	—	—	—	—	—	—	—	4
Horsegram (*Dolichos biflorus*)	16	—	16	—	—	—	—	—	—	—	
Navy bean[b] (*P. vulgaris*)	—	7	—	42	—	81	—	88	—	95	5
Peas[b] (*Pisum sativum*)	17	50	—	78	—	78	—	84	—	89	6

Hemagglutinating Activity

	1	2	3	4	5	6	7	8	9	10	Ref.
Navy bean[a] (*P. vulgaris*)	—	77	—	89	—	96	—	97	—	98	5
Horsegram (*Dolichos biflorus*)	0	—	77	—	—	—	—	—	—	—	4
Moth bean (*P. vulgaris*)	0	—	0	—	—	—	—	—	—	—	4

[a] Loss or gain in activity (%) by days of germination.
[b] Germination was carried out in dark.

REFERENCES

1. Kakade, M. L. and Evans, R. J., Effect of soaking and germinating on the nutritive value of navy beans, *J. Food Sci.*, 31, 781—783, 1966.
2. Collins, J. L. and Sanders, G. G., Changes in trypsin inhibitor activity in some soybean varieties during maturation and germination, *J. Food Sci.*, 41, 168—172, 1976.
3. Palmer, R., McIntoch, A., and Pusztai, A., The nutritional evaluation of kidney beans (*Phaseolus vulgaris*): the effect on nutritional value of seed germination and changes in trypsin inhibitor content, *J. Sci. Food Agric.*, 24, 937—944, 1973.
4. Subbulakshmi, G., Ganeshkumar, K., and Venkataraman, L. V., Effect of germination on the carbohydrates, proteins, trypsin inhibitor, amylase inhibitor, and hemahemagglutinin in horsegram and mothbean, *Nutr. Rep. Int.*, 13, 19—31, 1976.
5. Kakade, M. L. and Liener, L. E., unpublished observation, 1967.
6. Hobday, S. M., Thruman, D. A., and Barber, D. J., Proteolytic and trypsin inhibitory activities in extracts of germinating *Pisum sativum*, *Phytochemistry*, 12, 1041—1046, 1973.

Table 4
EFFECT OF HEAT ON THE TRYPSIN INHIBITOR (TI) ACTIVITY OF SOYBEANS *(GLYCINE MAX)*

Soybean or soybean product	Heat treatment	Loss of TI activity (%)	Ref.
Whole soybeans	Steam at 100°C for 20 min	21	1
	Same as above, except beans were tempered at 25% moisture before steaming	97	
Cotyledons	Steam at 100°C for 20 min	60	
	Steam at 100°C for 30 min	91	
Chips	Steam at 100°C for 20 min	74	
	Steam at 100°C for 30 min	96	
Full fat flakes	Steam at 100°C for 15 min	95	
Defatted flakes	Steam at 100°C for 1 min	23	2
	Steam at 100°C for 2 min	41	
	Steam at 100°C for 3 min	54	
	Steam at 100°C for 4 min	69	
	Steam at 100°C for 6 min	71	
	Steam at 100°C for 7 min	83	
	Steam at 100°C for 9 min	79	
	Steam at 100°C for 10 min	87	
	Steam at 100°C for 20 min	90	
	Steam at 100°C for 30 min	92	
Whole soybeans	Steam at 120°C for 30 min	100	3
	Dielectric heating at 127°C for 0.83 min	32	
	Dielectric heating at 132°C for 1 min	62	
	Dielectric heating at 146°C for 1.67 min	68	
	Dielectric heating at 168°C for 2 min	72	
Soy milk	Cooked at 93°C for 15 min	34	4
	Cooked at 93°C for 30 min	44	
	Cooked at 93°C for 60 min	86	
	Cooked at 93°C for 120 min	96	
	Cooked at 93°C for 240 min	100	
	Cooked at 121°C for 2 min	64	
	Cooked at 121°C for 4 min	80	
	Cooked at 121°C for 8 min	84	
	Cooked at 121°C for 16 min	88	
	Cooked at 121°C for 32 min	95	

REFERENCES

1. **Rackis, J. J.**, Soybean trypsin inhibitors: their inactivation during meal processing, *Food Tech. (Chicago)*, 20, 102—104, 1966.
2. **Rackis, J. J., McGhee, J. E., and Booth, A. N.**, Biological threshold levels of soybean trypsin inhibitors by rat bioassay, *Cereal Chem.*, 52, 85—92, 1975.
3. **Borchers, R., Manage, L. D., Nelson, S. O., and Stetson, L. E.**, Rapid improvement in nutritional quality of soybeans by dielectric heating, *J. Food Sci.*, 37, 331—334, 1972.
4. **Hackler, L. R., Van Bureon, J. P., Steinkraus, K. H., El-Rawi, I., and Hand, D. B.**, Effect of heat treatment on nutritive value of soymilk protein fed to weanling rats, *J. Food Sci.*, 30, 723—728, 1965.

Table 5
EFFECT OF COMMERCIAL PROCESSING ON THE TRYPSIN INHIBITOR (TI) ACTIVITY OF COMMERCIAL SOYBEAN PRODUCTS

Commercial product	Loss of TI activity (%)	Ref.
Toasted soy flour	92	1
Soy protein concentrate	90	
Soy protein isolate	92	
	82	
Specially processed soy flour for calf feeding	96	
Soybean spun fiber	86	2
Chicken analogues	92	
Ham analogue	88	
Beef analogue	93	
Soy infant formulas	92	3
Extruded soy flour	94	4

REFERENCES

1. **Kakade, M. L., Rackis, J. J., McGhee, J. E., and Puski, G.,** Determination of trypsin inhibitor activity of soy products: collaborative analysis of an improved procedure, *Cereal Chem.,* 51, 376—382, 1974.
2. **Liener, I. E.,** Toxic factors in protein foods, in *Proteins in Human Nutrition,* Porter, J. W. G. and Rolls, B. A., Eds., Academic Press, London, 1973, 481—499.
3. **Churella, H. R., Yao, B. C., and Thomson, W. A. B.,** Soybean trypsin inhibitor activity of soy infant formulas and its nutritional significance for the rat, *J. Agric. Food Chem.,* 24, 393—397, 1976.
4. **Kakade, M. L. and Martin, S. M.,** Unpublished observation, 1976.

EFFECT OF PROCESSING ON MICROBIAL CONTAMINANTS IN FEEDS

M. van Schothorst and A. W. M. Brooymans

INTRODUCTION

Feeds are intended to support the life of animals. Sometimes, however, feeds or feed constituents may be cont.. ~inated with pathogenic microorganisms which may endanger the animals' health. Feed processing may reduce the health hazard intentionally or unintentionally. This chapter will describe the effect of certain processes on the microbial contaminants, but will also point to the problem of recontamination of feeds and feed ingredients. Main emphasis will be laid on feeds used for animals kept for food production (cattle, pigs, poultry), but foods for pets and laboratory animals will also be touched upon.

THE MICROBIOLOGY OF FEEDS

Unprocessed feed constituents of vegetable origin (grains, etc.) normally carry 10^6 to 10^8 of all kinds of microorganisms.[1,2] These organisms originate from the soil or in some cases from natural fertilizers. The majority of these microorganisms is harmless to animal or human health, but sometimes a limited number of potentially harmful organisms is present.

Feed constituents of animal origin (meat and bonemeal, fishmeal, etc.) are mostly heat processed, but not all processes eliminate all harmful microorganisms and these feeds are nearly always recontaminated during further processing, storage, and distribution. These meals can therefore be contaminated with all kinds of microorganisms including those of animal or human health significance.[3-5] The occurrence of potentially harmful microorganisms cannot be predicted, since the processing conditions and the contamination of the environment of fishmeal and rendering plants differ all over the world.

Processed meals of vegetable origin, like byproducts of the oil and fat industry (soymeal, peanutmeal, etc.) can also harbor all kinds of microbes. During extraction of the vegetable oils a reduction of vegetative bacteria will be obtained in most cases but, during further processing of the cakes or extracted material, recontamination, also with potential pathogens, normally occurs.[6]

The dominant flora in mixed feeds mostly reflects the flora of unprocessed vegetable meals and is of no concern to animal or public health microbiologists. This flora, however, plays an important role in the deterioration of feeds when the moisture content, more specifically the water activity (a_w),* rises above a certain level.[7] The relationship of moisture level and a_w in various feeds is presented in Figure 1. Depending on the a_w certain groups of microorganisms may start to grow while others are still inhibited. Yeasts and molds are the microorganisms growing at the lowest a_w, and with increasing a_w various cocci and lactobacilli start to multiply followed by Gram negative and other bacteria.[8,9] Moisture level is in most cases the only limiting factor for microbial growth since feeds contain all necessary nutrients and feeds are in most cases not stored under refrigeration. In the literature spoiled feeds have been mentioned as the cause of disease in animals, but only seldom have other specific causative agents than

---

* Water activity is defined as the ratio of the equilibrium vapor pressure and the maximum vapor pressure under condition of the same temperature.

Relationship of moisture level and wateractivity in various
feeds.

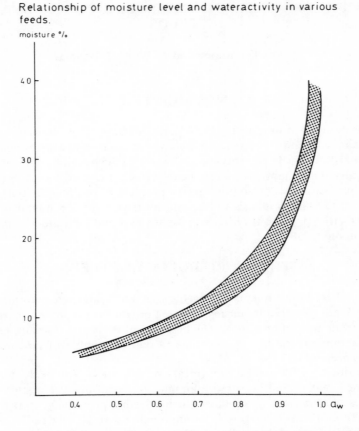

FIGURE 1.   Relationship of water activity and moisture level in various
feeds based on data of Carlson and Snoeyenbos[73] and Doesburg et al.[36,47]
Determinations were made in fishmeal, feathermeal, and poultry feed.

mycotoxins been described.[10,11] In the next part of this chapter the various groups of
microorganisms which may in principle provoke disease through ingestion of the feed
will be briefly mentioned and some of these will be selected to describe the effects of
processing.

MICROBIAL CONTAMINANTS

The role of feed as a causative vehicle of infectious diseases of animals is not well-
documented. Feeds may serve as transmitters of infectious agents or they may cause
disease through preformed bacterial or fungal metabolites. The formation of toxin in
feeds takes place only during or after growth of microorganisms in feeds. Sometimes
multiplication is necessary to reach the minimal infective dose, in other instances
growth of microorganisms is not necessary to provoke disease while in the case of
viruses multiplication is even impossible. In this chapter all microorganisms which are
of concern to animal or public health microbiologists are in principle regarded as contaminants. It would, however, be impossible to make a comprehensive list of all microorganisms which may be present as contaminants in feeds. Moreover, there is hardly
any information on the effect of processing on most of these microorganisms. Therefore, only a few representatives of the various groups of microorganisms having more
or less the same resistance to the different treatments will be discussed. The most im-

portant groups are viruses, sporeforming bacteria, nonsporeforming bacteria, and fungi. In the spread of animal diseases of viral origin feeding stuffs may play a role, but no information is available on the occurrence of viruses in commercial feeds. Representatives of pathogenic sporeforming bacteria occurring in feeds are *Clostridium perfringens*[12,13] and *Bacillus anthracis*,[14] isolated from meat and bone meal and *Clostridium botulinum* found in silage.[15,16] Most information is available on the occurrence of *Salmonella* as one of the nonsporeforming bacteria.[3,17-23] Salmonellae are not only of interest as potential pathogens for animals, but also as human pathogens. Ingestion with the feed does not frequently cause a disease, in most cases the animals only become symptomless excreters. Also, the metabolites of fungi are not only endangering animal but also human health through the residues in foods of animal origin.[24-26]

HEATING

Feed constituents of animal origin have mostly undergone a cooking process in which temperatures around 100°C or higher have been applied. The moisture levels during these processes are increased by the addition of water or the injection of steam. The purpose of the heating may be either to obtain tallow and fat or to destroy microorganisms. Sometimes the objective is to obtain a protein rich animal feed constituent like in the fishmeal industries, where the fish solubles are merely a byproduct. Destruction of pathogenic microorganisms is the main purpose of the rendering of condemned carcasses and animal offal as carried out in most European countries. Many different pieces of equipment and ways of processing are in use, but remarkably little information on heat penetration, and D values* of the various pathogenic microorganisms in this kind of material is available. Practical problems like the sampling of 15 tons (15,000 kg) of material in all sizes and shapes, as encountered in rendered carcasses, may be one of the reasons for the scarcity of information, another the difficulties in determining the composition of the initial flora. In the literature quite different information is found[3-5,27] regarding the flora of rendered material. Frequently it is not clear whether the bacteria are survivors or postprocess contaminants. Moist heating at 120 to 140°C for 20 min to 1 hr or even longer is recommended to obtain a "sterile" product. In the author's experience a moist heating of crushed carcasses during 20 min at 130°C together with the warming up and cooling down period leads to a reduction of 7 to 8 decimals of all the sporeforming bacteria normally encountered in condemned carcasses and heavily contaminated offal of animal origin. Absence of sporeformers may sometimes also be obtained at lower temperatures without prolonging the heating period, depending on fat content, particle size, and intial flora of the raw material.[28,29] While the heat processes used to destroy pathogenic microorganisms may lead to virtually sterile products, all other rendering processes will lead in principle to absence of vegetative microorganisms, because the raw material has to be heated, frequently with the addition of steam or water, to separate the fat or tallow. All the available information indicates that in the production of fishmeal all vegetative microorganisms are also destroyed.[30-32] To produce feathermeal the raw feathers are "hydrolyzed" under high pressure and temperatures, before further processing and also in the production of bloodmeal temperatures of approximately 100°C are used, leading to the absence of microorganisms with the exception of bacterial spores.

Whereas temperatures of 100°C and higher are required to kill spores of pathogenic bacteria, lower temperatures are sufficient to inactivate nonsporeforming microorganisms in feeds like Salmonellae. Especially with regard to Salmonellae, several experi-

---

* D value or decimal reduction value is defined as the time needed for a tenfold reduction of a certain microorganism or group of microorganisms at a certain temperature.

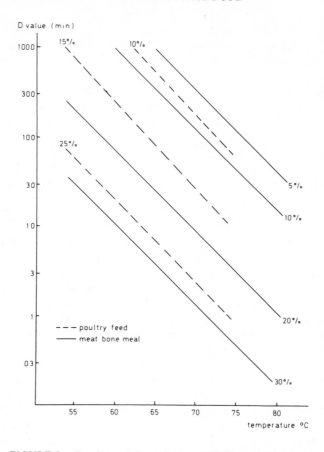

FIGURE 2.   D values of *S. senftenberg* 775W in meat and bone meal and poultry feed by heating at various moisture levels (Liu et al.[33]). Sterilized feeds were moistened with water and stored at room temperature for 2 weeks before inoculation. *Salmonella senftenberg* was grown in meat and bone meal slurry, dried and mixed with moisture stabilized feeds. The inoculated feeds were stored for 5 days before the heat resistance determinations were carried out in TDT tubes. Counts were made on trypticase soy agar.

ments have been carried out to determine time-temperature combinations to reduce them to safe levels. Vegetative microorganisms have a relatively high heat resistance at intermediate a_w (0.70 to 0.85) but with an increase of the water activity the D values decrease rapidly.[33] As a consequence, feeds have to be moistened with water or steam to eliminate Salmonellae successfully. The D values which have been published for the various temperatures and moisture levels for different types of feeds and *Salmonella* strains show a large variation. In Figures 2 and 3 some of the best documented data are presented. Reasons for these differences are manyfold since heat resistance may depend on experimental design, age of the inoculum, way of inoculation, water activity, method of moistening, equipment (TDT tubes), to name only a few of the variables (see also "Irradiation"). Also, the method of detection or enumeration of the survivors may influence the D values. Bacteria may survive a heat treatment in a more or less injured state and then should not be considered as dead. Selective methods should be used for the determination of a number of specific groups of microorganisms like Salmonellae and other Enterobacteriaceae. The differing degree of injury of a part of

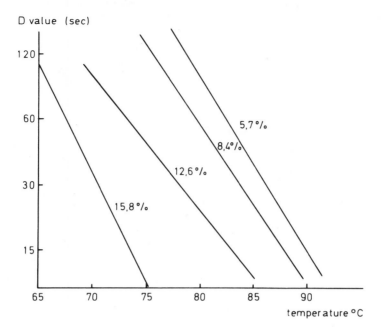

FIGURE 3. D values of *S. oranienburg* in fishmeal by heating at various moisture levels.[36] Full anchovy meals with 400 ppm ethoxiquin were moistened with water and stored at 0 to 3°C for a few days. *S. oranienburg* was grown in sterile stickwater and sprayed on fishmeal. After drying at 37°C to the required moisture level, the contaminated meals were mixed with the corresponding uninfected fishmeals and stored at 0 to 3°C for a few days. The heat resistance determinations were carried out in TDT plates of aluminum. The most probable number (MPN) of Salmonellae and their presence or absence (P/A) were determined with selenite broth (incubated at 37°C for 18 to 24 hr) and brilliant green agar plates.

the population and the counting methods applied may influence the D values more than has been recognized in the past.[34] When feeds are inoculated with Salmonellae a certain proportion of the inoculum dies rapidly in a short period of time (see "Storage"). The D values calculated on these data differ considerably from data which are based on the heat resistance of bacteria which have "survived" this initial decline[35] or from the D values which have been calculated using the so-called endpoint technique.[36] In the last method D values are calculated from results of heat treatments which reduce the *Salmonella* population to very low or nondetectable levels and therefore include the "tailing" effect.

To avoid problems of inoculation some investigators have used Enterobacteriaceae as indicator organisms, since these bacteria are normally present in sufficiently high numbers and since the heat resistance of most other Enterobacteriaceae is considered to be nearly the same as of most Salmonellae.[37] Some *Salmonella* strains (*S. senftenberg* 775 W), however, were found to have a higher heat resistance than others[38] and also some strains of the *Enterobacter-Citrobacter* group were found to have a higher heat resistance than other species of the Enterobacteriaceae.[39] Still these experiments are very useful. The results of one study on the reduction of Enterobacteriaceae in mixed feed are presented in Figure 4.

Fungi have more or less the same heat resistance as the more resistant nonsporeforming bacteria and viruses. D values (minimum) for *Aspergillus niger* at various temperatures and a_w values are 105 (90°C, 0.10 a_w), 210 to 216 (80°C, 0.20 to 0.30 a_w), 100

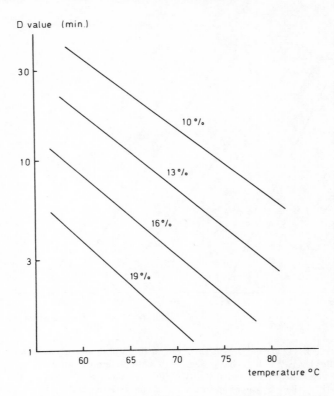

FIGURE 4.   D values of Enterobacteriaceae in mixed feed by heating at various moisture levels.[35] Naturally contaminated swine feed was moistened with water. The heat resistance determinations were carried out in TDT tubes. The most probable number (MPN) of Enterobacteriaceae was determined with EE broth incubated at 37°C.

(70°C, 0.60 a_w), 6 (55°C, 1.00 a_w). Increased D values were found for the extremely heat resistant *Humicola fuscoatra*.[40]

Yeasts are less heat resistant, D values (minutes) at 60°C for *Saccharomyces cerevisieae* are according to Hsieh et al.[41]: 0.04 (0.30 a_w), 0.09 (0.45 a_w), 0.76 (0.65 a_w), 2.3 (0.75 a_w), 1.5 (0.85 a_w), 0.4 (0.95 a_w).

When all the above-mentioned parameters are taken into account, D values are useful to design heat treatments, but one of the largest problems still remains to be solved, namely how many decimal reductions are necessary to obtain a safe product. In most instances the initial contamination level is not known, and the minimal infection dose of microbial contaminants is also not determined. Practical experiments with pigs have, however, demonstrated that a reduction of the Enterobacteriaceae in feeds with four to six decimals is sufficient to raise *Salmonella*-free animals, when other routes of infection have been excluded.[42,43]

PELLETING

Depending on animal species and region of the world, a varying percentage of the mixed feeds is pelletized. The pelleting process applied may differ, but the end product always consists of compressed feed particles of various sizes and shapes. During compression some heat may be generated which has an adverse effect on the microbial contaminants. The extent of the inactivation depends on a variety of factors, but tem-

Table 1
REDUCTION OF ENTEROBACTERIACEAE IN A MIXED SWINE FEED
DURING VARIOUS CONDITIONING AND PELLETING PROCEDURES[44]

| Pellet size (mm) | Moisture content (%) | | | Temperature (°C) | | Conditioning time (sec) | Enterobacteriaceae count (log 10)[a] | | |
	Meal[b]	Conditioned[c] meal	Pellets	Conditioned meal	Pellets		Meal	Conditioned meal	Pellets
5	12.2	14.2	14.4	51	73	458	5.6	5.1	2.1
8	12.2	12.2	12.3	28	78	437	5.7	5.7	<1.0
10	11.7	11.8	11.8	25	45	210	5.3	5.1	3.8
10	11.7	13.7	13.6	47	55	103	5.3	5.3	3.7
10	11.7	14.7	14.5	52	60	87	5.3	4.3	3.6
10	11.7	15.3	15.7	59	65	73	5.3	4.3	3.3
10	11.7	15.3	16.0	66	70	61	5.3	<1.0	<1.0
10	12.2	16.1	16.6	71	75	?	5.3	<1.0	<1.0
10	11.7	16.5	16.8	75	80	95	5.3	<1.0	<1.0
12	12.2	14.1	14.6	48	76	278	5.6	5.1	2.1

[a] Counts were determined with violet red bile glucose agar pour plates incubated at 37°C.
[b] Mean particle size 730 μ.
[c] Steam was injected during continuous mixing.

perature and moisture levels are the most important ones. Viruses, vegetative forms of bacteria, and molds may be affected, bacterial spores hardly ever. Pelleting may be applied for economic reasons (increase of feed efficiency, reduction of transport cost, etc.), but also to eliminate microbial contaminants like *Salmonella*. Especially with regard to the latter, data have been gathered on the decontaminating effect of various pelleting processes. Some decontamination may be obtained during pelleting of feed with a large particle size when a die with a mesh diameter of 5 mm is used without raising the moisture level. When other dies are used, leading to larger pellets, and when feeds are pelletized with smaller initial particle sizes (<845 μ), conditioning of the meal with steam prior to extrusion is needed to reduce Salmonellae and other Enterobacteriaceae in sufficient numbers to obtain a safe product. In Table 1 some data on the reduction of Enterobacteriaceae during conditioning and subsequent pelleting are presented. In Table 2 some of the results of extensive studies of Reinders[44] on the decontaminating effect of various pelleting procedures under commercial circumstances are presented. The overall conclusion is that a reduction of Enterobacteriaceae of at least three decimals can be obtained under most conditions of particle size and composition of the meal, when during conditioning a temperature is reached of 70°C or higher, the moisture content is raised with a minimum of 4% and the conditioning time is at least 70 sec. Other experiments have indicated that conditioning for 2 to 3 sec at 85°C reduced the Enterobacteriaceae with 5 decimals.[21]

Most publications on the effect of pelleting of Salmonellae do not describe the above-mentioned parameters in full detail. In most cases, however, a significant reduction of Salmonellae has been found.[45-48] It may be assumed that the findings for the whole family of Enterobacteriaceae can also be applied to Salmonellae. Moreover, field experiments using pelletized and nonpelletized meal have clearly demonstrated the "sanitizing" effect of pelleting mixed feed.[41,42] Molds can be inactivated by pelleting,[49-51] but data on the inactivation of viruses are not available. Most laboratory animal diets are pelletized, but they still contain 10^3 to 10^6 of microorganisms and can therefore not be used for germfree or even specific pathogen free (SPF) animals.[52-54]

Table 2

REDUCTION OF ENTEROBACTERIACEAE IN VARIOUS
MIXED FEEDS DURING PELLETING UNDER VARIOUS
CONDITIONS[44]

Mean particle size (μ)	Temperature (°C)	Time (sec)	Percent moisture increase	Enterobacteriaceae count (log 10)[a]	
				Meal	Pellets[b]
845	65	70	3.7[c]	3.3	<1.0
710	65	70	4.0	4.5	<1.0
475	65	70	3.2	5.0	2.1
845	70	70	3.6	3.3	<1.0
845	70	40	3.4	3.8	<1.0
845	70	7	2.5	3.6	<1.0
710	70	70	4.9	5.1	<1.0
710	70	40	3.8	4.7	<1.0
710	70	7	2.9	4.5	2.5
475	70	70	3.3	4.9	<1.0
475	70	40	3.5	4.9	<1.0
475	70	7	3.0	4.6	2.7
845	75	7	3.1	3.8	<1.0
710	75	7	2.6	4.5	<1.0
475	75	7	2.0	4.8	2.4

[a]   Counts were determined with violet red bile glucose agar pour plates incubated at 37°C.
[b]   Pellet size 10 mm.
[c]   Initial moisture level 11.8 to 12.9%.

EXPANSION — EXTRUSION

This process is almost only applied in the pet food industry. Extruded pet foods which are characterized by an open or "expanded" texture are produced by means of cooker-extruders. The more sophisticated cooker-extruders are segmented, double-walled, barrel-like machines with a centrally located screw which mixes and conveys the ingredient mix or slurry. The temperature of the segments and thus indirectly the temperature of the slurry can be controlled with steam or water. Measurement of the actual slurry temperature is virtually impossible in extruders, but is possible at the extruder exit. Extruder exit temperatures are necessarily in excess of 90°C and often between 100 and 130°C as otherwise the desired degree of expansion would not be obtained. The speed of the screw can usually be varied and thus the residence time in the extruder. However, as a result of laminar flow, the residence time of a given mix is not constant even at a constant screw speed, but is estimated as being generally less than 1 min under optimal process conditions. Published[55] and unpublished data strongly suggest a reduction of up to six to eight decimals of vegetative forms of microorganisms when heavily contaminated ingredients were used, leading to absence of Salmonellae. Additional factors which may affect the survival of microorganisms including spores are the steep pressure gradient occurring at the extruder exit which largely controls the degree of product expansion and the shear forces in the extruder. Following extrusion, products are cooled and dried to a moisture level of usually less than 10%.

Pet foods which are extruded but not dried to this moisture level are the intermediate-moisture pet foods (IMPF). These foods are usually formulated on a basis of fresh meat and meat byproducts or fish to which soy, protein concentrates, cereals, and minor nutrients are added. Despite a typical moisture content of between 20% and 40%, usually relatively mild thermo-processing and inexpensive protective wrapping in cellophane or similar air and moisture permeable materials, IMPFs show a virtually unlimited microbiological storage stability and do not require refrigeration. The microbial condition of IMPF has been designated as excellent[56] and in a recent study, Salmonellae could not be isolated from the limited number of samples studied.[23] The IMPF production process is basically similar to that of extruded dry pet foods, though the extruder exit temperature is lower as expansion is only rarely required. For nonexpanded IMPFs the extruder exit temperature must necessarily be somewhat above 80°C as otherwise a desirable degree of starch gelatinization would not be obtained, but cannot be much above 90°C at which temperature expansion would begin to occur. Higher extruder exit temperatures are obviously necessary if an expanded product is required. As in dry pet foods, the basic purpose of heat processing IMPF is twofold. Thermal processing is obviously needed to impart to the product the desired texture and other characteristics, but also to reduce the initial microbial load to a level at which storage stability can be maintained through the factors listed below and the product is safe both for the pet animal and its environment.

The storage stability of IMPF is obtained through a combination of factors which include manipulation of a_w and pH, the use of plasticizing humectants with bacteriostatic and germicidal properties (e.g., 1,2-propanediol-1,3-butanediol) and the use of preserving agents, in particular antimycotics.[57]

The a_w of IMPF is often above the level at which aerobic growth of *S. aureus* is possible (0.86). However, it is feasible to formulate IMPF-systems which react effectively to challenging with *Staphylococcus aureus*.[9] In an unpublished study of 21 samples of expanded and nonexpanded IMPF Staphylococci counts were below 10^2/g both in recently produced products and after several months storage. In the same study Enterobacteriaceae were also below 10^2/g.

Evidently postprocessing contamination should be avoided also in pet foods and at all costs. It is likely that postprocessing contamination was the cause of a reported finding of *Salmonella havana* in dry dog food.[58]

IRRADIATION

Irradiation of feeds has been advocated to eliminate Salmonellae, to sterilize feeds for laboratory animals and to eliminate *B. anthracis*. One of the advantages of irradiation above heat treatments is that no heat is generated and that as a consequence no cooling is required. Morever, feeds can be irradiated in closed bags so that recontamination is avoided. The D values* of various groups of microorganisms in the various feeds have not been studied extensively; some data are presented in Table 3. D values of Salmonellae have been determined more frequently and some of the findings are listed in Table 4. The differences in D values can be explained by differences in experimental conditions, like moisture level, partial oxygen pressure, temperature, pH, substrate, method of inoculation, method of bacterial analysis, etc.[59] (see also "Heating"). Doses of 0.5 to 0.8 Mrad are recommended to reduce Salmonellae in sufficient numbers to obtain a safe product.[60,61] A combination of pelleting (to reduce the initial contamination) and irradiation may reduce the required dose. Due to the fact that the initial contamination is not known, and also that a certain "tailing effect" occurs when inactivation rates are determined, the usefulness of D values to calculate "safe" doses is limited.[62] To inactivate fungi doses up to 1 Mrad may be necessary.[54]

---

* D value in irradiation is defined as the dose needed for a tenfold reduction of a certain microorganism or group of microorganisms.

Table 3
MEAN NUMBERS OF BACTERIA (LOG 10) IN FISHMEAL IRRADIATED WITH VARIOUS DOSES[76]

	Dose (Mrad)						
	0.0	0.2	0.4	0.5	0.6	0.7	0.8
Aerobic bacteria	5.7	3.8	3.5	3.5	3.2	3.0	2.8
Lancefield D streptococci	2.2	1.0	<1.0	<1.0	<1.0	<1.0	<1.0
Sulfite reducing clostridia	1.8	1.7	1.3	1.2	1.1	1.0	<1.0
Catalase-positive cocci	4.5	2.1	2.0	<2.0	<2.0	<2.0	<2.0

Table 4
D VALUES OF SALMONELLAE IN VARIOUS FEEDSTUFFS BY IRRADIATION

Material	Serotype, strain	D value (Mrad)	Ref.
Fishmeal	*S. binza*	0.09	60
	S. oranienburg	0.09	60
	S. senftenberg 1502	0.15	53
	S. senftenberg 56	0.19	84
	S. senftenberg 775W	0.19	84
Bone meal	*S. senftenberg* 775W	0.05	53
	S. typhi murium	0.09	53
Meat and bone meal	*S. typhi murium*	0.05	63
	S. give	0.05	63
	S. senftenberg	0.05	63
Mixed feed	*S. binza*	0.09	60
	S. oranienburg	0.10	60
	S. anatum	0.09	62
	S. enterititis	0.08	62
	S. give	0.08	62
	S. infantis	0.06	62
	S. typhi murium	0.10	62
	S. typhi murium SII	0.13	62

To reduce *B. anthracis* in meat and bonemeal with 5 to 6 decimals a dose of 1.25 Mrad may be required. To be on the safe side 2.0 Mrad has been advised.[63] To obtain a sterile laboratory animal diet doses of 2.5 to 5 Mrad are required.[51,53]

CHEMICAL TREATMENTS

Fumigation of feeds with gases like ethylene oxide and formaldehyde is used to "sterilize" or to "decontaminate" feeds. Bruhin[64] recommends sterilization of laboratory animal diets by fumigation with ethylene oxide under pressure (5.5 atm). For the treatment 1200 mg/l ethylene oxide is necessary at a temperature of 55°C during 60 min. Satisfactory sterilization of dried pelleted fish foods was achieved with an ethylene oxide carbon dioxide gas mixture (20:80).[65] To reduce the number of Salmonellae in meat and bone meal (7 to 15% moisture) with 5 decimals, Slavkov et al.[66] used approximately 1400 mg/l ethylene oxide at approximately 20°C. Tucker et al.[67] eliminated Salmonellae from poultry feed with methyl bromide (800 mg h/l at 25°C,

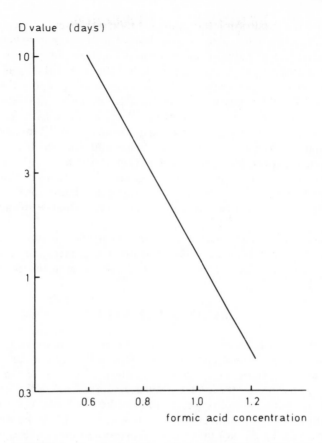

FIGURE 5. D values of Enterobacteriaceae in mixed feed by formic acid treatment.[69] Swine feed was mixed with various concentrations (%) of formic acid. The moisture content was 13 to 15%, the temperature 20°C. The most probable number (MPN) of Enterobacteriaceae was determined with EE broth incubated at 37°C.

relative humidity 70%). Fumigation with formaldehyde also proved to be effective for this purpose. A 5 min fumigation period during continuous mixing of poultry feed at 37°C and 60% relative humidity reduced the Salmonellae with more than 5 decimals.[68] The efficacy of most of these processes was checked by feeding experiments.

The effectiveness of the addition of organic acids to decontaminate feeds has also been studied. Addition of formic acid to mixed feeds at 20°C, 13 to 15% moisture proved to be more effective to reduce the number of Enterobacteriaceae than propionic, acetic, or lactic acid.[69] At a 1% concentration the D values in days were 1.5 for formic acid and more than 10 for the other organic acids. A reduction in D values was obtained by raising the temperature at which the acid was applied. The influence of the formic acid concentration on D values is presented in Figure 5. In feeding experiments a concentration of 0.9% formic acid was most efficient as regards the performance of the pigs and reduction of Salmonellae.[70]

Propionic and other organic acids can be useful to inhibit molds in high moisture feeds.[11,71]

STORAGE (EFFECT OF LOW WATER ACTIVITY)

It is well known that many bacteria may die off during storage. The rate of inactivation depends on the kind of microorganisms, moisture level (a_w), temperature, presence of oxygen, nature of the substrate, method of analysis, etc. Several studies were done on the inactivation of Salmonellae in various substrates at various a_w and temperatures.[47,71,73] As a general rule it can be stated that the higher the a_w and the temperature are, the greater the decrease in viability will be, unless temperature and a_w allows for multiplication. D values may vary from 1 day to several months, depending on the conditions during storage. The presence of highly unsaturated fatty acids in meat and fishmeal may accelerate the inactivation of Salmonellae[74,75] but Mossel and de Groot[76] demonstrated that the dying-off in fishmeal, meat and bonemeal, and caseine at the same a_w (0.46) was identical during a storage period of 6 months at 14°C. A significant loss of viability occurs when "unprotected" bacteria (i.e., grown in culture media) are brought in contact with feeds,[47,77,78] a phenomenon sometimes referred to as "osmotic shock". This effect does not occur in nature, but may influence storage stability and inactivation experiments.

The loss of viability of for instance Salmonellae in the various feeds may add an additional effect to irradiation, pelleting, and other treatments to control Salmonellae.[60,62] Storage, however, can in no way be relied upon as a means to reduce microbial contaminants to safe levels.

RECONTAMINATION

As stated before microbial contaminants may be present in mixed feeds due to contamination of unprocessed feed constituents of vegetable origin, to survival in feed constituents of animal origin, or to recontamination of processed feeds or feed constituents. In a discussion of the effect of processing on microbial contaminants in feeds, the problem of recontamination should not be overlooked. Carefully designed time-temperature schedules calculated on well-documented D values are virtually meaingless when sufficient care is not taken to avoid recontamination. Suspected processes for the production of feeds or feed constituents are dehydration, cooling, milling, sieving, sorting, mixing, transporting, and bagging since during these processes the feed may come into contact with contaminated air.[4,17,28,79-81] However, not only contaminated dust particles in the air may find access to the product, also flies, human hands, and excrements of birds and rodents may contaminate the product.[3,82] Dust especially has to be looked upon with concern since bacteria can multiply when the moisture level is raised through leakage of condensation water, leakage through open windows, cyclones, etc., and high numbers of microbial contaminants may find their way to the product.[83]

In conclusion, processed feeds cannot be regarded as safe unless adequate measures have been taken for the prevention of recontamination.

REFERENCES

1. Obi, S. C. Von, Die Bedeutung des Keimgehaltes in Versuchstiernahrung, *Berl. Münch. Tierärztl. Wochenschr.*, 84, 472—476, 1971.
2. Milanovic, A. and Beganović, H. A., Microflora of fodder of plant origin and mineral additives, *Veterinaria*, 23, 477—482, 1974.
3. Loken, K. I., Culbert, K. H., Solee, R. E., and Pomeroy, B. S., Microbiological quality of protein feed supplements produced by rendering plants, *Appl. Microbiol.*, 16, 1002—1005, 1968.
4. Tittiger, F., Studies on the contamination of products produced by rendering plants, *Can. J. Comp. Med.*, 35, 167—173, 1971.

5. Milanović, A. and Beganović, H. A., Microflora of fodder of animal origin, *Veterinaria,* 23, 467—476, 1974.

6. Rutqvist, L., Vorkommen von *Salmonella* in Futermitteln Vegetabilischen Ursprunges, *Zentralbl. Vet. Med.,* 10, 1016—1024, 1961.

7. Gedek, B., Futtermittelverderb durch Bacterien und Pilze und Seine Nachteiligen Folgen, *Obers. Tierernährg,* 1, 45—56, 1973.

8. Thalmann, A., Veränderungen der Mikroflora von Futtermitteln bei der Lagerung, *Landwirtsch. Forsch.,* 26(2), 58—62, 1971.

9. Haas, G. J. and Herman, E. B., Bacterial growth in intermediate moisture food systems, *Lebensm. Wiss. Technol.,* 11, 74—78, 1978.

10. Hoflund, S., Animal diseases associated with the use of deteriorated feeding stuffs under Swedish conditions, *Vet. Bull.,* 37, 701—717, 1967.

11. Harwig, J. and Munro, I. C., Mycotoxins of possible importance in diseases of Canadian farm animals, *Can. Vet. J.,* 16, 125—141, 1975.

12. Catsaras, M., Les *Clostridium* dans les aliments du betail, *Ann. Inst. Pasteur Lille,* 19, 55—61, 1968.

13. Jansen, M. M., Keimgehalt und Keimarten in Handelsfuttermitteln Pflanzlicher Herkunfst, *Übers. Tierernährg.,* 3, 277—289, 1975.

14. Davies, D. G. and Harvey, R. W. S., Anthrax infection in bone meal from various countries of origin, *J. Hyg.,* 70, 455—457, 1972.

15. Fjolstad, M. and Klund, T., An outbreak of botulism among ruminants in connection with ensilage feeding, *Nord. Veterinaer. Med,* 21, 609—613, 1969.

16. Haagsma, J. and Ter Laak, E. A., Type B botulism in cattle, caused by feeding grass silage. Report of a case, *Tijdschr. Diergeneesk.,* 103, 910—912, 1978.

17. Timoney, J., The sources and extent of *Salmonella* contamination in rendering plants, *Vet. Rec.,* 83, 541—543, 1968.

18. Tittiger, F. and Alexander, D. C., Recontamination of products with Salmonellae after rendering in Canadian plants, *Can. Vet. J.,* 12, 200—203, 1971.

19. Skovgaard, N. and Brest Nielson, B., *Salmonellas* in pigs and animal feeding stuffs in England and Wales and in Denmark, *J. Hyg.,* 70, 127—140, 1972.

20. Tompkin, R. B. and Kueper, T. V., Factors influencing detection of Salmonellae in rendered animal by-products, *Appl. Microbiol.,* 25, 485—487, 1973.

21. Stott, J. A., Hodgson, J. E., and Chaney, J. C., Incidence of Salmonellae in animal feed and the effect of pelleting on content of Enterobacteriaceae, *J. Appl. Bacteriol.,* 39, 41—46, 1975.

22. Reusse, U., Hafke, A., and Geister, R., Feststellung des Enterobacteriaceen-Gehaltes von Fischmehl zur Beurteilung der Salmonellen-Freiheit. *Zentralbl. Bakteriol. Hyg., I. Abt. Orig. B,* 162, 288—306, 1976.

23. D'Aoust, J. Y., *Salmonella* in commercial pet foods, *Can. Vet. J.,* 19, 99—100, 1978.

24. Armbrecht, B. H., Aflatoxin residues in food and feed derived from plant and animal sources in *Residue Rev.,* Vol. 41, Gunther, F. A., Ed., Springer-Verlag, Berlin, 1972, 13—54.

25. Elling, F., Hald, B., Jacobsen, C., and Krogh, P., Spontaneous toxic nephropathy in poultry associated with Ochratoxin A, *Acta Path. Microbiol. Scand.,* 83A, 739—741, 1975.

26. Krogh, P., Ochratoxin A residues in tissues of slaughter pigs with nephropathy, *Nord. Veterinaermd.,* 29, 402—405, 1977.

27. Hahn, G. von, Kondfeld, C. A., Hilburg, W., Sonnenschein, B., Flashoff, F. G., and Bisping, W., Hygienische Untersuchungen zur Kontrolle des Sterilisationsprozesses in Tierkörperbeseitigungsanstalt, *Dtsch. Tierärztl. Wochenschr.,* 84, 22—26, 1977.

28. Riedinger, O., Strauch, D., and Böhm, R., Die Abtötung von pathogenen und nichtpathogenen Sporenbildnern bei der Hitzesterilization von Schlachtabfallen, *Zentralbl. Vet. Med. B,* 22, 860—865, 1975.

29. von Bisping, W. and Bugl, G., Hygienische Forderungen an Einrichtungen und Betrieb von Anlagen zur Beseitigung von Schlachtäbfallen, *Berl. Münch. Tieräertzl. Wochenschr.,* 91, 238—241, 1978.

30. Quevedo, F., Les Enterobacteriaceae dans la farine de poisson, *Ann. Inst. Pasteur Lille,* 16, 157—162, 1965.

31. van Schothorst, M., Mossel, D. A. A., Kampelmacher, E. H., and Drion, E. F., The estimation of the hygienic quality of feed components using an Enterobacteriaceae enrichment test, *Zentralbl. Vet. Med., B.,* 3, 273—285, 1966.

32. Hauge, S., Fish meal and its bacteriological problems, in *The Microbiology of Dried Foods,* Kampelmacher, E. H., Ingram, M., and Mossel, D. A. A., Eds., Grafische Industrie, Haarlem, Netherlands, 1969, 446—450.

33. Liu, T. S., Snoeyenbos, G. H., and Carlson, V. L., Thermal resistance of *Salmonella senftenberg* 775W in dry animal feeds, *Avian Diseases,* 13, 611—631, 1969.

34. van Schothorst, M., Resuscitation of injured bacteria in foods, in *Inhibition and Inactivation of Vegetative Microbes, Series No. 5,* Skinner, F. A. and Hugo, W. B., Eds., Academic Press, London, 1976, 317—328.

35. Carroll, B. J. and Ward, B. Q., Control of Salmonellae in fish meal, *Fish. Ind. Res.,* 4, 29—36, 1967.

36. Doesburg, J. J., Lamprecht, E. C., and Elliott, M., Death rates of Salmonellae in fishmeals with different water activities, *J. Sci. Food Agric.,* 21, 636—640, 1970.

37. Kampelmacher, E. H., Guineé, P. A. M., van Schothorst, M., and Willems, H. M. C. C., Experimental studies to determine the temperature and duration of heat treatment required for decontamination of feed meals, *Zentralbl. Vet. Med.,* 1, 50—54, 1965.

38. Rossebø, L., Wet heat resistance in strains of *Salmonella senftenberg* isolated from herring meal, *Nord. Veterinaer Med.,* 22, 631—633, 1970.

39. Pietzsch, O. and Bulling, E., Die Futtermittel-Pelletierung, Ein Beitrag zur Salmonellose-Bekämpfung, *Fortschr. Veterinärmed.,* 25, 249—251, 1976.

40. Lubieniecki-von Schelhorn, M. and Heiss, R., The influence of relative humidity on the thermal resistance of mold spores, in *Water Relations of Foods,* Duckworth, R. B., Ed., Academic Press, London, 1975.

41. Hsieh, F., Acott, K., Elizondo, H., and Labuza, T. P., The effect of water activity on the heat resistance of vegetative cells in the intermediate moisture range, *Lebensm. Wiss. Technol.,* 8, 78—81, 1975.

42. Edel, W., Guineé, P. A. M., van Schothorst, M., and Kampelmacher, E. H., *Salmonella* infections in pigs fattened with pellets and unpelleted meal, *Zentralbl. Vet. Med.,* 14, 393—401, 1967.

43. Edel, W., Schothorst, M. van, Guineé, P. A. M., and Kampelmacher, E. H., Effect of feeding pellets on the prevention and sanitation of *Salmonella* infections in fattening pigs, *Zentralbl. Vet. Med.,* 17, 730—738, 1970.

44. Reinders, M. E., Technologisch onderzoek op het gebied van de Salmonelladecontaminatie van mengvoeders door middel van persen. Report No. 73—18, Instituut voor Graan, Meel en Brood TNO, Wageningen, Netherlands, 1973.

45. Christie, D. R., Cook, G. T., Dixon, J. M. S., Hobbs, B. C., Jones, H. C., King, G. J. G., McCoy, J. H., Norton, R., Pilsworth, R., Taylor, J., Tomlinson, A. J. H., Walker, J. H. C., Harvey, R. W. S., Jellard, C. H., Parker, M. T., and Tee, G. H., *Salmonella* organisms in animal feeding stuffs, *Mon. Bull. Minist. Health Public Health Lab. Serv.,* 20, 73—85, 1961.

46. von Kielstein, P., Bathke, W., and Schimmel, D., Der Einfluss der Pelletierung auf den Salmonellengehalt im Futter, *Mh. Vet. Med.,* 26, 12—17, 1970.

47. Doesburg, J. J., Lamprecht, E. C., and Elliott, M., Death rates of *Salmonellae* in fishmeals with different water activities, *J. Sci. Food Agric.,* 21, 632—635, 1970.

48. Slavkov, I. L., Milev, M., and Kolev, K., A study on the decontamination of *Salmonella*-contaminated forage mixtures, *Vet. Med. Nauki,* 13, 3—8, 1976.

49. Hörter, R., Kulturelle Keimzählungen an pelletierten Futtermittelproben, *Zentralbl. Bakt. Hyg. I. Abt. Orig.,* 207, 248—253, 1968.

50. Abily, B., Moreau, C., and Bougoin, M., Analyse mycologique d'un aliment pour volailles et de ses composants influence de la mise en pellets sur la mycoflore, *Bull. Inf. Stn. Exp. Avic. Ploufragan,* 14, 77—85, 1974.

51. Obi, S. K. C., The effects of pelleting and gamma irradiation on the microbial population of animal diets, *Zentralbl. Vet. Med.,* 25, 178—185, 1978.

52. von Hörter, R., Untersuchungen an Pelletierten Futtermitteln und ihre Eignung für SPF-Versuchstiere, *Berl. Münch. Tierärztl. Wochenschr.,* 17, 339—340, 1968.

53. Ley, F. J., Bleby, J., Goates, M. E., and Paterson, J. S., Sterilization of laboratory animal diets using gamma radiation, *Lab. Anim.,* 3, 221—254, 1969.

54. Münzner, R., Untersuchungen zur Strahlenbehandlung von Tierfutter, *Zentralbl. Bakt. Hyg., I. Abt. Orig. B.,* 158, 588—592, 1974.

55. Crane, F. M., Hansen, M., Yoder, R., Lepley, K., and Cox, P., Effect of processing feeds on molds, *Salmonella* and other harmful substances in feeds, *Feedstuffs,* 44, 34—36, 1972.

56. Mossel, D. A., Discussion contribution, in *Intermediate Moisture Foods,* Davies, R., Birch, G. G., and Parker, K. J., Eds., Applied Science, London, 1976, 258.

57. Acott, K., Sloan, A. E., and Labuza, T. P., Evaluation of antimicrobial challenge study for an intermediate moisture dog food, *J. Food Sci.,* 41, 541—546, 1976.

58. Pace, P. J., Silver, K. J., and Wisniewski, H. J., *Salmonella* in commercially produced dried dog food: possible relationship to a human infection caused by *Salmonella enteritidis* serotype Havana, *J. Food Protection,* 40, 317—321, 1977.

59. Mossel, D. A. A., van Schothorst, M., and Kampelmacher, E. H., Prospects for the *Salmonella* radicidation of some foods and feeds with particular reference to the estimation of the dose required, in *Elimination of Harmful Organisms from Food and Feed by Irradiation,* International Atomic Energy Agency, Vienna, 1967, 43—57.

60. Mossel, D. A. A., van Schothorst, M., and Kampelmacher, E. H., Comparative study on decontamination of mixed feeds by radicidation and by pelletisation, *J. Sci. Food Agric.,* 18, 362—367, 1967.

61. Reusse, U., Bischoff, J., Fleischhauer, G., and Geister, R., Pasteurisierung von Fischmehl durch Bestrahlung, *Zentralbl. Veterinaermed. B,* 23, 158—170, 1976.

62. Epps, N. A. and Idziak, E. S., Poultry feed radicidation. I. Microbial aspects of poultry feed irradiation, *Poult. Sci.,* 51, 277—282, 1972.

63. Hansen, P. I. E., Radiation treatment of meat products and animal by-products, in *Food Irradiation,* International Atomic Energy Agency, Vienna, 1966, 411—426.

64. von Bruhin, H., Futtersterilisation mit Aethylenoxid, *Z. Versuchstierk.,* 16, 315—321, 1974.

65. Trust, T. J. and Wood, A. J., An initial evaluation of ethylene oxide for the sterilization of formulated and pelleted fish feeds, *J. Fish. Res. Board Can.,* 30, 269—247, 1973.

66. Slavkov, I. L., Delchev, H., and Genchev, P., Attempts at the sterilization with ethylene oxide of bone-meal contaminated with Salmonellae, in *Salmonella and Salmonelloses in Bulgaria,* 2 Nats. Konf. Sofia, 1974, 411—414.

67. Tucker, J. F., Brown, W. B., and Goodship, G., Fumigation with methyl bromide of poultry foods artificially contaminated with Salmonella, *Br. Poult. Sci.,* 15, 587—595, 1974.

68. Duncan, M. S. and Adams, A. W., Effects of a chemical additive and of formaldehyde-gas fumigation on *Salmonella* in poultry feeds, *Poult. Sci.,* 51, 797—802, 1972.

69. Wieringa, G. W. and Viering, J., Het afsterven van Enterobacteriaceae in Mengvoer, *Mededelingen 395,* Instituut voor Bewaring en Verwerking van Landbouwprodukten, Wageningen, Netherlands, 1972, 1—4.

70. van der Wal, P., *Salmonella* control of feedstuffs by pelleting or acid treatment. Paper presented at the World Poultry Association meeting, December 12—14, Zichron, Yaacov, Israel, 1976.

71. Wieringa, G. W. and Viering, J., Invloed van organische zuren op gisten en schimmels in Mengvoer, *Mededelingen 416,* Instituut voor Bewaring en Verwerking van Landbouwprodukten, Wageningen, Netherlands, 1973, 1—3.

72. Liu, T. S., Snoeyenbos, G. H., and Carlson, V. L., The effect of moisture and storage temperature on a *Salmonella senftenberg* 775W population in meat and bone meal, *Poult. Sci.,* 48, 1628—1655, 1969.

73. Carlson, V. L. and Snoeyenbos, G. H., Effect of moisture on *Salmonella* populations in animal feeds, *Poult. Sci.,* 49, 718—726, 1970.

74. Khan, M. and Katamay, M., Antagonistic effect of fatty acids against *Salmonella* in meat and bone meal, *Appl. Microbiol.,* 17, 402—404, 1969.

75. Lamprecht, E. C. and Elliott, M. C., Death rate of *Salmonella oranienburg* in fish meals as influenced by autoxidation treatment, *J. Sci. Food Agric.,* 25, 1329—1338, 1974.

76. Mossel, D. A. A. and de Groot, A. P., The use of pasteurizing doses of gamma radiation for the destruction of Salmonellae and other Enterobacteriaceae in some foods of low water activity, in *Radiation Preservation of Foods,* National Academy of Sciences, National Research Council Publication 1273, Washington, D.C., 1965, 233—263.

77. Mossel, D. A. A. and Koopman, M. J., Losses in viable cells of Salmonellae upon inoculation into dry animal feeds of various types, *Poult. Sci.,* 44, 890—892, 1965.

78. Riemann, H., Effect of water activity on the heat resistance of *Salmonella* in "dry" materials, *Appl. Microbiol.,* 16, 1621—1622, 1968.

79. Magwood, S. E., Fung, J., and Byrne, J. L., Studies on *Salmonella* contamination of environment and product of rendering plants, *Avian Dis.,* 9, 302—308, 1965.

80. Orthoefer, J. G., Schrieber, M., Nichols, J. B., and Schneider, N., *Salmonella* contamination in a rendering plant, *Avian Dis.,* 12, 303—310, 1968.

81. Hess, G. W., Moulthrop, J. I., and Norton, H. R., New decontamination efforts and techniques for elimination of *Salmonella* from animal protein rendering plants, *J. Am. Vet. Med. Assoc.,* 157, 1975—1980, 1970.

82. Quevedo, F. and Carranza, N., Le rôle des mouches dans la contamination des aliments au Perou, *Ann. Inst. Pasteur Lille,* 17, 119—202, 1966.

83. Clise, J. D. and Swecker, E. E., Salmonellae from animal by-products, *Public Health Rep.,* 80, 899—905, 1965.

84. Underdal, B. and Rossebø, L., Inactivation of strains of *Salmonella senftenberg* by gamma irradiation, *J. Appl. Bacteriol.,* 35, 371—377, 1972.

85. Wieringa, G. W. and Viering, J., Uber die Einfluss des Wassergehaltes im Kraftfutter auf die Hitzeresistenz der Enterobacteriaceae, *Kraftfutter,* 53, 102—106, 1970.

Diets for Specific Animals

EFFECT OF PROCESSING ON NUTRITIVE VALUE OF DIETS FOR PIGS*

T. L. J. Lawrence

INTRODUCTION

Most of the raw materials suitable for including in pig diets, and the complete diets themselves, must be subjected to one or more processing techniques if their intake, and the utilization of nutrients contained within them, are to be optimal. In this context there is much variation in the techniques commonly used. Some are sophisticated and have to be applied in specialist plants (e.g., pelleting of diets). Others are relatively simple and can be applied at farm level (e.g., soaking diets in water or adding water to diets before feeding). In addition, some are physical in nature (e.g., grinding and rolling of cereals) while others may rely on the effect of a chemical substance which has been added (e.g., adding organic acids to high moisture cereal grains). The most important processing techniques may be summarized thus:

Physical	Chemical
Grinding	Soaking in water before feeding
Crimping	Adding water at feeding
Rolling	Adding enzymes
Dehulling	Adding salts of metals as detoxicants
Dry heating	Adding organic acids to preserve damp cereal grains and in ensiling processes
Wet heating	
Application of pressure	

CHEMICAL PROCESSING

Water Treatment

Soaking diets in water before feeding or adding water at feeding are processing techniques which are commonly used. Soaking has been the subject of much study in the past and is the source of much interest at present with the ever-increasing use of pipeline feeding systems dependent on the prior mixing and soaking of feeds at a central point. The general thesis is that soaking improves the nutritive value of the diet by its action on the cereal fraction, damaged cereal starch being converted to dextrins and reducing sugars by either natural cereal enzyme action and/or by enzymes produced by the growth of microorganisms.[1-4] Such improvements in the nutritive value of barley have been reflected in increased growth responses in the chick.[3-5] In the pig, however, there is little evidence to suggest that soaking improves nutritive value, most results indicating either no effect or, at the most, a very slight improvement.[6-12] On the other hand there is little evidence that soaking deleteriously affects either intake or resulting pig response.

In the last quarter of a century much effort has been devoted to the study of the effects of mixing food with water at feeding time compared with feeding the diet dry. A review in 1972[13] of 44 papers published during the previous 16 years found that

---

* Article written 1977.

while wet feeding had improved growth rate and efficiency of food conversion in 29 and 25 cases, respectively, approximately one third of the experiments showed no differences between the two systems of feeding. Often, advantages of wet over dry feeding in efficiency of food conversion may be more apparent than real due to the adding of water to feed preventing a certain amount of wastage of the dry feed and therein giving the apparent improvement in efficiency of food conversion. Support for this lies in the fact that in many,[14-16] though not all,[17] experiments no differences in digestibility between wet and dry feeding have been found.

Given free access to water the growing pig may regulate its intake to consume approximately 2.5 parts of water for each part of dry food which it eats.[18] If wet feeding is practiced then any ratio of water to dry food between 1.5:1 and 4:1 will be unlikely to affect performance.[16,19-21] Higher ratios than 4:1 may affect growth rate, but not always food conversion efficiency, by causing a decreased food intake.[14,15,22] This may be important in the young growing pig, immediately after weaning, with a small total alimentary tract capacity. Ratios of water to dry feed of less than 1.5:1 may also affect nutrient utilization.[23] At narrow (e.g., 2.5:1) ratios water temperature may have little deleterious effect[24] but the converse may be the case if the liquid intake is high.[25]

Enzyme Treatment

If α-amylase (EC 3.2.1.1) is added to cereals, in particular some varieties of barley, then the nutritive value is improved for the chick.[1,4,26,27] Similar additions to pig diets have so far failed to elicit a similar beneficial response.[8,28]

Salts of Metals

It is now well established that whole unextracted cottonseeds and raw soybeans are relatively unpalatable to the pig, are poorly utilized and cause growth depressions. In the cottonseed, the toxin gossypol is probably most responsible for this effect. Steam cooking of the whole cottonseed will reduce this toxicity by rupturing the resin glands to free the gossypol which then becomes attached to the proteins to form large complex molecules which are poorly absorbed. However, the steam cooking damages the proteins which are present. In a similar manner if oil is extracted from the seeds by the expeller process then the heat generated also damages the proteins which are present. Solvent extraction of oil leaves the proteins undamaged but leaves the resultant meals with fairly high levels of gossypol. The soaking of the whole or dehulled seeds in cold water to which calcium or iron (particularly the sulfate) salts have been added, followed by drying before feeding, have proved to be successful techniques for reducing toxicity without damaging the proteins.[30-33] The possible mode of action is one by which there is selective binding between the iron and the gossypol so that the latter cannot react with the epsilon amino group of lysine.

With whole raw soybeans, the growth depressions are probably due mostly to the presence of the trypsin inhibitor soyin or urease (EC 3.5.1.5) which causes a decreased food consumption in the first instance. These effects can to a large extent be ameliorated by either cooking the beans in solutions of or adding to the dry diet, copper sulfate.[34,35]

Organic Acids

If damp grain is stored aerobically it will heat. This may have a marked effect on its composition. Often reducing sugars will increase, the levels of some amino acids will decrease and this will cause a decrease in nitrogen digestibility and retention and reduce net energy availability and growth.[36] In addition fungal growth will be accelerated and the fungi of greatest concern include *Aspergillus flavus, Fusarium* and *Penicillium,* of which if there is a 3 to 5% invasion of the former, the cereal for the pig becomes potentially toxic.[37] To overcome these problems without having to resort to anaerobic storage, there has been an increasing tendency to treat cereals with organic acids. Propionic acid is probably the most widely used and possibly the most effective[38] although the successful treatment of cereals has been achieved with acetic, lactic, and formic acids either alone or in combination with each other and/or propionic acid. Compared with anaerobic storage fungal growth is kept at near zero levels[39,40] and the changes involving the production of lactic and acetic acids from carbohydrate fermentation, with accompanying increases in ammonia and soluble nitrogen levels, are inhibited.[41-44] Indeed there is some evidence to indicate that the protein quality of acid treated grain is better than that of ensiled or dried grain.[45,46] On the debit side organic acids may accelerate the natural storage decline in tocopherol levels.[47-49] Depending on circumstances this may[50] or may not[47,49] have a practical significance. The pig would appear to be able to tolerate higher levels of acid in the diet than those commonly used in grain preservation (up to a maximum of 2.5%) and there is evidence of an efficient absorption from the gut.[51,52] However, free fatty acid levels may increase in acid treated grain[53] and while up to approximately eight months postharvesting this may have little effect on voluntary consumption[54-56] after this time it may cause a decrease in intake.[49] In many experiments where the physical forms of the cereal have been the same, the acid treated cereal has given better growth and efficiency of conversion responses than the anaerobically stored[40,57,58] or dried[59] cereal and there is evidence[60] that a partial predigestion of the starch molecules and an easier penetration of the damp grain by digestive enzymes may be responsible for these responses.

Further potential uses of organic acids may be in storing undried field beans (*Vicia faba* L) in inducing an acceleration in enzyme hydrolysis in making fish silage and in treating whole rapeseeds (*Brassica* sp.). With field beans it has been shown that nitrogen digestibility in undried field beans treated with 2% propionic acid can equal that in soybean meal and exceed that in dried field beans.[61] In fish silage the liquefaction process which takes place can be accelerated by adding acids, organic acids being preferable to inorganic in that they do not need neutralizing before feeding. Formic acid may be best because it allows liquefaction to proceed at a high pH and because it is less corrosive than others.[62] The process is still imperfectly understood but it is probable that there is a rapid breakdown of proteins to low molecular weight peptides and free amino acids. Depending on solubility[63] fish silage can be a useful source of protein for the pig.[64,65] Last, it is possible that organic acids added to whole rapeseeds may improve their nutritive value although conflicting results have so far been produced.[66,67]

PHYSICAL PROCESSING

Dehulling

The potentially valuable components of many foodstuffs are relatively unavailable to the pig because of the fibrous coat which surrounds them. For example, in barley and oats the lemma and palea fuse to form two glumes which in turn fuse with the ovary to give the husk or hull which adheres to the endosperm at threshing. This contrasts to the situation with wheat, maize, and sorghum in which the lemma and palea are loose and separate at threshing leaving the naked kernel. These morphological differences are reflected in differences in nutrient availability for the pig and when the hull has been removed from barley,[68,69] whole rapeseeds (*Brassica* sp.)[70,71] and horse beans (*Vicia faba.* Var. *equina* Pers.)[72,73] the nutritive value has been improved.

Grinding and Rolling of Dried and Undried Dietary Components

In view of the fact that cereals form a high percentage of most pig diets this section will deal, almost exclusively, with the effects of grinding and rolling procedures on the nutritive value of dried and undried cereals. Other dietary ingredients usually form small percentages of the diet and the influence of particle size resultant from varying grinding procedures may be of small consequence.[74] With cereals, particle sizes resultant from differing grinding and rolling procedures would appear to be able to exert a profound effect on nutritive value and pig response. Thus, the pig may show a reluctance to eat barley, oats, and maize in very coarsely ground or whole forms.[75-78] If whole sorghum (*Sorghum vulgare* Pers)[79,80] maize,[81] barley,[8,75,82] or wheat[83] grains are eaten they are poorly utilized. Very coarse grinding or cracking of sorghum,[84] maize,[77,84,85] and barley[82,84,86] grains also gives poor utilization. It is difficult, however, to ascertain at what stage differences in particle sizes of smaller magnitudes than those above will cease to be of importance in affecting nutritive value and pig response. With barley it would appear likely that particle sizes resultant from using screen sizes up to 5.25 mm diameter are unlikely to impair utilization significantly[82,87] but that screen sizes of approximately 9.5 mm or above will affect utilization deleteriously.[82] However, the use of screen size may have serious limitations because nominally identical grinding procedure may give, ultimately, very different results because of hammer mills being operated differently by different individuals and because apparently identical screen sizes may not, in fact, be identical because of differences in the extent to which they have worn with normal usage. Instead, the use of "modulus of fineness of grinding",[88-90] offers a greater degree of accuracy in defining the overall particle size distribution. In this context a tentative conclusion is that a grinding process giving a modulus of fineness of grinding up to 2.2 will be unlikely to affect utilization. Compared with this, cold rolling, if carried out efficiently so that each grain is thoroughly flattened, is capable of eliciting similar responses.[82]

Wheat (because of a high glutelin content) and oats (because of a high crude fiber content), however, present more complex problems. Wheat is probably best prepared by efficient rolling.[16,83,91] Oats may best be prepared by fine grinding.[76,92,93]

Damp grain which has been stored anaerobically or with organic acids added as preservatives is difficult to grind and the alternative processing techniques available are those of plate milling and rolling. Experiments conducted on the nutritive value of barley processed by these two techniques have given very variable results.[40,47,50,54,55,94-98] A tentative conclusion from these experiments and work at this center[59] is that within any storage method rolling will give similar responses to dried ground or rolled barley providing the rolling is efficiently carried out and the moisture level is not above 22% approximately. With maize similar conclusions may be drawn.[49,99]

Heat Treatments

Heat is applied to many feedstuffs for the purpose of drying, reducing the levels of potentially toxic components and enzyme inhibitors, improving palatability and improving the nutritive value of cereals by disrupting the protein and starch matrices to render the carbohydrate fraction more available to the animal. These effects can be measured in vitro[100-104] but there is currently a need to standardize in vitro measurements so that in vivo responses can be measured against a standardized in vitro background.

Dry Heat

Because of wet harvesting conditions cereals have often to be dried and the evidence of the effects of drying on nutritive value is conflicting. With maize, drying temperatures between 60 and 150°C have in some cases had no effect on nutritive value.[105,106] In other cases drying temperatures between 90 and 150°C have given quadratic effects on pig nitrogen retention,[107,108] or (at 143°C cf 60°C or 88°C) deleteriously affected pig response.[109] In yet other cases drying temperatures between 371 and 482°C for periods of time between 2½ and 7 min have had no effect on nutritive value.[110] Some work[86] with barley has shown that high drying temperatures depress nitrogen retention, the depression being associated with a lower content of total and available lysine. Other work[111] has shown that a drying temperature of 90°C (cf 55, 67, or 80°C) need not necessarily affect lysine availability but may affect deleteriously growth rate, food conversion efficiency, and nitrogen digestibility.

Deliberate subjection of cereals, whole soybeans and other feedstuffs to dry heat (roasting) is often practiced with the aim of improving nutritive value per se or indirectly by removing enzyme inhibitors. Roasting soybeans[112,113] and field beans[114] has been shown to successfully destroy their inherent enzyme inhibitors. In the soybean evidence[115] suggests that a low (115°C) roasting temperature should be aimed at but that the growth response to temperatures between 110 and 160°C is quadratic with the greatest response occurring between 130 and 150°C.[74,116] In the rapeseed, heat treatment can successfully inactivate the native enzyme myrosinase,[117] which is thought to be largely responsible for the formation of isothiocyanates and oxazolidinethiones from the glucosinolates present. Similarly, in certain tropical legumes, roasting at 121°C for 15 min has been shown to improve the apparent digestibility of nitrogen and energy and to significantly improve growth rate.[118] Direct application of infrared heat to soybeans leaves satisfactory levels of available lysine and urease but gives a relatively poor response from the pig.[119] Micronizing (which involves some infrared heating) of field beans satisfactorily reduces the levels of trypsin inhibitors.[120]

Dry heating procedures applied to feedstuffs to improve, directly, nutritive value include roasting, popping, infrared heating (including micronization), and extrusion. Roasting of maize at 80, 100, 120, 140, and 160°C has been shown to have a quadratic effect on its metabolizable energy content and the nitrogen retained from it.[121] At 100°C there was no effect on growth rate but a significant improvement in food conversion efficiency.[121] Extrusion of full-fat soybeans and rapeseeds has been shown to improve their nutritive value.[122] With wheat, barley, and maize extrusion temperatures of 120°C may increase the initial amylolysis rate above that of popping and flaking procedures[123] and some evidence[124] suggests that the nutritive value of a barley based diet extruded at 40°C is superior to the nutritive value of the same diet extruded at 58°C after steam heating. The nutritive value of maize may, however, be unaltered by this processing procedure.[125] Although smaller than the effects from extrusion and flaking, popping increases the dry matter and nitrogen digestibilities of wheat but decreases the biological value of the protein.[125] The process may have little or no effect on the nutritive value of maize and barley[125] but does intensify the α-amylase activity in vitro and increase the long chain soluble carbohydrate formation.[126] Micronization of cereals has recently received much attention.[11,127-132] The measurement in vitro of the effects have been discussed[127] and although significant improvements in nutritive value have been found in maize, wheat and barley[11,128-130,132] the importance of physical processing after heating is clearly important for maximal responses to be achieved.[11,132] Results from some work[11,128] suggest that the nutritive value of maize is enhanced more than that of barley with improvements in wheat being very small and temperature dependent. Other work shows larger improvements in wheat.[132] Growth and efficiency of food conversion responses have been significantly improved from feeding micronized maize and barley diets.[128,132]

Wet Heat

The most important wet heating processes are those of expansion, flaking, and pelleting (and regrinding). With the latter process the feedstuff is also subjected to the pressure of the extrusion process. All processes may alter the external physical characteristics of the feedstuff but the primary aim, particularly with cereals, is to modify the structure of the starches present through the process of gelatinization. (This applies to a large extent to the dry heating processes mentioned previously). In expansion a high degree of starch gelatinization in cereals is usually achieved by softening the ground grain by steam heating and then forcing this material through a steel tube by an auger so as to increase heat and pressure until the material is eventually forced through the cone shaped holes of the expander head. On passing through these holes there is a sudden release of pressure and the escaping steam expands (gelatinizes) the grain. This process has been used mostly for processing maize and with this cereal and others there is no evidence that the nutritive value is improved.[133-136]

The flaking process may vary according to a number of factors and generally speaking it is a process which is applied most frequently to maize grain. Recently, however, its application to the potato to produce potato flakes has received attention,[137-141] and improvements in nutritive value have been found due partially to improved nitrogen utilization.[141] The process of flaking is applied to cereals in an attempt to improve their palatability and their nutritive value. In many diets flaked cereals (maize in particular) are added to keep the diet "open" and give it texture. If high levels of flakes are included it is possible, however, and dependent on the method of feeding, that palatability and nutritive value may be depressed.[142] Early work with maize[6,143] suggested that the flaking process improved digestibility appreciably and, largely as a result of this work, it was assumed that flaked maize contained approximately 8% more digestible energy than ground maize meal. Contrary to these findings some later work[144] showed that flaking had little or no effect on digestibility and that growth responses from flaked cereal based diets were similar to those from ground cereal-based diets.[144,145] Further work[146] showed flaking to improve the digestible energy content of maize by 3% (and of barley and wheat by 3 and 1%, respectively,) and to significantly depress nitrogen retention, growth and efficiency of food conversion responses (similar responses were found for barley and wheat). More recently French work[126,147] showed that flaking increased the long chain soluble carbohydrate formation, intensified in vitro α-amylase action upon starch, improved organic matter digestibility by approximately 3% but had no significant effect on nitrogen retention. These effects were accompanied by a higher water content in the gut, by a slower water flow from the stomach where the α-amylase activity was higher and the pH decreased later and more markedly. Also, in the small intestine, starch degredation was more intensive and could have explained the higher water retention. The improvement in digestible energy content was 6%. Other recent French work[148] has shown improvements in organic matter and nitrogen digestibilities of 3% for maize and 3 and 5%, respectively, for barley. These values were obtained with 21-day-old pigs and higher voluntary intakes of the flaked cereals were obtained. These variable responses to flaking may be explicable on the basis of variations in techniques used. It is impossible to obtain detailed information on the earliest work[143,144] but it is perhaps of some importance to note that some of the maize[143] had been degermed before cooking. In other work more details of the flaking process are available. In some work[142,146] the grain has first been cracked and then soaked in water for 48 hr before being passed through a steam chamber at 100°C and then rolled (flaked). In yet other work[126] no soaking was carried out and the maize was steam heated at 130°C for 45 to 55 min.

In the pelleting process the dietary components are subjected to pressure on being forced through the die of the cubing machine. There is a resulting frictional heat and prior to this, in most instances, the diet is steam heated. In some cases, however, no steam heating is applied ("cold" pelleting) and it is convenient to discuss this process here. The effects of these processes have been studied in many experiments. Two extensive reviews of most of these experiments have pointed overwhelmingly to an improvement in nutritive value and response in pigs from various pelleting procedures applied to a wide range of diets. In the first of these reviews[149] 117 experiments were considered and it was concluded, from all methods of feeding, that pelleting had improved growth rate by 6.6% and food conversion efficiency by 7.9%, this latter effect being largely a reflection of a decrease in food intake of 2.1%. In the second review,[13] in 57 published papers, growth rate and efficiency of food conversion were reported as having been improved in 39 and 48 cases, respectively. There is diversity of opinion as to whether all or some of the apparent advantages of cubing are retained when the pellet is reground or crumbled.[124,150-153]

The reasons behind the above improvements may be partially apparent in that there may be less feed wastage when pelleted diets are fed[154,155] and this may be reflected in the improved efficiency of conversion responses obtained. However, there is other evidence to suggest that the improvements may be real due to changes in the physical and chemical characteristics of the diet. Pelleting may reduce crude fiber[155-157] and moisture levels,[124] improve dry matter[124,159] and energy[124,159,160] digestibility, and improve amino acid[161] and phosphorus availability.[162] A combination of increased dry matter and improved digestibility of energy in the dry matter has been estimated[124] to mean that cubed diets, on restricted scales of feeding, may give up to 10% more digestible energy in an entire fattening period than diets of identical compositions in meal forms. However, the type of pelleting process may affect the degree of response achieved. Thus it is possible,[124,163] but not certain,[164] that the physical pressure of the process may be of greater importance than the steam heating. This would mean that cold pelleting could be more beneficial than hot pelleting. In one of the experiments cited above[124] this would imply a die extrusion temperature of 40°C compared with 70°C after steam heating at 58°C.

REFERENCES

1. **Willingham, H. E., Jensen, L. S., and McGinnis, J.,** Studies on the role of enzyme supplements and water treatment for improving the nutritional value of barley, *Poult. Sci.,* 38, 539—44, 1959.
2. **Thomas, J. M., Jensen, L. S., Leong, K. C., and McGinnis, J.,** Role of microbial fermentation in the improvement of barley by water treatment, *Proc. Soc. Exp. Biol. Med.,* 103, 198—200, 1960.
3. **Willingham, H. E., McGinnis, J., Nelson, F., and Jensen, L. S.,** Relation of superiority of water treated barley over enzyme supplements to antibiotics, *Poult. Sci.,* 39 (Abstr.), 1307, 1960.
4. **Burnett, G. S.,** The effect of damaged starch, amylolytic enzymes and proteolytic enzymes on the utilization of cereals by chickens, *Br. Poult. Sci.,* 3, 89—103, 1962.
5. **Fry, R. E., Allred, J. B., Jensen, L. S., and McGinnis, J.,** Influence of water treatment on the nutritional value of barley, *Proc. Soc. Exp. Biol. Med.,* 95, 249—251, 1957.
6. **Woodman, H. E.,** Digestion trials with swine. II. Comparative determinations of the digestibility of dry-fed maize, soaked maize, cooked maize and flaked maize, *J. Agric. Sci.,* 15, 1—25, 1925.
7. **Barber, R. S., Braude, R., and Mitchell, K. G.,** Effect of soaking the meal ration of growing pigs in water or skim milk, *Anim. Prod.,* 4, 313—18, 1962.
8. **Gill, D. R.,** Effect of barley pretreatment on feeding behaviour rate and efficiency of gains in swine, *Diss. Abstr.,* 25, 6862, 1965.
9. **King, J. O. L.,** Influence of the water treatment of rations supplemented with copper sulphate and an antibiotic on the growth rate of pigs, *Expl. Husb.,* 18, 25—31, 1969.
10. **Aumaitre, A., Henry, Y., Mercier, C., Ivorec-Szylit, O., and Thivend, P.,** The influence of soaking and precooking on the nutritive value of wheat and maize, *Ann. Zootech.,* 21, 133—137, 1972.
11. **Lawrence, T. L. J.,** An evaluation of the micronization process for preparing cereals for the growing pig. I. Effects on digestibility and nitrogen retention, *Anim. Prod.,* 16, 99—107, 1973.
12. **Smith, P.,** A comparison of dry, wet and soaked meal for fattening bacon pigs, *Expl. Husb.,* 30, 87—94, 1976.

13. **Braude, R.,** Feeding methods, in *Pig Production,* Cole, D. J. A., Ed., Butterworths, London, 1972, 279—291.
14. **Kornegay, E. T. and Vander Noot, G. W.,** Digestibility and nitrogen retention of swine fed rations with added water, *J. Anim. Sci.,* 24 (Abstr.), 892, 1965.
15. **Kornegay, E. T. and Vander Noot, G. W.,** Performance, digestibility of diet constituents and nitrogen retention of swine fed diets with added water, *J. Anim. Sci.,* 27, 1307—1312, 1968.
16. **Lawrence, T. L. J.,** High level cereal diets for the growing/finishing pig. I. The effect of cereal preparation and water level on the performance of pigs fed diets containing high levels of wheat, *J. Agric. Sci.,* 68, 269—274, 1967.
17. **Weissbach, F. and Laube, W.,** Effect of different moisture contents of the feed on digestibility of a mixed ration by growing pigs, *Arch. Tierz.,* 9, 333, 1963.
18. **Braude, R., Mitchell, K. G., Cray, A. S., Franke, A., and Sedgewick, P. H.,** Comparison of different levels of all meal feeding for fattening pigs, *J. Agric. Sci.,* 52, 223—229, 1959.
19. **Barber, R. S., Braude, R., and Mitchell, K. G.,** Further studies on the water requirement of the growing pig, *Anim. Prod.,* 5, 277—282, 1963.
20. **Bowland, J. P.,** Water restriction of market pigs, *44th Annual Feeders Day, Department of Animal Science, University of Alberta,* 1965, 18—20.
21. **Braude, R. and Rowell, J. G.,** Comparison of dry and wet feeding of growing pigs, *J. Agric. Sci.,* 68, 325—330, 1967.
22. **Rerat, A. and Fevrier, C.,** The influence of proportions of water in the diet on the growth and body composition of pigs, *Ann. Zootech.,* 14, 39—51, 1965.
23. **Cunningham, H. M. and Friend, D. W.,** Studies of water restriction on nitrogen retention and carcass composition of pigs, *J. Anim. Sci.,* 25, 663—67, 1966.
24. **Forbes, T. J. and Walker, N.,** The utilization of wet feed by bacon pigs with special reference to pipe-line feeding, *J. Agric. Sci.,* 71, 145—151, 1968.
25. **Holmes, C. W.,** Growth of pigs fed cool whey at two ambient temperatures, *Anim. Prod.,* 13, 1—6, 1971.
26. **Jensen, L. S., Fry, R. E., Allred, J. B., and McGinnis, J.,** Improvement in the nutritional value of barley for chicks by enzyme supplementation, *Poult. Sci.,* 36, 919—921, 1957.
27. **Herstad, O. and McNab, J. M.,** The effect of heat treatment and enzyme supplementation on the nutritive value of barley for broiler chicks, *Br. Poult. Sci.,* 16, 1—8, 1975.
28. **Burnett, G. S. and Neil, E. L.,** The influence of processing and of certain crude enzyme preparations on the utilization of cereals by pigs, *Anim. Prod.,* 6, 237—244, 1964.
29. **Baird, J. R., Cromwell, G. L., and Hays, V. W.,** Effects of enzyme supplementation and presoaking of diet on performance and nutrient digestibility in early weaned pigs, *J. Anim. Sci.,* 43 (Abstr.), 249, 1976.
30. **Stevenson, J. W., Cabell, C. A., and Kincaid, C. M.,** Methods of reducing cottonseed meal toxicity for swine, *J. Anim. Sci.,* 24 (Abstr.), 290, 1965.
31. **Clawson, A. J., Maner, J. H., Gomez, G., Mejia, O., Flores, Z., and Buitrago, J.,** Unextracted cottonseed in diets for monogastric animals. I. The effect of ferrous sulphate and calcium hydroxide in reducing gossypol toxicity, *J. Anim. Sci.,* 40, 640—647, 1975.
32. **Clawson, A. J., Maner, J. H., Gomez, G., Flores, Z., and Buitrago, J.,** Unextracted cottonseed in diets for monogastric animals. II. The effect of boiling and oven vs. sun drying following pretreatment with a ferrous sulphate solution, *J. Anim. Sci.,* 40, 648—654, 1975.
33. **Rincon, R. and Clawson, A. J.,** Detoxication of gossypol with iron, *J. Anim. Sci.,* 43 (Abstr.), 258, 1976.
34. **Young, L. G., Brown, R. G., Ashton, G. C., and Smith, G. C.,** Effect of copper on the utilization of raw soyabeans by market pigs, *Can. J. Anim. Sci.,* 50, 717—726, 1970.
35. **Young, L. G. and Smith, G. C.,** Processing soybeans with sodium hydroxide and copper sulphate for pigs, *Can. J. Anim. Sci.,* 53, 587—593, 1973.
36. **Matre, T., Nordum, E., Thorjørnsrud, B., and Homb, B. T.,** Experiments with self heated barley for pigs, *Beret. Inst. Husdyrernaering Foringslaere, Nor. Landbrukshøgsk.,* 151, 1—17, 1972.
37. **Jones, G. M., Mowat, D. N., Elliot, J. I., and Moran, E. T.,** Organic acid preservation of high moisture corn and other grains and the nutritional value: a review, *Can. J. Anim. Sci.,* 54, 499—517, 1974.
38. **Rao, C. S., Knake, R. P., Deyoe, C. W., and Allee, G. L.,** Preserving grain with organic acids, *Feedstuffs,* 46, 41—43, 1974.
39. **Jones, G. M., Donefer, E., and Elliot, J. I.,** Feeding value for dairy cattle and pigs of high moisture corn preserved with propionic acid, *Can. J. Anim. Sci.,* 50, 483—489, 1970.
40. **Livingstone, R. M., Dennerley, H., Stewart, C. S., and Elsley, F. W. H.,** Moist barley for growing pigs: some effects of storage method and processing, *Anim. Prod.,* 13, 547—556, 1971.
41. **Dumay, C., Delort-Laval, J., and Zelter, S. Z.,** Quality of wet maize grain silage preserved with propionic acid, *Journ. Rech. Porcine Fr.,* 127—129, 1972.

42. Darley, M. R. and Vetter, R. L., Artificially altered corn grain harvested at three moisture levels. I. Dry matter and nitrogen losses and changes in the carbohydrate fractions, *J. Anim. Sci.*, 38, 417—423, 1974.

43. Darley, M. R. and Vetter, R. L., Artificially altered corn grain harvested at three moisture levels. II. Changes in nitrogen fractions, *J. Anim. Sci.*, 38, 424—430, 1974.

44. Darley, M. R. and Vetter, R. L., Artificially altered corn grain harvested at three moisture levels. III. In vitro utilization of the carbohydrate and nitrogen fractions, *J. Anim. Sci.*, 38, 430—436, 1974.

45. Fevrier, C., Bourdon, D., and Chambolle, M., Feeding value of propionic acid treated high moisture maize for sows and piglets and of maize silage for growing-finishing pigs, *Journ. Rech. Porcine Fr.*, 135—142, 1972.

46. Gaye, M., Effect of propionic acid treatment of maize on the fattening of growing-finishing pigs, *Journ. Rech. Porcine Fr.*, 143—147, 1972.

47. Madsen, A., Mortensen, H. P., Larsen, A. E., Laursen, B., Keller-Nielsen, E., Welling, B., and Jensen, A., Moist barley preserved with propionic acid in the diet for bacon pigs, *Beret. Forsøgslab. Statens Husdrybrugsudualg.*, 1973, No. 407.

48. Allen, W. M., Parr, W. H., Bradley, R., Swannock, K., Barton, C. R. Q., and Tyler, R., Loss of vitamin E in stored cereals in relation to a myopathy of yearling cattle, *Vet. Rec.*, 94, 373—375, 1974.

49. Lawrence, T. L. J., High moisture maize grain for growing pigs: some effects on acceptability, digestibility, nitrogen retention and performance of physical form and tocopherol supplementation, *J. Agric. Sci.*, 86, 315—324, 1976.

50. Lawrence, T. L. J. and Boyd, J. W., Growth responses and serum enzyme activities in growing pigs resultant from supplementing diets based on aged barley with synthetic δ-α- tocopherol, *J. Agric. Sci.*, 88, 233—235, 1977.

51. Bowland, J. P., Young, B. A., and Milligan, L. P., Volatile fatty acids in pig diets, *49th Annual Feeders Day, Department of Animal Science, University of Alberta*, 1970, 14—16.

52. Lawrence, T. L. J., Volatile fatty acids as sources of energy in the diet of the growing pig, *Int. Res. Commun. Syst.*, (73-7) 45-6-1, 1973.

53. Young, L. G., Lun, A., and Forshaw, R. P., Selenium and α-tocopherol in stored corn, *J. Anim. Sci.*, 35 (Abstr.), 1112, 1972.

54. Perez-Alemen, S., Dempster, D. G., English, P. R., and Topps, J. H., Moist barley preserved with propionic acid in the diet of the growing pig, *Anim. Prod.*, 13, 271—278, 1971.

55. English, P. R., Topps, J. H., and Dempster, D. G., Moist barley preserved with propionic acid in the diet of the growing pig, *Anim. Prod.*, 17, 75—84, 1973.

56. Lynch, P. B., Hall, G. E., Hill, L. D., Hatfield, E. E., and Jensen, A. H., Chemically preserved high-moisture corns in diets for growing-finishing swine, *J. Anim. Sci.*, 40, 1063—1069, 1975.

57. Cole, D. J. A., Dean, G. W., and Luscombe, J. R., Single cereal diets for bacon pigs. II. The effect of methods of storage and preparation of barley on performance and carcass quality, *Anim. Prod.*, 12, 1—6, 1970.

58. Young, L. G., Brown, R. G., and Sharp, B. A., Propionic acid preservation of corn for pigs, *Can. J. Anim. Sci.*, 50, 711—715, 1970.

59. Lawrence, T. L. J., Rolled barley for the bacon pig: some effects on performance of grain moisture content and feeding method, *J. Sci. Food Agric.*, 22, 407—411, 1971.

60. Holmes, J. H. G., Bayley, H. S., and Horney, F. D., Digestion and absorption of dry and high moisture maize diets in the small and large intestine of the pig, *Br. J. Nutr.*, 30, 401—410, 1973.

61. Whittemore, C. T. and Taylor, A. G., Digestibility and nitrogen retention in pigs fed diets containing dried and undried field beans treated with propionic acid, *J. Sci. Food Agric.*, 24, 1133—1136, 1973.

62. Tatterson, I. N. and Windsor, M. L., Fish silage, *J. Sci. Food Agric.*, 25, 369—379, 1974.

63. Seve, B., Aumaitre, A., and Tord, P., Feeding value of soluble white fish meals prepared according to various technological procedures: incorporation into milk replacers of piglets weaned at 12 days, *Ann. Zootech.*, 24, 21—42, 1975.

64. Whittemore, C. T. and Taylor, A. G., Nutritive value for the growing pig of deoiled liquified herring offal preserved with formic acid (fish silage), *J. Sci. Food Agric.*, 27, 239—243, 1976.

65. Smith, P. and Adamson, A. H., An evaluation of liquified fish as a protein source for fattening pigs, *Anim. Prod.*, 22 (Abstr.), 161, 1976.

66. Bowland, J. P., Unprocessed rapeseed treated with propionic acid as an energy and protein supplement for market pigs, *51st Annual Feeders Day, Department of Animal Science, University of Alberta*, 1972, 1—3.

67. Bowland, J. P. and Newell, J. A., Ground rapeseed from low erucic acid (Lear) cultivars Span and Zephyr with or without organic acid treatment as a dietary ingredient for growing-finishing pigs, *Can. J. Anim. Sci.*, 54, 455—464, 1974.

68. Henry, Y. and Bourdon, D., Energy value and utilization of two types of barley (regular and hulless) by the growing-finishing pig, *Journ. Rech. Porcine Fr.*, 71—80, 1975.

69. Mitchell, K. G., Bell, J. M., and Sosulski, F. W., Digestibility and feeding value of hulless barley for pigs, *Can. J. Anim. Sci.,* 56, 505—511, 1976.

70. Leslie, A. J., Summers, J. D., and Jones, J. D., Nutritive value of air-classified rapeseed fractions, *Can. J. Anim. Sci.,* 53, 153—156, 1973.

71. Seth, P. C. C. and Clandinin, D. R., Metabolizable energy and composition of rapeseed meal and of fractions derived therefrom by air classification, *Br. Poult. Sci.,* 14, 499—505, 1973.

72. Pastuszewska, B., Duce, P. H., Henry, Y., Bourdon, D., and Jung, J., Utilization of whole and shelled horse beans by the growing pig: digestibility and amino acid availability, *Ann. Zootech.,* 23, 537—554, 1974.

73. Henry, Y. and Bourdon, D., Apparent digestibility of energy and protein in horse beans, with or without dehulling, as compared to soybean oil meal, *Journ. Rech. Porcine Fr.,* 105—114, 1973.

74. Olsen, E. M., Effect of roasting, particle size and dietary protein level on the utilization of soybeans by pigs, *Diss. Abstr.,* 32, 4695, 1972.

75. Haugse, C. N., Dinusson, W. E., Erickson, D. O., and Bolin, D. W., Effect of physical form of barley in rations for fattening pigs, *N. D. Agric. Exp. Stn. Bull.,* 17, 1966.

76. Crampton, E. W. and Bell, J. M., The effect of fineness of grinding on the utilization of oats by market hogs, *J. Anim. Sci.,* 5, 200—210, 1946.

77. Clawson, A. J., Effect of fineness of grinding on the growth of pigs raised in confinement and on pasture, *J. Anim. Sci.,* 21 (Abstr.), 377, 1962.

78. Maxwell, C. V., Reimann, E. M., Hoekstra, W. G., Kowalczyk, T., Benevenga, N. J., and Grummer, R. H., Effect of dietary particle size on lesion development and on the contents of various regions of the swine stomach, *J. Anim. Sci.,* 30, 911—922, 1970.

79. Baker, M. L. and Reinmiller, C. F., Feeding sorghum grain to growing and fattening pigs, *Nebr. Agric. Exp. Stn. Res. Bull.,* 323, 1939.

80. Aubel, C. E., The comparative value of corn and whole and ground milo with antibiotics as swine fattening feeds, *Kansas Agric. Exp. Stn. Bull.,* 308, 1954.

81. Young, L. G., Moisture content and processing of corn for pigs, *Can. J. Anim. Sci.,* 50, 705—709, 1970.

82. Lawrence, T. L. J., Some effects of including differently processed barley in the diet of the growing pig. I. Growth rate, food conversion efficiency, digestibility and rate of passage through the gut, *Anim. Prod.,* 12, 139—150, 1970.

83. Ivan, M., Giles, L. R., Alimon, A. R., and Farrell, D. J., Nutritional evaluation of wheat. I. Effect of preparation on digestibility of dry matter energy and nitrogen in pigs, *Anim. Prod.,* 19, 359—366, 1974.

84. Lawrence, T. L. J., High level cereal grains for the growing/finishing pig. II. The effect of cereal preparation on the performance of pigs fed diets containing high levels of maize, sorghum and barley, *J. Agric. Sci.,* 69, 271—281, 1967.

85. Moal, J. and Castaing, J., Effect of grinding fineness of feed mixtures (maize — soybean) on pig performances and gastric alterations, *Journ. Rech. Porcine Fr.,* 75—78, 1973.

86. Delort-Laval, J., Effect of grinding and heat treatment on the nutritional value of mature barley grain for growing pigs, *Journ. Rech. Porcine Fr.,* 115—119, 1972.

87. Simonsson, A. and Bjorklund, N. E., Differently processed barley for the growing pig. Growth rate, feed conversion efficiency, digestibility, carcass quality and development of esophogogastric lesions, in *Proc. Soviet-Swedish Symp. Ultuna,* 1970.

88. Hebblethwaite, P. and Hepherd, R. Q., A detailed procedure of testing for hammer mills and other farm grinding mills, *Nat. Inst. Agric. Eng. Tech. Memorandum,* 129, 1956.

89. Hebblethwaite, P., A brief survey on the fineness of grinding of barley for pigs, *J. Agric. Eng. Res.,* 3, 350—352, 1958.

90. Haigh, P. M. and Eden, A., Studies on particle size distribution in pig dietary constituents, *J. Agric. Sci.,* 81, 353—360, 1973.

91. Braude, R., Townsend, J., Harrington, G., and Rowell, J. G., A comparison of wheat and fine wheat offal in the rations of fattening pigs, *J. Agric. Sci.,* 57, 257—266, 1961.

92. Woodman, H. E., Evans, R. E., Menzies, A. W., and Kitchen, T., The value of oats in the nutrition of swine, *J. Agric. Sci.,* 22, 657—675, 1932.

93. Calder, A. F., Davidson, J., Duckworth, J., Hepburn, W. R., Lucas, I. A. M., Sokarovski, J., and Walker, D. M., Utilization by pigs of diets containing oats and oat husks ground to different degrees of fineness, *J. Sci. Food Agric.,* 10, 682—691, 1959.

94. Forbes, T. J., Holme, D. W., and Robinson, K. L., Further trials on feeding hermetically stored high moisture barley to pigs and cattle, *North Irel. Minist. Agric. Rec. Agricultural Res.,* 13, 151—159, 1964.

95. Livingstone, R. M. and Livingston, D. M. S., The use of moist barley in diets for growing pigs, *Anim. Prod.,* 12, 561—568, 1970.

96. Forbes, T. J., The utilization of chilled and hermetically stored undried barley by bacon pigs, *North Irel. Minist. Agric. Rec. Agricultural Res.,* 21, 47—55, 1973.

97. Thomke, S. and Tiden, A., Moist barley treated with propionic, acetic or formic acid in rations for growing pigs, *Swed. J. Agric. Res.,* 3, 145—151, 1973.

98. Cole, D. J. A., Brooks, P. H., English, P. R., Livingstone, R. M., and Luscombe, J. R., Propionic acid treated barley in the diets of bacon pigs, *Anim. Prod.,* 21, 295—302, 1975.

99. Young, L. G., Moisture content and processing of corn for pigs, *Can. J. Anim. Sci.,* 50, 705—709, 1970.

100. Hale, W. H., Grain processing as it affects beef cattle, *Feedstuffs,* 43, 71, 1971.

101. Hale, W. H., Influence of processing on the utilization of grains (starch) by ruminants, *J. Anim. Sci.,* 37, 1075—1080, 1973.

102. Mercier, C., Effects of various U.S. grain processes on the alteration and the in vitro digestibility of starch granule, *Feedstuffs,* 43, 33, 1971.

103. Croka, D. C. and Wagner, D. G., Micronized sorghum grain. II. Influence on in vitro digestibility, in vitro gas production and gelatinization, *J. Anim. Sci.,* 40, 931—935, 1975.

104. Prasad, D. A., Morrill, J. L., Melton, S. L., Dayton, A. D., Arnett, D. W., and Pfost, H. B., Evaluation of processed sorghum grain and wheat by cattle and by in vitro techniques, *J. Anim. Sci.,* 41, 578—587, 1975.

105. Jensen, A. H., Terrill, S. W., and Becker, D. E., Nutritive value of corn dried at 140°, 180° and 220°F for swine at different ages, *J. Anim. Sci.,* 19, 629—638, 1960.

106. Moal, J. and Castaing, J., Feeding value of maize subjected to different drying treatments: Two-stages drying or drying followed by delayed slow cooling, *Journ. Rech. Porcine Fr.,* 193—197, 1974.

107. Costa, P. M. A., Baker, D. H., Harmon, B. G., Norton, H. W., and Jensen, A. H., Studies on the nutritive value of roasted corns for growing and finishing swine, *J. Anim. Sci.,* 37 (Abstr.), 278, 1973.

108. Costa, P. M. A., Jensen, A. H., and Owens, F. N., The effect of drying temperature on the nutritive value of corn for swine, *J. Anim. Sci.,* 37 (Abstr.), 278, 1973.

109. Taylor, M. E., Pickett, R. A., Issacs, G. W., and Foster, G., Effects of dried corn on growing-finishing swine, *J. Anim. Sci.,* 23 (Abstr.), 894, 1964.

110. Jensen, A. H., Becker, D. E., and Harmon, B. G., Nutritive value of corn dried at different temperatures, in *Proc. Ill. Nutr. Conf.,* 1964.

111. Jones, A. S., Nutritional developments in the feeding of cereals to pigs, in *Cereal Supply and Utilization,* London U.S. Feed Grains Council, London Office, 1974, 77—84.

112. Combs, G. E., Conness, R. G., Berry, T. H., and Wallace, H. D., Effect of raw and heated soyabeans on gain, nutrient digestibility, plasma amino acids and other blood constituents of growing swine, *J. Anim. Sci.,* 26, 1067—1071, 1967.

113. Yen, J. T., Hymowitz, T., and Jensen, A. H., Effects of soyabeans of different trypsin inhibitor activities on performance of growing swine, *J. Anim. Sci.,* 38, 304—309, 1974.

114. Wilson, B. J., McNab, J. M., and Bently, H., Trypsin inhibitor activity in the field bean (*Vicia fabia* L.), *J. Sci. Food Agric.,* 23, 679—684, 1972.

115. Seerley, R. W., Emberson, J. W., McCampbell, H. C., Burdick, D., and Grimes, L. W., Cooked soybeans in swine and rat diets, *J. Anim. Sci.,* 39, 1082—1091, 1974.

116. Olsen, E. M., Young, L. G., Ashton, G. C., and Smith, G. C., Effects of roasting and particle size on the utilization of soybeans by pigs and rats, *Can. J. Anim. Sci.,* 55, 431—440, 1975.

117. Josefsson, E., Effects of variation of heat treatment conditions on the nutritional value of low glucosinolate rapeseed meal, *J. Sci. Food Agric.,* 26, 157—164, 1975.

118. Dividich, J. Le. and Seve, B., Energy and protein value of *Vigna sinensis* for the pig: effects of cooking and comparison with soyabean oil meal, *Ann. Zootech.,* 24, 13—20, 1975.

119. Faber, J. L. and Zimmerman, J., Evaluation of infrared-roasted and extruder-processed soyabeans in baby pig diets, *J. Anim. Sci.,* 36, 902—907, 1973.

120. McNab, J. M. and Wilson, B. J., Effects of micronizing on the utilization of field beans (*Vicia fabia* L.) by the young chick, *J. Sci. Food Agric.,* 25, 395—400, 1974.

121. Costa, P. M. A., Jensen, A. H., Harmon, B. G., and Norton, H. W., The effects of roasting temperature on the nutritive value of corn for swine, *J. Anim. Sci.,* 42, 365—374, 1976.

122. Bayley, H. S. and Summers, J. D., Nutritional evaluation of extruded full-fat soybeans and rapeseeds using pigs and chickens, *Can. J. Anim. Sci.,* 55, 441—450, 1975.

123. Delort-Laval, J. and Mercier, C., Efficiency of various treatments of cereals. I. Selection of treatments and study of their influence on the carbohydrate fraction of wheat, barley and maize, *Ann. Zootech.,* 25, 3—12, 1976.

124. Lawrence, T. L. J., Cubing the diet of the growing pig: some effects on nutritive value of temperature, pressure and physical form, *J. Sci. Food Agric.,* 22, 403—406, 1971.

125. Essatara, M., Saintaurin, M. A., and Abraham, J., Efficiency of various treatments of cereals. III. Effect of popping, extrusion and flaking on the feeding value of wheat and maize for growing rats, *Ann. Zootech.,* 25, 31—39, 1976.

126. Borgida, L. P., Maize and barley flaking and popping. I. Processing. II. Flaked maize digestion in the growing pig, *Journ. Rech. Porcine Fr.,* 99—104, 1975.

127. Papasalamontas, S. A., Putting a measure to micronization, in *Feed Grains,* U.S. Feed Grains Council, London Office, No. 18, 1975.

128. Lawrence, T. L. J., An evaluation of the micronization process for preparing cereals for the growing pig. II. Effects on growth rate, food conversion efficiency and carcass characteristics, *Anim. Prod.,* 16, 109—116, 1973.

129. Lawrence, T. L. J., An evaluation of the micronization process for preparing cereals for the growing pig. III. A note on the effect of micronization temperature on the nutritive value of wheat, *Anim. Prod.,* 20, 167—170, 1975.

130. Shiau, S. Y., Yang, S. P., Tribble, L. F., and Williams, I. L., Effect of micronizing of sorghum for swine, *J. Anim. Sci.,* 43 (Abstr.), 258, 1976.

131. Fernandes, T. H., Hutton, K., and Smith, W. C., A note on the use of micronized barley for growing pigs, *Anim. Prod.,* 20, 307—310, 1975.

132. Tardif, H. and Leuillet, M., Effects of micronizing of maize and wheat on the performance of *ad libitum* fed pigs, *Journ. Rech. Porcine Fr.,* 17—24, 1976.

133. Riker, J. T., Perry, T. W., Pickett, R. A., Featherstone, W. R., Beeson, W. M., and Curtin, T. M., Influence of gelatinized barley, corn, milo or wheat on the occurrence of gastric ulcers in swine, *J. Anim. Sci.,* 23 (Abstr.), 1218, 1964.

134. Riker, J. T., Perry, T. W., Pickett, R. A., and Curtin, T. M., Influence of various grains on the incidence of esophogastric ulcers in swine, *J. Anim. Sci.,* 26, 731—735, 1967.

135. Nuwer, A. J., Perry, T. W., Pickett, R. A., and Curtin, T. M., Expanded or heat-processed fractions of corn and their relative ability to elicit esophogogastric ulcers in swine, *J. Anim. Sci.,* 26, 518—525, 1967.

136. Maxson, D. W., Stanley, G. R., Perry, T. W., Pickett, R. A., and Curtin, T. M., Influence of various ratios of raw and gelatinized corn, oats, oat components and sand on the incidence of esophogogastric lesions in swine, *J. Anim. Sci.,* 27, 1006—1010, 1968.

137. Whittemore, C. T., Taylor, A. G., and Elsley, F. W. H., The influence of processing on the nutritive value of the potato, *J. Sci. Food Agric.,* 24, 539—545, 1973.

138. Whittemore, C. T., Taylor, A. G., and Crooks, P., The nutritive value for young pigs of cooked potato flake in comparison with maize meal, *J. Agric. Sci.,* 83, 1—5, 1974.

139. Hillyer, G. M. and Whittemore, C. T., Intake by piglets of diets containing cooked potato flake, *J. Sci. Food Agric.,* 26, 1215—1217, 1975.

140. Whittemore, C. T., Moffat, I. W., and Taylor, A. G., Influence of cooking upon the nutritive value of potato and maize in diets for growing pigs, *J. Sci. Food Agric.,* 26, 1567—1576, 1975.

141. Whittemore, C. T., Taylor, A. G., Moffat, I. W., and Scott, A. J., Nutritive value of raw potato for pigs, *J. Sci. Food Agric.,* 26, 255—260, 1975.

142. Lawrence, T. L. J., High level cereal diets for the growing/finishing pig. III. A comparison with a control diet of diets containing high levels of maize, flaked maize, sorghum, wheat and barley, *J. Agric. Sci.,* 70, 287—297, 1968.

143. Woodman, H. E. and Evans, R. E., The value of degermed maize meal (cooked) in the nutrition of swine, *J. Agric. Sci.,* 22, 670—675, 1932.

144. Sheehy, E. J. and Senior, B. J., The comparative feeding value for pigs of cereals prepared in the flaked and ground forms, *J. Dep. Agric. Repub. Ire.,* 36, 230—245, 1939.

145. Burnett, G. S. and Neil, E. L., The influence of processing and of certain crude enzyme preparations on the utilization of cereals by pigs, *Anim. Prod.,* 6, 237—244, 1964.

146. Lawrence, T. L. J., High level cereal diets for the growing finishing pig. IV. An evaluation of flaked maize, wheat and barley when included at high levels in the diet of the weaned pig grown to cutter weight (160 lb), *J. Agric. Sci.,* 79, 155—160, 1972.

147. Borgida, L. P., Effect of flaking on the digestive pattern and nutritive value of maize in growing pigs, *Ann. Zootech.,* 25, 337—349, 1976.

148. Aumaitre, A., Efficiency of various treatments of cereals. IV. Effect of flaking and popping of barley and maize on the performance of piglets weaned at 21 days; influence on the digestibility of the dietary constituents, *Ann. Zootech.,* 25, 41—51, 1976.

149. Vanschoubroek, F., Coucke, L., and van Spaendonck, R., The quantitative effect of pelleting feed on the performance of piglets and fattening pigs, *Nutr. Abstr. Rev.,* 41, 1—9, 1971.

150. Dinusson, W. E. and Bolin, D. W., Comparisons of meal, crumbles, pellets and repelleting feeds for swine, *N.D. Agric. Exp. Stn. Bull.,* 20, 16—20, 1958.

151. Laird, R. and Robertson, J. B., A comparison of cubes and meal for growing and fattening pigs, *Anim. Prod.*, 5, 97—103, 1963.

152. Jensen, A. H. and Becker, D. E., Effect of pelleting diets and dietary components on the performance of young pigs, *J. Anim. Sci.*, 24, 392—397, 1965.

153. Pettersson, A. and Bjorklund, N. E., Crumbles contra meal for bacon pigs. Effect on daily gain, feed efficiency, carcass quality and on the oesophageal part of the stomach, *Acta Agric. Scand.*, 26, 130—136, 1976.

154. Braude, R. and Rowell, J. G., Comparison of meal and pellets for growing pigs fed either in troughs or off the floor, *J. Agric. Sci.*, 67, 53—57, 1966.

155. Baird, D. M., Influence of pelleting swine diets on metabolizable energy, growth and carcass characteristics, *J. Anim. Sci.*, 36, 516—521, 1973.

156. Bohman, V. R., Kidwell, J. F., and McCormick, J. A., High levels of alfalfa in the rations of growing-fattening swine, *J. Anim. Sci.*, 12, 876—880, 1953.

157. Cameron, C. D. T., Effects of high fiber and pelleted and nonpelleted high fiber — high fat rations on the performance and carcass characteristics of bacon pigs, *Can. J. Anim. Sci.*, 40, 126—133, 1960.

158. Meade, R. J., Dukelow, W. R., and Grant, R. S., Influence of percentage oats in diet, lysine and methionine supplementation and of pelleting on rate and efficiency of gain of growing pigs and on carcass characteristics, *J. Anim. Sci.*, 25, 58—63, 1966.

159. Gorrill, A. D. L., Bell, J. M., and Williams, C. M., Ingredient and processing interrelationships in swine feed. I. Effects of antibiotics, protein source and wheat bran on the responses to pelleted feed, *Can. J. Anim. Sci.*, 40, 83—92, 1960.

160. Seerley, R. W., Hoefer, J. A., and Miller, E. R., Digestible energy and nitrogen and rate of passage of food in pigs fed meal and pellets, *J. Anim. Sci.*, 19 (Abstr.), 1291, 1960.

161. Yen, J. T., Baker, D. H., Harmon, B. G., and Jensen, A. H., Corn gluten feed in swine diets and effect of pelleting on tryptophan availability to pigs and rats, *J. Anim. Sci.*, 33, 987—991, 1971.

162. Bayley, H. S., Pos, J., and Thomson, R. G., Influence of steam pelleting and dietary calcium level on the utilization of phosphorus by the pig, *J. Anim. Sci.*, 40, 857—863, 1975.

163. Mercier, C. and Guilbot, A., Effect of pelleting processing on the physiochemical characteristics of starch, *Ann. Zootech.*, 23, 241—251, 1974.

164. Melcion, J. P., Vaissade, P., Valdebouze, P., and Viroben, G., Effect of pelleting conditions on some physico-chemico characteristics of a piglet diet, *Ann. Zootech.*, 23, 149—160, 1974.

EFFECT OF PROCESSING ON NUTRITIVE VALUE OF DIETS FOR POULTRY

James M. McNab

INTRODUCTION

Increasing pressures on world food resources and increasing sophistication in the nutritional information applied in the formulation of livestock diets have resulted in the expenditure of considerable effort to enhance the feeding value of particular ingredients by processing. When consideration is being given to processing feedstuffs for poultry, unlike foods for human consumption, economic considerations are paramount and organoleptic factors of relatively minor importance. Processing costs invariably can be tolerated only if they are accompanied by a more than commensurate increase in the value of yield from the animal being fed. Unfortunately, defining an improvement in economic terms is an almost impossible task, for three major reasons.

First, poultry are offered feedstuffs of almost infinite variability; for instance, it is well-known that appreciable chemical differences can occur between different batches of what is known as the same raw material. These differences can be caused, if the feedstuff is of plant origin for example, by breed, climate, soil conditions, or fertilizer application. For practical reasons such variations are generally ignored by animal nutritionists, whose experimental work is usually carried out on a single, ill-defined sample of the material; the responses obtained are then considered typical for that feedstuff.

Second, feedstuffs are often subjected to processes which are difficult to define by physical parameters in a way which allows the repetition of a particular set of circumstances. For example, many of the more common feed ingredients, e.g., the oilseed meals, are secondary or even tertiary products and have already been subjected to processing. Formerly, because these pretreatments had little bearing on improving the nutritive value of the secondary products, the demands of the animal feed compounder for a uniformly high quality product were seldom given much consideration. The use of trichloroethylene as a solvent for extracting oil from seeds is a good example of the type of problem which can emerge under these circumstances. Trichloroethylene-extracted soya bean meal was found to induce aplastic anemia when included in chick diets[1] and, although the chemistry of the reaction has not been completely elucidated, some of the symptoms can be reproduced by feeding S-dichlorovinyl-L-cysteine.[2,3] Interaction between the soya beans and the trichloroethylene presumably resulted in the formation of this highly toxic amino acid. It is now probably true to say that, due to the supreme importance of soybean meal as a constituent in livestock diets, events following that sequence are unlikely to recur. The early emphasis on maximum yield and quality of the oil at the expense of those of the residual meal has been modified so as to produce a high quality meal for the animal feed industry. Care must always be taken, however, especially with less familiar ingredients or solvents, to ensure that interactions between solvent and substrate do not result in the formation of a toxic substance.[4]

Finally, some processes are actually a combination of treatments. For example, *water treatment* commonly includes hot air drying and grinding, and *micronizing* consists not only of a heat treatment, but also of flaking and possibly grinding and water addition. In these respects poultry are no different from other livestock. However, poultry differ from other stock in two areas which are of particular relevance to processing. Poultry are fed on a mixed diet complex in the number of its constituents where it is generally assumed, again for practical reasons, that any effects observed experimentally when individually processed ingredients are tested will be present when the whole mix is so treated; such assumptions are not always valid. Finally, most birds raised for meat production are offered their diet in the form of pellets made by a process which is applied to the whole mix of ingredients. If the pellets are nutritionally superior to the parent mash it can prove difficult to establish which component(s) of which ingredient(s) in the diet have been affected and whether the effect is characteristic of the ingredient irrespective of the circumstances or only applies under certain dietary conditions. These drawbacks to proper process evaluation have not, however, prevented a vast amount of research being carried out on the pretreatment of diets and dietary ingredients for poultry.

STORAGE

Although storage cannot properly be described as a pretreatment it is a state undergone by most feedstuffs and processes can occur which may ultimately affect their feeding value. Despite this, storage has received fairly scant attention and tends to be associated with spoilage, particularly of materials which are rich in polyenoic fatty acids such as fish and meat meals. Although the precise mechanisms which result in the deterioration are even now not completely understood, both fatty acid polymerization and the formation of condensation products between lipids and proteins are considered to be the causative reactions; the consequence is a general reduction in the digestibility of both fat and amino acids. However, more severe damage to certain amino acids can occur and the digestibility of methionine has been shown to be lowered by 60% in oxidized anchovy meal[5] and by 16% in oxidized herring meal.[6] Many vegetable protein sources are also rich in polyunsaturated fatty acids and, although there is little published evidence of their involvement in the reduction in nutritive value of these materials during storage, oxidative polymerization could entangle proteins and thereby lower the digestibility of the amino acid residues.[7]

In many cases the reduction in nutritive value of protein-rich feedstuffs during storage has been attributed to the reaction of free amino groups of proteins with reducing compounds, the most prevalent of which are the carbohydrate carbonyl groups. Because these newly-formed bonds resist the action of the digestive enzymes, the effect is to reduce the digestibility of both protein and carbohydrate. Known as the browning or Maillard reaction[8] it can take place at relatively low temperatures but is accelerated by heat, which can be a feature of feedstuffs in storage. The nature and the form of the feedstuff can affect the extent of heating occurring under store but the principal factor appears to be the moisture content.[9] The early findings on the Maillard reaction have been comprehensively summarized by Reynolds,[10-11] and more recently Sulser and Büchi,[12] Adrian,[13] Finot,[14] Sulser,[15] Baumann and Gierschner,[16] and Feeney et al.[17] have covered various aspects of nutritional importance.

Spoilage of feedstuffs during storage can also be caused by reactions of nutrients with polyphenols.[18-19] Tannins, which are known to impair the digestion of protein,[20] can be formed during storage and quinone production by atmospheric oxygen may be responsible for further nutritional damage to certain materials.[21] Despite the effects of oxidation on the nutritive value of feedstuffs having been under investigation for more than 30 years, much still remains to be done to elucidate completely the detrimental reactions and to devise means for their prevention. The introduction of antioxidants such as ethoxyquin and butylated hydroxytoluene[22-23] revolutionized the shelf life of products such as fish and meat meals without having any unfavorable side effects on the well-being of stock fed on diets containing the treated products.[24-25]

Reactions initiated by oxidation are not responsible for all the damage occurring during storage. Sellam and Christensen[26] have shown that moisture transfer can occur in maize kept in laboratory containers designed to simulate conditions during transport. It was found that fungal invasion and the deterioration in feedstuff quality that invariably accompanies such contamination was affected by the moisture content of the maize, the magnitude of the temperature differentials within the container, and the duration of storage. To overcome spoilage of feedstuffs by fungi, bacteria, insects, and rodents an impressive array of preservation treatments has been developed.[27] However, the effects that preservation, essentially another form of processing, can have on the nutritive value of the material to poultry have been rather less thoroughly studied.[28-29]

Storage after treatment with 0.5% formalin did not appear to affect the nutritive value of wheat, as judged by the weight gain of broilers fed on diets containing 75% of the treated grain.[30] However, if the wheat was dried at 100°C for 20 hr before diet preparation, the food intake of the birds was reduced; this was attributed to the presence of irritating odors which had been released during the drying process. Moran et al.[31] reported that gamma irradiation of wheat bran at rates of up to 5 Mrad had no effects on its metabolizable energy (M.E.) content for the chick, but improved net phosphorus retention. This effect was attributed to increased availability of phytic acid phosphorus, probably as a result of cleavage of the ester linkage of the myo-inositol hexaphosphate.[32]

In certain parts of the world high-moisture maize has become a popular dietary ingredient for livestock, partly because it can be harvested early with less grain loss. With maize, however, moisture levels in excess of 16% are known to favor mold formation, with implied reductions in its nutritive value.[33] Application of 1% propionic acid inhibits mold formation[34] and, although Moran et al.[35] reported that performance by broilers fed on diets containing 52.75% maize treated in this way tended to be variable, the treatment appears to have little effect when the maize is fed to broilers at inclusion levels of 65%[36] or to laying hens.[37] Christensen[38] has reported that the addition of a 1% (weight to weight) mixture of acetic and propionic acids (60/40, volume to volume) immediately after harvest prevents mold growth on high moisture maize. Addition of this acid mixture at 1.5% (weight to weight) prevented maize (moisture content 27%) spoiling even after 10 months storage[39] and reduced the mold count by 10,000. Feeding experiments with broiler chicks revealed that the treated high moisture maize had the same M.E. value as the untreated maize, dried to 12% moisture, and that the performance of birds fed on diets containing 42% maize was unaffected by the treatment.

Reactions which occur during storage generally result in a downgrading in the nutritive value of the stored product even when care is taken to prevent it. Thus, soya beans, when stored as the whole, raw seed (6.6% moisture) at 25.5°C for 34 months lost 25% of their nutritive value[40] and the sum of ten essential amino acids fell by 20% when dried peas were stored in a ventilated warehouse for 12 months;[41] the methionine content, which fell by 32%, showed the greatest change. Examples where storage per se has proved beneficial to the nutritive value of feedstuffs are few. It is considered good husbandry not to include field beans (*Vicia faba* L.) in poultry diets during the year of their harvest and the observation in our laboratory that the trypsin inhibitor content of the beans decreases during storage may offer the scientific explanation for this practice. Finally in this context, Chitre et al.[42] reported that the riboflavin content of pulses stored in sacks for 6 months increased from 132.5 to 186.6 g/100 g. They also documented that the riboflavin content of rice and other cereals increased during storage but, unfortunately, the types of pulse stored, the temperature of storage and the moisture content of the stored materials are not detailed.

GRINDING

Current practice in the manufacture of poultry diets makes it essential that the component raw materials are ground to a meal before compounding, yet the effects of grinding and particle size have tended to be neglected. A common finding appears to be that finely ground particles (0.5 mm diameter) cause "pressure necrosis" of the beak in young birds.[43-46] Eley and Bell,[47] and Eley and Hoffman[48] reported an inverse relationship between the size of food particles and water intake. Waste from the food trough is greater with large particles, but more food is recovered from the water trough with feedstuffs consisting of fine particles, which have a tendency to adhere to the beak. Davis et al.[49] offered birds a choice of diets of different particle size and reported a preferential selection of the largest (2000—1410 μm). Smith et al.[50] found considerable variation in the elemental composition of different sized fractions of ground feedstuffs and preferences for particles within a particular size range may cause bias in nutrient intake. Feeding management, however, should prevent such an effect from being anything other than a short-term phenomenon.

There is remarkably little published evidence on the effect of grinding on nutritive value. Fritz[51] found that the digestibilities of the fat from whole, cracked, and ground maize were 0.75, 0.87, and 0.91, respectively, although those of the other components remained unchanged. More recently, Mitchell et al.[52] in a study to investigate whether the fineness of grind affected the digestion of full fat soybeans, found that fat digestibility increased from 0.577 to 0.900 and nitrogen retention from 0.528 to 0.617 as the particle size of the soya beans decreased. It is now generally accepted that the more finely ground are fat-containing raw materials, the more highly digestible the fat fraction and the higher the materials M.E. value. For example, it has been shown that fats extracted from certain foodstuffs, e.g., soya beans, are more digestible than when they are present in the parent material.[53-54] This effect is attributed to the physical disruption of the fat-containing cells, thereby allowing the lipolytic enzymes readier access to their substrates.

Summers[55] claimed that reduction in the particle size of grains by grinding improves their digestibility. He attributed the effect to cell wall breakage and release of nutrients to the digestive enzymes which, until disruption, had no means of access. Examples of this, however, are not numerous although grinding wheat shorts three times raised the digestibility of the protein fraction.[56] Any increase in digestibility might be expected to raise the M.E. value and it is perhaps significant that grinding and cold pelleting barley, wheat, and oats had only small effects on their true M.E. values.[57] Milling has been claimed to improve the nutritive value of certain materials by mechanical release of intracellular starch with a consequential increase in its digestibility.[58] Although increasing the digestibility of cereal starches does not seem a likely proposition in view of the observations that starch in unprocessed cereals is almost completely digested by poultry, it could explain the recent report that fineness of grind influences the M.E. value of field beans for broilers.[59] The M.E. values of each of two varieties of field bean rose by more than 10% as screen size decreased from 3.5 to 1 mm; protein digestibility was not affected.

These observations may possibly offer an explanation why feeding grit improves dietary digestibility coefficients.[46] The shear forces created by the grit and the action of the gizzard are presumably sufficient to disrupt the cell walls and release their contents to the digestive processes. Although a recent report must cast doubts on the value of grit feeding,[60] its effect could still depend on the nature of the dietary components. The possibility of the introduction of choice feeding systems for poultry where part of the choice is whole cereal may necessitate a reevaluation of the practice of offering dietary grit.[61,62]

Finally, grinding may induce detrimental effects as a result of exposing the previously protected components of the interior to atmospheric oxygen. It has been observed in our laboratory that field beans have a much better shelf life stored whole, because the meal appears to be prone to mold infestation.

WATER TREATMENT

It has been claimed for many years that the feeding value of certain dietary materials can be improved by pretreatment with water. Thus, McGinnis and Polis[63] have reported that the nutritive value of linseed meal is improved by moistening and incubating overnight at 37°C. Chick growth is improved as a result of feeding diets containing water-treated whey[64] and soaking salseed cake for 24 hr improved its nutritive value to the young chick.[65] The improvement of linseed and salseed meals was attributed to the removal or inactivation of toxic substances. Many other plant products respond similarly to water treatment and the nutritive value of various beans, rapeseed, crambe seed, cycads, peas, and leafy vegetables are improved after soaking.[66]

Because of their universal importance in poultry diets most research on water treatment has been directed at the cereals. Barley has received greatest attention, probably because its low digestibility coefficients offer the greatest scope for improvement.[67-69]

Although barley is similar in composition to maize and wheat, its feeding value is generally considered inferior. Despite the report by Hamm,[70] who noted a growth depression in chicks when water-treated barley replaced the untreated cereal in their diets, there is convincing evidence from many sources that water treatment of barley enhances its nutritive value for the young chick. Very little work appears to have been carried out with mature birds, although Berg[71] has recorded that the inclusion of water-treated barley in laying hen diets resulted in a small reduction in the amount of feed required to produce a dozen eggs compared with diets containing untreated barley.

Despite having been under investigation for more than 20 years,[72] it is still unclear which components of barley are affected by the water treatment. It has been established that the methods used in wetting and drying influence the nutritional value of the final product, but the importance of the ratio of barley to water, the time and temperature of the soaking and drying stages and the possible interaction with other dietary ingredients have yet to be established.[73] Drying temperatures of between 85 to 90°C are claimed to be optimal[74] and soaking temperatures of that same order are beneficial, hotter conditions reducing the nutritive value.[75] The improvement does not depend on the variety of barley[76] or on its proximate composition,[77] but barley grown in the western U.S. responded more than eastern or midwestern grown grain.[77] It appears possible that the warm, humid conditions prevailing in the midwest and east during grain maturation induced the same changes as the soaking process. Recently it has been shown that the digestibility coefficients of the crude protein of barleys grown in different Swedish locales decreased significantly as the latitude of the site increased.[78] The variation was attributed to the differences in their tannin content.

Novacek and Petersen[79] showed that the husk, pericarp, germ, aleurone, and endosperm fractions all responded slightly to water treatment, although the improvement with unfractionated barley was greater than could be accounted for by the sum of the improvements in the various anatomical parts. Because the method used to isolate the endosperm influenced the extent of the response from that fraction, the greatest effect being measured with intact endosperm cells, the effect of soaking was attributed to the rupture of these cells. This could explain the increased availability of energy which is an outstanding feature of water-treated barley[79-81] and which was thought to be due to the breakdown of cellular carbohydrate and/or the alteration in the structure of intracellular starch.[72,81,82] Lawrence[83] has recently suggested that soaking improves the nutritive value of barley, for the pig at least, by converting damaged starch into maltodextrins and reducing sugars, which are more digestible. Gohl,[84] on the other hand, has presented convincing evidence that the reason for at least part of the improvement in the value of water-treated barley for the young chick lies in the reduction in the hydrocolloid content of the barley by the action of endogenous enzymes. The presence in barley of a hydrocolloidal β-glucan has been blamed for the poor digestibility of barley by the chick.[85,86] The similarity in the response of chicks fed on diets containing water-treated barley to those fed on barley-based diets containing cellulolytic or hemicellulolytic enzymes appears to suggest that the water is affecting the carbohydrate fraction. However, the fact that digestibility coefficient of the nitrogen-free extract component of barley shows less response to water treatment (0.75 to 0.78) than those of either the fat (0 to 0.84) or protein (0.45 to 0.84) components,[81] appears to lend weight to those who believe that the similarity in response to water soaking and enzyme addition occurs merely by chance and has been misleading when attempts have been made to devise explanations.[76,77]

Bacterial action during soaking, with the production of a factor (probably an enzyme) which alters some component of the barley, is another possible explanation of the beneficial effect of soaking and its similarity to enzyme supplementation. Water treated barley contains larger numbers of bacteria than untreated barley, even though the drying temperature was 70°C[87,88] and rats fed on diets containing water-treated barley had greatly increased numbers of bacteria in the gastrointestinal tract compared with those fed on diets containing untreated barley.[84] Further evidence for the involvement of an antibiotic is that soaking barley interfered with the improvement in nutritional value.[89] However, studies showing that neither the addition of hydrochloric acid to the water[74,90,91] nor autoclaving the barley[88] significantly altered the response to soaking imply that the bacteria must be stable to heat and to 0.1 N acid or possibly that they originate from the water used in soaking. Microbial involvement during soaking does, therefore, appear to be a possible explanation in certain circumstances although the mechanism of the improvement is still unknown.

It is noteworthy that barley which has been soaked has been shown to contain bacitracin-like activity.[92] Antibiotics may control the gut microflora and increase intestinal permeability with a parallel increase in nutrient absorption.[80] The report that water soaking destroys a growth inhibitor in barley could well be explained in terms of antibiotic reaction.[93] There seems to be little doubt that soaking improves the nutritive value of barley but the extent of the improvement appears to vary. Until the improvement mechanism is conclusively identified and the economic benefits precisely quantified, commercial adoption of soaking barley will remain a process dependent on circumstances.

It can be inferred from the lack of any extensive study with other cereals that water treatment has little effect on their nutritive value. Nevertheless, water-treated wheat was superior to the untreated cereal when included in young turkey[82] or broiler diets.[74,80,94] The improvement, which has been calculated to be as much as 19% in the complete cereal, is seen in the bran, germ, shorts, middlings, and mill feed fractions.[95] Adams and Naber[90] reported that the utilization of hardwheat, its flour and gluten were all improved after treatment with either water or 0.1 N hydrochloric acid. Naber and Touchburn[96] also reported that a diet containing water-treated ground hard red wheat and one containing its long patent flour milling fraction (65% of the whole grain) significantly increased chick growth compared with those from diets containing untreated materials; dietary dry matter retention and M.E. were also improved.

Neither the efficiency nor the growth of broilers was affected by feeding diets containing maize which had been soaked at 85°C for 45 min.[97] Longer soaking in water, 16 to 20 hr at room temperature or at 38°C, occasionally resulted in improved chick performance[74] but it was insufficient to justify the inconvenience of the process. In general, the variability of the response to water treatment and the high cost of drying the wet material make it a most unattractive process for feedstuffs.

ENZYME TREATMENT

Enzyme addition has been a well-tried means of improving poultry diets but enzyme pretreatment of individual ingredients is rare. Saunders et al.[98] pretreated wheat bran with cellulase and estimated that in vitro its protein digestibility coefficient had risen from 0.7 to 0.9.

GERMINATION

Germination is known to result in an increase in the nutritive value of various seeds, e.g., soya beans[99,100] although the reason for the improvement does not appear to have been investigated. It is known that sprouted wheat can be fed to poultry with safety[101] and that the nutritive value of maize and wheat but, surprisingly, not barley were increased after germination.[74]

HEAT TREATMENT

Most processes which involve heat treatment are similar and, despite differences in the sophistication of the equipment, tend to differ only in the degree and time of heating and whether or not moisture is involved. The classical ways of applying heat to poultry diets, or ingredients intended for inclusion in diets, are steam pelleting, roasting, toasting, mashing, and flaking. Autoclaving or pressure cooking, although a popular laboratory tool, has been applied commercially only to the production of meals from animal sources.

It was Osborne and Mendel[102] who first recorded that the nutritional values of certain protein-rich foods were improved by heat treatment. This observation has led to the recognition of a wide range of compounds which, if ingested in a raw state, result in reduced performance.[103] Many of these so-called antinutritive compounds are sensitive to heat and this property has promoted a vast amount of research into the processing conditions necessary for their inactivation. The inhibitors of the proteolytic enzymes trypsin and chymotrypsin have received greatest attention, partly because of their ubiquity and partly because of their presence at relatively high concentrations in the commercially important soya bean. Other factors which are known to affect the nutritive value of feedstuffs are hemagglutinins, amylase inhibitors, tannins, gossypol, phytates, goitrogens, cyanides, and saponins. The relevance of most of them to commercial poultry nutrition unfortunately still awaits evaluation. Many, however, are destroyed by heat but, since recent reviews detailing their modes of action are available, they will not be dealt with here.[104-106]

Preoccupation with attempts to correlate the beneficial effects of heat treatment with the levels of one or more of the antinutritive factors has tended to obscure the beneficial effects that heat can have on the digestion of the protein and carbohydrate components of many feedstuffs. Structural changes induced by heat may well result in an increase in the efficiency of digestion. Kakade[107] has argued that the improved nutritive value of many protein-containing feedstuffs after the application of heat may be explained, at least in part, by the rupture of the three dimensional structure of the protein. For example, Fukushima[108] has shown that the folded conformations adopted by the proteins in soya beans resist breakdown until their form has been disrupted by heat. Heat alone, however, is less likely to be able to disrupt tertiary protein structure formed by hydrophobic bonds, a feature of many globular proteins,[109] and soaking before the application of heat is essential if the digestion of the protein is to be improved.[110,111] It seems predictable that heat will also break enzyme-resistant salt bonds, glycoprotein, and peptide linkages with consequent improvements in the digestion of the involved amino acids.

Mercier et al.[112] have shown that the physical changes brought about by heating make starch six times more accessible to amylases and Mercier[113] has shown that, in vitro at least, the digestibilities of barley, maize, and sorghum starches can be increased by various forms of heat. Since most cereal and vegetable starches are highly digestible by poultry without any form of heat application, as judged by the absence of starch from feces, these findings are of doubtful significance in practical situations. However, the increased susceptibility of dietary starch to enzymic breakdown might alter its site of digestion in the bird and Burt[114] has argued that this could result in more starch being digested to glucose, and less to energetically inferior products, by the bacteria in the lower gut; this might be expected to increase the net energy of the diet and result in improved food conversion efficiency. Starch in less conventional dietary ingredients may behave differently during digestion. For instance, it has long been known that the digestibility of potato starch is greatly affected by pretreatment. Halnan[115] showed that the digestibility of raw potato starch by chickens fell within the range 12 to 35% whereas the corresponding values in potatoes which had been boiled or mashed were 83.5 and 90%, respectively. Earlier Hock[116] had shown that for optimal digestion of potato starch some form of pregelatinization by heating was necessary and Hakansson and Lindgren[117] have concluded that the digestibility of potato starch is directly proportional to the extent to which the starch granules have been gelatinized before ingestion. Although other heat-labile factors present in potatoes complicate the interpretation of the effect of heat treatment on the starch,[118-123] there is little doubt that its structure is not readily penetrated by the digestive enzymes without some form of pretreatment. Potato starch differs from the cereal starches in granule size and X-ray pattern but the significance of these differences has yet to be understood in nutritional terms.[124] The increased energy metabolism by chicks of certain tropical legumes after autoclaving could possibly be explained in terms of increased starch digestion.[125]

Moderate heat treatment is, therefore, generally acknowledged to enhance the nutritive value for poultry of many dietary ingredients and many explanations for the improvements are possible. This conclusion is perhaps less than surprising since, like man, the chicken is monogastric and might be expected to respond similarly. It is claimed that the introduction of cooking has in no small way contributed to the evolutionary superiority that man now enjoys.[66] Improvements in taste by the formation of aromatic substances can be a further favorable consequence of heating,[126] although its relevance to poultry feeding must be considered doubtful.

Heating ingredients for poultry diets commercially is labor intensive, time consuming, and costs money. Steam pelleting, which will be discussed later, is more readily applicable to finished diets rather than to individual ingredients. Some industrial by-products are marketed as pellets, although it is invariably done more for reasons of bulk reduction by the seller than for reasons of improved feedstuff value for the customer. Despite all the drawbacks to heat treatment much effort has been expended investigating methods and effects. Many in the feed and poultry industries appear convinced that heating does increase a feedstuff's value and research in this area has been encouraged. Micronizing is a recently introduced heating process which has been applied to a number of raw materials with promising results. A micronizer consists of a battery of heated ceramic tiles below which passes a belt carrying the grains or seed to be processed. The material is cooked very quickly from the inside by a rapid increase in the water pressure within the kernel, then rolled and ground. The whole process,

Table 1
THE EFFECT OF MICRONIZING FIELD
BEANS ON BROILERS FED FROM 7 TO
14 DAYS ON DIETS CONTAINING 300 g
BEANS PER kg

	Raw	Micronized
Weight gain (g)	103	112
Food intake (g)	149	142
F.C.E. (g gain per g food)	0.69	0.79
Pancreas size (g pancreas per g body weight)	0.58	0.49
Diet digestibility (%)	62	67
Nitrogen retention (%)	52	58

Table 2
THE EFFECT OF MICRONIZING ON THE
COMPOSITION OF FIELD BEANS (g/kg
DRY MATTER)

Component	Raw	Micronized	% Change
Dry matter (as received)	835	868	+ 4.0
Ash	37.7	40.1	+ 6.3
Ether extract	13.2	17.3	+ 31.1
Crude protein	316	327	+ 3.5
Available carbohydrate	406	502	+ 23.6
Trypsin inhibitor activity (units/mg)	2.49	0.21	−91.6
Metabolizable energy (MJ/kg)	10.59	11.67	+ 10.2

which can be controlled by the speed of the belt, takes less than 2 min. The nutritive value of field beans for poultry is significantly improved as a result of micronizing (Table 1). Food conversion efficiency, apparent protein digestibility and nitrogen retention are increased and pancreas size of the birds reduced when beans constitute 30% of the diet.[127] Shannon and Clandinin,[128] studying similar beans grown in Canada, obtained responses of a similar order after autoclaving whole beans for 30 min at 121°C. This may indicate that there is little special about micronizing other than being a commercially viable process, a not unimportant consideration. The principal chemical change in the beans appears to be a substantial increase in their available carbohydrate content (Table 2); the M.E. value is also significantly increased. The nutritive value of field peas (*Pisum sativum* L.) for the laying hen has also been shown to be improved as a result of micronization.[129]

Full-fat soya beans and whole rapeseed appear to be converted into safe high energy feedstuffs after micronization and are suitable for inclusion in poultry diets. The effect of micronizing on the M.E. values of several feedstuffs for poultry is shown in Table 3. Improved shelf life (field beans did not develop molds during storage) and reduced moisture contents are further features of micronized products.

Jet-sploding is an even more recent innovation as a commercially feasible means of applying heat to feedstuffs for short periods of time. The material under process is "popped" in a heated chamber, rolled, and ground to a meal. Because the heat can be recycled jet-sploding is claimed to be energetically more efficient than micronizing,

Table 3
EFFECT OF MICRONIZING ON THE METABOLIZABLE ENERGY VALUES OF SOME FOODSTUFFS (MJ/kg DRY MATTER)

	Untreated	Micronized	Increase (%)
Barley	10.7	11.3	5.6
Wheat	13.2	13.6	3.0
Maize	14.0	14.6	4.3
Full fat soya beans	11.3	16.9	49.6
Full fat rapeseed	8.1	15.1	86.4

Note: All values were estimated with broiler chicks between 1- and 2-weeks old and are corrected to zero nitrogen retention.

with implied cost benefits. The effect of the process on maize, wheat, barley, field beans, and soya beans has been under study and preliminary data indicate that the results are similar to those obtained after micronization.

Heat can, of course, reduce nutritive value and the occurrence and consequences of the Maillard reaction have already been discussed. In addition to encouraging the formation of enzyme-resistant carbohydrate-protein bonds, heat can aid other deleterious reactions, such as the creation of new linkages both within and between peptide chains.[7,130,131] Such bonds are also thought to resist breakage by the digestive enzymes and the result can be impaired availability of both the involved and adjacent amino acid residues. Some of the nutritional consequences of heat have been studied by Carpenter and his colleagues[132-137] and a review has appeared.[138]

Under certain circumstances and with certain feedstuffs heat can result in the destruction of certain amino acids. For example, the cystine contents of soya bean meal[139] and of wheat germ[140] were lower after quite moderate heating. Autoclaving rapeseed meal at 121°C for 6 hr, admittedly, fairly extreme treatment, reduced the lysine content by 60%, the arginine by 45%, and those of both phenylalanine and tyrosine by 8%.[141] Care must therefore be taken, when heat is applied to dietary ingredients, to ensure that the detrimental effects do not outweigh the beneficial. General recommendations on the treatment conditions necessary to produce optimal nutritive values for all of the wide range of ingredients which go into poultry diets are unfortunately not available. The presence of water, fat, and carbohydrate, the structure of the protein and the nature of the air movement during heating are only some of the factors which can influence the nutritive value of the final product. Thus, autoclaving muscle protein has little effect on its quality whereas autoclaving in the presence of glucose reduced the availability of the lysine markedly after only 20 min.[142] Feedstuffs containing keratinous proteins often benefit from what appears to be quite severe treatment. For example, processing temperatures of between 142 and 153°C (28 to 42 Mg/m²) for 1 hr result in the highest quality feather meals.[143-145]

Removal of moisture from feedstuffs, particularly cereals, invariably involves the application of heat although its effect on the nutritive value of the raw material has not always been quantified. A grain temperature of 90°C did not affect the nutritive value of barley when it was being dried in a concurrent flow drier where the air inlet temperature was 171°C.[146] Drying wheat at 89°C did not affect its feed value to poultry[147] and the quality of maize dried by air with an air inlet temperature of 177°C was unaffected;[148] only at 232°, where a deliberate attempt was made to damage the maize, was there any indication of a reduction in its nutritive value. This confirms the earlier finding that drying temperatures of up to 90°C do not affect the nutritive value of maize,[149] although it has since been shown that the efficiency of utilization of maize by poultry progressively decreased as the length of time in storage before heating increased.[150] An extensive study in France[151] investigated the economics of drying maize at 80°C or 140°C after three different storage treatments. Chemical estimations indicated that the availability of lysine was reduced at the higher temperature[152] but this could not be confirmed with evaluations with chicks in vivo.[153] There was, however, an interaction between drying temperature and lysine availability because with no supplementary dietary lysine, drying at 80°C gave the best results whereas with 0.4% added lysine the 140°C dried maize gave better growth. When the maize was tested under practical-type conditions it was shown that an interaction existed between storage conditions and drying temperature with ventilated maize dried at 140°C having the highest quality and with maize stored in a bin for 6 days and dried at 80°C the poorest.[154] These examples serve to illustrate some of the wide variety of factors which can influence the quality of a heated feedstuff and demonstrate the extreme difficulty in predicting conditions which will result in the highest quality product.

STEAM PELLETING

The formation of diets into pellets or crumbles is routine in many types of poultry production and the effects of pelleting on both individual feedstuffs and complete diets has been a topic of considerable research. Early reports on the subject in the context of poultry feeding have been comprehensively reviewed by Calet[155] and certain more recent aspects have been discussed.[55,114,156-159] Even now, however, after the expenditure of vast amounts of research effort, the effects of pelleting on dietary nutritive values are seldom known. Steam pelleting is still generally regarded as a production rather than processing procedure and pelleting machinery is usually run under conditions which will produce the greatest tonnage of suitable material per hour rather than under conditions which will produce a diet of greater nutritive value. Easier handling, less separation of the dietary ingredients and less waste of feed by the birds are readily seen advantages of pellets; improved palatability and enhanced nutritional qualities are less readily seen and hence seldom justified as reasons for pelleting except in the research laboratory.

It has been concluded that at least part of the improvement observed as a result of pelleting diets for poultry was a consequence of increased density.[160,161] Certainly the physical effects of pelleting feedstuffs can be considerable. Hussar and Robblee[162] found that diet density increased from 0.57 to 0.71 weight per unit volume on pelleting whereas regrinding gave a density of 0.60 weight per unit volume. Density changes are likely to be greatest with fibrous materials, such as cereal byproducts, which are bulky and may have awkward flow characteristics. Moisture contents seem little affected by steam pelleting, being between 10 and 14% before and between 9 and 15% after cooling.[163] The amount of moisture absorbed is apparently a function of the time of exposure to and the pressure of the steam, more moisture being taken up if the steam is at low pressure with time to condense.[164] The impact of other physical factors which are part of the pelleting process are discussed in MacBain's[164] excellent review. Bayley et al.[163] noted that steam preconditioning for 3 to 5 sec at a pressure of 45.7 g/mm² heated the diet to a temperature of 90°C whereas Hussar and Robblee[162] found that pellet temperatures lay in the range 65 to 70°C. Their finding that maximum die temperature was reached only after 15 to 25 min of pelleting should be borne in mind, particularly when pelleting short-run experimental diets. In most studies the process is not adequately described in terms of physical parameters to allow more general conclusions to be drawn.

The factors which affect the physical and chemical changes which take place during pelleting are even less well-documented. Melcion et al.[165] described some of the effects of pelleting (2.5 mm diameter) for a dry diet (10% moisture), one with 4% added water, and one after steam conditioning which added about 1% moisture. The temperature rise in the die varied from 30°C with the steam treatment to 59°C with the dry diet and the temperatures of the pellets on extrusion were 77°, 58°, and 73°C, respectively. Available lysine was not affected by any of the treatments but vitamin A levels and reducing sugar contents were reduced and susceptibility of starch to amylase attack increased by the most severe treatment. Pellet throughput was reduced and the energy required for pelleting increased with the dry diet. Unfortunately, the diet was intended for piglets and was atypical in containing spray-dried skim milk, maize oil, denatured sugar, and cassava. One of the very few detailed studies carried out on poultry diets describes the parameters associated with the manufacture of diets containing either 22 or 26% protein as mash, 2 mm, and 3 mm pellets, and crumbles using meals conditioned by water addition, steam, or steam followed by double pressing.[166] The best performance was achieved by birds fed the 3 mm double pressed pellets prepared by steam conditioning, using the meal containing 26% protein.

There is considerable conflict among the reports of nutritive value changes caused by steam pelleting. Blakely et al.[167] found little effect of either dry or steam pelleting on the dietary M.E. content. It has also been claimed that in experiments with wheat, maize, and soya beans, M.E. differences could not explain observed growth responses.[168] McIntosh et al.[169] found that the M.E. of a diet containing 610 g wheat per kg was improved by about 9% after pelleting. However, it was subsequently reported that the M.E. values of cereals were not uniformly affected by the form of the feed.[46] Moran et al.[170] found the M.E. of field peas was improved from 2.48 to 2.70 kcal/g (8.9%) by pelleting, a response identical to that measured after autoclaving the peas.[170] Recently in a study involving three wheat fractions with fiber contents of 133, 94, and 36 g/kg Janssen[171] found that their M.E. contents were inversely proportional and improvements in M.E. and digestibilities of their protein and fat fractions were directly proportional to their fiber contents. The responses in M.E. are likely to reflect the scope for cellular disruption by the mechanical effects of pelleting. The importance of mechanical forces during pelleting was first emphasized by Carew and Nesheim,[172] who calculated that the absorbability of the oil in soya beans was raised from 0.73 to 0.78 after pelleting the beans by themselves, whereas it was raised to 0.90 when the beans were pelleted in a mixed feed. The difference was attributed to the higher pressures generated in pelleting the mixed diet. This example also illustrates the danger in assuming that a response obtained after treating an ingredient of a diet by itself will be reproduced exactly when the material is only part of the diet being processed.

That mechanical disruption is an important part of the pelleting process was demonstrated by Saunders et al.[56], who offered an explanation for the increase in the nutritive value of diets containing wheat bran as a result of pelleting in terms of aleurone cell rupture. It was shown by staining techniques and subsequent microscopic examination that the percentage of empty aleurone cells was higher in the feces from birds which had eaten the diets which had been pelleted. The protein content of the feces decreased proportionally as the number of empty cells increased and protein digestibility was always higher if the diets had been pelleted. It was also shown that the heat involved with the process was not responsible for the improvement and, although the steam may have aided the changes by softening the extracellular structure, the improvement was exclusively attributed to the pressure forces exerted by pelleting. Because the aleurone fraction of bran contains not only 90% of the protein but also 95% of the fat, 60% of the minerals, and almost all the vitamins, it seems likely that the digestibilities of these components will also have been increased. This may explain the observation that chicks fed on a pelleted diet containing 250 g wheat bran per kg had higher plasma inorganic phosphorus and bone ash contents than birds fed on the same diet

Table 4
EFFECT OF DIETARY FORM ON THE GROWTH OF WHITE
LEGHORN CHICKS AND THE METABOLIZABLE ENERGY
CONTENT OF DIETS CONTAINING EITHER STEAM
PELLETED OR UNTREATED BRAN, SHORTS, OR MAIZE

	Average Weight at 28 days (g)		M.E. (kcal/g dry matter)	
	Untreated	Steam pelleted	Untreated	Steam pelleted
Bran				
Mash	164	259	1.46	1.70
Dry pellets	288	296	1.48	1.85
Steam pellets	303	294	2.05	2.50
Shorts				
Mash	233	267	2.10	2.16
Dry pellets	324	318	2.09	2.13
Steam pellets	303	333	2.20	2.21
Maize				
Mash	231	248	3.45	3.51
Dry pellets	258	269	3.58	3.55
Steam pellets	314	183	3.61	3.61

in mash form.[173] A similar implied improvement in phosphorus availability as a result of steam pelleting has been observed by Summers[55] in diets for laying hens based on maize and soya bean meals and by Bayley et al.[168] in similar diets for broilers.

A good illustration of how the effects of pelleting depend on the conditions used and the ingredients is contained in an experiment of Summers et al.[174], who took samples of maize, wheat shorts, and wheat bran, subjected them to steam pelleting and then reground them. Either the processed material or the original meal was then mixed in equal amounts with a diet based on maize and soya bean meals and each of the six diets offered to 1-day-old white leghorn chicks for 28 days as mash, dry pellets, or steam pellets. Table 4 shows the mean weights of the birds fed on these diets and the determined M.E. values of the three ingredients. Because the diets were not isocaloric or isonitrogenous, performance comparisons can be made only within diets containing the same test ingredient. It can be seen that processing the ingredients before dietary incorporation generally resulted in improved bird weight and higher ingredient M.E. values when the diets were fed as mashes. Dry pelleting improved the growth of the birds fed on the diets containing untreated bran, shorts, and maize by 75, 28, and 12%, respectively, and those containing the corresponding pretreated cereals by 14, 19, and 8%. The response could not be explained in terms of improved ingredient M.E. values, which were altered little by dry pelleting. Steam pelleting generally resulted in improved performance except with the diet containing maize which had been pelleted. The imposition of two heat treatments was assumed to have reduced the availability of the amino acids. Steam pelleting improved the M.E. of the bran markedly (44%) but had generally little effect on those of the maize or shorts (both 3.5%). These results show that the feeding value of a fibrous low-energy ingredient (bran) can be enhanced by pelleting or by double pelleting whereas medium (shorts) and high energy (maize) ingredients respond less. Increasing the temperature and the time of heating is known to reduce the availability of lysine in maize[175] and could explain the poor performance of birds fed on steam pellets containing pretreated cereal.

Gürocak et al.[176] have shown similar effects. Both pellets and reground pellets had digestible energy contents 3.6% higher than that of the corresponding mash, which was formulated to contain 11.3 MJ M.E. and 204 g crude protein per kg whereas a diet formulated to contain 12.8 MJ M.E. and 230 g crude protein per kg gave no such responses to pelleting. Whether the response to pelleting is consistently greater with diets of lower energy content than with those of higher energy content within the range currently in practical use is clearly of considerable commercial importance and must warrant further investigation.

Despite variability in response there is widespread evidence of improved performance by poultry fed on diets which have been pelleted.[155] In view of this and of the almost universal commercial practice of feeding pellets or crumbles to broilers it is somewhat surprising that most of the experiments investigating the effect of nutrient density on the performance of broilers appear to have been carried out using mash diets. In reviewing the response of the growing chick to variations in dietary energy concentration Fisher and Wilson[177] could find only two series of experiments which met their particular criteria and the results of those differed so much that no general conclusions could be drawn about the effect of pelleting on the relationship between chick performance and dietary energy density.[178,179] Nevertheless, pelleting broiler diets is generally acknowledged to result in higher food intakes,[180] less food wastage,[48] lower mortality,[181] and better efficiencies of food conversion. The relative importance of the various factors which cause the bird to respond to the pelleted diet is not, however, clear. Increased food intake and less wastage would, in themselves, be expected to improve food conversion efficiency, but the consequences of increased nutrient density are not so predictable. Reddy et al.[182] incorporated cellulose into a pelleted diet to give it the same nutrient density as that of the same diet in mash form and entirely removed the growth response to pelleting that had been observed earlier.

Pelleting may result in other, less obvious, but no less important effects. For example, poults fed on mash or pellets spent 18.8 or 2.2% of their day, respectively, eating; the corresponding values for chicks were 14.3 and 4.7%.[183] Fujita[184] observed that, at similar food intakes, eating time was over 500 min/day for mash but less than 200 for pellets. A consequence of this is that, although M.E. responses may be small and erratic, there will be large — as much as 30% — responses in dietary productive energy (P.E.) levels after pelleting.[183,185] The decrease in activity and in the energy used to consume food must also result in an increased efficiency of utilization of M.E. (net energy) but the extent of this has not been quantified nor its relevance defined. Because the birds are less fully occupied eating pellets, an increase in vice may be expected[186] and increased downgrading due to feather loss has been reported among broilers fed on pellets.[187]

Defining the extent of the effect of pelleting may be further complicated by an interaction with the strain of bird. Savory[188] found that with Rhode Island Red × Light Sussex hybrids the mean apparent digestibilities of diets in the form of mash and pellets were similar but with brown leghorns the pellets appeared more digestible than the mash. This contradicted an earlier finding by Bolton,[189] who found that digestibility coefficients of protein, oil, or carbohydrate by brown leghorns were not affected by the form of the diet.

Apart from losses in the availability of amino acids, in particular lysine, as a result of the application of excessive heat, there are few reports of reduced performance from feeding pelleted diets. However, a broiler flock fed on pellets had an incidence of fatty liver and kidney syndrome twice that of a similar flock fed on the same diet as mash.[190] The effect was attributed partly to an increase in the net utilization of carbohydrate which had widened the dietary energy-to-protein ratio, and partly to the destruction of biotin.[191] Conversely, however, Scott[192] has suggested that steam pelleting may increase the availability of both biotin and nicotinic acid from plant feedstuffs by releasing them from bound forms.

CONCLUSIONS

To meet the protein and energy requirements of an ever increasing world population, less high quality feedstuffs are likely to be available for poultry diets. Increased use of waste products and materials with poorer nutritional properties seems inevitable. If the poultry industry is to survive it will have to make maximum use of these raw materials. It seems possible that processing these materials or diets containing them may well upgrade their qualities sufficiently to maintain high performance. It is unfortunately not possible from past experience to predict with complete certainty whether or not the potential improvement resulting from a process will justify its cost. This has arisen because, until now, the cost and availability of high quality feedstuffs have allowed most poultry producers in most parts of the world to formulate diets to supply in excess most of the nutrients required for optimal performance. In such circumstances it is simply not possible, however scientifically well-based the process, to improve, by increasing the nutritional quality of the diet, the performance of a stock whose genetic potential is already being realized. However, the predicted use of ingredients of lower nutrient specification may allow the improvements brought about by processing to be realized. Formulations must be capable of taking advantage of the improvements, which will have to be defined in chemical terms, by being more tightly constrained to meet more precisely the needs of the animal being fed.

REFERENCES

1. **Pritchard, W. R., Davis, O. S., Taylor, D. B., and Doyle, L. P.**, Aplastic anemia in chickens fed trichloroethylene extracted soybean oil meal and failure of the dietary meal to suppress the development of experimental lymphomatosis, *Am. J. Vet. Res.*, 17, 771—777, 1956.
2. **Kuiken, K. A.**, Effect of other processing factors on vegetable protein meals, in *Processed Plant Protein Foodstuffs*, Altschul, A. M., Ed., Academic Press, New York, 1958, 131—152.
3. **Friedman, L. and Shibko, S. I.**, Adventitious toxic factors in processed foods, in *Toxic Constituents of Plant Foodstuffs*, Liener, I. E., Ed., Academic Press, New York, 1969, 349—408.
4. **Pritchard, J. L. R., Farmer, S. N., and Wong, D. R.**, Solvent extraction of oilseeds. Estimation of solvent in residual meal, *Chem. Ind. (London)*, 2062—2065, 1964.
5. **Bürke, R. P. and Maddy, K. H.**, Verbesserte Fishmehlqualität durch Stabilisierung mit Santoquin. II. Einfluss der Stabilisierung auf den Fütterungswert von Fishmehl, *Kraftfutter*, 49, 66—74, 1966.
6. **Opstvedt, J.**, Nutritional significance of residual lipids in fish meal, *Feedstuffs*, 46(23), 22—28, 1974.
7. **Mauron, J.**, Le comportement chimique des protéines lors de la préparation des aliments et ses incidences biologiques, *J. Int. Vitaminol.*, 40, 209—229, 1970.
8. **Maillard, L. C.**, Action des acides amines sur les sucres. Formation de melanoidines par voie methodique, *C.R. Acad. Sci.*, 154, 66—72, 1912.
9. **Halick, J. V., Richardson, L. R., and Cline, M.**, Studies on feed spoilage — heating in feed ingredients and mixtures containing molasses and added fat, *Tex. Agric. Exp. Stn. Bull.*, 860, 1—7, 1957.
10. **Reynolds, T. M.**, Chemistry of non-enzymic browning. I. The reaction between aldoses and amines, *Adv. Food Res.*, 12, 1—52, 1963.
11. **Reynolds, T. M.**, Chemistry of non-enzymic browning. II, *Adv. Food Res.*, 14, 167—283, 1965.
12. **Sulser, H. and Büchi, W.**, Abbauprodukte von Fructoselysin in pflazlichen Trockenlebensmitteln und in einem Modellgemisch nach thermischer Handlung, *Lebensm. Wiss. Technol.*, 2, 105—112, 1969.
13. **Adrian, J.**, La réaction de Maillard vue sous l'angle nutritionnel. II. Comportement des matières alimentaires, *Ind. Aliment. Agric.*, 89, 1713—1720, 1972.
14. **Finot, P. A.**, Non-enzymic browning, in *Proteins in Human Nutrition*, Porter, J. W. G. and Rolls, B. A., Eds., Academic Press, London, 1973, 501—514.
15. **Sulser, H.**, Die bedeutung des Fructose — lysin und seiner Abbauprodukte Furosin und Pyridosin für die Qualitätsbeurteilung von Lebensmitteln, *Lebensm. Wiss. Technol.*, 6, 66—75, 1973.

16. **Baumann, G. and Gierschner, K.**, Die Bedeutung aminogruppen — haltiger Verbindungen, insbesondere der freien Amino-saüren, für planzliche Lebensmittel, vor allem für Fruchterzeugnisse, *Dtsch. Lebensm. Rundsch.*, 70, 273—279, 1974.

17. **Feeney, R. E., Blankenhorn, G., and Dixon, H. B. F.**, Carbonyl-amine reactions in protein chemistry, *Adv. Protein Chem.*, 29, 135—203, 1975.

18. **Synge, R. L. M.**, Interactions of polyphenols with proteins in plants and plant products, *Qualitas Plant Plant Foods Hum. Nutr.*, 24, 337—351, 1975.

19. **Van Sumere, C. F., Albrecht, J., Dedonder, A., De Pooter, H., and Pe, I.**, Plant proteins and phenolics, in *The Chemistry and Biochemistry of Plant Proteins,* Harborne, J. B. and Van Sumere, C. F., Eds., Academic Press, London, 1975, 221—270.

20. **McLeod, M. N.**, Plant tannins — their role in forage quality, *Nutr. Abstr. Rev.*, 44, 803—815, 1974.

21. **Synge, R. L. M.**, Damage to nutritional value of plant proteins by chemical reactions during storage and processing, *Qualitas Plant Plant Foods Hum. Nutr.*, 26, 9—27, 1976.

22. **Aure, L.**, Oxidation and stabilisation of herring meal by BHT on a technical scale, *Arsberet. Vedkomm. Nor. Fisk.*, 3, 17—24, 1957.

23. **Meade, T. L.**, A new development in fish meal processing, *Feedstuffs,* 28(20), 15—22, 1956.

24. **Marsh, B. E., Biely, J., Tarr, H. L. A., and Claggett, F.**, The effect of antioxidant treatment on the metabolizable energy and protein value of herring meal, *Poult. Sci.,* 44, 679—685, 1965.

25. **Combs, G. F.**, Influences of vitamin A and other reducing compounds on the selenium-vitamin E nutrition of the chick, in *Proc. 1976 Distillers Feed Research Council Conf.,* Distillers Feed Research Council, Cincinnati, 1976, 40—48.

26. **Sellam, M. A. and Christensen, C. M.**, Temperature differences, moisture transfer and spoilage in stored corn, *Feedstuffs,* 48(36), 28—29, 1976.

27. **Romoser, G. L.**, The role of chemical compounds in preserving feed and feed ingredients, *Feedstuffs,* 48(26), 18—19, 1976.

28. **Whitehead, C. C.**, The effects of pesticides on production in poultry, *Vet. Rec.,* 88, 114—117, 1971.

29. **Whitehead, C. C.**, Pesticides and poultry production, *DVM Newsmagazine,* 4, 5, 1972.

30. **Bragg, D. B., Sharma, H. R., and Ingalls, J. R.**, Effect of formalin treated wheat in the diet on the performance of growing chicks, *Can. J. Anim. Sci.,* 50, 601—604, 1970.

31. **Moran, E. T., Summers, J. D., and Bayley, H. S.**, Effect of cobalt 60 gamma-irradiation on the utilization of energy, protein and phosphorus from wheat bran by the chicken, *Cereal Chem.,* 45, 469—479, 1968.

32. **Chung, O., Finney, K. F., and Pomeranz, Y.**, Lipids in flour from gamma-irradiated wheat, *J. Food Sci.,* 32, 315—317, 1967.

33. **Wyatt, R. D.**, Mycotoxins and animal feeds — detection and control, in Proc. 1975 Georgia Nutr. Conf. Feed Industry, University of Georgia, Atlanta, 1975, 41—46.

34. **Jones, G. M.**, Preservation of high moisture corn with volatile fatty acids, *Can. J. Anim. Sci.,* 52, 73—79, 1970.

35. **Moran, E. T., Longworth, D. M., and Carlson, H. C.**, High moisture corn for the broiler chicken and a spoilage provoked vitamin E-selenium deficiency, *Feedstuffs,* 46(7), 27—45, 1974.

36. **Sunde, K. L., Din, M. G., and Holm, G. P.**, Feeding value of low bushel weight corn and propionic acid treated corn for broiler chicks, *Feedstuffs,* 48(15), 18—20, 1976.

37. **Moran, E. T. and Leslie, A. J.**, High moisture corn for the laying hen, *Can. Poult. Rev.,* 95, 26—28, 1971.

38. **Christensen, C. M.**, Tests with propionic and acetic acids as grain preservatives, *Feedstuffs,* 45(9), 37, 1973.

39. **Garlich, J. D., Wyatt, R. D., and Hamilton, P. B.**, The metabolizable energy value of high moisture corn preserved with a mixture of acetic and propionic acids, *Poult. Sci.,* 55, 225—228, 1976.

40. **Mitchell, H. H. and Beadles, J. R.**, The impairment in nutritive value of corn grain damaged by specific fungi, *J. Agric. Res. (Washington, D.C.),* 61, 135—141, 1940.

41. **Schupan, W.**, Essentielle Aminosaüren und B-vitamine als Qualitätskriterien bei Nahrungspflanzen under besonderer Berücksichtigung tropischer Leguminosen, *Qualitas Plant. Mater. Veg.,* 10, 187—203, 1963.

42. **Chitre, R. G., Desai, D. B., and Raut, V. S.**, The nutritive value of pure bred strains of cereals and pulses. The effect of storage on thiamine, riboflavin and nicotinic acid in cereals and pulses, *Indian J. Med. Res.,* 43, 585—589, 1955.

43. **Poley, W. E.**, The utilization of wheat and wheat byproducts in feeding young chickens. I. The effect of the fineness of grinding wheat, *Poult. Sci.,* 17, 331—337, 1938.

44. **Berg, L. R. and Bearse, G. E.**, The effect of the size and shape of feed particles on the development of growing poults, *Poult. Sci.,* 26, 532, 1947.

45. **Nikolaiczuk, N.**, The adverse effect of texture upon the feeding value of linseed oil meal for ducks, *Poult. Sci.,* 29, 773—774, 1950.

46. **McIntosh, J. I., Slinger, S. J., Sibbald, I. R., and Ashton, G. C.**, Factors affecting the metabolisable energy content of poultry feeds. VII. The effects of grinding, pelleting and grit feeding on the availability of the energy of wheat, corn, oats and barley, *Poult. Sci.*, 41, 445—456, 1962.

47. **Eley, C. P. and Bell, J. C.**, Particle size of broiler feed as a factor in the consumption and excretion of water, *Poult. Sci.*, 27, 660—661, 1948.

48. **Eley, C. P. and Hoffman, E.**, Feed particle size as a factor in water consumption and elimination, *Poult. Sci.*, 28, 215—222, 1949.

49. **Davis, R. L., Hill, E. G., Sloan, H. J., and Briggs, G. M.**, Detrimental effect of corn of coarse particle size in rations for chicks, *Poult. Sci.*, 30, 325—328, 1951.

50. **Smith, J. H., Carter, D. L., Brown, M. J., and Douglas, C. L.**, Differences in chemical composition of plant sample fractions resulting from grinding and screening, *Agron. J.*, 60, 149—151, 1968.

51. **Fritz, J. C.**, Effect of grinding on digestibility of Argentine flint corn, *Poult. Sci.*, 14, 267—279, 1935.

52. **Mitchell, R. J., Waldroup, P. W., Hillard, C. M., and Hazen, K. R.**, Effects of pelleting and particle size on utilization of roasted soybeans by broilers, *Poult. Sci.*, 51, 506—510, 1972.

53. **Renner, R. and Hill, F. W.**, Studies of the effect of heat treatment on the metabolizable energy value of soybeans and extracted soybean flakes for the chick, *J. Nutr.*, 70, 219—225, 1960.

54. **Carew, L. B., Hill, F. W., and Nesheim, M. C.**, The comparative value of heated ground unextracted soybeans and heated dehulled soybean flakes as a source of soybean oil and energy for the chick, *J. Am. Oil Chem. Soc.*, 38, 249—253, 1961.

55. **Summers, J. D.**, Effect of processing on the nutritive value of poultry feeds, in Proc. 1975 Georgia Nutrition Conf. Feed Industry, University of Georgia, Atlanta, 1975, 113—130.

56. **Saunders, R. M., Walker, H. G., and Kohler, G. O.**, Aleurone cells and the digestibility of wheat mill feeds, *Poult. Sci.*, 48, 1497—1503, 1969.

57. **Sibbald, I. R.**, The effect of cold pelleting on the true metabolizable energy values of cereal grains fed to adult roosters and a comparison of observed and predicted metabolizable energy values, *Poult. Sci.*, 55, 970—974, 1976.

58. **Jones, C. R.**, The production of mechanically damaged starch in milling as a governing factor in the diastatic activity of flour, *Cereal Chem.*, 17, 133—169, 1940.

59. **Totsuka, K., Tajima, M., Saito, T., and Shoji, K.**, Studies on the energy and protein values of faba beans for poultry rations, *Jpn. Poult. Sci.*, 14, 109—114, 1977.

60. **Sibbald, I. R. and Gowe, R. S.**, Effects of insoluble grit on the productive performance of ten White Leghorn strains, *Br. Poult. Sci.*, 18, 433—442, 1977.

61. **Blair, R., Dewar, W. A., and Downie, J. N.**, Egg production responses of hens given a complete mash of unground grain together with concentrate pellets, *Br. Poult. Sci.*, 14, 373—377, 1973.

62. **Auckland, J. N. and Wilson, S. B.**, Mineral and vitamin supplementation of a whole wheat diet for developing pullets, *Br. Poult. Sci.*, 18, 121—127, 1977.

63. **McGinnis, J. and Polis, H. L.**, Factors affecting the nutritive value of linseed meal for growing chicks, *Poult. Sci.*, 25, 408, 1946.

64. **Kratzer, F. H., Vohra, P., Atkinson, R. L., and Davis, P. N.**, Treatment of dried whey with various solvents; effects upon the growth of chicks, *Proc. Soc. Exp. Biol. Med.*, 89, 273—274, 1955.

65. **Panda, B., Jayaram, M., Ramamurthy, M. S., and Mair, R. B.**, Processing and utilization of salseed *(Shorea robusto)* as a source of energy in poultry feeds, *Indian Vet. J.*, 46, 1073—1077, 1969.

66. **Kakade, M. L. and Liener, I. E.**, The increased availability of nutrients from plant foodstuffs through processing, in *Man, Food and Nutrition*, Rechcigl, M., Ed., CRC Press, Boca Raton, Fla., 1973, 231—241.

67. **Jakobsen, P. E., Gertov, K., and Nielsen, S. H.**, Fordojelighedsforshog med fjerksae, 322. Beretning fra Forsogslaboratoriet, Copenhagen, 1960.

68. **Vogt, H. and Stute, K.**, Über die verdaulichkeit eineger Kohlenhydratfraktionen (Zucker, Stärke, Pentosane, Rohcellulose, Lignin) im Hühnerfutter, *Arch. Geflüegelkd.*, 35, 29—35, 1971.

69. **Petersen, V. E.**, The properties and value of the various feed grains in poultry nutrition, in *Cereal Processing and Digestion*, U.S. Feed Grains Council, London, 1972, 67—75.

70. **Hamm, D.**, Pelleting, soaking and adding of enzymes to feeds, in Proc. 1958 Arkansas Formula Feed Conf., University of Arkansas, Fayetteville, 1958.

71. **Berg, L. R.**, Enzyme supplementation of barley diets for laying hens, *Poult. Sci.*, 38, 1132—1139, 1959.

72. **Fry, R. E., Allred, J. B., Jensen, L. S., and McGinnis, J.**, Influence of water treatment on nutritive value of barley, *Proc. Soc. Exp. Biol. Med.*, 95, 249—251, 1957.

73. **Lepkovsky, S. and Furuta, F.**, The effect of water treatment of feeds upon the nutritional value of feeds, *Poult. Sci.*, 39, 394—398, 1960.

74. **Adams, O. L. and Naber, E. C.**, Effect of physical and chemical treatment of grains on growth and feed utilization by the chick. I. The effect of water and acid treatments of corn, wheat, barley and expanded or germinated grains on chick performance, *Poult. Sci.*, 48, 853—858, 1969.

75. Anderson, J. O., Dobson, D. C., and Wagstaff, R. K., Studies on the value of hulless barley in chick diets and means of increasing this value, *Poult. Sci.,* 40, 1571—1584, 1961.

76. Willingham, H. E., Jensen, L. S., and McGinnis, J., Studies on the role of enzyme supplements and water treatment for improving the nutritional value of barley, *Poult. Sci.,* 38, 539—544, 1959.

77. Willingham, H. E., Leong, K. C., Jensen, L. S., and McGinnis, J., Influence of geographical area of production on response of different barley samples to enzyme supplements or water treatment, *Poult. Sci.,* 39, 103—108, 1960.

78. Gohl, B. and Thomke, S., Digestibility coefficients and metabolizable energy of barley diets for layers as influenced by geographical area of production, *Poult. Sci.,* 55, 2369—2374, 1976.

79. Novacek, E. J. and Petersen, C. F., Metabolizable energy of the anatomical parts and other fractions of Western barley and the effect of enzymes and water treatment, *Poult. Sci.,* 46, 1008—1015, 1967.

80. Leong, K. C., Jensen, L. S., and McGinnis, J., Effect of water treatment and enzyme supplementation on the metabolizable energy value of barley, *Poult. Sci.,* 41, 36—39, 1962.

81. Potter, L. M., Stutz, K. W., and Matterson, L. D., Metabolizable energy and digestibility coefficients of barley for chicks as influenced by water treatment or by presence of fungal enzyme, *Poult. Sci.,* 44, 565—573, 1965.

82. Fry, R. E., Allred, J. B., Jensen, L. S., and McGinnis, J., Influence of enzyme supplementation and water treatment on the nutritional value of different grains for poults, *Poult. Sci.,* 37, 372—375, 1958.

83. Lawrence, T. L. J., Some effects of processing on the nutritive value of feedstuffs for growing pigs, *Proc. Nutr. Soc.,* 35, 237—243, 1976.

84. Gohl, B., Influence of water-treatment of barley on the digestion process in rats, *Z. Tierernäehr. Futtermittelkd,* 39, 57—67, 1977.

85. Rickes, E. L., Ham, E. A., Moscatelli, E. A., and Ott, W. H., The isolation and biological properties of the β-glucanase from *B. subtillis, Arch. Biochem. Biophys.,* 96, 371—375, 1962.

86. Burnett, G. S., Studies of viscosity as the probable factor involved in the improvement of certain barleys for chicks by enzyme supplementation, *Br. Poult. Sci.,* 7, 55—75, 1966.

87. Thomas, J. M., Jensen, L. S., Leong, K. C., and McGinnis, J., Role of microbial fermentation in improvement of barley by water treatment, *Proc. Soc. Exp. Biol. Med.,* 103, 198—200, 1960.

88. Thomas, J. M., Jensen, L. S., and McGinnis, J., Further studies on the role of microbial fermentation in the nutritional improvement of barley by water treatment, *Poult. Sci.,* 40, 1209—1213, 1961.

89. Thomas, J. M., Jensen, L. S., and McGinnis, J., Interference with the nutritional improvement of water-treated barley by antibiotics, *Poult. Sci.,* 40, 1204—1208, 1961.

90. Adams, O. L. and Naber, E. C., Effect of physical and chemical treatment of grains on growth of and feed utilization by the chick. II. Effect of water and acid treatments of grains and grain components on chick growth, nitrogen retention and energy utilization, *Poult. Sci.,* 48, 922—928, 1969.

91. Herstad, O. and McNab, J. M., The effect of heat treatment and enzyme supplementation on the nutritive value of barley for broiler chicks, *Br. Poult. Sci.,* 16, 1—8, 1975.

92. Willingham, H. E., McGinnis, J., Nelson, F., and Jensen, L. S., Relation of superiority of water-treated barley over enzyme supplements to antibiotics, *Poult. Sci.,* 39, 1307, 1960.

93. Arscott, G. H., Rose, R. J., and Harper, J. A., An apparent inhibitor in barley influencing efficiency ···tilization by chicks, *Poult. Sci.,* 39, 268—270, 1960.

94. Willingham, H. E., Leong, K. C., McGinnis, J., and Jensen, L. S., Nutritional improvement of cereal grains for chicks, *Poult. Sci.,* 40, 1470, 1961.

95. Leong, K. C., Jensen, L. S., and McGinnis, J., Improvement of the feeding value of wheat fractions for poultry, *Poult. Sci.,* 39, 1269, 1960.

96. Naber, E. C. and Touchburn, S. P., Effect of water treatment of components of hard red wheat on growth and energy utilization by the chick, *Poult. Sci.,* 48, 2052—2058, 1969.

97. Aumaitre, A., Henry, Y., Mercier, C., Ivorec-Szylit, O., and Thivend, P., Etude préliminaire de l'influence d'un traitment hydrothermique sur la valeur alimentaire du blé et du maïs, *Ann. Zootech.,* 21, 133—137, 1972.

98. Saunders, R. M., Connor, M. A., Edwards, R. H., and Kohler, G. O., Enzymatic processing of wheat bran: effects on nutrient availability, *Cereal Chem.,* 49, 436—442, 1972.

99. Everson, G., Steenbock, H., Cederquist, D. C., and Parsons, A. J., The effect of germination, the stage of maturity, and variety upon the nutritive value of soybean protein, *J. Nutr.,* 27, 225—232, 1944.

100. Viswanatha, T. and De, S. S., Relative availability of cystine and methionine in raw, germinated and autoclaved soybeans and soybean milk, *Ind. J. Physiol. Allied Sci.,* 5, 51—59, 1951.

101. Falen, L. F. and Petersen, C. F., Comparison of sprouted versus normal wheat when fed to White Leghorn cockerel chicks, *Poult. Sci.,* 48, 1772—1774, 1969.

102. Osborne, J. B. and Mendel, L. B., The use of soyabean as food, *J. Biol. Chem.,* 32, 369—387, 1917.

103. Liener, I. E., Ed., *Toxic Constituents of Plant Foodstuffs,* Academic Press, New York, 1969.

104. Couch, J. R. and Hooper, F. G., Antitryspin factors, in *Newer Methods of Nutritional Biochemistry*, Vol. 5, Albanese, A. A., Ed., Academic Press, London, 1972, 183—195.

105. Liener, I. E., Toxic factors in protein foods, in *Proteins in Human Nutrition*, Porter, J. W. G. and Rolls, B. A., Eds., Academic Press, New York, 1973, 481—500.

106. Liener, I. E., Endogenous toxic factors in oilseed residues, in *Animal Feeds of Tropical and Subtropical Origin*, Halliday, D., Ed., Tropical Products Institute, London, 1975, 179—188.

107. Kakade, M. L., Biochemical basis for the differences in plant protein utilization, *J. Agric. Food Chem.*, 22, 550—555, 1974.

108. Fukushima, D., Internal structure of 7S and 11S globulin molecules in soyabean proteins, *Cereal Chem.*, 45, 203—224, 1968.

109. Seidl, D., Jaffé, M., and Jaffé, W. G., Digestibility and proteinase inhibitory action of a kidney bean globulin, *J. Agric. Food Chem.*, 77, 1318—1321, 1969.

110. Jaffé, W. G., Toxicity of raw kidney beans, *Experientia*, 5, 81, 1949.

111. Honavar, P. M., Shih, C. V., and Liener, I. E., Inhibition of the growth of rats by purified hemagglutinin fractions isolated from *Phaseolus vulgaris*, *J. Nutr.*, 7, 109—114, 1962.

112. Mercier, C., Ivorec-Szylit, O., Guilbot, A., and Calet, C., Influence du traitment subi par le maïs sur la dégradation de son amidon dans le jabot du coq, *C.R. Acad. Sci.*, 263, 2033—2036, 1966.

113. Mercier, C., Effects of various U.S. grain processes on the alteration and *in vitro* digestibility of starch granule, *Feedstuffs*, 43(50), 33, 1971.

114. Burt, A. W. A., Effect of processing on the nutritive value of cereals in animal feeds, *Proc. Nutr. Soc.*, 32, 31—39, 1973.

115. Halnan, E. T., Digestibility trials with poultry. XI. The digestibility and metabolizable energy of raw and cooked potatoes, potato flakes, dried potato slices and dried potato shreds, *J. Agric. Sci.*, 34, 139—154, 1944.

116. Hock, A., Über die Verdauung verschiedener Stärkearten, *Tierernährung*, 10, 3—20, 1938.

117. Hakansson, J. and Lindgren, E., The ability of laying hens to digest raw potato starch, *Swed. J. Agric. Res.*, 4, 191—194, 1974.

118. Ryan, C. A. and Balls, A. K., An inhibitor of chymotrypsin from *Solanum tuberosum* and its behaviour towards trypsin, *Proc. Nat. Acad. Sci. U.S.A.*, 48, 1839—1844, 1962.

119. Pusztai, A., Trypsin inhibitors of plant origin, their chemistry and potential role in animal nutrition, *Nutr. Abstr. Rev.*, 37, 1—9, 1967.

120. D'Mello, J. P. F., Whittemore, C. T., and Elsley, F. W. H., The influence of processing upon the nutritive value of the potato. Studies with young chicks, *J. Sci. Food Agric.*, 24, 533—538, 1973.

121. Whittemore, C. T., Moffat, I. W., and Mitchell-Manson, J., Performance of broilers fed on diets containing cooked potato flake, *Br. Poult. Sci.*, 15, 225—230, 1974.

122. Whittemore, C. T., Moffat, I. W., and Taylor, A. G., The effect of dietary cooked potato on performance of broilers and on litter quality, *Br. Poult. Sci.*, 16, 115—120, 1975.

123. D'Mello, J. P. F. and Whittemore, C. T., Nutritive value of cooked potato flakes for the young chick, *J. Sci. Food Agric.*, 26, 261—266, 1975.

124. French, D., Chemical and physical properties of starch, *J. Anim. Sci.*, 37, 1048—1061, 1973.

125. Oluyemi, J. A., Fetuga, B. L., and Endeley, H. K. L., The metabolizable energy value of some feed ingredients for young chicks, *Poult. Sci.*, 55, 611—618, 1976.

126. Hertz, W. J., Some aromas produced by simple amino acid sugar reaction, *Food Res.*, 25, 491—495, 1960.

127. McNab, J. M. and Wilson, B. J., Effects of micronising on the utilisation of field beans *(Vicia faba L.)* by the young chick, *J. Sci. Food Agric.*, 25, 395—400, 1974.

128. Shannon, D. W. F. and Clandinin, D. R., Effect of heat treatment on the nutritive value of faba beans *(Vicia faba)* for broiler chicks, *Can. J. Anim. Sci.*, 57, 499—507, 1977.

129. Davidson, J., Attempts to overcome anti-nutritive factors in field beans *(Vicia faba L)* and field peas *(Pisum sativum)* fed in diets to laying hens, *Proc. Nutr. Soc.*, 36, 51A, 1977.

130. Bjarnason, J. and Carpenter, K. J., Mechanisms of heat damage in proteins. I. Models with acylated lysine units, *Br. J. Nutr.*, 23, 859—868, 1969.

131. Bjarnason, J. and Carpenter, K. J., Mechanisms of heat damage in proteins. II. Chemical changes in pure proteins, *Br. J. Nutr.*, 24, 313—329, 1970.

132. Waibel, P. E. and Carpenter, K. J., Mechanisms of heat damage in proteins. III. Studies with E-(γ-L-glutamyl)-L-lysine, *Br. J. Nutr.*, 27, 509—515, 1972.

133. Hurrel, R. F. and Carpenter, K. J., Mechanisms of heat damage in proteins. IV. The reactive lysine content of heat-damaged material as measured in different ways, *Br. J. Nutr.*, 32, 589—604, 1974.

134. Varnish, S. A. and Carpenter, K. J., Mechanisms of heat damage in proteins. V. The nutritional values of heat-damaged and propionylated proteins as sources of lysine, methionine and tryptophan, *Br. J. Nutr.*, 34, 325—337, 1975.

135. **Varnish, S. A. and Carpenter, K. J.,** Mechanisms of heat damage in proteins. VI. The digestibility of individual amino acids in heated and propionylated proteins, *Br. J. Nutr.,* 34 339—349, 1975.

136. **Hurrell, R. F., Carpenter, K. J., Sinclair, W. J., Otterburn, M. S., and Asquith, R. S.,** Mechanisms of heat damage in proteins. VII. The significance of lysine-containing isopeptides and lanthionine in heated proteins, *Br. J. Nutr.,* 35, 383—395, 1976.

137. **Hurrell, R. F. and Carpenter, K. J.,** Mechanisms of heat damage in proteins. VIII. The role of sucrose in the susceptibility of protein foods to heat damage, *Br. J. Nutr.,* 38, 285—297, 1977.

138. **Carpenter, K. J.,** Damage to lysine in food processing: its measurements and its significance, *Nutr. Abstr. Rev.,* 43, 423—451, 1973.

139. **Iriarte, B. J. R. and Barnes, R. H.,** The effect of over-heating on certain nutritional properties on the proteins of soybeans, *Food Technol. (Chicago),* 20, 131—134, 1966.

140. **Moran, E. T., Summers, J. D., and Bass, E. J.,** Heat processing of wheat germ meal and its effect on utilization and protein quality for the growing chick: toasting and autoclaving, *Cereal Chem.,* 45, 304—318, 1968.

141. **Sarwar, G., Shannon, D. W. F., and Bowland, J. P.,** Effects of processing conditions on the availability of amino acids in soybean and rapeseed proteins when fed to rats, *J. Inst. Can. Sci. Technol. Aliment.,* 8, 137—141, 1975.

142. **Erbersdobler, H.,** Ergebnisse zur Qualitatbeurteilung der Trockenmagermilch, *Landwirtsch. Forsch.,* 19, 264—268, 1966.

143. **Moran, E. T., Summers, J. D., and Slinger, S. J.,** Keratin as a source of protein for the growing chick. I. Amino acid imbalance as the cause for inferior performance of feather meal and the implication of disulfide bonding in raw feathers as the reason for poor digestibility, *Poult. Sci.,* 45, 1257—1266, 1966.

144. **Morris, W. C. and Balloun, S. L.,** Effect of processing methods on the utilization of feather meal by broiler chicks, *Poult. Sci.,* 52, 858—866, 1973.

145. **Morris, W. C. and Balloun, S. L.,** Evaluation of five differently processed feather meals by nitrogen retention, net protein values, xanthine dehydrogenase activity and chemical analysis, *Poult. Sci.,* 52, 1075—1084, 1973.

146. **Woodham, A. A. and Bailey, P. H.,** The effect of drying temperatures upon the nutritive value of barley for growing chickens, *Proc. Nutr. Soc.,* 36, 50A, 1977.

147. **Milner, C. K. and Woodforde, J.,** The effect of heat in drying on the nutritive value of wheat for animal feed, *J. Sci. Food Agric.,* 16, 369—373, 1965.

148. **Emerick, R. J., Carlson, C. W., and Winterfeld, H. L.,** Effect of heat drying upon the nutritive value of corn, *Poult. Sci.,* 40, 991—995, 1961.

149. **Calet, C. and de Lambilly, H.,** Étude le la valeur alimentaire du maïs-grain séché artificiellement pour le poussin en crossance. I. Influence du mode de séchage sur la disponibilité des acides animés, *Ann. Zootech.,* 9, 181—184, 1960.

150. **Chavez, R., de Matheu, P. J., and Reid, B. L.,** Grain sorghums in laying hen diets, *Poult. Sci.,* 45, 1275—1283, 1966.

151. **Lasserán, J. C.,** Méthodes de preparation des lots, *Ann. Zootech.,* 20, 607—618, 1971.

152. **Godon, B. and Petit, L.,** Propriétes des protéines, *Ann. Zootech.,* 20, 641—644, 1971.

153. **Larbier, M., Guillame, J., and Calet, C.,** Mesure de la lysine libre du muscle chez le poulet, *Ann. Zootech.,* 20, 653—661, 1971.

154. **Baratou, J. and Vachel, J. P.,** Valeur alimentaire chez le poulet dans le conditions pratiques de l'élevage, *Ann. Zootech.,* 20, 683—689, 1971.

155. **Calet, C.,** The relative value of pellets versus mash and grain in poultry nutrition, *World's Poult. Sci. J.,* 21, 23—52, 1965.

156. **Pfost, H. B.,** Effect of feed processing on digestibility of animal feeds, in *University of Nottingham Nutr. Conf. Feed Manuf.,* Vol. 5, Swan, H. and Lewis, D., Eds., Churchill Livingstone, Edinburgh, 1971, 50—68.

157. **Hutton, K. and Armstrong, D. G.,** Cereal processing, in *Feed Energy Sources for Livestock,* Swan, H. and Lewis, D., Eds., Butterworths, London, 1976, 47—63.

158. **Burt, A. W. A.,** Processing to improve nutritive value, in *Digestion in the Fowl,* Boorman, K. N. and Freeman, B. M., Eds., British Poultry Science, Edinburgh, 1976, 235—311.

159. **Wilson, B. J. and McNab, J. M.,** The effect of pretreatment on the nutritive value of diets for poultry, *Proc. Nutr. Soc.,* 35, 231—236, 1976.

160. **Allred, J. B., Jensen, L. S., and McGinnis, J.,** Factors affecting the response of chicks and poults to feed pelleting, *Poult. Sci.,* 36, 517—523, 1957.

161. **Allred, J. B., Fry, R. E., Jensen, L. S., and McGinnis, J.,** Studies with chicks on improvement in nutritive value of feed ingredients by pelleting, *Poult. Sci.,* 36, 1284—1289, 1957.

162. **Hussar, N. and Robblee, A. R.,** Effects of pelleting on the utilization of feed by the growing chicken, *Poult. Sci.,* 41, 1489—1493, 1962.

163. Bayley, H. S., Summers, J. D., and Slinger, S. J., The influence of steam pelleting conditions on the nutritional value of chick diets, *Poult. Sci.*, 47, 931—939, 1968.
164. MacBain, R., *Pelleting Animal Feed*, American Feed Manufacturers Association, Arlington, Va., 1968, 1—28.
165. Melcion, J. P., Vassade, P., Valdebouge, P., and Viroben, G., Influence des conditions d'agglomeration sur quelques caracteristiques physicochimiques d'un aliment pour porcelet, *Ann. Zootech.*, 23, 149—160, 1974.
166. Härtel, H., Scholtyssek, S., and Friedrich, W., Beeinflussung der Mastlustung von Broilern durch die Futterform und das Pressvarfahren bei unterschiedlicher Proteinkonzentration des Futters, *Arch. Geflügelkd.*, 34, 23—37, 1970.
167. Blakely, R. M., Macgregor, H. I., and Hanel, D., The effect of type of pelleting on growth and metabolisable energy from turkey rations, *Br. Poult. Sci.*, 4, 261—265, 1963.
168. Bayley, H. S., Summers, J. D., and Slinger, S. J., The effect of steam pelleting feed ingredients on chick performance: effect on phosphorus availability, metabolizable energy value and carcass composition, *Poult. Sci.*, 47, 1140—1148, 1968.
169. McIntosh, J. I., Slinger, S. J., Sibbald, I. R., and Ashton, G. C., The effect of three physical forms of wheat on the weight gains and feed efficiencies of pullets from hatching to fifteen weeks of age, *Poult. Sci.*, 41, 438—445, 1962.
170. Moran, E. T., Summers, J. D., and Jones, G. E., Field peas as a major dietary protein source for the growing hen with emphasis on high-temperature steam pelleting as a practical means of improving nutritional value, *Can. J. Anim. Sci.*, 48, 47—55, 1968.
171. Janssen, W. M. M. A., Waanders, J., and Terpstra, K., Methods to predict the metabolisable energy of feedstuffs for poultry, in *Energy Metabolism of Farm Animals*, Vermorel, M., Ed., G. de Brussac, Clermont-Ferrand, France, 1976, 273—276.
172. Carew, L. B. and Nesheim, M. C., The effect of pelleting on the nutritional value of ground soybeans for the chick, *Poult. Sci.*, 41, 161—168, 1962.
173. Summers, J. D., Slinger, S. J., and Cisneros, G., Some factors affecting the biological availability of phosphorus in wheat by-products, *Cereal Chem.*, 44, 318—323, 1967.
174. Summers, J. D., Bentley, H. U., and Slinger, S. J., Influence of method of pelleting on utilization of energy from corn, wheat shorts and bran, *Cereal Chem.*, 45, 612—615, 1968.
175. Erbersdobler, H. and Gropp, J., Wachstum von Küken bei Futterung von Kornermais unterschiedlicher Proteinqualitat, *Arch. Geflügelk.*, 35, 36—40, 1971.
176. Gürocak, B., Stute, K., and Vogt, H., Untersuchunger über den Einfluss des Pressens auf die Vardaulichkeit des Geflügelmastfutters, *Arch. Geflügelk.*, 37, 81—84, 1973.
177. Fisher, C. and Wilson, B. J., Response to dietary energy concentration by growing chickens, in *Energy Requirements of Poultry*, Morris, T. R. and Freeman, B. M., Eds., British Poultry Science, Edinburgh, 1975, 151—184.
178. Pepper, W. F., Slinger, S. J., and Summers, J. D., Studies with chickens and turkeys on the relationship between fat, unidentified factors and pelleting, *Poult. Sci.*, 39, 66—74, 1960.
179. Auckland, J. M. and Fulton, R. B., The effects of dietary nutrient concentration, crumbles versus mash and age of dam on the growth of broiler chicks, *Poult. Sci.*, 51, p. 1968—1975, 1972.
180. Hamm, D. and Stephenson, E. L., The pelleting response in broiler feeding, *Poult. Sci.*, 38, 1211, 1959.
181. Shreck, P. K., Sterrit, G. M., Smith, M. P., and Stilson, D. W., Environmental factors in the development of eating in chicks, *Anim. Behav.*, 11, 306—309, 1963.
182. Reddy, C. V., Jensen, L. S., Merrill, L. H., and McGinnis, J., Influence of mechanical alteration of dietary density on energy available for chick growth, *J. Nutr.*, 77, 428—432, 1962.
183. Jensen, L. S., Merrill, L. H., Reddy, C. V., and McGinnis, J., Observations on eating patterns and rate of food passage of birds fed pelleted and unpelleted diets, *Poult. Sci.*, 41, 1414—1419, 1962.
184. Fujita, H., Quantitative studies on the variation in feeding activity of chickens. III. Effect of pelleting the feed on eating patterns and the rate of feed passage through the digestive tract in chicks, *Jpn. Poult. Sci.*, 11, 210—216, 1974.
185. Reddy, C. V., Jensen, L. S., Merrill, L. H., and McGinnis, J., Influence of pelleting on metabolizable and productive energy content of a complete diet for chicks, *Poult. Sci.*, 40, 1446, 1961.
186. Bearse, G. E., Berg, L. R., McClary, C. F., and Miller, V. L., The effect of pelleting chicken rations on the incidence of cannibalism, *Poult. Sci.*, 28, 756, 1949.
187. Merritt, E. S., Downs, J. H., Bordeleau, R., and Tinney, B. F., Growth, variability of growth, and market quality of broilers on mash and pellets, *Can. J. Anim. Sci.*, 40, 7—14, 1960.
188. Savory, C. J., Growth and behaviour of chicks fed on pellets or mash, *Br. Poult. Sci.*, 15, 281—286, 1974.
189. Bolton, W., The digestibility of mash and pellets by chicks, *J. Agric. Sci.*, 55, 141—142, 1960.

190. **Blair, R., Whitehead, C. C., and Teague, P. W.**, The effect of dietary fat and protein levels, form and cereal type on fatty liver and kidney syndrome in chicks, *Res. Vet. Sci.,* 18, 76—81, 1975.

191. **Whitehead, C. C.**, Extra biotin will prevent FLKS, *Poult. World,* 125(7), 30—31, 1974.

192. **Scott, M. L.**, Effect of processing on the availability and nutritional value of vitamins, in *Effect of Processing on the Nutritional Value of Feeds,* National Academy of Sciences, Washington, D.C., 1973, 119—130.

EFFECT OF PROCESSING ON NUTRITIVE VALUE OF DIETS FOR HORSES*.**

G. F. W. Haenlein

INTRODUCTION

The processing of feed is justifiable if the nutritional and economic benefits from feeding it to animals are increased. Animal feeds are bulky roughages, forages, hay, straw, silage or more dense grains, seeds, and industrial byproducts. There are physical and chemical processes which will change:

1. The density and properties that influence transportation cost and flow characteristics in conveyance.
2. The nutritional value of feeds.

Roughages, i.e., hay and straw, are bulky and the density of nutritional values per unit volume of roughage is low. Therefore, the interest in making transporation and conveyance of hay and straw more economical and more suitable for automation dominated engineering efforts to develop processing methods. Interest in effects on nutritional properties of such processed feed has been usually a subsequent, not a primary, objective. Not surprisingly, some processing methods do not improve the nutritional value of feeds or even decrease it, although cost benefits in handling and transportation are derived. There is a challenge to the nutritionist and to the ideal of cooperative development between engineers and nutritionists. For instance, the modern corn picker-sheller and the modern forage chopper are examples for the need of close cooperation between the two disciplines. Both machines minimize the cost of obtaining and handling the final product, corn kernels in the first and free-flowing, finely chopped forage in the latter. Disregarded are the nutritional values of corn cobs and corn stover in the first, which are left wasting on the field; disregarded in the latter machine are the consequences of finely chopping of forage upon the nutritional value of feed and upon the health of the animals fed such processed forages.

Modern domestic animals are in an additional position of conflict. Modern man has selected genetic strains during the last 150 years which outperform greatly the level of their ancestors and which have vastly increased nutritional requirements per unit volume of animal and per unit of time. This need for greater nutritional values per animal and thus per unit of feed coincides with engineering efforts to better package by increasing the density of feeds although primarily for the sake of reducing transportation cost and labor through processing methods.

Domestic animals are products of thousands of years of evolution in a particular geographic niche in relation to the available types and compositions of feeds which are mostly low in nutritive density and contain little fine, short fiber material, such as modern engineering produces with grinding and other processing methods. The challenge is to fit processing methods that suit transportation needs into the physiological demands of domestic animals, specifically the horse.

Together with ruminants and rabbits, horses are not in competitive conflict with

* Published with the approval of the Delaware Agricultural Experiment Station, Misc. Pap. No. 805, Contr. No. 38, Department of Animal Science and Agricultural Biochemistry, University of Delaware, Newark.

** Reprinted in part from *Landwirtsch. Forsch.*, 33, 227—235, 1980. With permission.

man's own food resources. They can maintain themselves through their unique ability of utilizing forages which are mostly inedible for man. Thus, they contribute greatly to the welfare of man. A growing world population depending on grains for its own food supply is looking more and more to the forage-consuming domestic animals for support. The world land area suitable for growing grain and other food crops is only 10% of the total world surface, while 20% of it is permanent grassland (i.e., forage), and 30% woodland.[1]

Forage consuming animals, if needed in greater numbers, require:

1.   Greater productivity of the forage producing lands.
2.   Greater net utilization of these forages.

Processing of forages could make a significant nutritional contribution to mankind if processing would increase the net utilization of forages. For instance, it is estimated for ruminants that a dairy cow can produce at present 5,000 kg of milk per year on forages alone.[2] Most available forages are only between 40 to 60% digestible, losing 60 to 40% of their nutritive values in the voided feces.[3] Processing methods of forages which would reduce this digestive waste could double the milk producing ability of forage-only fed cows.[2] Processes that break the lignocellulose bonds in forages which are the principal obstacles to higher net utilization, can improve digestibilities of forages up to 90%, which makes them comparable in nutritive value to grains.[3,4] Processing methods and their nutritional effects have been considered a subject of national significance by the Governing Board of the National Research Council and a comprehensive symposium was held in 1972.[5]

The nutrition of horses is of new interest since ownership of horses in the U.S. and much of the affluent world has shifted from the cavalry and farmers to suburbanites, race tracks, and city police, which all depend on purchased feeds of preferably little bulk and ease of handling. However, interest in processing methods to provide easily transported bulky feeds for horses is not new and can be traced back at least 125 years.[6-11]

Holbach, in Hamburg, Germany, then produced thin wafers containing concentrates and industry byproducts as a cheaper, easier handled, and better stored feed for his circus animals. During the Franco-Prussian War of 1870-71, and during the two World Wars, the German cavalry used feed mixtures which were wafered, e.g., 1 cm thick and 10 to 12 cm wide. Commercially, wafers, also named "biscuits", containing grains such as oatmeal, peaflour, rye and linseed, chopped hay, and straw, were produced and fed in Germany, England, France, Belgium, and Russia around 1880.

Between 1925 and 1937 more than 4000 Russian army horses were used in feeding trials involving "briquettes" that contained 40 to 50% chopped hay, 20 to 30% rolled oats, beet pulp, molasses, bran, ground corn, and salt.[7,8,11] It was reported that horses on such diets drank more water, maintained their weight better, had fewer digestive ailments, worked as well, but sweated more than horses on nonprocessed feed rations.

In 1952, Bruhn, of the Wisconsin Agricultural Experiment Station, pioneered the development of a wafering process from unground roughage.[12] Earlier it had been found that grinding of roughage prior to processing had subsequent physiological effects, e.g., bloat, ulcers, intestinal parakeratosis and, in cows, the depression of their milk composition toward abnormally low milk fat levels. These physiological effects were traced to effects on digestion and biochemical and microbiological changes during digestion.[13,14]

Thus, the different elements of processing such as compression, grinding, heating, gelatinization, etc. have to be evaluated separately as to their nutritional effects. This recognition also stimulated the search for a method of compression of feed without the need of grinding. In addition to compression, other processing methods for horses include cooking, toasting, steaming, expanding, extruding, dehydrating, extracting, malting, flaking, crimping, dehulling, hydrolyzing, micronizing, and fermenting, but respective nutritional literature is sparse, especially for horses.

Different names apply for compression depending on the density of the product, the least dense being bales, next wafers, and the densest are mostly called pellets. However, biscuits, briquettes, nuggets, crumbles, crackers, dice, and cubes are also used without specifically exclusive definitions. According to an American engineering definition:[15] hay pellets consist of agglomerated ground forage in which the particle size is generally less than 3/8 in. in length; hay wafers consist of agglomerated unground forage which has some fiber equal to or greater than the length of the minimum cross-section dimensions.

However, wafering machines can and have processed ground hay and pellets have been made from unground, chopped hay with fiber length that exceeded the diameter of pellets.[16,17] Also, the name "hay cubes" is used in place of "pellets" in the western U.S. and England although they may look like wafers but consist of ground compressed hay.[18,19]

EFFECTS ON COMPOSITION

It is obvious that processing methods which separate parts of the original feed affect the composition and nutritional value of the product accordingly. This applies to extracting, malting, dehulling, hydrolyzing, and fermenting. It is not generally realized that separation of feed components, although not intentionally, occurs also during baling, wafering, and pelleting. Shattering and loss of hay leaves is to varying degrees part of the baling process, resulting in higher fiber, lower protein, and lower total nutritive values. Back pressure expansion and disintegration is part of the processing of wafers, resulting in reduced fiber, higher protein contents, and greater apparent nutritive values, especially if no recycling of crumbles occurs. Disintegration and accumulation of fines is also part of the pelleting process. If it occurs after leaving the mill, the separations will again cause changes in nutritive values.

Compressing feeds into bales, wafers, or pellets diminishes the exposure to air and oxygen. The normal rate of decomposition in storage, e.g., of carotene is accordingly slowed, resulting in higher values compared to the nonprocessed feed in storage.[20,21] On the other hand, if wafers are not dried within 48 hr of processing below 14.5% moisture content, they will mold inside and outside within a few days, or heat and brown, unless the environmental relative humidity is low.[22]

Heat occurs during compression because of friction, but heat is also added because of at least four engineering reasons:[23] It cuts down power requirements, it reduces forage moisture, it relaxes and softens forage fibers, and it makes the feed particles stick together and the wafers or pellets more durable in subsequent handling (Table 1).

Alfalfa hay heated to 212°F was wafered at 500 ψ or 1.89 hp/t hr to a density of 57 lb./ft.^{3} The same hay, not heated, required a force of 2500 ψ or 9.60 hp/t hr to produce the same wafer. The nutritional consequences of heating during processing are several and depend on the degree and extent of heating: denaturation of portions of protein, vitamins and enzymes, gelatinization, dextrinization, rupture, and predigestion of carbohydrates, particularly starches, and carmelization of sugars.[24] These changes are not all necessarily detrimental to the nutritional value of the feed since they can mean enhanced flavors, digestibility, and palatability.

Table 1
COMPRESSION CONDITION IN PROCESSING FEEDS[6]

	Compression			
Feed mixture	Pressure (ψ)	°C	Time (min)	Condition of product
50% Chopped meadow hay 25% Rolled oats 12.5% Beet pulp 12.5% Molasses[a]	3500—5800	120	0.5	Too hard, soaking needed before feeding
40% Chopped alfalfa hay 40% Rolled oats 10% Beet pulp 5—8% Corn dextrin and steam[b]	500	120	2	Good
40% Chopped timothy hay 40% Rolled oats 10% Beet pulp	500	120	6	Disintegrated
5—19% Corn dextrin and steam[b]	3000	53	1	Good, if cooled under pressure

Note: All feeds preheated to 50 to 53°C.

[a]   Reference 7.
[b]   Reference 6.

Table 2
ADAPTATION OF HORSES TO DIFFERENT PHYSICAL FORMS OF ALFALFA HAY RATIONS[a]

	Average daily, voluntary dry matter consumption, kg in weeks				
Ration	1st	2nd	3rd	4th	Average
Pellets	3.3	3.6	3.7	3.9	3.6
Wafers	3.1	3.4	3.6	3.6	3.4
Loose	2.8	2.9[b]	3.0[c]	2.8[b]	2.9[b]

[a]   Averages of 6 horses from 3 × 3 replicated latin-square trials; wafered hay, 6.4 cm chopped, 20.7 lb/ft³; pelleted hay, 1.9 cm screen, 1.9 cm die.
[b] $P<0.01$.
[c] $P<0.05$.

From Haenlein, G. F. W., Holdren, R. D., and Yoon, Y. M., *J. Anim. Sci.*, 25, 740, 1966. With permission.

EFFECTS ON CONSUMPTION BY HORSES

Aside from the chemical changes of feeds during processing, the mere physical changes of bite size, unit density, and particle sizes, influence animal acceptance and performance. Horses adapt readily with a few exceptions to compressed feeds (Tables 2,3). The chewing of hay wafers by horses takes more time than when eating pellets.

Table 3
VOLUNTARY CONSUMPTION OF
HAY, LOOSE OR PELLETED[a]

Horse No.	Red clover hay intake/lb per day		
	Loose	Pelleted	Total
1	13.4	7.7	21.1
2	6.8	12.4	19.2
3	10.5	12.1	22.6
4	10.9	11.2	22.1
5	18.8	6.3	25.1
6	13.2	10.8	24.0
7	4.9	10.7	15.6
8	8.4	9.4	17.8
9	3.7	8.2	11.9
10	9.9	20.3	30.2
11	13.9	16.0	29.9
Average	10.4 ± 1.3	11.4 ± 1.2	

[a] Thoroughbreds; 25-day cafeteria trial; hay ground at ½ in. screen and pelleted through ¾ in. die.

From Mitchell, W. H. and Shropshire, J. H., *Proc. N. Atlantic Sec. Am. Soc. Anim. Prod.*, 1959, 1—12. With permission, American Dairy Science Association.

With their incisors they struggle to split and slice wafers parallel to the planes of fiber compression, knocking, and pushing the wafers against the sides and floors of mangers and spilling many wafers around the stall (Figures 1-3). A complete pelleted ration consisting of ⅓ alfalfa and ⅔ total corn plants, including ears, was eaten at the average rate of 2.1 lb/100 lb. body weight of four yearling ponies vs. 2.6 lb for loose alfalfa hay mixed with loose ground corn over 4-week trial periods.[26] The difference in consumption was due to preference for the free grain over loose hay or pellets.

However, even at the 2.1 lb rate, the ponies voluntarily overate at a 125% level of daily nutrient requirements according to the National Research Council standards and gained 3.5 ± 0.7 lb body weight per week. When comparing average consumption times of pelleted with nonpelleted complete rations, it took seven thoroughbred and three quarter horses an average 108 min vs. 77 min for the pellets.[27] Horses react like cattle and sheep, which have been studied more thoroughly, when offered pelleted or wafered feeds. They eat more nutrients per minute and per day in the compressed higher unit density form. It is not proven, however, that horses eat more pellets because they like them better tastewise than nonpelleted feeds. Greater palatability has been claimed but may not have been proven per se. Horses apparently eat more pellets than loose feed per day because first physiological signals for their intake regulation may not come from stomach fill or blood glucose levels but from unit metering sensors in the jaw, mouth, and pharynx, which signal reduced chewing and mastication when eating pellets.[28] This hypothesis needs testing but it is supported by the phenomena of gorging and wood chewing when feeding pellets to either horses or ruminants.

There is a high inverse correlation between unit density and average particle size in pellets and wafers.[17] This also means that grinding, i.e., the reduction of particle size of feeds, will increase consumption by horses if combined with compression.[21] However, grinding per se, if the product, e.g., ground grains are not dusty, will also im-

FIGURE 1.   Horses slice and shave wafers into small pieces parallel to the hay stems.

FIGURE 2.   Horses use the manger and other objects to aid in slicing wafers with their incisor teeth.

prove horse performance, by improved consumption and digestion.[29] The cost of grinding has to be related to the amount of extra gain and benefits. In order to avoid dustiness after grinding in the absence of compression, the procedure of rolling, crimping, and flaking of grain, with or without steam, has been used for a long time.[30,31] It is another particle reduction process and the response from horses thus fed is increased weight gains and better work performance.

EFFECTS ON DIGESTIBILITY IN HORSES

Pelleting will not affect the digestibility of horse rations much, except for decreased crude fiber digestibility (Table 4). This is more than compensated for by the higher

FIGURE 3. Horses chew wood progressively when on pelleted rations exclusively.

Table 4
DIGESTIBILITY OF NUTRIENTS IN ALFALFA HAY WHEN FED IN DIFFERENT PHYSICAL FORMS TO HORSES[a]

	Hay form		
Nutrient	Pellets	Wafers	Loose
Dry matter	52	53	52
Crude protein	68	69	67
Crude fiber	30[b]	35	36
N-free extract	68	70	67
Total digestible nutrients	47	50	48
Nutritive value index[c]	57	59	44[b]

[a]   Average of 6 ponies.
[b] $P < 0.01$.
[c]   Nutritive value index =

$$\frac{100 \times \text{observed intake g} \times \text{total digestible nutrients}}{80 \, (W_{kg}^{0.75})}$$

From Haenlein, G. F. W., Holdren, R. D., and Yoon, Y. M., *J. Anim. Sci.,* 25, 740, 1966. With permission.

voluntary consumption, resulting in a greater overall nutritive value index for pelleted over nonpelleted horse rations.[21] It is possible that the decrease in crude fiber digestion is a matter of decreased intestinal residence time. It has been shown in horses by the stained particle procedure and by the chromic oxide method that wafering and pelleting, even more, increases the speed of passage of hay through the gastrointestinal tract of horses.[27,32,33] The water content of feces is also increased.[17,27]

These findings in horses agree with similar results in sheep and cows, except that here the digestion of protein and carbohydrates is also depressed.

On the other hand, when the consumption levels of pelleted hay are restricted to those of loose hay, then the digestibilities of nutrients in the two physical forms of hay are equal.[16]

This also seems to indicate that increased consumption is not caused by decreased digestibility which is due to a faster rate of intestinal passage. On the other hand, it can be theorized that increased consumption causes faster rate of passage and therefore decreased digestibility, favored by the reduced particle size.

SIDE EFFECTS IN HORSES

Choking, bloating, scouring, and foundering have been observed in other species, while progressive and persistent wood chewing seems to be the major problem with horses when pellets make up the total ration.[21,25,26] No such problems have been observed when feeding wafers which do not contain ground particles, but the fiber length from only chopped hay is at least the length of the cross-section of the wafer, which then occupies considerably more time in chewing and mastication for horses than pellets do.[33] Wood chewing has been called a vice of horses caused by boredom and bad habit. This does not seem to be a satisfactory explanation in light of the discussed differences with different physical forms of the ration, and the physiology of wood chewing by horses deserves some research attention. Possibly the ethology of chewing as a form of displaced behavior, i.e., the stimulus for chewing, has not been satiated and is misdirected to another object, might help in the solution of this problem.[34]

Pelleting of feeds other than forages has the advantage of reducing fines and dust which are disliked by horses.[31] It also enables better mixing of feed ingredients and does not allow sorting out by horses. Thus pelleting has become a very acceptable form of feeding horses in recent years because it also reduces waste and increases nutritive values even with less body fill which is of interest to owners of racing horses. In addition, pelleting has labor, storage, flowability, and transportation advantages due to the increased bulk density from 5 lb/ft³ for loose hay, up to 20 lb/ft³ for wafers, and up to 45 lb/ft³ for pellets; besides reduced unit size and high durability on handling and in storage.

SUMMARY

Processing methods of grains and roughages are many, but research into their nutritional effects in horses has been limited to changes of physical forms of feeds and whole rations by reduction of particle size, agglomeration, and compression, such as grinding, rolling, pelleting, and wafering. The need to improve characteristics for transportation and automatic conveyance rather than the interest in improving the nutritive value of feeds has been a dominant factor in process development. Chemical changes in the composition of feeds occur during processes designed to change their physical form, because of heat generated during the process, steam and pressure added, and different surface areas being exposed now more or less to air and oxidation. Changes of physical form generally affect voluntary consumption by horses because some of the need for chewing has been reduced. Dustiness and large, dense shapes may have negative effects. Increased rate of consumption because of reduced particle size has physiological effects; overeating, weight gain, increased water consumption, increased fecal water content, increased rate of gastrointestinal passage, decreased crude fiber digestibility but increased overall nutritive value. Pelleting is the nondusty solution to reduction in particle size plus improved conveyance characteristics. Wafering is a modification of pelleting which preserves some fiber length for nutritional and behavioral benefits to horses.

REFERENCES

1. Campbell, J. R. and Lasley, J. F., *The Science of Animals that Serve Mankind,* 2nd ed., McGraw-Hill, New York, 1975, 20.
2. Reid, J. T., Potential for increased use of forages in dairy and beef rations, in *How Far with Forages for Meat and Milk Production?,* Matches, A. G. and Marks, J. J., Eds., Proc. 10th Res. Ind. Conf., Am. Forage and Grassland Coun., Lexington, Ky., 1977, 175.
3. Morrison, F. B., *Feeds and Feeding,* 22nd ed., Morrison, Ithaca, N.Y., 1956, 1000—1043.
4. Bull, L. S., Reid, J. T., and Johnson, D. E., Energetics of sheep concerned with the utilization of acetic acid, *J. Nutr.,* 100, 262—276, 1970.
5. Cunha, T. J., Baumgardt, B. R., Bell, J. M., Hale, W. H., Halver, J. E., Jacobson, N. L., Oltjen, R. R., Sunde, M. L., and Ullrey, D. E., *Effect of Processing on the Nutritional Value of Feeds,* National Academy of Sciences, Washington, D.C., 1973.
6. Earle, I. P., Compression of complete diets for horses, *J. Anim. Sci.,* 9, 255—260, 1950.
7. Dyakov, M. I., Oppel, V. V., Banaitis, S. I., Alexandrov, S. A., and Ivankin, V. K., Efficiency of combined feed No. 2 for working horses versus hay and oats rations, *Zap. Tsentr. N. I. Lab. Kormovoi Kombikormovoi Promishlenosti Zootekhn. Lab.,* (Leningrad), No. 18, 1937.
8. Sorokin, S. V., The preparation of mixed feed cakes, *Zap. Tsentr. N. I. Lab. Kormovoi Kombikormovoi Promishlenosti Zootekhn. Lab. Detskoje Selo,* (Leningrad), No. 14, 1936.
9. Minson, D. J., The effect of grinding, pelleting and wafering on the feeding value of roughages — a review, Contr. No. 84, Animal Research Institute, Canada Department of Agriculture, Ottawa, 1962.
10. Moore, L. A., Symposium on Forage Utilization, Nutritive Value of Forages as Affected by Physical Form. I. General principles involved with ruminants and effect of feeding pelleted or wafered forage to dairy cattle, *J. Anim. Sci.,* 23, 230, 1964.
11. Kuznetsov, P., Feeding of army horses with a balanced ration, *Veterinariya,* (Russ.), 8, 9, 1942.
12. Bruhn, H. D., Pelleting grain and hay mixtures, *Agric. Eng.,* 36, 330—331, 1955.
13. Tyznik, W. J., The Effect of the Amount and Physical State of the Roughage Upon the Rumen Fatty Acids and Milk Fat of Dairy Cows, Dissertation, University of Wisconsin, Department of Dairy Science, Madison, 1951.
14. Rodrigue, C. B. and Allen, N. N., The effect of fine grinding of hay on ration digestibility, rate of passage, and fat content of milk, *Can. J. Anim. Sci.,* 40, 23—29, 1960.
15. Anon., Hay pellets and wafers — definitions and methods for determining specific weight, durability and moisture content, *Agric. Eng. Yearbook,* 243—244, 1964.
16. Haenlein, G. F. W., Richards, C. R., and Mitchell, W. H., Effect of the size of grind and level of intake of pelleted alfalfa hay on its nutritive value in cows and sheep, *J. Dairy Sci.,* 45, 693, 1964.
17. Haenlein, G. F. W. and Holdren, R. D., Response of sheep to wafered hay having different physical characteristics, *J. Anim. Sci.,* 24, 810—818, 1965.
18. Bath, D. L., Nutritional value of alfalfa cubes and wafers for dairy and beef cattle, *Feedstuffs,* 38, 45—46, 1966.
19. Blaxter, K. L. and Graham, N. McC., The effect of the grinding and cubing process on the utilization of the energy of dried grass, *J. Agric. Sci.,* 47, 207—217, 1956.
20. Vander Noot, G. W. and O'Connor, J. J., Effect of wafering on alfalfa hay, in *Proc. Seminar on Wafering,* Chicago, 1962, 129—133.
21. Haenlein, G. F. W., Holdren, R. D., and Yoon, Y. M., Comparative response of horses and sheep to different physical forms of alfalfa hay, *J. Anim. Sci.,* 25, 740—743, 1966.
22. Hall, C. W., Operational observations on the M—F wafering machine, in *Proc. Seminar on Wafering,* Chicago, 1962, 73—79.
23. Kosch, M. A., Kosch, A. J., and Melcher, M. A., Role of heat in wafering, in *Proc. Seminar on Wafering,* Chicago, 1962, 88—97.
24. Pope, L. S., New methods of processing grain for fattening beef cattle, *Feedstuffs,* 34, 40—42, 67—69, 1962.
25. Mitchell, W. H. and Shropshire, J. H., Preference of various baled and pelleted hay mixtures by horses, in *Proc. N. Atlantic Sec. Am. Soc. Anim. Prod.,* 1959, 1—12.
26. Haenlein, G. F. W., Nutritive value of a pelleted horse ration, *Feedstuffs,* 41, 19—26, 1969.
27. Hintz, H. F. and Loy, R. G., Effects of pelleting on the nutritive value of horse rations, *J. Anim. Sci.,* 25, 1059—1062, 1966.
28. Haenlein, G. F. W., The Hypothalamus and Feed Intake Regulation, in Proc. Semin., Department of Dairy Science, University of Wisconsin, Madison, 1966.
29. Harper, M. W., Feeding work horses, *Cornell Agric. Exp. Stn., N.Y. Bull.,* 437, 1925.
30. Caine, A. B., The preparation of feed for colts, *Iowa Agric. Exp. Stn. Bull.,* 347, 1936.

31. Ott, E. A., Effect of processing feeds on their nutritional value for horses, in *Effect of Processing on the Nutritional Value of Feeds,* Cunha, T. J., Baumgardt, B. R., Bell, J. M., Hale, W. H., Halver, J. E., Jacobson, N. L., Oltjen, R. R., Sunde, M. L., and Ullrey, D. E., National Academy of Sciences, Washington, D.C., 373—382, 1973.
32. Haenlein, G. F. W., Smith, R. C., and Yoon, Y. M., Determination of the fecal excretion rate of horses with chromic oxide, *J. Anim. Sci.,* 25, 1091—1095, 1966.
33. Meyer, H., Ahlswede, L., and Reinhardt, H. J., Untersuchungen über Fressdauer, Kaufrequenz und Futterzerkleinerung beim Pferd, *Dtsch. Tierärztl. Wochenschr.,* 82, 54—58, 1975.
34. Scott, J. P., Introduction to animal behavior, in: *The Behavior of Domestic Animals,* 1st ed., Hafez, E. S. E., Ed., Williams & Wilkins, Baltimore, 1962, 3—20.

Index

INDEX

A

B

F

H

Q

R

T